Introduction to Wireless Systems

Introduction to Wireless Systems

Bruce A. Black

Philip S. DiPiazza

Bruce A. Ferguson

David R. Voltmer

Frederick C. Berry

Upper Saddle River, NJ • Boston • Indianapolis • San Francisco
New York • Toronto • Montreal • London • Munich • Paris • Madrid
Capetown • Sydney • Tokyo • Singapore • Mexico City

Visit us on the Web: www.informit.com/ph

Library of Congress Cataloging-in-Publication Data

Introduction to wireless systems / Bruce A. Black ... [et al.].

p. cm.

Includes bibliographical references and index.

ISBN 0-13-244789-4 (hardcover : alk. paper)

1. Wireless communication systems. I. Black, Bruce A.

TK5103.2.I62 2008

621.384—dc22

2008005897

ISBN-13: 978-0-13-244789-8
ISBN-10: 0-13-244789-4

Text printed in the United States on recycled paper at Courier in Westford, Massachusetts.
First printing, May 2008

This book is dedicated to Misty Baker, the Founding Director of the Global Wireless Education Consortium (GWEC), and to her successor, Susan Sloan, Executive Director of GWEC, for their tireless efforts in promoting the academic disciplines central to wireless communications. Their encouragement not only provided inspiration for this book, but also established the venue for the authors to launch its creation and to become sincere friends.

Contents

	Preface	*xiii*
	Acknowledgments	*xv*
	About the Authors	*xvii*
Chapter 1	**Introduction**	**1**
	Overview	1
	System Description	4
	What Is a Wireless System?	4
	General Architecture, Basic Concepts, and Terminology	6
	Historical Perspective	10
	Systems Engineering and the Role of the Systems Engineer	12
	Problem Statement	16
Chapter 2	**The Radio Link**	**17**
	Introduction	17
	Transmitting and Receiving Electromagnetic Waves	18
	Isotropic Radiation	20
	Antenna Radiation Patterns	22
	The Range Equation	28
	Thermal Noise and Receiver Analysis	34
	Characterizing Noise Sources	35
	Characterizing Two-Ports	47
	Optimizing the Energy Transmission System	61
	System-Level Design	61
	Receiver Sensitivity	62
	Top-Level Design	63
	An Example Link Budget	66
	Conclusions	70
	Problems	70

Chapter 3 Channel Characteristics **77**

Introduction 77

Macroscopic Models 1: Reflection from the Earth's Surface 79

Macroscopic Models 2: Empirical Models 86

The Hata Model 87

The Lee Model 90

Macroscopic Models 3: Log-Normal Shadowing 95

Microscopic Models 1: Multipath Propagation and Fading 100

Introduction 100

A Two-Ray Model for Multipath Propagation: Stationary Receiver 102

Microscopic Models 2: Statistical Models for Multipath Propagation 106

Rayleigh Fading 106

Coherence Bandwidth 115

Microscopic Models 3: A Two-Ray Model with a Moving Receiver 121

Microscopic Models 4: A Statistical Model with a Moving Receiver 129

Area Coverage 132

The Link Budget 137

Conclusions 139

Problems 141

Chapter 4 Radio Frequency Coverage: Systems Engineering and Design **149**

Motivation 149

Requirements Assessment and System Architecture 150

Cellular Concepts 153

Estimation of Interference Levels 167

Cochannel Interference 167

Adjacent-Channel Interference 171

Cellular System Planning and Engineering 173

The Key Trade-offs 173

Sectoring 175

Cell Splitting 179

Operational Considerations 183

The Mobile Switching Center 184

Dynamic Channel Assignment 185
Handoff Concepts and Considerations 185
Traffic Engineering, Trunking, and Grade of Service 187
Conclusions 194
Problems 196

Chapter 5 Digital Signaling Principles 203
Introduction 203
Baseband Digital Signaling 204
Baseband Digital Communication Architecture 204
Baseband Pulse Detection 207
The Matched Filter 212
Correlation 216
Correlation Receiver 220
Receiver Performance 222
Carrier-Based Signaling 226
Modulation Overview 226
Modulated Carrier Communication Architecture 227
Digital Modulation Principles 229
Binary Phase-Shift Keying (BPSK) 236
Differential Binary Phase-Shift Keying (DPSK) 239
Quadrature Phase-Shift Keying (QPSK) 243
Offset QPSK (OQPSK) 251
Frequency-Shift Keying (FSK) 254
Gaussian Frequency-Shift Keying (GFSK) 262
Minimum-Shift Keying (MSK) 264
Spread-Spectrum Signaling 267
Overview 267
Frequency-Hopping Spread Spectrum 268
Direct-Sequence Spread Spectrum 271
Conclusions 278
Problems 280

Chapter 6 Access Methods 287
Introduction 287
Channel Access in Cellular Systems 290
Frequency-Division Multiple Access 295
The AM Broadcasting Band 296

The AMPS Cellular Telephone System 297
Effect of Transmitted Signal Design 298
Frequency-Division Duplexing 299
Time-Division Multiple Access 300
The U.S. Digital Cellular (USDC) System 302
The GSM System 304
Time-Division Duplexing 305
Code-Division Multiple Access 306
Frequency-Hopping CDMA Systems 307
Direct-Sequence CDMA Systems 311
Contention-Based Multiple Access 325
The Aloha Multiple-Access Protocol 326
The Slotted Aloha Protocol 328
Carrier-Sense Multiple Access 330
Conclusions 335
Problems 337

Chapter 7 Information Sources 343
Introduction 343
Information Sources and Their Characterization 346
Speech 347
Music 348
Images 349
Video 350
Data 351
Quality of Service (QoS) 352
Smooth versus Chunky 354
Digitization of Speech Signals 355
Pulse Code Modulation 356
Differential PCM 367
Vocoders 371
Coding for Error Correction 376
Convolutional Codes 377
Conclusions 389
Problems 392

Chapter 8 Putting It All Together 397
Introduction 397
Looking Backward 399
The First Generation 399
The Second Generation 400
Toward a Third Generation 405
Generation 2.5 407
Contemporary Systems and 3G Evolution 411
Wideband CDMA (W-CDMA) 411
cdma2000 Radio Transmission Technology (RTT) 420
OFDM: An Architecture for the Fourth Generation 432
Conclusions 442

Appendix A Statistical Functions and Tables 443
The Normal Distribution 443
Function Tables 446

Appendix B Traffic Engineering 453
Grade of Service and the State of the Switch 453
A Model for Call Arrivals 454
A Model for Holding Time 456
The Switch State Probabilities 457
Blocking Probability, Offered Load, and Erlang B 460
Computational Techniques for the Erlang B Formula 462
Erlang B Table 465

Acronyms 477
Index 483

Preface

This text is intended to provide a senior undergraduate student in electrical or computer engineering with a systems-engineering perspective on the design and analysis of a wireless communication system. The focus of the text is on cellular telephone systems, as these systems are familiar to students; rich enough to encompass a variety of propagation issues, modulation techniques, and access schemes; and narrow enough to be treated meaningfully in a text that supports a single course. The presentation is limited to what cellular systems engineers call the "air interface" and what network engineers call the "physical layer."

The presentation is unique in a number of ways. First, it is aimed at undergraduate students, whereas most other textbooks written about wireless systems are intended for students either at the graduate level or at the community college level. In particular, the presentation combines a clear narrative with examples showing how theoretical principles are applied in system design. The text is based on ten years' experience in teaching wireless systems to electrical and computer engineering seniors. The lessons learned from their questions and responses have guided its development. The text not only presents the basic theory but also develops a coherent, integrated view of cellular systems that will motivate the undergraduate student to stay engaged and learn more.

Second, the text is written from a systems-engineering perspective. In this context a "system" comprises many parts, whose properties can be traded off against one another to provide the best possible service at an acceptable cost. A system with the complexity of a cellular network can be designed and implemented only by a team of component specialists whose skills complement one another. Top-level design is the responsibility of systems engineers who can translate market requirements into technical specifications, who can identify and resolve performance trade-off issues, and who can set subsystem requirements that "flow down" to the subsystem designers. The text introduces students to the concept that specialists from a wide range of engineering disciplines come together to develop a complex system. Theory and contemporary practice are developed in the context of a problem-solving discipline in which a divide-and-conquer approach is used to allocate top-level functional system requirements to lower-level subsystems. Standard analysis results are developed and presented to students in a way that shows how a systems engineer can use these results as a starting point in designing an optimized system. Thus an overlying systems-engineering theme ties together a wide variety of technical principles and analytical techniques.

This text comprises eight chapters. An introductory chapter sets out the systems-engineering story. Chapters 2 and 3 introduce the air interface by considering how to provide enough power over a wide enough area to support reliable communication. Chapter 2 introduces the free-space range equation and thermal noise. On completing this chapter, students should be aware of the dependence of received power on range and of the role of noise in determining how much power is enough for quality reception. Chapter 3 introduces the terrestrial channel and its impairments, including the effects of shadowing and multipath reception. Next, Chapter 4 introduces the principle of frequency reuse and the resulting cellular system structure. The goal of this chapter is to show how a communication system can be extended to provide service over a virtually unlimited area to a virtually unlimited number of subscribers.

Once a power link is established, information must be encoded to propagate effectively over that link. Chapter 5 introduces modulation. The emphasis is on digital techniques common to cellular systems. Of particular interest are frequency efficiency, power efficiency and bit error rate, bandwidth, and adjacent-channel interference. Chapter 5 also introduces spread-spectrum modulation, emphasizing the ability of spread-spectrum systems to provide robust communication in the presence of narrowband interference and frequency-selective fading.

On completion of Chapter 5, students will have an appreciation of the factors involved in designing a point-to-point data link between a single transmitter and a single receiver. Chapter 6 introduces methods for multiple access, including FDMA, TDMA, and an introduction to CDMA. The ability of spread-spectrum systems to support multiple users over a single channel is emphasized.

Wireless systems carry information from a wide variety of sources, from speech to music to video to short text messages to Internet pages. When digitized, information from various sources produces data streams with differing properties. Further, subscribers apply different criteria to assessing the quality of different kinds of received information. Chapter 7 distinguishes streaming from bursty information streams. As second- and subsequent-generation cellular systems are highly dependent on effective use of speech compression, examples are given showing traditional digitization of speech and a brief introduction to linear predictive coding. Chapter 7 concludes with presentations of convolutional coding for error control and the Viterbi decoding algorithm. The systems-engineering story is pulled together in Chapter 8.

This text has been written to support a one-term senior elective course. It is assumed that students taking the course will have completed conventional courses in signals and systems and an introduction to communication systems. The signals and systems background should include a thorough introduction to the Fourier transform. It is also assumed that readers of the text will have completed an introductory course in probability, including coverage of probability density functions, expectations, and exposure to several conventional probability distributions. The material included in this text should be more than sufficient to support a one-semester course. At Rose-Hulman Institute of Technology the book is used to support a one-quarter course that includes four instructional meetings per week for ten weeks. The course covers all of Chapter 2, selections from Chapter 3, all of Chapter 4, and most of Chapter 5. The CDMA material from Chapter 6 is included as time permits.

Acknowledgments

The authors would like to acknowledge the many individuals and organizations that helped in the preparation of this text. The National Science Foundation provided funding (DUE-0341209) to allow time to write. Special recognition is extended to the Global Wireless Education Consortium (GWEC) and its administrators, Misty Baker and Susan Sloan, for their support in establishing the concept for the book and for advice along the way. Special thanks is also due to Gerald DiPiazza and Richard Frenkiel, two early pioneers responsible for the development of cellular communications, for their contributions to the text and their insightful comments. Gerald provided a thermal noise analysis for an early cellular mobile unit and Richard provided the splendid overview of the history of wireless communications in Chapter 1. We wish to express our sincere appreciation for the enthusiastic support of the folks at Award Solutions, Inc. In particular, we thank Russ Edwards, Hooman Razani, Chris Reece, and Nishith Tripathi, for their helpful discussions on contemporary systems and their evolution, and Satyajit Doctor and Ramki Rajagopalan, for their support of GWEC and the creation of this book. We also thank Andy Groome, of Groome Technologies, for his helpful discussions on current practice for creating link budgets and for providing some real-world examples.

About the Authors

Bruce A. Black completed his B.S. at Columbia University, his S.M. at Massachusetts Institute of Technology, and his Ph.D. at the University of California at Berkeley, all in electrical engineering. Since 1983 he has been on the faculty of the Department of Electrical and Computer Engineering at Rose-Hulman Institute of Technology in Terre Haute, Indiana, where he has been adviser to Tau Beta Pi and is adviser to the Amateur Radio Club (W9NAA). His interests are in communications, wireless systems, and signal processing. He has developed a variety of courses and laboratories in the signal-processing and communications areas, including a junior-level laboratory in communication systems and a senior elective in wireless systems. In 2004 he was named Wireless Educator of the Year by the Global Wireless Education Consortium. He is a member of Tau Beta Pi, Eta Kappa Nu, and Sigma Xi.

Philip S. DiPiazza received a B.E.E. from Manhattan College in 1964, an M.E. in electrical engineering from New York University in 1965, and a Ph.D. (electrical engineering) from the Polytechnic Institute of New York in 1976. His career spans more than forty years of professional experience in industry, academe, and private practice. During the first ten years of his career, he was a systems engineer engaged in the development of advanced airborne radar systems at the Norden Division of United Technologies. He joined Bell Laboratories (AT&T) in 1977, where, as a systems engineer and technical manager, he was engaged in the development of cellular mobile telephone (AMPS) and later wireless PBX systems. Dr. DiPiazza was responsible for the system integration and test of the first North American deployment of AMPS. Since retiring from AT&T Labs in 1998, he has served as an industry management consultant, Executive Director at Rutgers WINLAB, and Vice President and General Manager of the Melbourne Division of SAFCO Technologies, Inc. As a Visiting Professor at the Florida Institute of Technology, he was founding director for its Wireless Center of Excellence and developed graduate programs in wireless. He is currently an Adjunct Professor at Rose-Hulman Institute of Technology and a Senior Consultant with Award Solutions, Inc. Dr. DiPiazza is an adviser and member of the Global Wireless Educational Consortium and a member of the IEEE.

Bruce A. Ferguson received the B.S., M.S., and Ph.D. degrees in electrical engineering from Purdue University, West Lafayette, Indiana, in 1987, 1988, and 1992 respectively. He is currently a Communication System Engineer with Northrop Grumman Space Technology. He has worked with space and ground communication systems and photonics at TRW Space and Electronics (now NGST) and taught at Rose-Hulman Institute of Technology and the University of Portland in Oregon. Dr. Ferguson is a member Eta Kappa Nu and IEEE.

David R. Voltmer received degrees from Iowa State University (B.S.), University of Southern California (M.S.), and the Ohio State University (Ph.D.), all in electrical engineering. During nearly four decades of teaching, Dr. Voltmer has maintained a technical focus in electromagnetics, microwaves, and antennas. His more recent efforts are directed toward the design process and project courses. He has served in many offices of the ERM division of ASEE and in FIE. Dr. Voltmer is an ASEE Fellow and a Life Senior Member of IEEE.

Frederick C. Berry received the B.S., M.S., and D.E. degrees from Louisiana Tech University in 1981, 1983, and 1988 respectively. He taught in the Electrical Engineering Department at Louisiana Tech University from 1982 to 1995. Currently Dr. Berry is Professor and Head of the Electrical and Computer Engineering Department at Rose-Hulman Institute of Technology. In 2007 he became Executive Director of the Global Wireless Education Consortium. He is a member of Tau Beta Pi, Eta Kappa Nu, and Sigma Xi.

CHAPTER 1

Introduction

Overview

On the night of April 14, 1912, the RMS *Titanic*, en route from Southampton, England, to New York, struck an iceberg and sank in the North Atlantic. Over fifteen hundred lives were lost when the ship went down, but fortunately for the more than seven hundred passengers and crew who were able to find accommodation in the ship's lifeboats, the *Titanic* was equipped with a wireless system. The *Titanic*'s wireless included a 5 kW rotary spark transmitter built by the Marconi Wireless Company. Distress calls were heard by a number of ships at sea, including the *Carpathia* that arrived on the scene of the disaster several hours later, in time to rescue the survivors.

The wireless traffic between the *Carpathia* and shore stations in North America was widely monitored. News was passed to the press even before the fate of the *Titanic*'s passengers was known. The widespread publicity given to this disaster galvanized public interest and propelled wireless communication into the forefront of attention. The age of wireless communication might be said to have begun with the sinking of the *Titanic*.

As social beings, humans have a fundamental need to communicate. As we grow and learn, so do our communication needs evolve. Dramatic advancements over the past century have made several facts about our evolving communication needs rather apparent: (1) The information that needs to be communicated varies widely; (2) the types and amount of information that needs to be communicated continuously change, typically toward higher complexity; and (3) current technology rarely meets communication demands, so technology evolves. These facts, along with a healthy worldwide economy, produced the wireless revolution in the late twentieth century. Wireless communication is here to stay, and the design principles used to create wireless technology differ enough from those used to create wired communication systems that a separate treatment is necessary.

In this text the process of designing a wireless communication system is presented from the perspective of a systems engineer. Two main goals of the text follow immediately: (1) to present the concepts and design processes involved in creating wireless communication systems, and (2) to introduce the process of systems engineering and the role of a systems engineer to provide an organizing framework under which to introduce the wireless system concepts. In the industrial world, the design process flows in an organized manner from problem definition, through conceptual and detailed design, to actual deployment. In this text, information from first principles to advanced topics is presented in a fashion compatible with systems-engineering design processes, which are required to manage the development of complex systems.

In Chapter 1 the problem of moving information wirelessly from any point A to any point B is introduced. In every engineering endeavor it is important to have a clear understanding of the problem to be solved before beginning, and so the system and its requirements are defined. The role of a systems engineer and the methods of systems engineering are introduced as the perspective for conducting our study and design.

Chapter 2 presents the most fundamental element of our wireless system, the radio link that connects points A and B. This chapter addresses two issues: how radio waves propagate in space, and how much power must be provided at point B to ensure a desirable quality of communication service. This chapter focuses on propagation in free space and the development of the **range equation**, a mathematical model familiar to both radio and radar engineers. We introduce the antenna as a system element and the antenna design engineer as a member of the design team. To answer the question of how much power is enough, we develop models and analysis tools for thermal noise and describe how thermal noise limits performance. Signal-to-noise ratio (SNR) is introduced as a measure of system performance. Finally, the concept of link budget, a fundamental tool for radio frequency (RF) systems engineering, is presented and supported with examples.

Chapter 3 focuses on signal propagation in the real world. Obstacles in the signal path and indeed the very presence of the Earth itself modify a signal as it travels between endpoints. Various terrestrial propagation models are presented and discussed. These models provide methods for predicting how a signal will propagate in various types of environments. The phenomena associated with shadow fading, Rayleigh fading, and multipath propagation are described as well as the effects of relative motion of a receiver and of objects in the environment. Statistical methods are developed that allow engineers to create robust designs in an unstable and changing environment. Receiver design and channel modeling are added to the list of design functions that a systems engineer must understand to competently interact with team members who specialize in these design disciplines.

Given a basic understanding of methods to ensure that an adequate signal can be conveyed between two endpoints, we discuss the concepts and complexities involved in allowing many users in a large area to share a common system. Geographic diversity and frequency reuse are discussed and used as the basis for developing the "cellular" concept in Chapter 4. The cellular concept is the fundamental basis for designing and deploying most wireless communication systems that must provide service for many users over a large geographic area. The chapter describes how engineers using cellular engineering techniques plan for growth in user capacity and coverage area. Traffic engineering and the use of the Erlang formula as tools for predicting and designing a system for user capacity are demonstrated. At this stage of the design, system-level concerns are well above the device or subsystem level.

In Chapter 5 we describe the methods used to convey information over the wireless link. The process for conveying information using a radio signal, called modulation, is described from a trade-off perspective. The characteristics of several digital modulation schemes are developed and their attributes are compared. The key design parameters of data throughput, error rate, bandwidth, and spectral efficiency are contrasted in the context of optimizing a system design. Also in this chapter we introduce spread-spectrum signaling. Spread spectrum is a modulation technique that broadens the bandwidth of the transmitted signal in a manner unrelated to the information to be transmitted. Spread-spectrum techniques are very effective in making signals resilient in the presence of interference and frequency-selective fading. Our study of spread-spectrum techniques continues in Chapter 6, as these techniques provide an important basis for multiple-access communications.

The first five chapters provide all of the fundamental elements of system design for providing radio coverage to many users over a large area and for designing the components that support the conveying of information at a given quality of service (QoS) across a wireless link between individual users. Chapter 6 introduces various methods that allow many users to access the system and to simultaneously use the resources it provides. In this chapter we introduce the classical methods of frequency-division and time-division multiple access, as well as spread-spectrum-based code-division multiple access which allows independent users to share the same bandwidth at the same time. In providing a multiple-access capability, a systems engineer unifies a variety of system-level design activities to make the system accessible to a varying number of users.

People wish to communicate different types of information, and the information they want to communicate comes from a variety of sources. Chapter 7 discusses several of the types and sources of information commonly communicated in contemporary wireless systems. The required QoS that is to be provided to a system's users must be accounted for in nearly every aspect of a system design. Users' perceptions of what constitutes good quality vary for different types and sources of information and always depend on how well a signal representing the information is preserved in the communication process. Chapter 7 discusses some of the fundamental relationships between the perceptual measures of QoS and specific system design parameters. Understanding these relationships allows a systems engineer to design for predictable QoS at minimum cost. As modern wireless systems are designed to carry information in digital form, a major part of this chapter is about efficient digitization of speech. We discuss two general categories of speech "coding": waveform coding and source coding. As an example of the waveform coding technique we examine traditional pulse code modulation (PCM). Our example of source coding is linear predictive coding (LPC). This latter technique has been extremely successful in providing high-quality, low-bit rate digitization of voice signals for cellular telephone applications. Following the discussion of speech coding, the chapter concludes with an example of coding for error control. Convolutional coding is used for this purpose in all of the digital cellular telephone systems. We introduce the coding method and the widely used Viterbi decoding algorithm.

Chapter 8 wraps up the presentation with a review of the lessons developed in the preceding chapters. This is followed by an overview of the generations of cellular telephone systems and a look into the future at the way wireless systems are evolving to provide an increasing array of services at ever-higher quality. As wireless systems evolve, they tend to become more complicated. Thus the role of the systems engineer in managing the design process and in understanding the myriad of design trade-offs and interactions becomes ever more important.

System Description

What Is a Wireless System?

In the most general sense, a wireless system is any collection of elements (or subsystems) that operate interdependently and use unguided electromagnetic-wave propagation to perform some specified function(s). Some examples of systems that fit this definition are

- Systems that convey information between two or more locations, such as personal communication systems (PCS), police and fire department radio systems, commercial broadcast systems, satellite broadcast systems, telemetry and remote monitoring systems
- Systems that sense the environment and/or objects in the environment, including radar systems that may be used for detecting the presence of objects in some region or volume of the environment and measuring their relative motion and/or position, systems for sensing or measuring atmospheric conditions, and systems for mapping the surface of the Earth or planets
- Systems that aid in navigation or determine the location of an object on the Earth or in space

Each of these systems contains at least one transmitting antenna and at least one receiving antenna. In the abstract, an antenna may be thought of as any device that converts a guided signal, such as a signal in an electrical circuit or transmission line, into an unguided signal propagating in space, or vice versa. We note in passing that some systems do not need to transmit and receive simultaneously. For example, the WiFi local area network computer interface uses a single antenna that is switched between transmitter and receiver. Specifically, a pulse of energy is transmitted, after which the antenna is switched to a receiver to detect the response from the network access point.

As the examples show, some systems may be used to convey information, whereas others may be used to extract information about the environment based on how the transmitted signal is modified as it traverses the path between transmitting and receiving antennas. In either case, the physical and electromagnetic environment in the neighborhood of the path may significantly modify the signal. We define a **channel** as the physical and electromagnetic environment surrounding and connecting the endpoints of the transmission path, that is, surrounding and connecting the system's transmitter and receiver. A channel may consist of wires, waveguide and coaxial cable, fiber, the Earth's atmosphere and surface, free space, and so on. When a wireless system is used to convey information between endpoints, the environment often corrupts the signal in an unpredictable[1] way and impairs the system's ability to extract the transmitted information accurately at a receiving end. Therein lies a major difference between wired and wireless systems. To provide a little further insight, we compare some of these differences.

The signal environment or channel characteristics of a single-link wired system are rather benign.

- At any instant of time, the path between endpoints is well known and many of its degrading effects upon a signal can be measured and compensated for.
- Signal dropout (signal loss), momentary or otherwise, is very rare.
- Random effects such as "thermal noise" and "interference" are fairly predictable and controllable and therefore less likely to corrupt the signal to the extent of unintelligibility.
- The signal environment does not change or changes very slowly with time.
- The endpoints do not move.

In contrast, the signal environment of a wireless system is hostile.

- The direction of the signal cannot be completely controlled, and the path between endpoints is not unique.
- The path between endpoints is time-varying.
- Signal dropouts are frequent.
- Noise and interference levels are often difficult to predict and time-varying.
- Objects in the path between and surrounding the endpoints affect the signal level and its content.
- Variations in the signal environment change with geographic location, seasons, and weather.
- For mobile systems, as in cellular and PCS systems, at least one of the endpoints may be moving at an unknown and sometimes significant speed.

As an everyday example, the differences between wired and wireless systems may be compared to the difference between carrying on a conversation with someone in the environment of your living room versus conversing in the environment of a busy airport runway. The same principles of communication theory apply to the design of both wired and wireless communication systems. In addition to those specific functions associated with the unguided propagation of signals, however, the most profound differences between the implementations of wired and wireless communication systems relate to overcoming the signal impairments introduced by a changing wireless channel and, for mobile systems, compensating for the possible motion of the endpoints.

1. The term *unpredictable* is used in the sense that the signal cannot be perfectly determined at any point in space. As we will see, however, we can infer a great deal about a signal using statistical modeling. These models are a fundamental basis for system design.

In addition to providing the fundamental basis for the design of wireless communication systems, the principles of communication theory, RF engineering, and propagation in real-world environments also apply to a host of other applications. As examples, these principles apply to a multitude of radar applications, including object or target detection, location and ranging, speed/velocity measurement, terrain mapping, weather monitoring, and navigation. In fact, many of the techniques used to develop modern personal communication systems were originally developed and proved for radar applications. In contrast to wireless communication systems that convey information between endpoints, radar systems analyze the way transmitted signals are reflected and modified by the presence of objects or variations along the signal path to extract information about the objects or the environment that the signal traverses. As a simple example, consider that a narrow pulsed-RF signal is transmitted in a given direction. Objects within the transmission path reflect some fraction of the signal incident upon them. If a receiver colocated with the transmitter detects an approximate replica of the transmitted signal sometime after the transmitted signal is sent, it is reasonable to assume that an object is located in the direction of transmission and the distance to the object is proportional to the time delay between transmitted and received signals. If no signal is detected within a specified period of time, it is assumed that there are no reflecting objects in the path of the signal, over a given range.

Clearly our general definition of a wireless system fits a vast range of seemingly unrelated applications. It is profoundly important, however, to recognize that all of these applications are founded on a common set of enabling principles and technologies encompassing communication theory, RF engineering, and RF propagation. Although the focus of this text is personal communication systems, the principles and techniques to be presented provide a strong foundation for study of other wireless system applications.

General Architecture, Basic Concepts, and Terminology

At a high level, every communication system is described by a common block diagram. In this section we present a basic functional description of each of the blocks to introduce some of the terminology of wireless systems and to help motivate later discussion of each of the functions.

We begin by considering the general block diagram of a wireless system for a generic application as shown in Figure 1.1. Many of the blocks and their functions apply to both wired and wireless communication systems. Note, however, that the blocks contained within the dashed outline are fundamental and necessary to wireless systems. With the exception of the antennas, all of the remaining blocks may also be found in wired system applications.

The box labeled "Information Source" includes all functions necessary to produce an electrical signal that adequately represents the actual information to be communicated between end users. The term **end user** refers to a person or device that is the source or recipient (sink) of the information to be communicated. The term **endpoint** refers to the location of the transmitters and receivers in the communication path. End users may or may not be colocated with the endpoints. The functions of the Information Source box might include

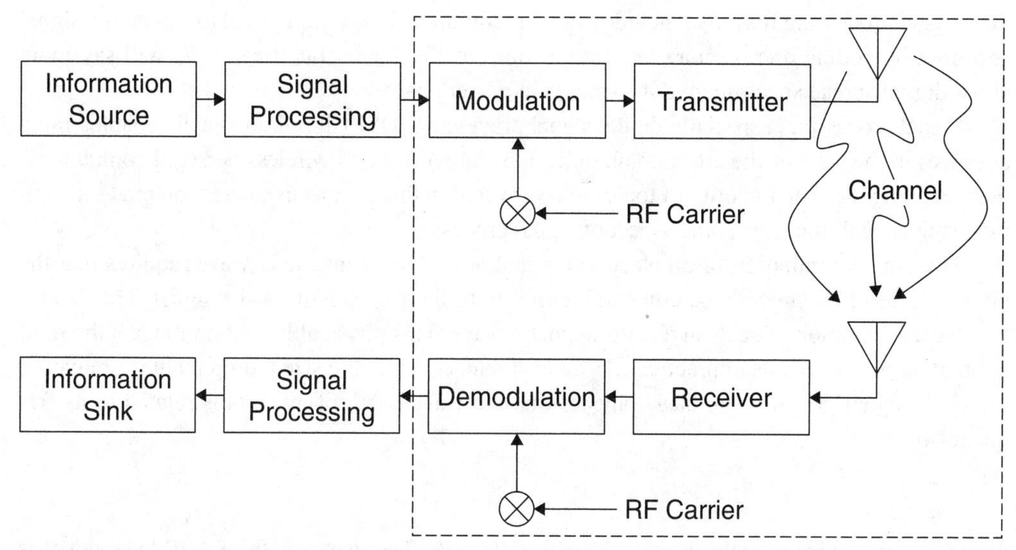

Figure 1.1 A Wireless System

- Creation of an analog waveform representing speech derived from a microphone, or creation of a digital bit stream resulting from sampling of an analog waveform
- Formatting digital information such as data, text, sampled audio, images, video, and so forth

The signals from information sources are typically bandlimited; that is, they contain frequencies from DC (or near DC) to some nominal cutoff frequency. They are termed **baseband signals**.

The box labeled "Signal Processing" encompasses all operations necessary to convert information signals into waveforms designed to maximize system performance. Signals may be processed to increase capacity, throughput, intelligibility, and accuracy and to provide other auxiliary functions. In modern wireless systems, many of the signal-processing functions are aimed at improving signal reception by mitigating the corrupting effects of the transmission medium or environment. Signal-processing functions on the transmitting end may include

- Converting analog signals to digital signals of a specific type
- Shaping signals to minimize the corrupting effects of the environment or transmission medium
- Compressing and coding signals to remove redundancies and improve throughput
- Coding signals to aid in the detection and correction of errors caused by the environment
- Encryption of signals for privacy
- Multiplexing information from several sources to fully utilize the channel bandwidth
- Adding information that simplifies or enhances access and control for the endpoints or end users

Signal processing may also include digital modulation, a technique used to spread the signal spectrum by coding one or more bits into a substantially longer bit stream. We will say more about digital spread spectrum and its benefits in a later chapter.

Signal processing, especially digital signal processing (DSP), has dramatically enabled rapid advances in the state of the art of communications in general and wireless personal communications in particular. The majority of topics to be covered in this text, as in any text on modern communications, will focus on some aspect of signal processing.

The efficient radiation of an electrical signal as an electromagnetic wave requires that the physical size of the antenna be comparable in size to the wavelength of the signal. This is also true for the reception of such an electromagnetic wave. This physical limitation renders the radiation of baseband signals impractical. As an example, consider the size requirement for radiating a 10 kHz signal. Recall from basic physics that the wavelength of a signal is related to its frequency by

$$\lambda = c/f, \tag{1.1}$$

where c is the speed of light in free space, 3×10^8 m/s. The wavelength of a 10 kHz signal is about 98,000 feet. If a typical quarter-wavelength ($\lambda/4$) antenna were used, it would be 24,600 feet or 4.7 miles in length. In contrast, $\lambda/4$ antennas in the cellular (900 MHz) or PCS (2 GHz) bands are 3.3 inches and 1.5 inches long, respectively. For this reason, practical wireless systems employ high-frequency or radio frequency sinusoidal signals called **carriers** to transport (or carry) information between endpoints.

The laws and regulations of the countries in which the systems are to be deployed govern and constrain the radiation of electromagnetic waves. Various frequency bands are allocated by law for specific applications; for example, there are AM, FM, and TV broadcast bands; public safety bands; airport communication, radar, traffic control, and maritime applications bands; and others. Furthermore, the laws may regulate transmitted power, transmitted spectrum and spectrum characteristics, modulation method, geographic location, tower height, and so on. Figure 1.2 shows some of the spectrum allocations in the ultra-high-frequency (UHF) band from 300 MHz to 3 GHz. A detailed chart of spectrum allocations in the United States is available from the National Telecommunications and Information Administration (NTIA).[2] In the United States, the Federal Communications Commission (FCC) is the agency entrusted with the responsibility for administering the use of the radio spectrum, granting licenses, and working with government and private industry to develop fair and equitable regulatory rules and standards.

Information signals are imposed upon a carrier signal by modulating (varying) its amplitude, frequency, and/or phase in direct relation to the variations of the information signal. At the receiving end, an information signal is extracted from the carrier by a process of demodulation. The boxes labeled "Modulation" and "Demodulation" refer to any of a wide range of techniques

2. National Telecommunications and Information Administration. Source: www.ntia.doc.gov/osmhome/allochrt.html, accessed August 8, 2006.

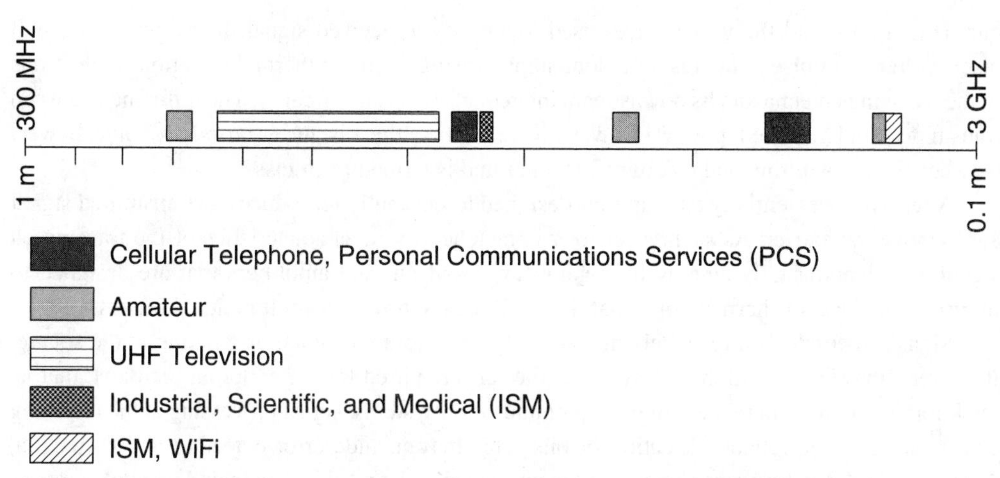

Figure 1.2 Some Spectrum Allocations in the UHF Band (January 2008)

that may be used to impose/extract an information signal upon/from a carrier. As we will discuss later, the choice of modulation scheme is strongly influenced by a number of factors, including available frequency spectrum, spectrum rules and regulations, required throughput, channel characteristics, and QoS requirements. In the context of a wireless system (or a "broadband" wired system that employs coaxial cable, waveguide, or fiber as a transmission medium), a modulator translates the spectrum of a baseband signal to a bandpass spectrum centered about some high "radio" frequency appropriate for the intended application and consistent with spectrum regulations.

Many wired systems (for example, "plain old telephone service" [POTS]) use transmission media that allow the system to operate effectively at baseband. For such systems, a modulator translates an information signal into waveforms (sometimes called line codes) that are optimized for the given transmission medium and application. For example, a line code may convert a binary bit stream (1s and 0s) into a bipolar or multilevel voltage waveform, or it may convert a bit stream to positive and negative voltage transitions.

For wireless systems, a transmitter is essentially an RF power amplifier and appropriate bandpass filter. A transmitter drives a transmitting antenna (often through a coaxial cable or waveguide) and ensures that the modulated RF signal is radiated at a power level, and within a bandwidth, specific to the application and applicable regulations. Wired systems, on the other hand, often use transmitters termed **line drivers** that ensure that transmitted signals have sufficient energy to overcome the line losses in the path to the receiving end.

The power intercepted and absorbed by a receiving antenna is usually much lower than the transmitted power. For example, when a cellular base station transmits with a power of one watt, the received signal two miles away may be only a few tenths of a nanowatt. In fact, a receiver may be located so far from the transmitter that the signal level is comparable to system noise. **System noise** is a random signal that arises from a number of sources such as galactic radiation, engine ignitions, and the very devices used to amplify a received signal. In particular, we will discuss **thermal noise**, which is a random signal arising from the thermal agitation of electrons in the receiving antenna and its downstream interconnections and circuitry. The difference between transmitted and received power is inversely related to the distance (raised to some power) between the transmitting and receiving antennas and is termed **path loss**.

A receiver is essentially an amplifier designed to optimally reproduce the transmitted signal and remove the carrier. As such, a receiver is matched to the characteristics of the transmitted signal. Receivers usually employ high-gain, low-loss front-end amplifiers that are designed to minimize the level of thermal noise that they will pass to downstream functional blocks.

Signal processing on the receiving end seeks to restore the signal originating at the source. It converts the signal from the receiver into the form required for the endpoint recipient, that is, the Information Sink. In modern digital communication systems, signal processing at the receiving end is aimed at the reliable detection of bits. This may include error detection and correction, depending on the coding used to transmit the original signal, and also may include digital demodulation of a spread-spectrum signal.

Historical Perspective

A hundred years ago, a radio "system" was a transmitter, a receiver, and a path that could be successfully traversed by the miracle of radio waves. Even then there were broader issues to resolve—trade-offs that could be made between one element of the configuration and another. A more powerful transmitter or a more sensitive receiver; a higher mast or a directive antenna—these were some of the potential design improvements that could extend the range of the system when needed. Which of these to adopt became an important question, affecting cost and having performance implications in other dimensions. Was the power demand excessive? Was range being limited by circuit noise within the receiver or by external environmental noise? Was it limited by a physical obstruction over which one might radiate?

Radio had evolved from the design of general-purpose transmitters and receivers to a variety of "systems" with specific applications. Broadcast systems created the wildly popular phenomena of radio and television entertainment, by creating a way to deliver that entertainment inexpensively to a mass market. The trade-offs shifted again; base transmitters could be extremely powerful and expensive, sited on tall buildings or hills, using tall masts and elaborate gain antennas, but the millions of home receivers had to be low-cost consumer products.

"Propagation engineers" now had a more difficult problem; rather than designing a single path from one radio to another, they were concerned with an "area of coverage" in which signal quality was likely (but not guaranteed) to be acceptable. Moreover, the demand for channels required channels to be reused in nearby areas, so that interference needed to be predicted and controlled in the service areas of systems. Probability and statistics had joined the sciences that contributed to system design.

The first mobile telephone systems emerged in the 1940s, quickly became popular, and introduced a number of new trade-offs. The mobile equipment, carried in the trunks of cars and powered from the car battery, needed to be smaller and lower in power (as well as cheaper) than the base station equipment; but coverage areas needed to be large, since cars would need to operate throughout large urban areas. A single high-powered base station could serve an entire urban area of more than a thousand square miles, but the lower-powered return paths from the vehicles could not, and satellite base station receivers became necessary. The higher cost of the (relatively few) satellite base stations could now be traded off for the (smaller) savings in power in the more numerous mobile units. This trade-off of expensive base equipment against more numerous mobile radios is characteristic of such systems.

In the major urban areas, a mobile telephone system would now consist of several radio channels, serving several hundred customers. This aggregation of radios and customers led to the incorporation of telephone traffic-handling probabilities into mobile system design—designers would now calculate the probability that an idle channel would be available. Because of the shortage of channels, however, service was very poor before the days of cellular systems. In the 1950s mobile operators who set up calls were replaced by equipment to automatically select idle channels, allowing the dialing of telephone calls in both directions. Signaling had been added to voice communication on radio channels, together with the first steps toward complex logic.

As early as the 1940s, when the first crude mobile telephone systems were going into service, AT&T had begun to propose a new concept in mobile radio system design. Rather than using a single high-powered base station to cover an entire urban area, they proposed to create a service area from a grid of smaller coverage areas, called "cells." This had several important advantages. It allowed both base and mobile radios to operate at lower power, which would reduce radio costs. It also allowed larger service areas, since additional small coverage areas could always be added around the periphery to expand the system. Most importantly, although nearby cells required different channels to prevent interference, farther cells could reuse the same channels. In this way each channel could handle tens or even hundreds of calls in the same urban area, overcoming the limitations on capacity that were a result of spectrum shortages. These new systems would require a few hundred channels to get started, however, and the needs of the broadcasters were more persuasive in that period.

In 1968 the FCC finally opened the inquiry that ultimately led to cellular systems in the 1980s. For the advantages they provided, however, these systems demanded a new level of complexity. This time, the major complexity was not in the radio design, which saw few radical changes. With the introduction of small cells, calls could cross many cells, requiring mobile locating, channel switching during calls, and the simultaneous switching of wireline connections from one cell to another. Mobiles had to identify systems and find the channels on which calls could be received or originated, which required the introduction of microcomputers in mobile radios and made the technology of telephone switching machines an important element of radio system design. Moreover, with the introduction of so many disciplines in the design of a single system, and a variety of new trade-offs to be made, it was no longer practical for the many engineers to mediate these trade-offs, and the practice of systems engineering became a new and important discipline.

The introduction of cellular systems also marked the continuation of a long-term trend, in which spectrum shortages drove system designs to higher and higher frequencies. Frequencies such as 900 MHz (and later, 2 GHz) were used for such applications for the first time, and it became necessary to understand the propagation characteristics at these frequencies in real-world environments. Moreover, the old methods of the propagation engineer, in which terrain elevations were plotted to determine coverage, were no longer practical for hundreds of cells in a single system, and statistical coverage methods were developed to assure an acceptable quality of coverage. This trend has reversed once again more recently, as computers have allowed detailed terrain studies to be carried out for many cells.

Even as the first analog systems such as the Advanced Mobile Phone Service (AMPS) were being deployed in the early 1980s, efforts were under way to provide significant performance and capacity enhancements enabled by digital communications, advancements in digital signal-processing technology, and speech encoding. The Global System for Mobile Communications (GSM) was a cooperative effort of European countries to define an evolutionary system that provided increased capacity (or equivalently improved spectral efficiency), improved the quality of service, and allowed seamless roaming and coverage across the continent, and eventually around the world. The GSM standard was the first to encompass both the radio elements and the interconnection of serving areas to provide a holistic approach to ubiquitous service. As the first commercial digital cellular system, GSM demonstrated the power of digital signal processing in providing spectrally efficient, high-quality communications. GSM systems began deployment in the early 1990s.

By the mid-nineties, digital spread-spectrum systems were being introduced in North America under standard IS-95. Introduced by Qualcomm, Inc., a U.S.-based company, this system allows all cells to use the same frequency. Each channel is distinguished not by a distinct frequency or time slot but by a spreading code. The fundamental basis for this system is a technique called code-division multiple access (CDMA), a technique that has become the universal architecture for third-generation systems and beyond. CDMA systems have provided another technological leap in complexity, bringing additional enhancements to capacity, information bandwidth, quality of service, and variety of services that can be provided.

Each generation of wireless systems builds upon the technological advances of the prior generation. For each step in this evolution, the classical tools of the engineer remain, but they are honed and reshaped by each subsequent generation. The importance of system design and the role of systems engineering have grown substantially with each new technological generation. The continuing demand for new services and increased capacity, interacting with ongoing technological advancement, leads to new opportunities for system design, new problems to solve, and even the development of new engineering disciplines.

Systems Engineering and the Role of the Systems Engineer

Wireless communications, and communications in general, are specializations in the discipline of systems engineering. Our approach to the study of wireless communications from the perspective of a systems engineer is therefore a study in a specialized field of systems engineering. It is fitting, then, that we begin our study by discussing systems engineering at a general level.

Some type of system supports nearly every aspect of our daily life. Systems help us to travel anywhere on the Earth (and beyond), create memos, solve complex problems, store and retrieve vast arrays of information, cook our food, heat and light our homes, entertain ourselves and our friends, and of course communicate with each other. There is no universally accepted standard definition of a "system," but the purposes of this discussion are served by the working definition: A **system** is any collection of elements or subsystems that operate interdependently to perform some specified function or functions.

An automobile, an airplane, a personal computer, a home, a television or radio, an ordinary telephone, and a microwave oven are all common examples of systems. But at a lower level, an automobile engine or transmission, an airplane's hydraulic system, a computer chip or microprocessor, a home air-conditioning or heating unit, or the video circuitry of a television or audio circuitry of a radio are also systems by definition. Depending on the context of discussion, a system may often be referred to as a subsystem, since it may perform only a few of the intended or auxiliary functions of the overall system of which it is a part. For example, an automobile is a system that conveys passengers and/or objects to arbitrary geographic locations. Its subsystems are the engine, transmission, braking system, steering system, chassis, dashboard, and so on, all of which are necessary for an automobile to perform the functions we have come to expect. Likewise, an engine is a collection of other subsystems such as the ignition system, fuel system, and emission control system. At some level the terms *system* and *subsystem* become less relevant. For example, a passive circuit may be considered a system, but considering resistors, capacitors, and inductors as subsystems has little relevance. In such instances we may choose to use the term *component* or *element.*

In the previous sections we introduced a simplified block diagram for a generic wireless system. Each block may be considered a subsystem of the overall system. Furthermore, each block performs some direct or auxiliary function needed to meet the general requirements of the system application. Regardless of the intended wireless application, the designs of the various blocks and their functions are founded on principles derived from distinct specialty areas. To identify a few of these specialty areas:

- Antennas and electromagnetic wave propagation
- Microwave circuit theory and techniques
- Signals, systems, and signal processing
- Noise and random processes
- Statistical nature of the environment and its effects on a propagating signal
- Communication theory
- Traffic theory
- Switching and networking theory

Depending on complexity and scale, the design and development of a system usually require knowledge and professional expertise in a number of distinctly different disciplines. For example, the development of large-scale wireless systems, such as personal communication systems or advanced radar systems, requires the participation of specialists in such diverse disciplines as

- Antenna design
- RF propagation and radio environment modeling
- Microwave circuit design
- Transmitter design
- Low-noise amplifier (LNA) design
- Modulator/demodulator (modem) design
- Digital circuit/system design
- Signal processing
- Real-time, non-real-time, and embedded software development
- Power systems and power management
- Switching, networking, and transmission
- Mechanical structures and packaging
- Human factors engineering
- Manufacturing engineering
- Reliability and quality engineering
- And, last but not least, systems engineering

Successful development companies usually have processes (sequences of well-defined steps and procedures) and information systems that allow development teams to work and communicate effectively; track progress; manage schedule, budget, and other resources; control changes; and maintain and improve product quality. Highly successful companies also review their processes, constantly seeking ways to reduce development costs and schedule time while improving product quality, customer satisfaction, and cost competitiveness. In fact, process engineering and improvement is an important area of specialization. A strong and continuously improving development process is often vital to a company's ability to compete in a given market.

Development processes may vary among companies, but they all possess common phases of particular emphasis, for example,

- Product/system definition
- Design/development
- Integration and system test
- Manufacture
- Product life-cycle management

The specific activities in each phase may vary significantly, and many of the phases may, and often do, run concurrently.

One of the most important factors contributing to the successful development of a system is a complete, well-directed, and stable product definition. The product definition, sometimes called "functional product requirements" (FPR), is usually developed by a marketing or market research organization in concert with members of the technical development community, especially systems engineering. In addition to specifying the required attributes of the system from a customer perspective, an FPR also defines all the information necessary to ensure a viable financial return for the investors, including cost to manufacture the product, time to market, development budget, projected manufacturing ramp-up and life-cycle volumes, key competitive attributes, and so forth.

The design and development phase of any system usually begins with a system design. It is one of the most important products of a systems-engineering effort. A system design consists of all the requirements, specifications, algorithms, and parameters that a development team uses to design and develop the hardware and software necessary to implement and manufacture a product in accordance with an agreed-upon product definition. System-level documentation may include

- System-level requirements—a high-level technical document that translates the needs expressed from a customer perspective into technical constraints on system functions, performance, testing, and manufacture
- System architecture—a specification that defines all of the parameters and subsystem functions necessary to ensure interoperability among subsystems and meet system requirements, including distribution of system-level functions among the subsystems, definition of subsystem interfaces, and specification of system-level controls
- Supporting analyses—appropriate documentation of all analyses, simulations, experimentation, trade-off studies, and so on, that support the choice of key technical parameters and predict and/or verify system-level performance

As it relates to a system design, the responsibilities of a systems engineer are to

- Translate customer-level functional requirements into technical specifications at a system level
- Develop a system architecture and determine specific parameters to ensure that the system will meet the desired level of functionality and performance within specified constraints
- Perform trade-off analyses among the system elements to ensure that the implementation requirements can be met within the specified constraints and technology limitations
- Develop and negotiate specific requirements for each of the subsystems based on analysis, modeling, experimentation, and simulation

These functions are the focus of this text and are the basis for many other functions that systems engineers perform. These other functions might include

- Interacting with potential customers
- Developing human-interface specifications

- Developing plans, methods, and criteria for system integration and verification
- Interfacing with government and legal entities
- Specifying deployment, maintenance, and operations procedures
- Competitive analysis
- Supporting regulatory and standards development

Depending on the complexity of the system being developed, a team of systems engineers, each of whom has a particular area of expertise, may be required to fully perform all of the systems-engineering functions of a development.

Problem Statement

This text develops the major systems aspects of personal communication systems while demonstrating the application of relevant theory and principles and introducing students to some of the real-world aspects of the wireless systems-engineering profession. It is fitting, therefore, that the subject matter be presented in the context of a solution to a general systems-engineering problem. To be specific, we seek to design a wireless telecommunication system that will

- Support the communication of information of various types, including speech, text, data, images, and video, in urban, suburban, and rural environments and with quality approximating that of wired communications
- Be capable of expanding in geographic coverage
- Allow for virtually limitless growth in the number of users
- Support endpoints that are not geographically fixed and, in fact, may be moving at vehicular speeds

Many of the attributes of this system, as stated previously, were in fact the major objectives underlying the development of the very first cellular mobile telephone systems. Our discussions of principles and concepts are presented as motivation for solving this systems-engineering problem in particular. In view of the continued advances in digital technologies and the directions of modern communication systems, our emphasis will be on digital wireless communications, although many of the principles apply to both analog and digital systems.

Since the advent of the first mobile phone systems, the meanings of some commonly used terms have become blurred by marketing and advertising efforts to provide some level of distinction between early first and later generations of systems. Specifically, the terms *cellular* and *PCS* are often used to identify the cellular frequency (850 MHz) or personal communication systems (or services) (1.9 GHz) frequency bands. The term *cellular,* however, originally referred to the technology underlying nearly all of the systems that may be classified as personal communication systems. We will endeavor to ensure that the meaning is always clear in context; however, it is important to recognize that most modern systems are capable of operating in either band.

CHAPTER 2

The Radio Link

Introduction

What is involved in the engineering of a wireless system? When your mobile phone rings, a set of actions has already been put in motion to allow you to stay effortlessly connected to a vast communication network not dreamed of by your parents. Several disciplines, theoretical studies, technologies, and engineering developments have converged to create this wonder, and it starts with the invisible radio frequency (RF) link connecting your mobile phone to one of the ubiquitous base station towers around your town. To begin our study of wireless systems engineering, we first investigate RF links. The engineering of these connections will introduce the process of systems engineering that will take us through most of the physical system design.

Two fundamental and interrelated design considerations for a wireless communication system are

- The distance, area, or volume over which the system must meet specified performance objectives, and
- The capacity of the system, as related to the number of users that can be served by the system and the uses to which the system will be put

Note that capacity may be expressed in a number of ways. It may be expressed as the number of simultaneous two-way conversations that can take place, the average data rate available to a user, the average throughput for data aggregated over all users, or in many other ways, each related to the nature of the information being communicated. It should become apparent in later discussions that area and capacity criteria are interdependent from both technical and economic viewpoints. For wireless systems, coverage area and capacity are intimately related to the attributes of the **RF link** (or, in cellular telephone terminology, the "air interface"), and thus we begin our study with an overview of this type of link.

Most people have experienced the degradation and eventual loss of a favorite broadcast radio station as they drive out of the station's coverage area. This common experience clearly demonstrates that there are practical limits to the distance over which a signal can be reliably communicated. Intuition suggests that reliable reception depends on how strong a signal is at a receiver and that the strength of a signal decreases with distance. Reception also depends on the strength of the desired signal relative to competing phenomena such as noise and interference that might also be present at a receiver input. In fact, the robustness of a wireless system design strongly depends on the designers' abilities to ensure an adequate ratio of signal to noise and interference over the entire coverage area. A successful design relies on the validity of the predictions and assumptions systems engineers make about how a signal propagates, the characteristics of the signal path, and the nature of the noise and interference that will be encountered.

Among the many environmental factors systems engineers must consider are how a signal varies with distance from a transmitter in the specific application environment, and the minimum signal level required for reliable communications. Consideration of these factors is a logical starting point for a system design. Therefore, to prepare for a meaningful discussion of system design, we must develop some understanding of how signals propagate and the environmental factors that influence them as they traverse a path between a transmitter and a receiver.

In this chapter we investigate how an RF signal varies with distance from a transmitter under ideal conditions and how the RF energy is processed in the early stages of the receiver. Specifically, we first introduce the concept of **path loss** or **free-space loss**, and we develop a simple model to predict the increase in path loss with propagation distance. Afterward, we investigate thermal noise in a receiver, which leads to the primary descriptor of signal quality in a receiver—the signal-to-noise ratio or SNR. This allows us to quantitatively describe the minimum signal level required for reliable communications. We end our discussion with examples of how quantitative analysis and design are accomplished for a basic RF link. In the next chapter we will discuss path loss in a real-world environment and describe other channel phenomena that may limit a receiver's ability to reliably detect a signal or corrupt the information carried by an RF signal regardless of the signal level.

Transmitting and Receiving Electromagnetic Waves

Electromagnetic theory, encapsulated in Maxwell's equations, provides a robust mathematical model of how time-varying electromagnetic fields behave. Although we won't deal directly with Maxwell's equations, our discussions are based on the behavior they predict. We begin by pointing out that electromagnetic waves propagate in free space, a perfect vacuum, at the speed of light c. The physical length of one cycle of a propagating wave, the wavelength λ, is inversely proportional to the frequency f of the wave, such that

$$\lambda = \frac{c}{f}. \tag{2.1}$$

For example, visible light is an electromagnetic wave with a frequency of approximately 600 THz (1 THz = 10^{12} Hz), with a corresponding wavelength of 500 nm. Electric power in the United States operates at 60 Hz, and radiating waves would have a wavelength of 5000 km!

A typical wireless phone operating at 1.8 GHz emits a wavelength of 0.17 m. Obviously, the range of wavelengths and frequencies commonly dealt with in electromagnetic science and engineering is quite vast.

In earlier studies you learned how to use Kirchhoff's laws to analyze the effects of arbitrary configurations of circuit elements on time-varying signals. A fundamental assumption underlying Kirchhoff's laws is that the elements and their interconnections are very small relative to the wavelengths of the signals, so that the physical dimensions of a circuit are not important to consider. For this case, Maxwell's equations reduce to the familiar rules governing the relationships among voltage, current, and impedance. This case, to which Kirchhoff's laws apply, is commonly called **lumped-element analysis**. When the circuit elements and their interconnections are comparable in physical size to the signal wavelengths, however, distributed-element or transmission-line analysis must be employed. In such cases the position and size of the lumped elements are important, as are the lengths of the interconnecting wires. Interconnecting wires that are not small compared with a wavelength may in fact be modeled as possessing inductance, capacitance, and conductance on a per-length (distance) basis. This model allows the interconnections to be treated as a network of passive elements, fully possessing the properties of a passive circuit. It also explains the fact that any signal will be changed or distorted from its original form as it propagates through the interconnection in much the same way as any low-frequency signal is changed as it propagates through a lumped-element circuit such as a filter. Furthermore, when a circuit is comparable in size to a signal wavelength, some of the electromagnetic energy in the signal will radiate into the surroundings unless some special precautions are employed. When such radiation is unintentional, it may cause interference to nearby devices. In some cases, however, such radiation may be a desired effect.

An antenna is a device specifically designed to enhance the radiation phenomenon. A **transmitting antenna** is a device or **transducer** that is designed to efficiently transform a signal incident on its circuit terminals into an electromagnetic wave that propagates into the surrounding environment. Similarly, a **receiving antenna** captures an electromagnetic wave incident upon it and transforms the wave into a voltage or current signal at its circuit terminals. Maxwell's equations suggest that for an antenna to radiate *efficiently* its dimensions must be comparable to a signal wavelength (usually 0.1λ or greater). For example, a 1 MHz signal has a wavelength of 300 m (almost 1000 feet). Thus an antenna with dimensions appropriate for efficient radiation at that frequency would be impractical for many wireless applications. In contrast, a 1 GHz signal has a wavelength of 0.1 m (less than 1 foot). This leads to a practical constraint on any personal communication system; that is, it must operate in a frequency band high enough to allow the efficient transmission and reception of electromagnetic waves using antennas of manageable size.

An electromagnetic wave of a specific frequency provides the link or **channel** between transmitting and receiving antennas located at the endpoints, over which information will be conveyed. This electromagnetic wave is called a **carrier**, and as discussed in Chapter 1, information is conveyed by modulating the carrier using a signal that represents the information.

Isotropic Radiation

Development of the **range equation** is greatly facilitated by considering how a signal propagates in free space. By "free space" we mean a perfect vacuum with the closest object being infinitely far away. This ensures that the signal is not affected by resistive losses of the medium or objects that might otherwise reflect, refract, diffract, or absorb the signal. Although describing the propagation environment as free space may not be realistic, it does provide a useful model for understanding the fundamental properties of a radiated signal.

To begin, consider a transmitting antenna with power level P_t at its input terminals and a receiving antenna located at an arbitrary distance d from the transmitting antenna as shown in Figure 2.1. In keeping with our free-space simplification we consider that the transmitter and receiver have a negligible physical extent and, therefore, do not influence the propagating wave. We further assume that the transmitting antenna radiates power uniformly in all directions. Such an antenna is known as an **isotropic antenna**. (Note that an isotropic antenna is an idealization and is not physically realizable. Neither is an antenna that radiates energy only in a single direction.)

Imagine flux lines emanating from the transmitting antenna, where every flux line represents some fraction of the transmitted power P_t. Since radiation is uniform in all directions, the flux lines must be uniformly distributed around the transmitting antenna, as is suggested by Figure 2.2. If we surround the transmitting antenna by a sphere of arbitrary radius d, the number of flux lines per unit area crossing the surface of the sphere must be uniform everywhere on the surface. Thus, the power density p measured at any point on the surface of the sphere must also be uniform and constant. Assuming no loss of power due to absorption by the propagation medium, conservation of energy requires that the total of all the power crossing the surface of the sphere must equal P_t, the power being radiated by the transmitter. This total power must be the same for all concentric spheres regardless of radius, although the power density becomes smaller as the radius of the sphere becomes larger. We can write the power density on the surface of a sphere of radius d as

$$p = \frac{\text{Transmitted Power}}{\text{Area of Sphere}} = \frac{P_t}{4\pi d^2}, \tag{2.2}$$

where p is measured in watts per square meter.

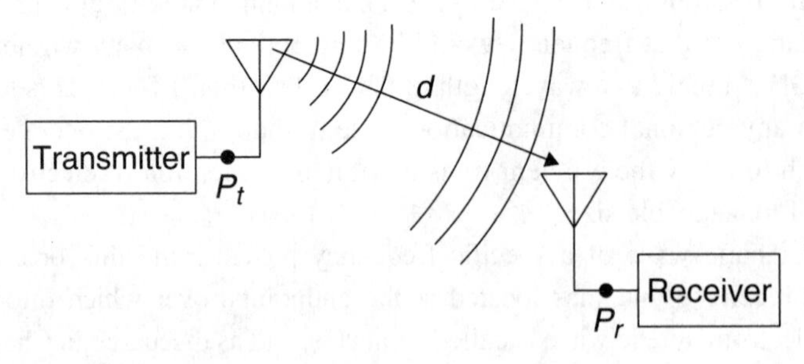

Figure 2.1 Free-Space Link Geometry

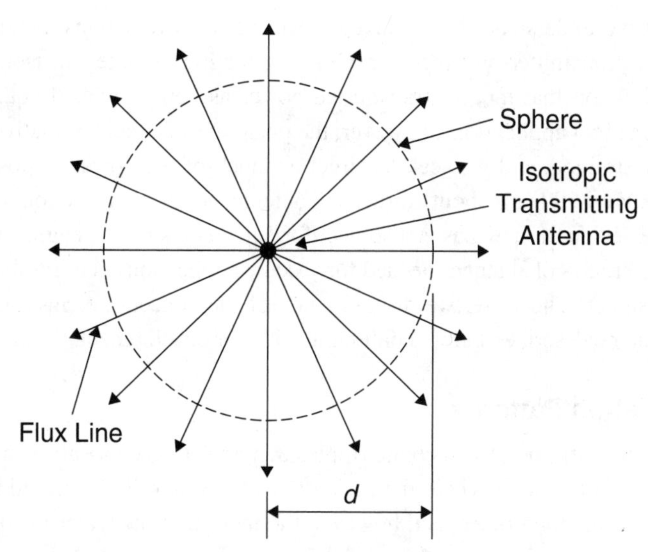

Figure 2.2 Isotropic Radiation Showing Surface of Constant Power Density and Flux Lines

---

Example

Suppose the isotropic radiator of Figure 2.2 is emitting a total radiated power of 1 W. Compare the power densities at ranges of 1, 10, and 100 km.

Solution

The calculation is a straightforward application of Equation (2.2), giving the results $p = 80 \times 10^{-9}$ W/m^2, 800×10^{-12} W/m^2, and 8×10^{-12} W/m^2, respectively. There are two things to note from these results. First, the power density numbers are very small. Second, the power density falls off as $1/d^2$, as expected from Equation (2.2). Obviously transmission range is a very important factor in the design of any communication link.

A receiving antenna located in the path of an electromagnetic wave captures some of its power and delivers it to a load at its output terminals. The amount of power an antenna captures and delivers to its output terminals depends on its **effective aperture (area)** A_e, a parameter related to its physical area, its specific structure, and other parameters. Given this characterization of the antenna, the power intercepted by the receiver antenna may be written as

$$P_r = pA_{er} = \frac{P_t A_{er}}{4\pi d^2}, \tag{2.3}$$

where the r subscript on the antenna aperture term refers to the receiving antenna.

At this point we understand that received power varies inversely as the square of the distance between the transmitter and receiver. Our formulation assumes an isotropic transmitter antenna, an idealization that models transmitted power as being radiated equally in all directions. For most practical applications, however, the position of a receiver relative to a transmitter is much more constrained, and we seek to direct as much of the power as possible toward the intended receiver. We will learn about "directive" antennas in the next section.

Although the design of antennas is a subject for experts, a systems engineer must be able to express the characteristics of antennas needed for specific applications in terms that are meaningful to an antenna designer. Therefore, we discuss some of the concepts, terms, and parameters by which antennas can be described before continuing with our development of the range equation.

Antenna Radiation Patterns

A rock dropped into a still pond will create ripples that appear as concentric circles propagating radially outward from the point where the rock strikes the water. Well-formed ripples appear at some distance from the point of impact; however, the motion of the water is not so well defined at the point of impact or its immediately surrounding area.

A physically realizable antenna launches both electric and magnetic fields. At a distance sufficiently far from the antenna we can observe these fields propagating in a radial direction from the antenna in much the same fashion as the ripples on the surface of the pond. The region of well-defined radially propagating "ripples" is known as the **far-field radiation region**, or the **Fraunhofer region**. It is the region of interest for most (if not all) communication applications. Much nearer to the antenna there are capacitive and inductive "near" fields that vary greatly from point to point and rapidly become negligible with distance from the antenna. A good approximation is that the far field begins at a distance from the antenna of

$$d_{far\,field} = \frac{2l^2}{\lambda}, \tag{2.4}$$

where l is the antenna's largest physical dimension.[1] Often in communication applications, the antenna's physical dimensions, though greater than $\lambda/10$, are less than a wavelength, so that the far field begins at a distance less than twice the largest dimension of the antenna. For cellular telephone applications, the far field begins a few centimeters away from the handset.

A polar plot of the far-field power density as a function of angle referenced to some axis of the antenna structure is known as a **power pattern**. The peak or lobe in the desired direction is called the **main lobe** or **main beam**. The remainder of the power (i.e., the power outside the main beam) is radiated in lobes called **side lobes**. "Directive" antennas can be designed that radiate most of the antenna input power in a given direction.

Figure 2.3 illustrates the beam pattern of a directive antenna with a conical beam. The antenna is located at the left of the figure from which the several lobes of the pattern emanate.

1. C. A. Balanis, *Antenna Theory*, 3rd ed. (New York: John Wiley & Sons, 2005), 167.

Figure 2.3 Typical Antenna Power Pattern Showing Main Lobe and Side Lobes

The plot represents the far-field power density measured at an arbitrary constant distance d from the antenna, where $d > d_{far\,field}$. The shape of the beam represents the power density as a function of angle, usually referenced to the peak of the main beam or lobe. For the antenna represented in the figure, most of the power is radiated within a small solid angle directed to the right in the plot. In general, a convenient coordinate system is chosen in which to describe the power pattern. For example, one might plot the power density versus the spherical coordinates θ and ϕ for an antenna oriented along one of the coordinate axes as shown in Figure 2.4. For each angular position (θ, ϕ), the radius of the plot represents the value of power density measured at distance d. Since d is arbitrary, the graph is usually normalized by expressing the power density as a ratio in decibels relative to the power density value at the peak of the pattern. Alternatively, the normalization factor can be the power density produced at distance d by an isotropic antenna, see Equation (2.2). When an isotropic antenna is used to provide the reference power density, the units of power density are dBi, where the *i* stands for "isotropic reference." Figure 2.5 shows an

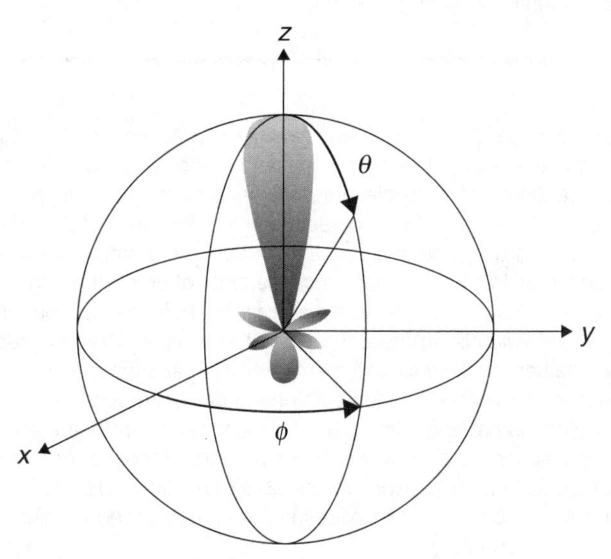

Figure 2.4 An Antenna Pattern in a Spherical Coordinate System

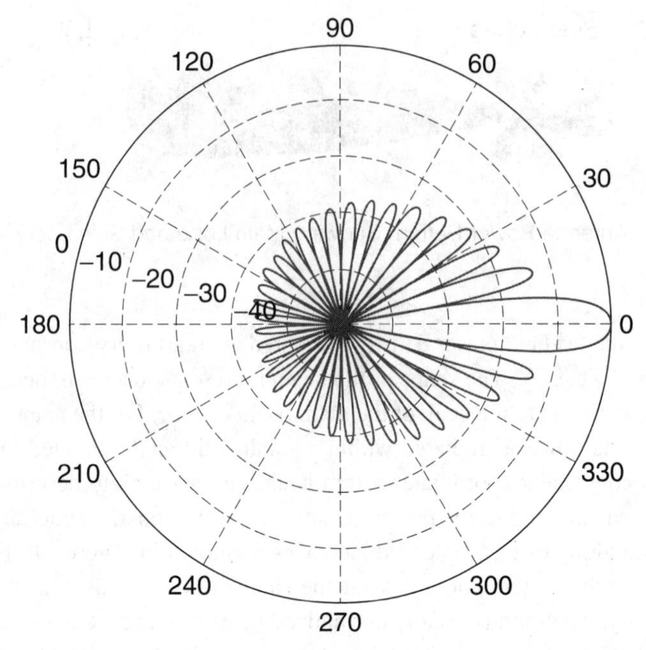

Figure 2.5 An Antenna Power Pattern Plotted versus θ for a Fixed Value of ϕ

antenna power pattern plotted versus one coordinate angle θ with the other coordinate angle ϕ held constant. This kind of plot is easier to draw than the full three-dimensional power pattern and is often adequate to characterize an antenna.

Example

A "half-wave dipole" or just "dipole" is a very simple but practical antenna. It is made from a length of wire a half-wavelength long, fed in the center by a source. Figure 2.6 shows a schematic representation of a dipole antenna. A slice through the power pattern is shown in Figure 2.7. The dipole does not radiate in the directions of the ends of the wire, so the power pattern is doughnut shaped. If the wire were oriented vertically, the dipole would radiate uniformly in all horizontal directions, but not up or down. In this orientation the dipole is sometimes described as having an "omnidirectional" pattern. The peak power density, in any direction perpendicular to the wire, is 2.15 dB greater than the power density obtained from an isotropic antenna (with all measurements made at the same distance from the antenna). Dipole antennas are often used for making radiation measurements. Because a dipole is a practical antenna, whereas an isotropic antenna is only a hypothetical construct, power patterns of other antennas are often normalized with respect to the maximum power density obtained from a dipole. In this case the power pattern will be labeled in dBd. Measurements expressed in dBd can easily be converted to dBi by adding 2.15 dB.

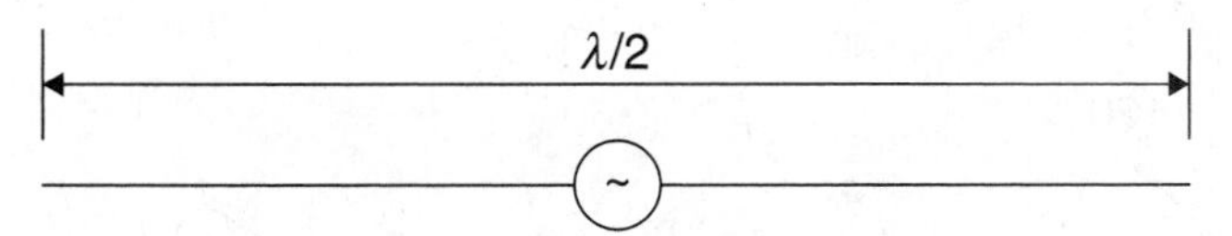

Figure 2.6 A Dipole Antenna

The antenna **beamwidth** along the main lobe axis in a specified plane is defined as the angle between points where the power density is one-half the power density at the peak, or 3 dB down from the peak. This is known as the "3 dB beamwidth" or the "half-power beamwidth" as shown in Figure 2.8. This terminology is analogous to that used in describing the 3 dB bandwidth of a filter. The "first null-to-null beamwidth," as shown in the figure, is often another parameter of interest.

Since a power pattern is a three-dimensional plot, beamwidths can be defined in various planes. Typically the beamwidths in two orthogonal planes are specified. Often the two planes of interest are the azimuth and elevation planes. In cellular telephone systems, where all of the transmitters and receivers are located near the surface of the Earth, the beamwidth in the azimuth plane is of primary interest. The azimuth-plane beamwidth is shown in Figure 2.8.

The physics governing radiation from an antenna indicates that the narrower the beamwidth of an antenna, the larger must be its physical extent. An approximate relationship exists between the beamwidth of an antenna in any given plane and its physical dimensions in that same plane. Typically beamwidth in any given plane is expressed by

$$Beamwidth \cong k\frac{\lambda}{L}, \tag{2.5}$$

where L is the length of the antenna in the given plane, λ is the wavelength, and k is a proportionality constant called the "beamwidth factor" that depends on the antenna type. A transmitting

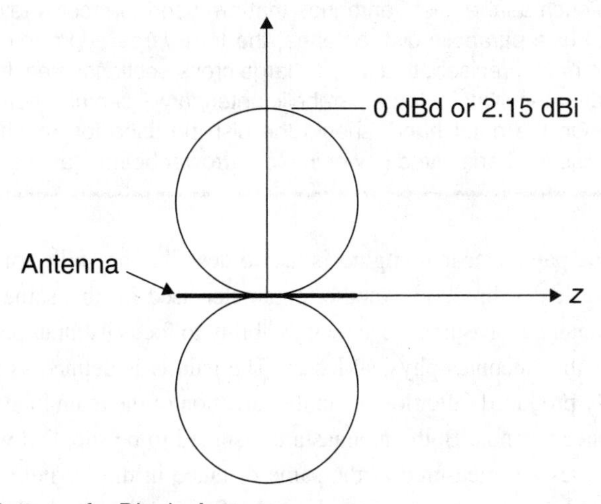

Figure 2.7 Power Pattern of a Dipole Antenna

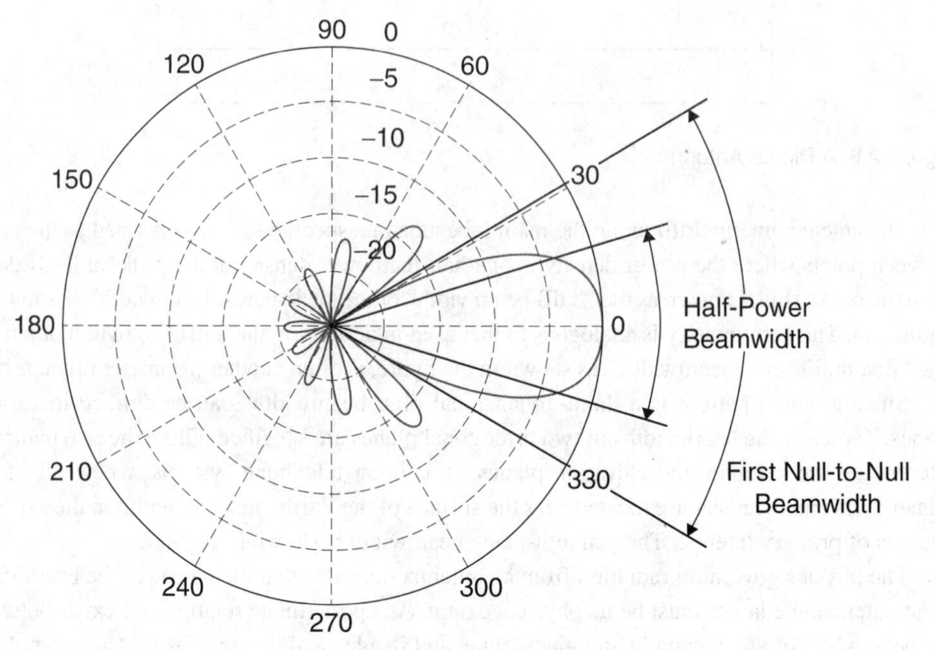

Figure 2.8 Half-Power and First Null-to-Null Beamwidths of an Antenna Power Pattern

antenna with a narrow "pencil" beam will have large physical dimensions. This implies that if the same antenna were used for receiving, it would have a large effective aperture A_e.

Example

The relation between beamwidth and effective aperture is easiest to visualize for parabolic antennas such as the "dish" antennas that are used to receive television signals from satellites. For a parabolic dish antenna, the term *large aperture* means that the dish is large, or, more precisely, that it has a large cross-sectional area. It is easy to see that when used for receiving, a large parabolic antenna will capture more power than a small one will. On the other hand, should the dish be used for transmitting, a large parabola will focus the transmitted power into a narrower beam than a small one.

Beamwidth is one parameter that engineers use to describe the ability of an antenna to focus power density in a particular direction. Another parameter used for this same purpose is **antenna gain**. The gain of an antenna measures the antenna's ability to focus its input power in a given direction but also includes the antenna's physical losses. The gain G is defined as the ratio of the peak power density actually produced (after losses) in the direction of the main lobe to the power density produced by a reference antenna. Both antennas are assumed to be supplied with the same power, and both power densities are measured at the same distance in the far field. A lossless isotropic antenna is a common choice for the reference antenna. To visualize the relationship between gain and beamwidth, consider the total power radiated as being analogous to a certain volume of water enclosed within a flexible balloon. Normally, the balloon is spherical in shape, corresponding to an antenna that radiates isotropically in all directions. When pressure is applied to the balloon, the regions of greater pressure are compressed, while the regions of lesser pressure expand; but the total volume remains unaltered. Similarly, proper antenna design alters the antenna power pattern by increasing the power density in certain directions with an accompanying decrease in the power density in other directions. Because the total power radiated remains unaltered, a high power density in a particular direction can be achieved only if the width of the high-power-density beam is small.

Beamwidth, antenna gain, and effective aperture are all different ways of describing the same phenomenon. (Technically, gain also takes into account ohmic losses in the antenna.) It can be shown that the gain is related to the effective antenna aperture A_e in the relationship[2]

$$G = 4\pi\eta\frac{A_e}{\lambda^2}, \tag{2.6}$$

where the losses are represented by the parameter η. For most telecommunications antennas the losses are small, and in the following development we will usually assume $\eta = 1$. The gain given by Equation (2.6) is referenced to an isotropic radiator.

Example

Suppose an isotropic radiator is emitting a total radiated power of 1 W. The receiving antenna has a gain described by Equation (2.6) with $\eta = 1$. Compute the required effective aperture of a receiving antenna at 10 km that will produce the same power P_r at the receiver input as would be produced by a receiving antenna with an effective aperture of $A_{er1} = 1\ \text{cm}^2$ at a distance of 1 km.

Solution

The power density of the transmitted wave is given by Equation (2.2). As we showed in a previous example, $p_1 = 80 \times 10^{-9}\ \text{W/m}^2$ at $d = 1$ km and $p_2 = 800 \times 10^{-12}\ \text{W/m}^2$ at $d = 10$ km. According to Equation (2.3), the total power produced at the receiver input is the power density multiplied by the effective receiving antenna aperture. Thus, for the antenna at 1 km,

$$P_r = p_1 A_{er1} = 80 \times 10^{-9}\ \frac{\text{W}}{\text{m}^2} \times (10^{-2})^2\ \text{m}^2 = 8\ \text{pW}. \tag{2.7}$$

Now at $d = 10$ we have

$$P_r = 8\ \text{pW} = p_2 A_{er2} = 800 \times 10^{-12}\ \frac{\text{W}}{\text{m}^2} \times A_{er2}. \tag{2.8}$$

2. Balanis, *Antenna Theory*, 92–94.

Solving gives $A_{er2} = 0.01\ \text{m}^2$. This equals an area of 100 cm^2, a much larger antenna. Since gain is proportional to effective area, the second antenna requires 100 times the gain of the first in this scenario, or an additional 20 dB. Note that we could have predicted these results simply by realizing that the power density falls off as the square of the range.

Example

Continuing the previous example, suppose the communication link operates at a frequency of 10 GHz. Find the gains of the two receiving antennas.

Solution

From Equation (2.1), the wavelength is $\lambda = \frac{c}{f} = \frac{3 \times 10^8}{10 \times 10^9} = 0.01$ m. Now using Equation (2.6) with $\eta = 1$ gives $G_{r1} = 4\pi \frac{10^{-4}\ \text{m}^2}{(0.03\ \text{m})^2} = 1.40$ or $G_{r1}\big|_{\text{dB}} = 10\log(1.40) = 1.4$ dBi. Similarly, $G_{r2} = 4\pi \frac{10^{-2}\ \text{m}^2}{(0.03\ \text{m})^2} = 140$ or $G_{r2}\big|_{dB} = 10\log(140) = 21.4$ dBi.

It is very important to keep in mind that although an antenna has "gain," the antenna does not raise the power level of a signal passing through it. An antenna is a passive device, and when transmitting, its total radiated power must be less than or equal to its input power. The gain of an antenna represents the antenna's ability to focus power in a preferred direction. This is always accompanied by a reduction in the power radiated in other directions. To a receiver whose antenna lies in the direction of the transmitter's main beam, increasing the transmitting antenna's gain is equivalent to increasing its power. This is a consequence of improved focusing, however, and not a consequence of amplification. Increasing the receiving antenna's gain will also increase the received power. This is because antenna gain is obtained by increasing the antenna's effective aperture, thereby allowing the antenna to capture more of the power that has already been radiated by the transmitter antenna.

It should be apparent from our discussion to this point that both transmitting and receiving antennas have gain, and both receiving and transmitting antennas have effective apertures. In fact, there is no difference between a transmitting and a receiving antenna; the same antenna can be used for either purpose. An important theorem from electromagnetics shows that antennas obey the property called **reciprocity**. This means that an antenna has the same gain when used for either transmitting or receiving, and consequently, by Equation (2.6), it has the same effective aperture when used for either transmitting or receiving as well.

The Range Equation

Earlier we stated that under ideal conditions, and assuming an isotropic transmitting antenna, the power received at the output of the receiving antenna is given by

$$P_r = pA_{er} = \frac{P_t A_{er}}{4\pi d^2}. \tag{2.9}$$

This relationship assumes that all of the power supplied to a transmitting antenna's input terminals is radiated and all of the power incident on a receiving antenna's effective aperture is captured and available at its output terminals. If we now utilize a directional transmitting antenna with gain G_t, then the power received in the direction of the transmitting antenna main lobe is given as

$$P_r = pA_{er} = \frac{P_t G_t A_{er}}{4\pi d^2}. \tag{2.10}$$

The term **effective isotropic radiated power** (EIRP) is often used instead of $P_t G_t$. EIRP is the power that would have to be supplied to an isotropic transmitting antenna to provide the same received power as that from a directive antenna. Note that a receiver cannot tell whether a transmitting antenna is directive or isotropic. The only distinction is that when a directive antenna is used, the transmitter radiates very little power in directions other than toward the receiver, whereas if an isotropic antenna is used, the transmitter radiates equally in all directions.

Example

A transmitter radiates 50 W through an antenna with a gain of 12 dBi. Find the EIRP.

Solution

Converting the antenna gain from decibels, $G_t = 10^{\frac{12}{10}} = 15.8$. Then $EIRP = 50 \times 15.8 =$ 792 W.

Using the definition of EIRP, we can write Equation (2.10) as

$$P_r = pA_{er} = \frac{(EIRP)A_{er}}{4\pi d^2}. \tag{2.11}$$

Since the gain and effective aperture of an antenna are related by Equation (2.6), and since the gain and effective aperture are properties of the antenna that do not depend on whether the antenna is used for transmitting or receiving, it is not necessary to specify both parameters. Antenna manufacturers frequently list the antenna gain on specification sheets. If we replace A_{er} in Equation (2.11) using Equation (2.6) we obtain

$$P_r = \frac{P_t G_t G_r \lambda^2}{(4\pi)^2 d^2}. \tag{2.12}$$

In applying Equation (2.6) we have set the loss parameter $\eta = 1$ for simplicity. We will include a term later on to account for antenna losses as well as losses of other kinds. It is instructive to write Equation (2.12) in the following way:

$$P_r = P_t G_t G_r \frac{\lambda^2}{(4\pi)^2 d^2} = P_t G_t G_r \frac{1}{(4\pi)^2 (d/\lambda)^2}. \tag{2.13}$$

This form suggests that the received power can be obtained by starting with the transmitted power and multiplying by three "gain" terms. Since the reciprocal of gain is "loss," we can interpret the term $(4\pi)^2 (d/\lambda)^2$ as a loss term. Let us define the free-space path loss L_{path} as

$$L_{path} = (4\pi)^2 (d/\lambda)^2. \tag{2.14}$$

It should be observed that the free-space path loss L_{path} varies directly with the square of the distance and inversely with the square of the wavelength. We can interpret the ratio d/λ in Equation (2.14) to mean that path loss is a function of the distance between transmitting and receiving antennas measured in wavelengths. It is important to note that the path loss is not an ohmic or resistive loss that converts electrical power into heat. Rather, it represents the reduction in power density due to a fixed amount of transmitted power spreading over an increasingly greater surface area as the waves propagate farther from the transmitting antenna; that is, just as G represents focusing and not actual "gain," L_{path} represents spreading and not actual "loss."

Equation (2.12), the so-called **range equation**, can be written in any of the alternate forms

$$P_r = \frac{P_t G_t G_r \lambda^2}{(4\pi)^2 d^2} = \frac{(EIRP) G_r \lambda^2}{(4\pi)^2 d^2} = \frac{P_t G_t G_r}{L_{path}} = \frac{(EIRP) G_r}{L_{path}}. \tag{2.15}$$

The first of these forms is known commonly as the Friis transmission equation in honor of its developer, Harald T. Friis[3] (pronounced "freese"). Subsequently we will see that there is a minimum received power that a given receiver requires to provide the user with an adequate quality of service. We call that minimum power the **receiver sensitivity** p_{sens}. Given the receiver sensitivity, we can use Equation (2.15) to calculate the maximum range of the data link for free-space propagation, that is,

$$d_{max} = \sqrt{\frac{P_t G_t G_r \lambda^2}{(4\pi)^2 p_{sens}}}. \tag{2.16}$$

In our analysis so far we have ignored any sources of real loss, that is, system elements that turn electrical power into heat. In a realistic communication system there are a number of such elements. The transmission lines that connect the transmitter to the antenna and the antenna to the receiver are slightly lossy. Connectors that connect segments of transmission line and connect transmission lines to antennas and other components provide slight impedance mismatches that lead to small losses. Although the loss caused by one connector is usually not significant, there may be a number of connectors in the transmission system. We have mentioned above that antennas may produce small amounts of ohmic loss. There may also be losses in propagation, where signals pass through the walls of buildings or into or out of automobiles. In mobile radio systems there can be signal absorption by the person holding the mobile unit. And so on. Fortunately it is easy to amend the range equation to include these losses. If $L_1, L_2, \ldots, L_n$ represent

3. H. T. Friis, "A Note on a Simple Transmission Formula," *Proceedings of the Institute of Radio Engineers* 34 (1946): 254–56.

real-world losses, then we can modify Equation (2.15) to read

$$P_r = \frac{P_t G_t G_r}{(4\pi)^2 (d/\lambda)^2 \prod_{i=1}^{n} L_i} = \frac{P_t G_t G_r}{L_{path} \prod_{i=1}^{n} L_i} = \frac{P_t G_t G_r}{L_{path} L_{sys}}, \tag{2.17}$$

where in the last form we have lumped all of the real-world losses together into a single term L_{sys}, which we call "system losses."

In typical applications, signal levels range over many orders of magnitude. In such situations it is often very convenient to represent signal levels in terms of decibels. Recall that a decibel is defined in terms of a power ratio as

$$10 \log\left(\frac{P_2}{P_1}\right), \tag{2.18}$$

where P_1 and P_2 represent power levels, and the logarithm is computed to the base 10. For example, if the signal power is 1 W at 1 mile, and 10 mW at 10 miles, then the ratio in decibels is

$$10 \log\left(\frac{10 \text{ mW}}{1 \text{ W}}\right) = 10 \log\left(\frac{10 \times 10^{-3}}{1}\right) = -20 \text{ dB}. \tag{2.19}$$

Absolute power levels (as opposed to ratios) can also be expressed in decibels, but a reference power is needed for P_1 in Equation (2.18). Reference values of 1 W and 1 mW are common. When a 1 W reference is used, the units are written "dBW" to reflect that fact; when a 1 mW reference is used, the units are written as "dBm." Expressing quantities in decibels is particularly useful in calculating received power and in any analysis involving the range equation. This is because the logarithm in the decibel definition converts multiplication into addition, and as a result the various gain and loss terms can be added and subtracted. Converting Equation (2.17) to decibels gives

$$\begin{aligned}
P_r\big|_{dB} &= 10 \log\left(\frac{P_r}{1 \text{ mW}}\right) \\
&= 10 \log\left(\frac{P_t G_t G_r}{L_{path} L_{sys} (1 \text{ mW})}\right) \\
&= 10 \log\left(\frac{P_t}{1 \text{ mW}} \times G_t \times G_r \times \frac{1}{L_{path}} \times \frac{1}{L_{sys}}\right) \\
&= 10 \log\left(\frac{P_t}{1 \text{ mW}}\right) + 10 \log(G_t) + 10 \log(G_r) \\
&\quad - 10 \log(L_{path}) - 10 \log(L_{sys}) \\
&= P_t\big|_{dB} + G_t\big|_{dB} + G_r\big|_{dB} - L_{path}\big|_{dB} - L_{sys}\big|_{dB}.
\end{aligned} \tag{2.20}$$

Notice particularly the way in which the 1 mW reference power is handled in Equation (2.20). The reference power is needed as a denominator to convert the transmitted power P_t into decibels. The gain and loss terms are all ratios of powers or ratios of power densities and can be converted to decibels directly, without need for additional reference powers.

To make the dependence of received signal power on range explicit, let us use Equation (2.14) to substitute for L_{path} in Equation (2.20). We have

$$L_{path} = \left(\frac{4\pi}{\lambda}\right)^2 d^2. \tag{2.21}$$

The exponent of distance d in Equation (2.21) is called the **path-loss exponent**. We see that the path-loss exponent has the value of 2 for free-space propagation. The path-loss exponent will play an important role in the models for propagation in real-world environments that will be discussed in detail in Chapter 3. Converting Equation (2.21) to decibels gives

$$\begin{aligned} L_{path}\big|_{dB} &= 10\log\left(\left(\frac{4\pi}{\lambda}\right)^2\right) + 10\log(d^2) \\ &= 10\log\left(\left(\frac{4\pi}{\lambda}\right)^2\right) + 20\log(d). \end{aligned} \tag{2.22}$$

We see that in free space, the path loss increases by 20 dB for every decade change in range. Substituting Equation (2.22) into Equation (2.20) gives

$$\begin{aligned} P_r\big|_{dB} &= P_t\big|_{dB} + G_t\big|_{dB} + G_r\big|_{dB} - L_{sys}\big|_{dB} - 10\log\left(\left(\frac{4\pi}{\lambda}\right)^2\right) - 20\log(d) \\ &= K - 20\log(d), \end{aligned} \tag{2.23}$$

where all of the constant terms have been lumped together as K. Note that Equation (2.23) is a straight line when plotted on semilog scale.

Example

A transmitter at 1900 MHz produces a power of 25 W. The transmitting antenna has a gain of 15 dBi and the receiving antenna has a gain of 2.15 dBi. System losses are 3 dB. If the range between transmitting and receiving antennas is 10 km, find the received signal power.

Solution

In decibels the transmitted power is $P_t\big|_{dB} = 10\log\left(\frac{25\text{ W}}{1\text{ mW}}\right) = 44.0$ dBm. The wavelength at 1900 MHz is given by Equation (2.1) to be $\lambda = \frac{3\times10^8}{1900\times10^6} = 0.158$ m. The path loss can then be found from Equation (2.22) as $L_{path}\big|_{dB} = 10\log\left(\left(\frac{4\pi 10\times 10^3}{0.158}\right)^2\right) = 118$ dB. Now Equation (2.20) gives

$$\begin{aligned} P_r\big|_{dB} &= 44.0 \text{ dBm} + 15 \text{ dB} + 2.15 \text{ dB} - 118 \text{ dB} - 3 \text{ dB} \\ &= -59.9 \text{ dBm}. \end{aligned} \tag{2.24}$$

Note that system losses of 3 dB mean that *half* of the transmitted power is turned into heat before it can reach the receiver.

Example

Suppose a transmitting base station emits an EIRP of 100 W. The handset receiver has a sensitivity of $p_{sens}\big|_{dB} = -90$ dBm. What is the maximum range for the mobile unit given the following selected antennas: a dipole having gain of 2.15 dBi, and an omnidirectional having gain of 0 dBi? The operating frequency is 860 MHz. Neglect system losses.

Solution

When working problems with the range equation, it is important to take the proper approach. In this problem, most of the data is given in decibel units, so it would make sense to start with a form of the range equation expressed in decibel units, such as Equation (2.20). Expressed in decibels, the EIRP is $EIRP\big|_{dB} = 10\log\left(\frac{100 \text{ W}}{1 \text{ mW}}\right) = 50$ dBm. Recall that $EIRP = P_tG_t$, so in decibels, $EIRP\big|_{dB} = P_t\big|_{dB} + G_t\big|_{dB}$. Substituting into Equation (2.20) gives

$$\begin{aligned} -90 \text{ dBm} &= 50 \text{ dBm} + G_r\big|_{dB} - L_{path}\big|_{dB} \\ &= \begin{cases} 50 \text{ dBm} + 2.15 \text{ dB} - L_{path}\big|_{dB}, & \text{for the dipole, and} \\ 50 \text{ dBm} + 0 \text{ dB} - L_{path}\big|_{dB}, & \text{for the omnidirectional.} \end{cases} \end{aligned} \tag{2.25}$$

Solving for path loss gives

$$L_{path}\big|_{dB} = \begin{cases} 142.15 \text{ dB}, & \text{for the dipole, and} \\ 140 \text{ dB}, & \text{for the omnidirectional.} \end{cases} \tag{2.26}$$

Now the wavelength λ is given by Equation (2.1):

$$\lambda = \frac{3\times 10^8}{860\times 10^3} = 0.349 \text{ m}.$$

Substituting in Equation (2.22) gives

$$\begin{cases} 142.15\ \text{dB} = 10\log\left(\left(\dfrac{4\pi d}{0.349}\right)^2\right), & \text{for the dipole, and} \\ 140\ \text{dB} = 10\log\left(\left(\dfrac{4\pi d}{0.349}\right)^2\right), & \text{for the omnidirectional.} \end{cases} \tag{2.27}$$

Solving for range gives

$$\begin{aligned} d &= 356\ \text{km}, && \text{for the dipole, and} \\ d &= 278\ \text{km}, && \text{for the omnidirectional.} \end{aligned} \tag{2.28}$$

A 2.15 dB difference in antenna gain can make a big difference in range (28% in this free-space example).

Thermal Noise and Receiver Analysis

All communication systems, wired or unwired, are affected by unwanted signals. These unwanted signals are termed either **noise** or **interference**. There are no clear and universal definitions that distinguish noise from interference. Most often the term *interference* refers to unwanted signals entering the passband of the desired system from other systems that intentionally radiate electromagnetic waves. An intentional radiator is any radio system that uses electromagnetic waves to perform its function. The term *noise* often refers to unwanted signals arising from natural phenomena or unintentional radiation by man-made systems. Examples of natural phenomena that produce noise are atmospheric disturbances, extraterrestrial radiation, and the random motions of electrons. Examples of man-made unintentional radiators that produce noise are power generators, automobile ignition systems, electronic instruments, and microwave ovens. The distinction between noise and interference will become more evident when we discuss interference management in cellular systems.

The effects of interference and certain types of noise can often be mitigated and sometimes eliminated by appropriate engineering techniques and/or the establishment of rules to restrict both intentional and unintentional radiation. One particular type of electrical noise, however, is ubiquitous insofar as it arises in the very components used to implement a system. This noise, called **thermal noise**, arises from the thermal motion of electrons in a conducting medium and is present in any circuit consisting of resistive elements such as wires, semiconductors, and, of course, resistors. It is, therefore, present in any system that uses these components. The presence of thermal noise limits the sensitivity of all electronic wireless systems. Sensitivity is a measure of a system's ability to reliably detect a signal.

In addition to assuming that waves propagate in free space, our development of the range equation tacitly assumed ideal transmit and receive antennas and a lossless connection between the receive antenna and an ideal receiver. Under these optimal conditions there is no minimum limit to a receiver's ability to detect a signal, and therefore the operating range of such a system is limitless. Real antennas, receivers, and interconnecting circuits, however, all have resistive (lossy) elements and electronic components. These elements and components introduce thermal noise throughout a receiver and in particular in its front end, that is, the first stages immediately following the receiving antenna. When the information-carrying signal is comparable to the noise level, the information may become corrupted or may not even be detectable or distinguishable from the noise. The maximum range is, therefore, constrained by the need to maintain the information-carrying signal at some level relative to the noise level at the input to the receiver. This noise level is often called the **noise floor**.

Characterizing Noise Sources

Taking into account the thermodynamic nature of the phenomenon, the noise generated by a resistor has a normalized power spectrum[4] given by

$$S(f) = \frac{2h|f|R}{e^{\frac{h|f|}{kT}} - 1}, \tag{2.29}$$

where

R is the resistance of the noise source in ohms
T is the absolute temperature in kelvin
f is frequency in hertz
k is Boltzmann's constant (1.38×10^{-23} J/K)
h is Planck's constant (6.63×10^{-24} Js)

Example

If we take $T = 290$ K (that is, 62.6°F, a chilly "room" temperature), then

$$\frac{kT}{h} = \frac{(1.38 \times 10^{-23} \text{ J/K})(290 \text{ K})}{6.63 \times 10^{-24} \text{ Js}} = 6000 \text{ GHz}. \tag{2.30}$$

All radio frequencies (but not optical frequencies) in use today are significantly below this value. Then for $\frac{h|f|}{kT} << 1$ we can write

$$\begin{aligned} e^{\frac{h|f|}{kT}} &= 1 + \frac{h|f|}{kT} + \frac{1}{2!}\left(\frac{h|f|}{kT}\right)^2 + \cdots \\ &\cong 1 + \frac{h|f|}{kT}. \end{aligned} \tag{2.31}$$

4. This is actually a mean-square voltage spectrum. It represents the power per unit frequency that would be delivered to a 1-ohm resistor. This is discussed later in connection with Equation (2.40).

Thus at practical radio frequencies Equation (2.29) becomes

$$S(f) \cong \frac{2h|f|R}{1+\frac{h|f|}{kT}-1} = 2kTR, \qquad (2.32)$$

which is independent of frequency. The units of normalized power spectrum are JΩ or, more appropriately, V^2/Hz.

Noise having a power spectrum that is constant at all frequencies is called "white" noise, by analogy with white light, which has a constant power spectrum at all wavelengths. Equation (2.32) shows that thermal noise can be modeled as white noise for all radio frequencies of current practical interest. We often write the power spectrum of white noise as

$$S(f) = \frac{N_0}{2} \qquad (2.33)$$

when we do not want to imply that the noise was necessarily generated by a resistor. Figure 2.9 shows a signal $x(t)$ with Fourier transform $X(f)$ passing through a filter with frequency response $H(f)$. The output signal $y(t)$ has Fourier transform $Y(f)$. We know that

$$Y(f) = H(f)X(f). \qquad (2.34)$$

Taking the magnitude squared of both sides of Equation (2.34) gives a relation between the energy spectrum of the input signal and the energy spectrum of the output signal:

$$|Y(f)|^2 = |H(f)|^2|X(f)|^2 . \qquad (2.35)$$

Now noise does not have an energy spectrum, but it can be shown that an equation similar to Equation (2.35) applies to noise power spectra. Thus, if noise $x(t)$ having a power spectrum $Sx(f)$ is passed through the filter, then the power spectrum $S_y(f)$ of the output noise $y(t)$ is given by

$$S_y(f) = |H(f)|^2\, S_x(f). \qquad (2.36)$$

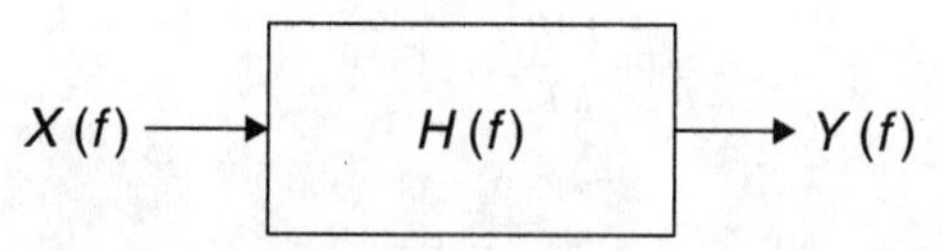

Figure 2.9 A Signal Is Passed through a Filter

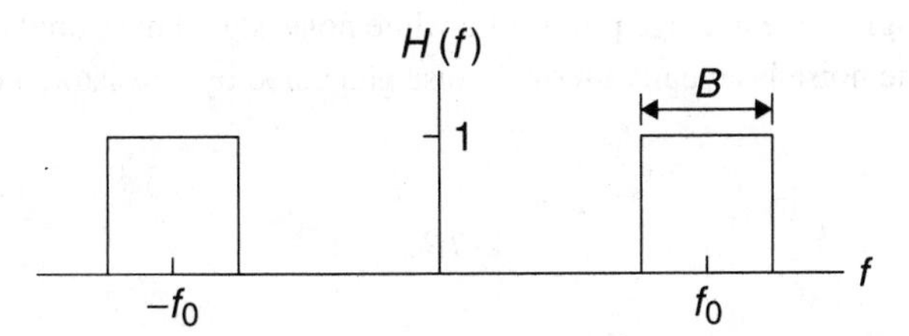

Figure 2.10 Frequency Response of an Ideal Bandpass Filter of Center Frequency f_0 and Bandwidth B

Example

Suppose $x(t)$ is white noise with power spectrum $S_x(f) = \frac{N_0}{2}$. Suppose the filter is a bandpass filter having the frequency response $H(f)$ shown in Figure 2.10. Then the output noise $y(t)$ has power spectrum $S_y(f)$ given by $S_y(f) = |H(f)|^2 \frac{N_0}{2}$. This power spectrum is shown in Figure 2.11.

The average noise power at the filter output is the area under the power spectrum, that is,

$$\begin{aligned} P_y &= \int_{-\infty}^{\infty} S_y(f)df \\ &= N_0 B. \end{aligned} \quad (2.37)$$

Equation (2.37) shows that the average power in a noise signal depends not only on the spectral level $N_0/2$ of the noise, but also on the bandwidth through which the noise is measured. It is important to emphasize that Equation (2.37) cannot be used to calculate the average noise power at the input to the filter. The unfiltered white noise at the filter input will appear to have an infinite power owing to its unlimited bandwidth. The fact that the noise power seems to be infinite is an artifact of the white noise model. As can be seen from Equation (2.29), the thermal noise power spectrum is not actually constant at all frequencies, and so thermal noise does not actually have infinite average power.

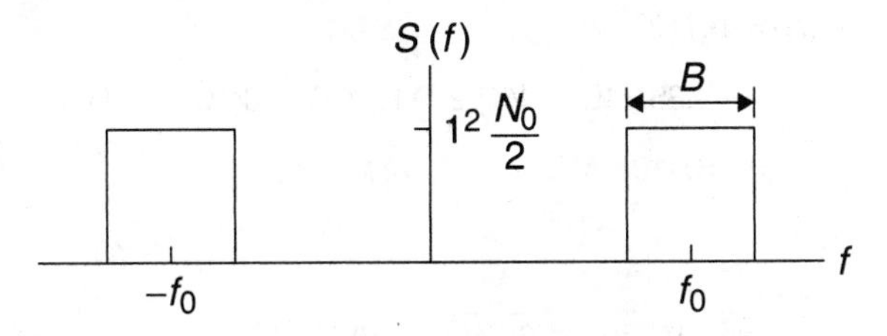

Figure 2.11 Power Spectrum of Noise at the Bandpass Filter Output

Equation (2.37) gives the average power of a white noise signal measured using an instrument of bandwidth B. If the noise is actually thermal noise generated by a resistor, then Equation (2.32) gives us

$$\frac{N_0}{2} = 2kTR \tag{2.38}$$

and

$$P_y = 4kTRB. \tag{2.39}$$

We need to be more precise about the meaning of the term *average power,* as several notions of average power will be used in the sequel. If $x(t)$ is a signal, the average power is given by

$$P_x = \lim_{T\to\infty}\frac{1}{T}\int_{-\frac{T}{2}}^{\frac{T}{2}} x^2(t)dt. \tag{2.40}$$

This is actually the mean square value of $x(t)$ or the average power that a voltage or current $x(t)$ would deliver to a 1-ohm resistor. The average power P_x is sometimes referred to as the "normalized" power in $x(t)$ and is sometimes written $\langle x^2(t)\rangle$ to emphasize the mean-square concept. We note that the square root of the mean-square value is the RMS value, so

$$X_{RMS} = \sqrt{\langle x^2(t)\rangle} = \sqrt{P_x}. \tag{2.41}$$

J. B. Johnson of the Bell Telephone Laboratories was the first to study and model thermal noise (also known as "Johnson" noise) in the late 1920s. In his model, thermal noise arising from a resistance of value R is represented as an ideal voltage source in series with a noiseless resistance of value R as shown in Figure 2.12. The mean-square open-circuit voltage of the ideal voltage source is given by Equation (2.39).

Example

Suppose we have an $R = 100\ \text{k}\Omega$ resistor at "room" temperature $T = 290$ K. Suppose we measure the open-circuit voltage across the resistor using a true-RMS voltmeter with a bandwidth of $B = 1$ MHz. Then

$$\begin{aligned}\langle v^2(t)\rangle &= 4kTRB \\ &= 4(1.38\times10^{-23}\ \text{J/K})(290\ \text{K})(100\times10^{3}\ \Omega)(10^{6}\ \text{Hz}) \\ &= 1.6\times10^{-9}\ \text{V}^2.\end{aligned} \tag{2.42}$$

This gives

$$V_{RMS} = \sqrt{1.6\times10^{-9}} = 40.0\ \mu\text{V}. \tag{2.43}$$

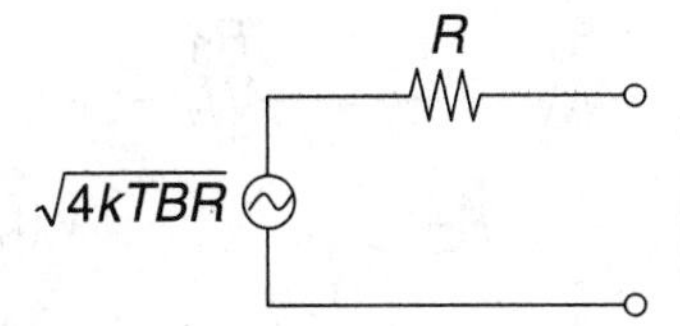

Figure 2.12 Thermal Noise Model of a Resistor

As was described earlier, thermal noise is generated by any lossy system, not just an isolated resistor. An interesting consequence of the thermal nature of the noise is that if we observe any passive system with a single electrical port, whatever noise there is appears to be generated by the equivalent resistance at the port. We will not attempt to prove this assertion, but we can illustrate it with several examples.

Example

The circuit shown in Figure 2.13 consists of three resistors all in thermal equilibrium with their surroundings at temperature T. We wish to find the average power $\langle v^2(t)\rangle$ in the noise voltage $v(t)$.

Method 1

Replace each resistor with a noise model as shown in Figure 2.12. The result is shown in Figure 2.14. This circuit can now be solved by standard circuit theory methods, say, superposition or the node-voltage method. We obtain

$$v(t)=\frac{R_2}{R_1+R_2}v_1(t)+\frac{R_1}{R_1+R_2}v_2(t)+v_3(t). \tag{2.44}$$

The mean-square value of $v(t)$ is given by

$$\begin{aligned}\langle v^2(t)\rangle &= \left\langle\left[\frac{R_2}{R_1+R_2}v_1+\frac{R_1}{R_1+R_2}v_2+v_3\right]^2\right\rangle\\ &=\left(\frac{R_2}{R_1+R_2}\right)^2\langle v_1^2\rangle+2\left(\frac{R_2}{R_1+R_2}\right)\left(\frac{R_1}{R_1+R_2}\right)\langle v_1v_2\rangle\\ &\quad+2\left(\frac{R_2}{R_1+R_2}\right)\langle v_1v_3\rangle+\left(\frac{R_1}{R_1+R_2}\right)^2\langle v_2^2\rangle+2\left(\frac{R_1}{R_1+R_2}\right)\langle v_2v_3\rangle\\ &\quad+\langle v_3^2\rangle,\end{aligned} \tag{2.45}$$

where we have used the fact that the average of a sum is the sum of the averages. Now when the voltages from two distinct resistors are multiplied and averaged, we have, for example,

$$\langle v_1v_2\rangle=\langle v_1\rangle\langle v_2\rangle=0, \tag{2.46}$$

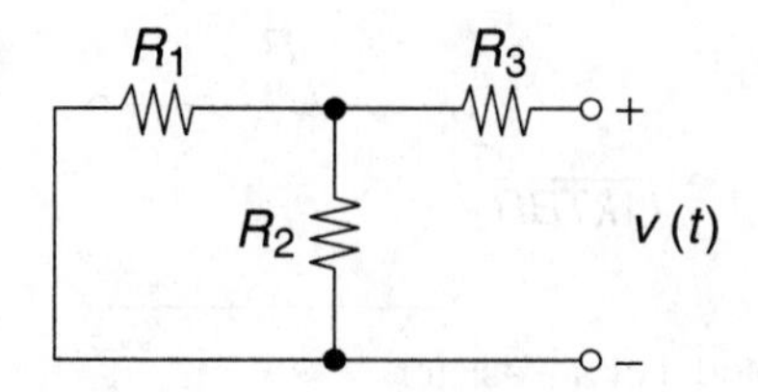

Figure 2.13 A Three-Resistor Noise Source

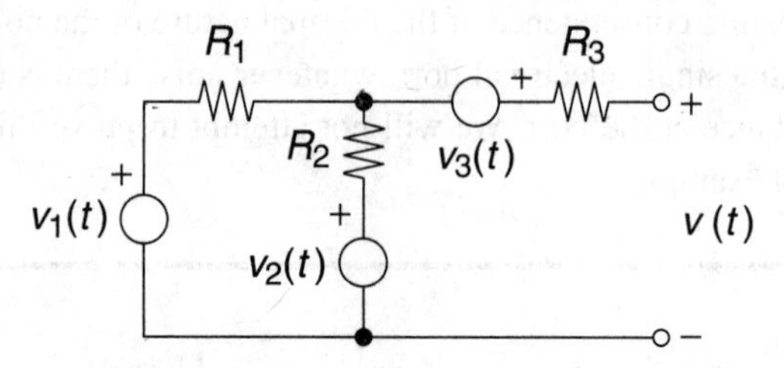

Figure 2.14 Resistors Replaced by Their Noise Models

where we have taken advantage of the statistical independence of the voltages $v_1(t)$ and $v_2(t)$ and also of the fact that thermal noise has a zero average value. Applying Equation (2.46) to Equation (2.45) gives

$$\langle v^2(t)\rangle = \left(\frac{R_2}{R_1+R_2}\right)^2 \langle v_1^2\rangle + \left(\frac{R_1}{R_1+R_2}\right)^2 \langle v_2^2\rangle + \langle v_3^2\rangle. \tag{2.47}$$

We can now substitute for the mean-square noise voltages using Equation (2.39). We obtain

$$\begin{aligned}\langle v^2(t)\rangle &= \left(\frac{R_2}{R_1+R_2}\right)^2 4kTR_1B + \left(\frac{R_1}{R_1+R_2}\right)^2 4kTR_2B + 4kTR_3B \\ &= 4kT\left(\frac{R_1R_2^2 + R_1^2R_2}{(R_1+R_2)^2} + R_3\right)B \\ &= 4kT\left(\frac{R_1R_2}{R_1+R_2} + R_3\right)B.\end{aligned} \tag{2.48}$$

Method 2

As an alternate approach to finding the average power in $v(t)$, let us replace the three resistors in Figure 2.13 with a single equivalent resistor R. To do this we first combine resistors R_1 and R_2 in parallel and then combine the result in series with R_3. The result is

$$R = \frac{R_1R_2}{R_1+R_2} + R_3. \tag{2.49}$$

The average thermal noise power generated by the equivalent resistor R is given by Equation (2.39) as

$$\begin{aligned}\langle v^2(t)\rangle &= 4kTRB \\ &= 4kT\left(\frac{R_1R_2}{R_1+R_2}+R_3\right)B.\end{aligned} \tag{2.50}$$

As expected, Equation (2.50) expresses the same result as Equation (2.48), but with considerably less effort. Note that the average noise power is proportional to the measurement bandwidth B. To avoid the vagueness of a result that contains an unspecified bandwidth, we can specify the noise power spectrum rather than the average noise power. Comparing Equation (2.32) and Equation (2.39) shows that the noise power spectrum at the output of the circuit is

$$S_v(f) = 2kT\left(\frac{R_1R_2}{R_1+R_2}+R_3\right). \tag{2.51}$$

Example

Figure 2.15 shows a one-port passive circuit that contains a reactive element as well as a resistor. As before, the circuit is at thermal equilibrium with its surroundings at temperature T. We wish to find the power spectrum $S_v(f)$ of the output voltage $v(t)$ and also to find the average power $P_v = \langle v^2(t)\rangle$. As in the previous example, we will solve the problem two ways. The second solution will illustrate the "equivalent resistance" assertion.

Method 1

Begin by replacing the resistor with its noise equivalent as shown in Figure 2.12. The result is the circuit shown in Figure 2.16. The voltage $v_R(t)$ is the resistor noise voltage. The power spectrum of this voltage is $S_R(f) = 2kTR$. The circuit of Figure 2.16 looks like an ideal voltage source driving an RC lowpass filter. We can, therefore, use Equation (2.36) to find the power spectrum of the noise at the output. First, the frequency response of the lowpass filter can be found by using a voltage divider. We obtain

$$\begin{aligned}H(f) &= \frac{\frac{1}{j2\pi fC}}{R+\frac{1}{j2\pi fC}} \\ &= \frac{1}{1+j2\pi fRC}.\end{aligned} \tag{2.52}$$

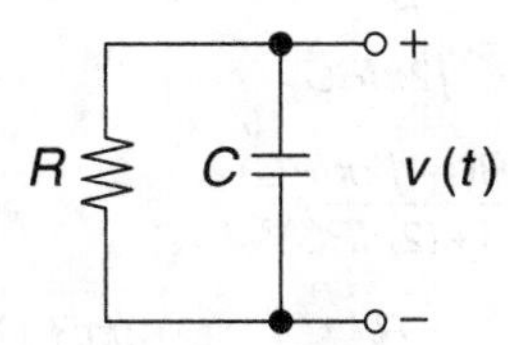

Figure 2.15 A Noise Source Having a Reactive Element

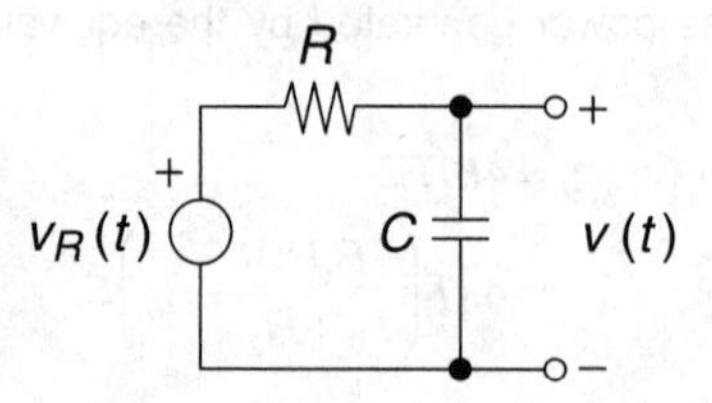

Figure 2.16 The Resistor Is Replaced with Its Noise Model

Next, the magnitude squared of the frequency response of the filter is

$$|H(f)|^2 = \frac{1}{1+(2\pi fRC)^2}. \tag{2.53}$$

Finally, we find the output-noise power spectrum, that is,

$$\begin{aligned} S_v(f) &= |H(f)|^2 S_R(f) \\ &= \frac{1}{1+(2\pi fRC)^2} 2kTR \\ &= 2kT\left(\frac{R}{1+(2\pi fRC)^2}\right). \end{aligned} \tag{2.54}$$

It is interesting to note that the noise at the output of the RC circuit is not white. At low frequencies the capacitor acts as an open circuit, and the noise power spectrum is essentially that of the resistor acting alone. At high frequencies the capacitor impedance becomes very small compared to the resistor and most of the noise voltage is dropped across the resistor rather than across the capacitor. The power spectrum reflects the capacitor voltage and decreases monotonically as frequency increases.

Method 2

The impedance of the circuit of Figure 2.15 as seen from the terminals is

$$\begin{aligned} Z(f) &= \frac{1}{\frac{1}{R} + j2\pi fC} \\ &= \frac{R}{1+j2\pi fRC} \\ &= \frac{R(1-j2\pi fRC)}{1+(2\pi fRC)^2} \\ &= \frac{R}{1+(2\pi fRC)^2} - j\frac{2\pi fR^2C}{1+(2\pi fRC)^2}. \end{aligned} \tag{2.55}$$

If we write

$$Z(f) = R(f) + jX(f), \tag{2.56}$$

then the resistive part of this impedance is given by

$$R(f) = \frac{R}{1+(2\pi fRC)^2}. \tag{2.57}$$

Now the thermal noise at the output of the circuit appears to be generated by the equivalent resistance at the port. Substituting in Equation (2.32) gives

$$\begin{aligned} S_v(f) &= 2kTR(f) \\ &= 2kT\left(\frac{R}{1+(2\pi fRC)^2}\right), \end{aligned} \tag{2.58}$$

which is precisely the same as the answer given in Equation (2.54).

To find the average power at the circuit terminals, we can calculate the area under the power spectrum. This will give a finite result since the noise generated by the RC circuit is not white. We find

$$\begin{aligned} P_v &= \int_{-\infty}^{\infty} S_v(f)df \\ &= \int_{-\infty}^{\infty} \frac{2kTR}{1+(2\pi fRC)^2} df \\ &= \frac{kT}{C}. \end{aligned} \tag{2.59}$$

There is an interesting interpretation to Equation (2.59). If we multiply and divide by $4R$, we obtain

$$\begin{aligned} P_v &= \frac{4kTR}{4RC} \\ &= \frac{4kTR}{\frac{2}{\pi}2\pi RC} \\ &= 4kTR\left(\frac{\pi}{2}\frac{1}{2\pi RC}\right) \\ &= 4kTR\left(\frac{\pi}{2}f_{3dB}\right), \end{aligned} \tag{2.60}$$

where f_{3dB} is the 3 dB bandwidth of an RC lowpass filter. Now refer to Figure 2.16. It appears that the average power of the noise generated by the resistor is being measured through an RC filter that sets the measurement bandwidth. If we compare

Equation (2.60) with Equation (2.39), we see that the measurement bandwidth is given by

$$B = \frac{\pi}{2} f_{3dB}. \tag{2.61}$$

What sort of bandwidth this is will be the subject of a subsequent discussion.

Not all sources of noise are passive. Allowing amplification of the noise can produce interesting effects. An example will illustrate the idea.

Example

Figure 2.17 shows an amplifier with a room-temperature resistive noise source providing the input signal. Suppose the amplifier has a high input resistance, an output resistance of 100Ω, and a voltage gain of 10. We wish to find the average power in the output-noise signal $v(t)$. We will consider only thermal noise in this example and will ignore any additional noise that may be contributed by transistors in the amplifier. Figure 2.18 shows the amplifier replaced by a circuit model.

Now

$$v(t) = v_{100}(t) + 10v_{1000}(t) \tag{2.62}$$

where $v_{100}(t)$ and $v_{1000}(t)$ are the noise voltages across the 100 Ω and 1kΩ resistors, respectively. The average power at the amplifier output is given by

$$\begin{aligned} \langle v^2(t)\rangle &= \langle v_{100}^2\rangle + 20\langle v_{100}v_{1000}\rangle + 100\langle v_{1000}^2\rangle \\ &= \langle v_{100}^2\rangle + 100\langle v_{1000}^2\rangle, \end{aligned} \tag{2.63}$$

where the product of $v_{100}(t)$ and $v_{1000}(t)$ averages to zero because the noise voltages generated by physically distinct resistors are statistically independent and thermal noise has zero average value. Substituting Equation (2.42) for the mean-square resistor noise voltages gives

$$\begin{aligned} \langle v^2(t)\rangle &= 4kT_0(100)B + 100\times 4kT_0(1000)B \\ &= 4kT_0(100100)B \end{aligned} \tag{2.64}$$

measured in bandwidth B.

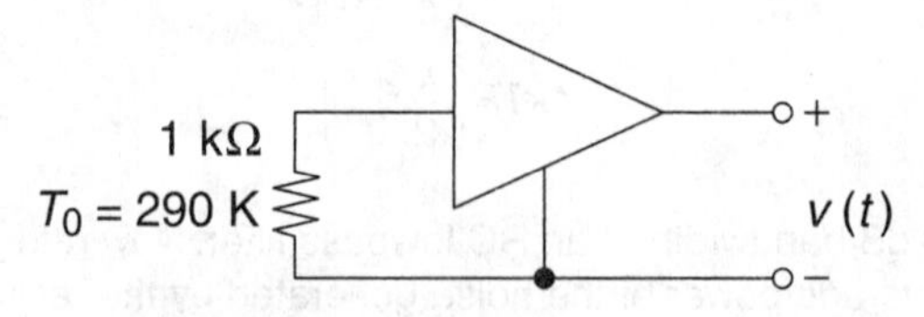

Figure 2.17 An Amplifier with a Noise Source as Input

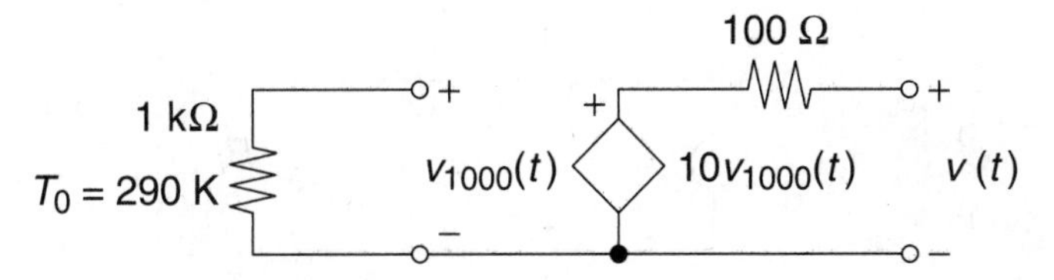

Figure 2.18 The Amplifier Is Replaced by a Circuit Model

A naive user looking back into the amplifier from the output terminals will see the Thevenin equivalent shown in Figure 2.19. The user will see a thermal noise voltage $v(t)$ that appears to be coming from a 100 Ω resistor. The user will interpret Equation (2.64) as

$$\langle v^2(t)\rangle = 4k(1001T_0)(100)B \tag{2.65}$$

and will conclude that the 100 Ω resistor must be at a temperature of $1001T_0 \cong 290{,}000$ K, but note details in the following discussion.

A white noise source is often characterized by a so-called **noise temperature**. This is the temperature at which a resistor would have to be to produce thermal noise of power (in a given bandwidth) equal to the observed noise power. The noise temperature of the circuit of Figure 2.17 is 290,000 K even though the circuit is physically at room temperature. Noise temperature is often used to characterize the noise observed at the terminals of an antenna. Antenna noise is usually a combination of noise picked up from electrical discharges in the atmosphere (static); radiation from the Earth, the sun, and other bodies that may lie within the antenna beam; cosmic radiation from space; and, of course, thermal noise from the conductive material out of which the antenna is made. At frequencies below about 30 MHz atmospheric noise is the principal component of antenna noise, and the noise temperature can be as high as 10^{10} K. Atmospheric noise becomes less important at frequencies above 30 MHz. A narrow-beam antenna at 5 GHz pointing directly upward on a dark night may pick up noise primarily from cosmic radiation and oxygen absorption in the atmosphere. The noise temperature in this case may be only 5 K.

Figure 2.20 shows a voltage source with a source resistance R driving a load resistance R_L. This simple model might represent the output of an antenna or the output of a filter or the output of an

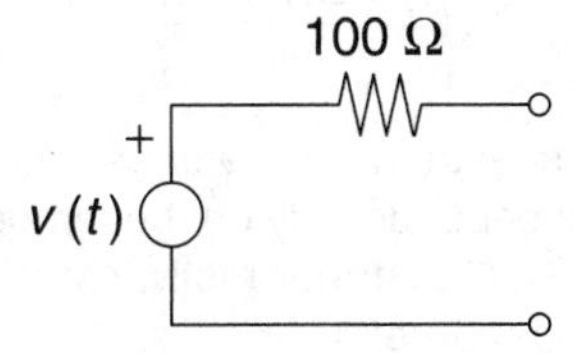

Figure 2.19 Thevenin Equivalent of the Amplifier

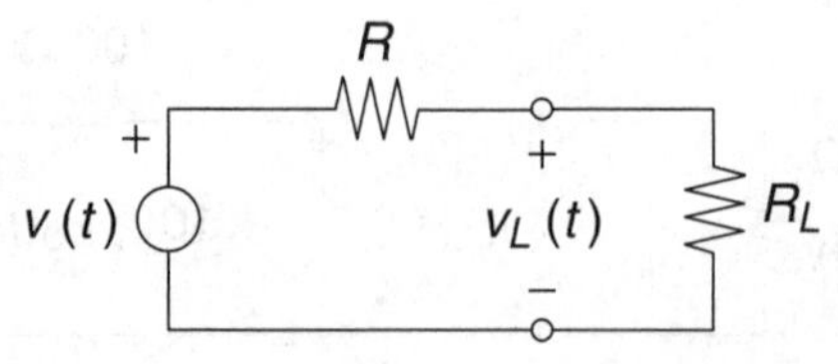

Figure 2.20 A Signal Source Driving a Load

amplifier. Two common situations arise in practice. In some circuits $R_L >> R$ and the load voltage is approximately equal to the source voltage. If the source and load voltages happen to be noise, it is meaningful to characterize these voltages by their mean-square values or, to use another term for the same quantity, by their normalized average power. A second important situation is the one in which we wish to transfer power, rather than voltage, from the source to the load. For a fixed value of source resistance R, maximum power transfer occurs when $R_L = R$. Under this condition we have

$$v_L(t) = \frac{v(t)}{2}, \tag{2.66}$$

and the actual average power transferred to the load is

$$\begin{aligned} P_a &= \frac{\langle (v(t)/2)^2 \rangle}{R_L} \\ &= \frac{\langle v^2(t) \rangle}{4R}. \end{aligned} \tag{2.67}$$

This quantity is called the **available power**.

Example

A resistor of value R at temperature T is a source of thermal noise. A model for the resistor has been given in Figure 2.12. Using Equation (2.67), we find the available noise power to be

$$P_a = \frac{\langle v^2 \rangle}{4R} = \frac{4kTRB}{4R} = kTB. \tag{2.68}$$

The units of available power are watts. We can also find the available power spectrum:

$$S_a(f) = \frac{2kTR}{4R} = \frac{kT}{2}. \tag{2.69}$$

The units of available power spectrum are watts per hertz. It is important to note that the available power spectrum depends only on the temperature of the source and not on the value of the resistor. This is further motivation for characterizing an arbitrary noise source by its noise temperature.

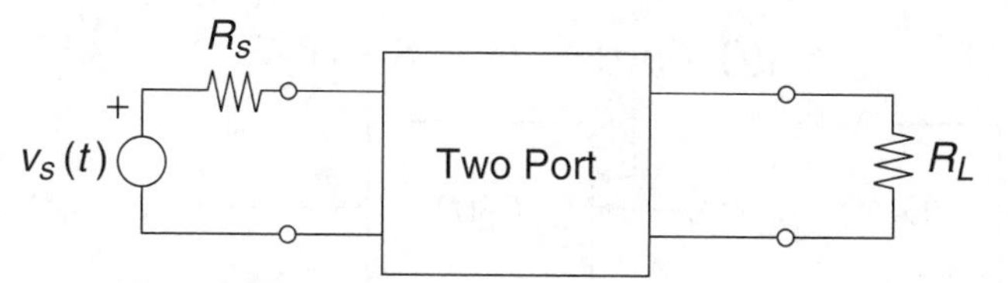

Figure 2.21 A Two-Port with a Source and a Load

Characterizing Two-Ports

A **two-port** is a subsystem with an input and an output. Two-ports model devices such as filters, amplifiers, transmission lines, and mixers. In this section we develop ways to characterize the noise performance of two-ports. We are interested in describing the effect that the two-port has on noise presented at its input and are also interested in identifying the effect of noise added by the two-port itself. We will find that for noise calculations a two-port device can be characterized by three parameters: the available gain, the noise bandwidth, and the noise figure.

Available Gain

Figure 2.21 shows a two-port connected in a circuit. The two-port is driven by a source having source resistance R_s and the two-port in turn drives a load of resistance R_L. Suppose that the source available power spectrum is $S_s(f)$. Note that this available power spectrum characterizes the power that could be drawn from the source into a matched load. Although communication systems are often designed so that the input impedance of a two-port such as that shown in the figure is matched to the source resistance, this is not necessarily the case. That means that the source available power spectrum may not represent the power that is actually absorbed by the two-port device. Let us suppose further that the load available power spectrum is $S_o(f)$. Again, the actual load resistor R_L may or may not be matched to the output resistance of the two-port. The load available power spectrum may not characterize the power actually delivered to the load. Now the **available gain** $G(f)$ is defined as

$$G(f) = \frac{S_o(f)}{S_s(f)}. \tag{2.70}$$

As the ratio of two power spectra, the available gain is a kind of power gain. Note that the available gain is a function of frequency. Some two-ports, such as transmission lines and broadband amplifiers, have available gains that are relatively constant over a broad frequency range. Other two-ports, such as filters and narrowband amplifiers, have available gains with a bandpass characteristic; that is, the available gains are constant over a narrow range of frequencies and very small otherwise. We will normally be interested only in the available gain in the passbands of such devices.

Figure 2.22 shows a cascade of three two-ports. Suppose the two-ports have individual available gains $G_1(f)$, $G_2(f)$, and $G_3(f)$ respectively, and we wish to find an expression for the

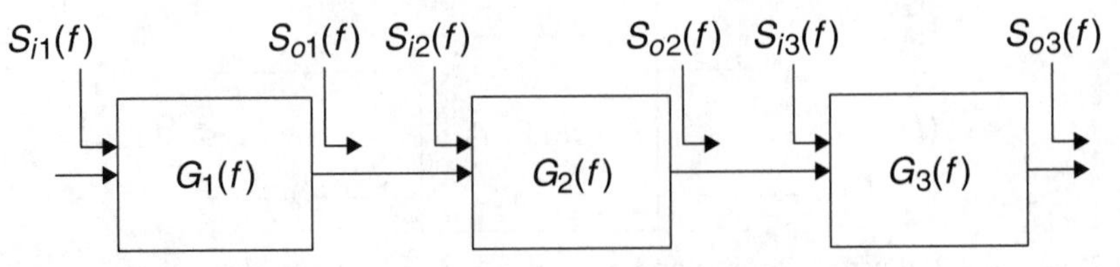

Figure 2.22 Cascaded Two-Ports

overall available gain $G(f)$ of the cascade. The available power spectrum at the input and output of each individual two-port is indicated in the figure. Note that the available power spectrum at the output of stage 1 is identical to the available power spectrum at the input to stage 2. Likewise, the available power spectrum at the output of stage 2 is identical to the available power spectrum at the input to stage 3. We have

$$\begin{aligned} G(f) &= \frac{S_{o3}(f)}{S_{i1}(f)} \\ &= \frac{S_{o1}(f)}{S_{i1}(f)} \frac{S_{o2}(f)}{S_{o1}(f)} \frac{S_{o3}(f)}{S_{o2}(f)} \\ &= \frac{S_{o1}(f)}{S_{i1}(f)} \frac{S_{o2}(f)}{S_{i2}(f)} \frac{S_{o3}(f)}{S_{i3}(f)} \\ &= G_{a1}(f) G_{a2}(f) G_{a3}(f). \end{aligned} \tag{2.71}$$

Thus, two-ports can be combined in cascade simply by multiplying their available gains. The usefulness of the available gain concept can be seen in the following example.

Example

A two-port with available gain $G(f)$ is driven by a white noise source having available power spectrum $S_s(f) = \frac{N_0}{2}$. At the output of the two-port we have

$$\begin{aligned} S_o(f) &= G(f) S_i(f) \\ &= G(f) \frac{N_0}{2}. \end{aligned} \tag{2.72}$$

The available noise power at the output is the area under the available power spectrum, that is,

$$P_{ao} = \int_{-\infty}^{\infty} S_o(f) df = \int_{-\infty}^{\infty} G(f) \frac{N_0}{2} df = \frac{N_0}{2} \int_{-\infty}^{\infty} G(f) df. \tag{2.73}$$

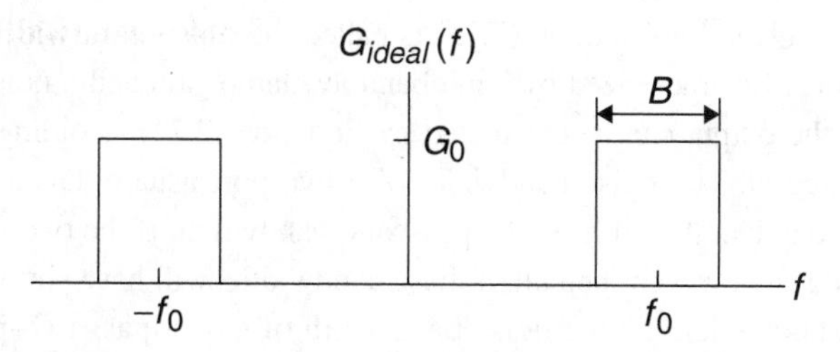

Figure 2.23 Available Gain of an Ideal Bandpass Filter

Noise Bandwidth

Many bandpass two-ports such as amplifiers and filters have an approximately rectangular passband. An idealization of such a passband is shown in the available gain plot of Figure 2.23. Suppose we have an actual two-port device and we model this device by adjusting the plot of Figure 2.23 so that the center frequency f_0 and the midband gain G_0 reflect parameter values of the actual device. Now suppose the actual device and the ideal model are both driven by identical white noise sources. The available power at the output of the actual device is given by Equation (2.73). To find the available power at the output of the ideal device we first find the available power spectrum at the output using Equation (2.72). This available power spectrum is shown in Figure 2.24. The available power at the output is the area under the available power spectrum, that is,

$$P_{ao} = 2G_0 \frac{N_0}{2} B = G_0 N_0 B \tag{2.74}$$

for the idealized model. Now we want the idealized model to predict the correct value for available power out, so Equation (2.74) and Equation (2.73) must give the same value of P_{ao}. To make this happen, the bandwidth parameter B in the idealized model must be given by

$$\begin{aligned} G_0 N_0 B &= \frac{N_0}{2} \int_{-\infty}^{\infty} G(f) df \\ B &= \frac{1}{2G_0} \int_{-\infty}^{\infty} G(f) df, \end{aligned} \tag{2.75}$$

where $G(f)$ is the available gain and G_0 is the midband available gain of the actual two-port.

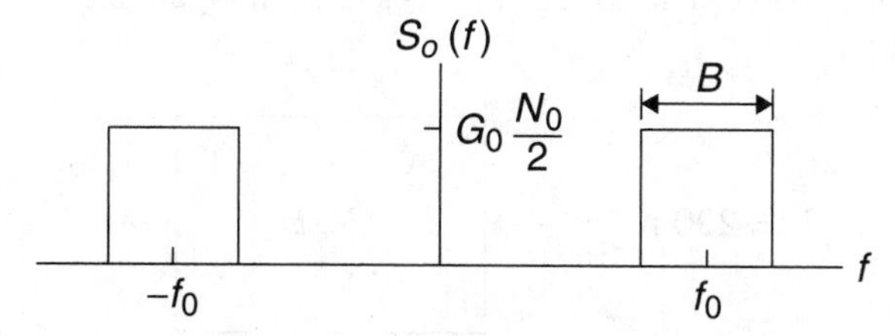

Figure 2.24 Available Power Spectrum at the Ideal Bandpass Filter Output

The bandwidth B given by Equation (2.75) is called the **noise bandwidth** of the actual two-port. When a two-port is characterized by a midband available gain and a noise bandwidth, available noise power at the output can be calculated by Equation (2.74), avoiding the evaluation of a sometimes messy integral. The noise bandwidth of a two-port usually has a somewhat different value from the 3 dB bandwidth and from the passband bandwidth. If the two-port has a sharp cutoff at its passband edges, however, then all of these bandwidths will have similar values.

The derivation above defined the noise bandwidth of a bandpass two-port. We can define the noise bandwidth for a lowpass two-port in an analogous way, as is shown in the following example.

Example

Consider an nth-order lowpass Butterworth filter, for which the available gain is

$$G(f) = \frac{1}{1+\left(\frac{f}{f_{3dB}}\right)^{2n}}, \tag{2.76}$$

where f_{3dB} is the 3 dB bandwidth. Note that the DC gain of this filter is unity. Then

$$\begin{aligned} B &= \frac{1}{2G_0}\int_{-\infty}^{\infty} G(f)df \\ &= \frac{1}{2}\int_{-\infty}^{\infty} \frac{1}{1+\left(\frac{f}{f_{3dB}}\right)^{2n}} df \\ &= \frac{\frac{\pi}{2} f_{3dB}}{n\sin\left(\frac{\pi}{2n}\right)}. \end{aligned} \tag{2.77}$$

For $n=1$ we have $B=\frac{\pi}{2}f_{3dB}$ (compare Equation (2.61)). As $n \to \infty$ we find $B \to f_{3dB}$.

Noise Figure

Now consider a physical amplifier with available gain G_0 and noise bandwidth B driven by a room-temperature noise source. The arrangement is shown in Figure 2.25.

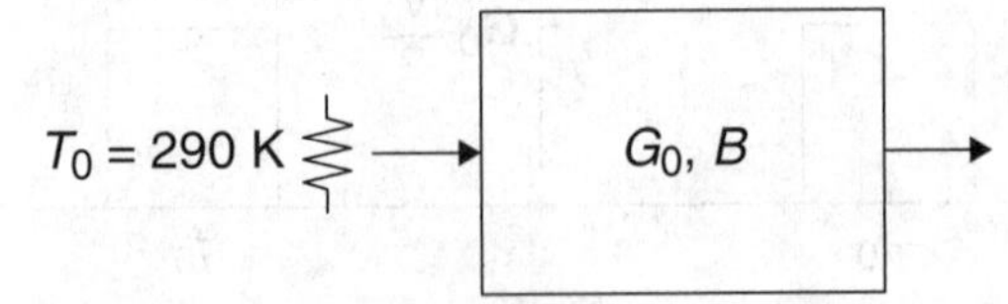

$T_0 = 290$ K → [G_0, B] →

Figure 2.25 A Two-Port with a Room-Temperature Noise Source

According to Equation (2.74), the available noise power at the amplifier output should be $P_{ao} = G_0 N_0 B$. In fact, the output power will be larger than this owing to noise generated within the amplifier. There are two widely used conventions for accounting for this additional noise, the **noise figure** and the **effective input-noise temperature**. Noise figure is the older of the two conventions; it is widely used in terrestrial wireless systems. Effective input-noise temperature is more common in specifying satellite and other space systems that operate with extremely low antenna noise temperatures. We will introduce noise figure first and then we will present the effective input-noise temperature alternative.

Suppose, in the setup shown in Figure 2.25, the available power actually obtained at the output of the two-port is P_{ao}. Then the noise figure is defined as

$$F = \frac{P_{ao}}{kT_0 G_0 B}, \tag{2.78}$$

where G_0 and B are the available gain and noise bandwidth respectively, that is,

$$P_{ao} = kT_0 G_0 BF. \tag{2.79}$$

It is important to note that Equation (2.79) is valid only when the noise source is at room temperature $T_0 = 290$ K. It is common practice to express the noise figure in decibels, where

$$F_{dB} = 10\log F. \tag{2.80}$$

In some texts, F is called the **noise factor** and the term **noise figure** is reserved for F_{dB}. Since a two-port can add noise, but not subtract noise, the noise figure F is always at least unity. This means that the noise figure in decibels F_{dB} is never negative. If we rewrite Equation (2.79) as

$$P_{ao} = kT_0 G_0 B + kT_0 G_0 B(F-1), \tag{2.81}$$

we recognize the first term $kT_0 G_0 B$ as the output-noise power caused by the room-temperature source. This means that the second term

$$N_r = kT_0 G_0 B(F-1) \tag{2.82}$$

must be the noise power contributed by the two-port. If the noise source were at some temperature T other than room temperature, then the first term in Equation (2.81) would change, but the second would not. In general, for a source of noise temperature T we have

$$P_{ao} = kTG_0 B + kT_0 G_0 B(F-1). \tag{2.83}$$

To help develop some intuitive feeling for the noise figure concept, let us consider the effect of two-port noise on signal-to-noise ratio. We shall see in subsequent chapters that signal-to-noise ratio is an important figure of merit for system performance. Suppose a two-port such as the one shown in Figure 2.25 receives a signal of available power P_{si} at its input. We will assume that the room-temperature white noise is also present. Assuming that the signal is within the passband of the two-port, the signal power at the output is given by

$$P_{so} = G_0 P_{si}. \tag{2.84}$$

Since the output-noise power is given by Equation (2.79), we can write the signal-to-noise ratio at the output as

$$\begin{aligned} SNR_o &= \frac{G_0 P_{si}}{kT_0 G_0 BF} \\ &= \frac{P_{si}}{kT_0 BF}. \end{aligned} \tag{2.85}$$

We would like to compare the signal-to-noise ratio at the two-port output with a similar ratio at the two-port input. At the input, however, the noise is white and the total noise power cannot be calculated. (Our white noise model would give a total input-noise power of infinity.) To allow the comparison, let us define an "input-noise power" that is the noise power in the bandwidth B. This is a plausible step, since only noise within the passband of the two-port will make its way through the two-port to the output. The input-noise power calculated this way is

$$P_{ai} = kT_0 B. \tag{2.86}$$

Using this definition of input-noise power, we can find the input signal-to-noise ratio,

$$SNR_i = \frac{P_{si}}{kT_0 B}. \tag{2.87}$$

We can calculate the degradation in signal-to-noise ratio caused by the two-port if we take the ratio of SNR_i to SNR_o. We obtain

$$\frac{SNR_i}{SNR_o} = \frac{\frac{P_{si}}{kT_0 B}}{\frac{P_{si}}{kT_0 BF}} = F \tag{2.88}$$

which is just the noise figure.

A low-noise amplifier will have a noise figure very close to unity. An alternative way of representing two-port noise that is useful in this case is the effective input-noise temperature T_e. In this formulation we refer again to Figure 2.25, only in this case we do not require the input-noise source to be at room temperature. We write the available noise power at the output as

$$P_{ao} = k(T + T_e) G_0 B, \tag{2.89}$$

where T is the noise temperature of the source. Comparing Equation (2.89) with Equation (2.81), we can write

$$P_{ao} = kTG_0B + N_r,$$

where kTG_0B is the available noise power at the output due to the input-noise source, and N_r is the available noise power at the output contributed by the two-port. We have

$$N_r = kT_eG_0B. \tag{2.90}$$

Example

Figure 2.26 shows a receiver "front end." This includes the amplifiers, mixers, and filters from the antenna to the demodulator. The antenna is shown as a source of noise having noise temperature $T_{ant} = 10$ K.

Now the available noise power spectrum from the antenna is

$$\begin{aligned} S_{ai} &= \frac{kT_{ant}}{2} \\ &= \frac{(1.38 \times 10^{-23}\text{ J/K})(10\text{ K})}{2} \\ &= 69.0 \times 10^{-24}\text{ W/Hz}. \end{aligned} \tag{2.91}$$

(We avoid calculating the available input-noise *power*.)

To include the amplifier noise, the output-noise power spectrum uses a temperature of $T_{ant} + T_e = 10 + 140 = 150$ K, that is,

$$S_{ao} = \frac{k(T_{ant} + T_e)G_0}{2}. \tag{2.92}$$

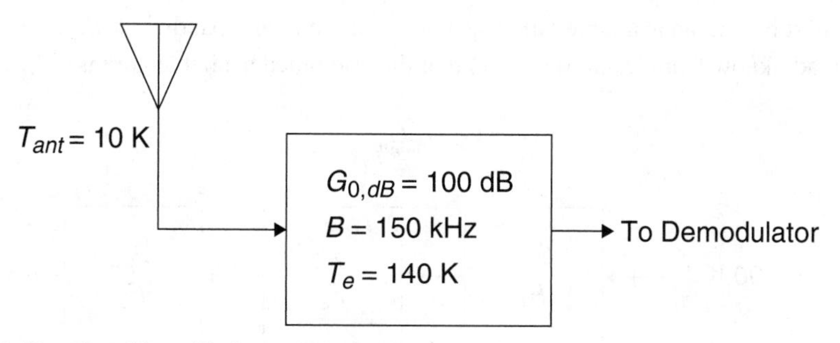

Figure 2.26 Receiver Front End

We must also convert the available gain from decibels: $G_0 = 10^{\frac{100}{10}} = 10^{10}$. Then the available output-noise spectrum is

$$S_{ao} = \frac{(1.38\times10^{-23}\ \text{J/K})(150\ \text{K})(10^{10})}{2} \tag{2.93}$$
$$= 10.4\times10^{-12}\ \text{W/Hz}.$$

The available output-noise power is

$$P_{ao} = k(T_{ant}+T_e)G_0B$$
$$= (1.38\times10^{-23}\ \text{J/K})(150\ \text{K})(10^{10})(150\times10^{3}\ \text{Hz}) \tag{2.94}$$
$$= 3.1\ \mu\text{W}.$$

Noise figure and available input-noise temperature are measures of the same quantity. Thus, Equation (2.82) and Equation (2.90) must give the same value for N_r. Then

$$N_r = kT_0G_0B(F-1) = kT_eG_0B$$
$$T_0(F-1) = T_e \tag{2.95}$$
$$F = 1+\frac{T_e}{T_0}.$$

Solving for T_e gives

$$T_e = T_0(F-1). \tag{2.96}$$

Cascade of Two-Ports

Figure 2.27 shows a system made up of a cascade of three noisy two-ports. In an actual receiver these might be, for example, a low-noise amplifier, a transmission line, and a high-gain amplifier-filter. Each of the individual stages is characterized by an available gain and a noise figure. We assume that one of the stages limits the bandwidth of the cascade, and we use the symbol B to represent that bandwidth. Our object is to combine the three stages and obtain an equivalent single two-port having an available gain G_0, a noise figure F, and bandwidth B.

We already know from Equation (2.71) that the combined available gain is

$$G_0 = G_{01}G_{02}G_{03}. \tag{2.97}$$

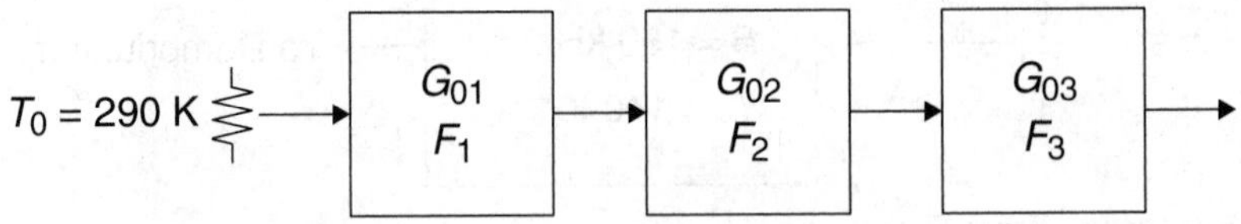

Figure 2.27 A Cascade of Noisy Two-Ports

To ensure the proper use of noise figure, the input to the cascade is a noise source at room temperature $T_0 = 290$ K.

The available noise power at the output of the cascade is the sum of the noise power contributed by the source plus the noise power contributed by each individual stage. We can write

$$P_{ao} = kT_0 G_{01} G_{02} G_{03} B + N_{r1} G_{02} G_{03} + N_{r2} G_{03} + N_{r3}, \tag{2.98}$$

where the noise from each stage is amplified by the stages that follow it. Using Equation (2.82), we obtain

$$\begin{aligned} P_{ao} &= kT_0 G_{01} G_{02} G_{03} B + (kT_0 G_{01} B(F_1 - 1)) G_{02} G_{03} + (kT_0 G_{02} B(F_2 - 1)) G_{03} + (kT_0 G_{03} B(F_3 - 1)) \\ &= kT_0 G_{01} G_{02} G_{03} B + kT_0 G_{01} G_{02} G_{03} B(F_1 - 1) + kT_0 G_{02} G_{03} B(F_2 - 1) + kT_0 G_{03} B(F_3 - 1) \\ &= kT_0 G_{01} G_{02} G_{03} BF_1 + kT_0 G_{02} G_{03} B(F_2 - 1) + kT_0 G_{03} B(F_3 - 1). \end{aligned} \tag{2.99}$$

Now for the equivalent single two-port,

$$P_{ao} = kT_0 G_0 BF, \tag{2.100}$$

with G_0 given by Equation (2.97). Equating Equation (2.99) to Equation (2.100) and dividing by $kT_0 G_0 B$ gives

$$F = F_1 + \frac{F_2 - 1}{G_{01}} + \frac{F_3 - 1}{G_{01} G_{02}}. \tag{2.101}$$

The extension to an arbitrary number of stages should be obvious.

If we wish to work with effective input-noise temperature rather than with noise figure, then we can substitute Equation (2.90) into Equation (2.98) to obtain

$$\begin{aligned} P_{ao} &= kT_0 G_{01} G_{02} G_{03} B + (kT_{e1} G_{01} B) G_{02} G_{03} + (kT_{e2} G_{02} B) G_{03} + (kT_{e3} G_{03} B) \\ &= kT_0 G_{01} G_{02} G_{03} B + kT_{e1} G_{01} G_{02} G_{03} B + kT_{e2} G_{02} G_{03} B + kT_{e3} G_{03} B, \end{aligned} \tag{2.102}$$

where T_{e1}, T_{e2}, T_{e3} are the effective input-noise temperatures of the three stages respectively. For the equivalent single two-port,

$$P_{a0} = k(T_0 + T_e) G_0 B. \tag{2.103}$$

As before, we equate Equation (2.102) to Equation (2.103) to obtain

$$T_e = T_{e1} + \frac{T_{e2}}{G_{01}} + \frac{T_{e3}}{G_{01} G_{02}}. \tag{2.104}$$

Equations (2.101) and (2.104) are sometimes known as Friis formulas, after Harald Friis, who developed the theory of noise figure at the Bell Telephone Laboratories in the 1940s. (This is the same Friis who brought you the Friis transmission formula!)

Equation (2.101) (or Equation (2.104)) contains an important message. If the three stages have comparable noise figures, then the dominant term in Equation (2.101) will be the first term. This suggests that if the first stage of a receiver has a low noise figure and a high gain, then the noise figures of the subsequent stages do not matter much. It is common practice for the first stage in a receiver to be a "low-noise" amplifier (LNA).

Example

Figure 2.28 shows the first two stages in a receiver. The first stage is a low-noise amplifier, and the second stage is a mixer. The available gains and noise figures are as given in the figure. The antenna noise temperature is also specified. No bandwidth is given in this example, as the bandwidth-limiting stage would normally be the intermediate-frequency (IF) amplifier that would follow the mixer.

The overall gain of the system is given by

$$G_{0,dB} = G_{01,dB} + G_{02,dB} = 10 + 9 = 19 \text{ dB}. \tag{2.105}$$

We must begin the noise analysis by converting the gains and noise figures from decibels. We obtain $G_{01} = 10$ and $F_1 = 2$ for the LNA, and $G_{02} = 8$ and $F_2 = 4.47$ for the mixer. Next we can find the overall noise figure from Equation (2.101):

$$F = 2 + \frac{4.47 - 1}{10} = 2.35, \tag{2.106}$$

or

$$F_{dB} = 10\log(2.35) = 3.71 \text{ dB}. \tag{2.107}$$

Notice how close the overall noise figure is to the noise figure of the LNA.

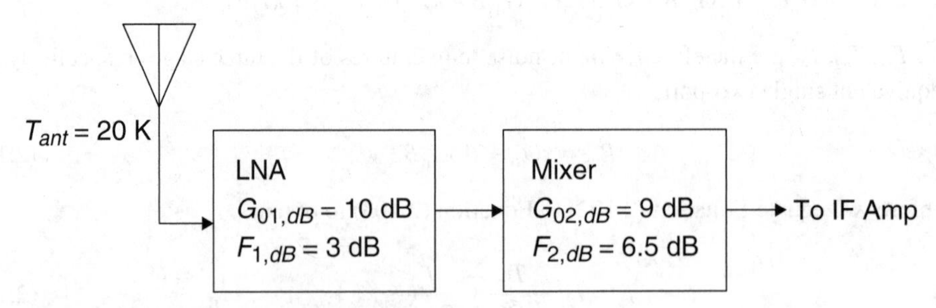

Figure 2.28 The First Two Stages in a Receiver

Since the input noise is not coming from a room-temperature source, we cannot use the overall noise figure to find the available output-noise power spectrum directly. One alternative approach is to use the effective input-noise temperature, which does not presume the use of a room-temperature noise source. From Equation (2.96) we obtain

$$T_e = 290(2.35-1) = 391\text{ K}. \tag{2.108}$$

Then

$$\begin{aligned} S_{ao} &= \frac{k(T_{ant}+T_e)G_0}{2} \\ &= \frac{(1.38\times10^{-23}\text{ J/K})(20\text{ K}+391\text{ K})(10\times8)}{2} \\ &= 227\times10^{-21}\text{ W/Hz}. \end{aligned} \tag{2.109}$$

Lossy Transmission Lines

A transmission line is a coaxial cable or waveguide used to connect a transmitter or a receiver to an antenna and sometimes to connect other stages in a communication system. Transmission lines are lossy, particularly at the frequencies used in cellular and personal communication systems. Consequently the losses of the transmission line must be taken into account in any system design. Losses in the receiving system are particularly damaging, as losses are a source of thermal noise.

Let us consider a length of transmission line of characteristic impedance Z_0. In most practical circumstances we can take Z_0 to be a real number. We will look at a transmission line from two points of view. First, a transmission line is a two-port, and therefore it has an available gain and a noise figure. We note, however, that the available gain will be less than unity. Expressed in decibels, the available gain will be a negative number. It is customary to drop the minus sign and describe the transmission line in terms of its loss; that is, letting G represent the gain and L represent the corresponding loss,

$$L = \frac{1}{G} \tag{2.110}$$

and

$$L_{dB} = 10\log(L) = 10\log\left(\frac{1}{G}\right) = -10\log(G) = -G_{dB}. \tag{2.111}$$

Now suppose the transmission line is terminated at its input end in a resistor of value $R = Z_0$. We will assume that the input resistor, and in fact the entire transmission line, is in thermal equilibrium at room temperature T_0. Figure 2.29 shows the setup. The available noise power at the output of the transmission line is

$$P_{ao} = kT_0GBF = kT_0\frac{1}{L}BF \tag{2.112}$$

measured in bandwidth B.

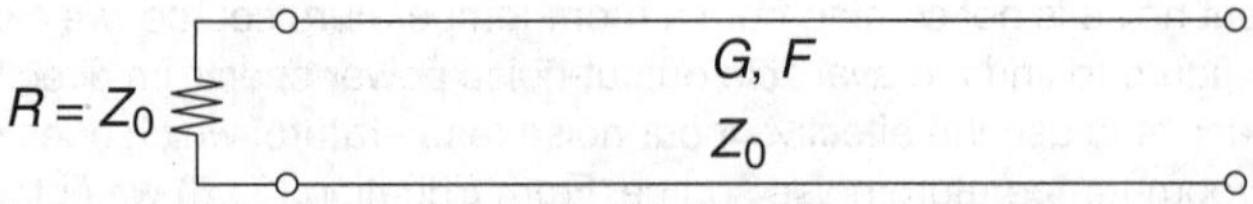

Figure 2.29 A Lossy Transmission Line at Room Temperature

For our second point of view we will take the transmission line with the input resistor connected to be a one-port circuit. We observe that since the input resistance equals the characteristic impedance of the transmission line, the impedance seen looking into the output end of the line is Z_0, which we are taking to be a real number. Now recall that for a passive circuit, whatever noise there is, appears to be generated by the equivalent resistance at the port. This means that the available power at the output of the transmission line is just the power generated by a room-temperature resistor. We have

$$P_{ao} = kT_0B. \tag{2.113}$$

Now Equation (2.112) and Equation (2.113) describe the same quantity. Equating these gives us the noise figure of the lossy line:

$$F = L. \tag{2.114}$$

As an intuitive description of the operation of a lossy transmission line, consider what happens when a signal in room-temperature noise is introduced at the input of the line. The signal is attenuated in passing down the transmission line. The input noise is attenuated also, but the transmission line adds noise of its own, to exactly make up the noise that is lost. At the output, then, we have a reduced signal that is still in room-temperature noise. The signal-to-noise ratio is reduced in the transmission line by exactly the loss in the signal. If the noise figure represents the degradation in signal-to-noise ratio, we are led immediately to Equation (2.114).

A Final Example

As a practical example of applying these concepts, we calculate the noise figure and effective input-noise temperature for the front end of a mobile telephone receiver. The receiver is shown in Figure 2.30. In the following analysis we derive expressions for the effective input-noise temperature at each interface. Most of the components shown in the figure are straightforward, but the mixer requires a comment. Since the mixer is specified as having loss L_C, a passive mixer is assumed. A passive mixer contains diodes, and so it is not a linear system. Consequently the analysis that we used to find the noise figure of a lossy transmission line cannot be applied to the mixer without some modification. It turns out, however, that for most practical mixers the noise figure is only very slightly higher than the loss. Consequently it is common practice to set $F = L_C$ for a passive mixer with only negligible inaccuracy.

Let us now proceed from right to left through the receiver system:

1. At this interface the effective input-noise temperature is given as $T_{e1} = T_{ex}$.
2. Using Equation (2.104), $T_{e2} = T_{ea2} + \frac{T_{ex}}{G_2}$.

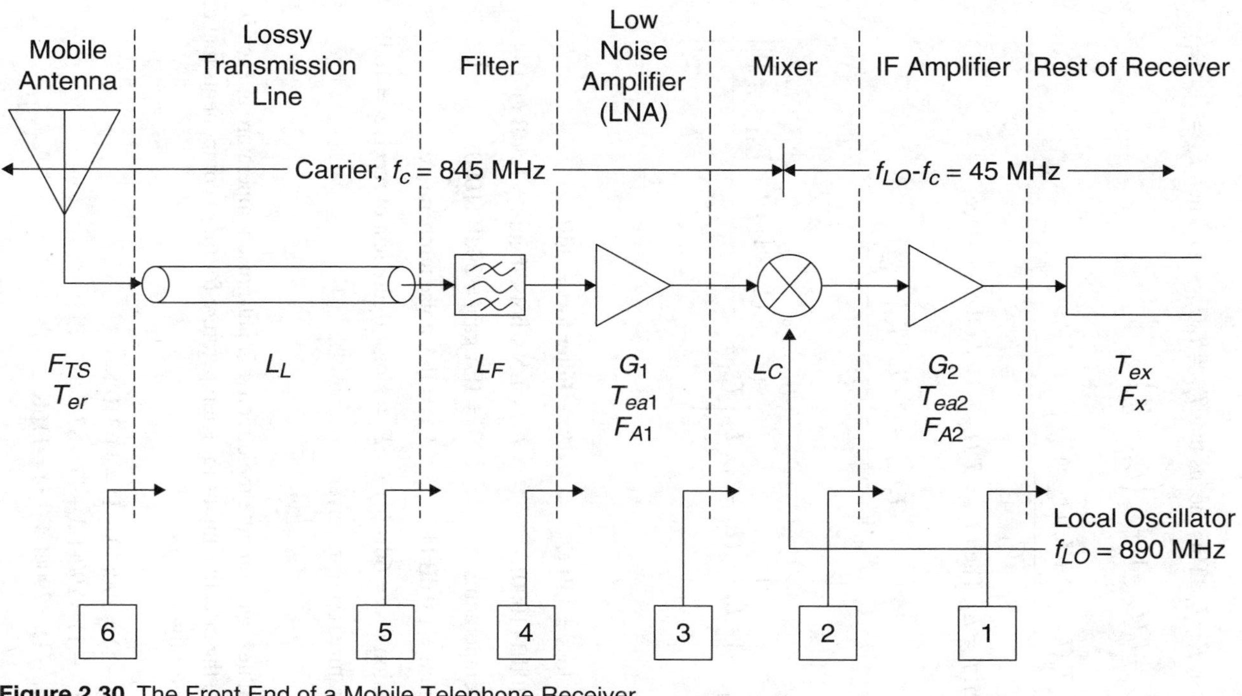

Figure 2.30 The Front End of a Mobile Telephone Receiver

3. For the mixer, $F = L_C$. This means that $T_{eC} = T_0(L_C - 1)$. Then $T_{e3} = T_{eC} + \frac{T_{e2}}{1/L_C}$. Substituting gives $T_{e3} = T_0(L_C - 1) + L_C\left(T_{ea2} + \frac{T_{ex}}{G_2}\right)$.

4. $$T_{e4} = T_{ea1} + \frac{T_0(L_C - 1) + L_C\left(T_{ea2} + \frac{T_{ex}}{G_2}\right)}{G_1}.$$

5. For the filter, $F = L_F$. Then $T_{eF} = T_0(L_F - 1)$. At the interface,

$$T_{e5} = T_0(L_F - 1) + L_F\left(T_{ea1} + \frac{T_0(L_C - 1) + L_C\left(T_{ea2} + \frac{T_{ex}}{G_2}\right)}{G_1}\right).$$

6. $$T_{e6} = T_0(L_L - 1) + L_L\left\{T_0(L_F - 1) + L_F\left(T_{ea1} + \frac{T_0(L_C - 1) + L_C\left(T_{ea2} + \frac{T_{ex}}{G_2}\right)}{G_1}\right)\right\}.$$

Typical Values

L_L = Antenna cable loss = 3dB (2)

L_F = Filter loss = 3dB (2)

G_1 = LNA gain = 30dB (1000)

F_{A1} = LNA noise figure = 1.5dB (1.4)

T_{eA1} = LNA noise temperature

G_2 = IFA gain = 30dB (1000)

F_{A2} = IFA noise figure = 1.0 dB (1.3)

T_{eA2} = IFA noise temperature

L_C = Mixer conversion loss = 6dB (4)

F_x = Noise figure rest of receiver = 16dB (40)

T_{ex} = Noise temperature rest of receiver

Using these values, we compute the effective input-noise temperature at each interface, but first let us calculate the effective input-noise temperature for those components that are specified in terms of their noise figure:

1. $T_{ex} = T_0\,(F_{ex} - 1) = 290(40 - 1) = 11{,}310$ K
2. $T_{ea2} = T_0\,(F_{A2} - 1) = 290(1.3 - 1) = 87$ K
3. $T_{ea1} = T_0(F_{A1} - 1) = 290(1.4 - 1) = 116$K

The effective input-noise temperature at each interface is computed:

1. $T_{e1} = T_{ex} = 11{,}310$ K
2. $T_{e2} = T_{ea2} + \frac{T_{ex}}{G_2} = 87 + \frac{11{,}310}{1000} = 98.31$ K

3. $T_{e3} = T_0(L_C - 1) + L_C(T_{e2}) = 290(4-1) + 4(98.31) = 1263 \text{ K}$
4. $T_{e4} = T_{ea1} + \dfrac{T_{e3}}{G_1} = 116 + \dfrac{1263}{1000} = 117 \text{ K}$
5. $T_{e5} = T_0(L_F - 1) + L_F(T_{e4}) = 290(2-1) + 2(117) = 525 \text{ K}$
6. $T_{e6} = T_0(L_L - 1) + L_L(T_{e5}) = 290(2-1) + 2(525) = 1339 \text{ K}$

The overall effective input-noise temperature of the receiver front end, T_{er}, is 1339 K. It is interesting to note how low the effective input-noise temperature is at interface 4. At this point T_{e4} is only one degree higher than the effective input-noise temperature of the low-noise amplifier alone. Having 6 dB of transmission line and filter loss ahead of the LNA has a very deleterious effect, increasing the effective input-noise temperature by more than an order of magnitude. Clearly if noise were the only design consideration, it would make sense to place the LNA right at the antenna, ahead of the transmission line and filter.

The overall receiver noise figure is given by Equation (2.95). We have

$$F_{TS} = 1 + \frac{T_{er}}{T_0} = 1 + \frac{1339}{290} = 5.62. \text{ Then } F_{TS,dB} = 10\log(5.62) = 7.5 \text{ dB}.$$

Optimizing the Energy Transmission System

System-Level Design

At the start of our discussion of thermal noise we stated, "When the information-carrying signal is comparable to the noise level, the information may become corrupted or may not even be detectable or distinguishable from the noise." It might at first seem plausible, then, that, given a fixed receiver noise level, a key objective of our radio link design should be to maximize the information-carrying signal power delivered to the receiver. On further reflection, however, we see that this is not precisely the case. For any communication system, the most important goal is to ensure that the user is provided with an adequate quality of service. For a telephone connection, the received voices should be intelligible, and the identity of the speaker should be discernible. Calls should not be interrupted or dropped. For a data connection the received data should be accurate and timely. Images should be clear, crisp, and with good color. Music should have good fidelity and there should be no annoying hisses or whines in the background. Stating design goals in terms of user perceptions, however, raises immediate issues about what is meant by "intelligible," "accurate," "clear," and "good fidelity." Not only must all of these "quality" terms be quantified, but they must also be expressed in terms of parameters that engineers can control. To guarantee intelligibility and fidelity, we must provide a certain bandwidth and dynamic range. To guarantee accuracy and timeliness, we must provide a low rate of data errors. Systems engineers describe the engineering parameters that directly affect user perceptions as **quality of service** (QoS) parameters. In a digital transmission system, several QoS parameters are related to a particularly important parameter of the data link, the **bit error rate** (BER), which is a measure of how often, on the average, a bit is interpreted incorrectly by the receiver. In a later chapter we will show that bit error rate is directly related to the ratio of the received signal level to the noise and interference level at the receiver's input terminals. Thus the design goals are more appropriately stated like this: To provide intelligible voice, or clear images, there will be a maximum acceptable bit error rate. To achieve this bit error rate, there will be a minimum acceptable signal-to-noise-and-interference ratio (SNIR), or just signal-to-noise in the absence of interference. We now see that a design objective that is more appropriate than maximizing received signal power is to ensure that an adequate SNR is achieved at the receiver input. Delivering more power than that required to meet this SNR objective is actually undesirable from a system perspective. In this section the reasons for using this design objective and its intimate relation to link design are investigated.

For discussion purposes, we consider our system as a basic radio link, that is, a system consisting of a transmitter and receiver, with intervening free space as the channel. To optimize the performance of this link, we must first determine the specific performance objectives to be optimized, and we must then identify all of the relevant parameters affecting these performance objectives. For a wireless system, particularly a mobile system such as a cellular telephone, cost and size, weight, and power are equally important as SNR. Optimization, then, has much broader scope than many other technical treatments might consider. From a systems-engineering point of view, optimization should be holistic, considering not only how well a system operates but also all of the other relevant attributes that might constrain a design. For example, designers should consider battery life and portability as constraints for optimizing the more common performance objectives such as data rate and bit error rate.

System design, then, is a process of optimizing a system's technical performance against desired and quantifiable constraints, such as cost, size, weight, and power. So, from a systems engineer's point of view, the design problem should be restated, "Given all of the system constraints, how do we guarantee that an adequate SNR is delivered to the receivers via the radio links?" (Yes, there are two links, forward and reverse, and they are not symmetric.) Given the importance of features such as size, weight, cost, and battery life, maximizing the power delivered to a receiver without considering constraints on these other features is not a well-founded design objective. The consequences of transmitting more power than necessary will certainly affect all of the system attributes as well as contribute to the overall background of RF energy, thereby causing unnecessary interference.

Receiver Sensitivity

As stated previously, the measures of goodness or QoS provided by a system are often related to the SNR. At this point in our study, however, we will not discuss the relationships quantitatively. For the present discussion, it is sufficient to assume that if the SNR is adequate, the relevant QoS objectives will be met. In later chapters we will show how the choice of modulation and signaling type is related to SNR. Understanding the relationships allows a link analyst to determine the SNR required to achieve a given level of performance. For a digital system, a typical performance parameter might be bit error rate at a particular transmission rate (e.g., a BER of 10^{-7} at 100 kbits/s), a parameter of particular interest at the most basic level of analysis.

A simplified description of a typical RF link design process is as follows. Once the key link performance parameters are identified and detailed objectives are established (bit error rate, data rate, etc.), a signaling scheme is chosen. Given a chosen signaling scheme, the required baseband SNR is calculated and adjusted for the implementation loss of the actual receiver. The result is a prediction of the actual SNR that the receiver requires to achieve the performance objectives. Noise power levels are calculated for typical operating environments and are then translated back to the receiver input to produce a prediction of the required SNR at the input to the receiver. (If interference is present, we also calculate the interference power levels to determine the required signal-to-noise-and-interference ratio at the input to the receiver.) Given the SNR and noise power, the corresponding minimum required power p_{sens} at the input to the receiver is then calculated. The power p_{sens} is known as the receiver sensitivity and is the minimum power required to meet the given performance objective (such as bit error rate or data rate). Systems engineers use the range equation to ensure that this power level is achieved over the range of interest and over all expected operating conditions.

Overdesigning the link by delivering more power than required could introduce difficulties in meeting objectives and constraints involved in the system implementation. For example, too much power radiated from an antenna could violate license parameters or even pose a health risk. Excessive power feeding the antenna will cause unnecessary battery draw in a portable unit or create excessive heat. Excessive transmitter power could also cause interference between the transmitter and receiving sections of a handheld unit, necessitating extra shielding and thus extra bulk and cost. Notwithstanding these considerations, a certain level of overdesign is usually mandated by good engineering practice: The link system engineer will provide for the delivery of the required minimum power to the receiver plus some extra power called "margin." Margin power ensures that the minimum required power is maintained in spite of design errors, aging effects, and manufacturing variations, as well as variations introduced by a real-world propagation environment, as will be discussed in the next chapter. Evidently, a systems engineer must have a good understanding of both the design objectives and system variables in order to optimize a system design to meet performance requirements within specific design constraints such as cost, size, weight, and battery life.

Top-Level Design

Given the discussion above, our design objective has been reduced to ensuring that some power P_r, which is the receiver sensitivity plus margin, is delivered to the receiver over the radio link. The range equation is the basic radio link model we developed earlier for predicting the variation in received power versus distance between a transmitter and a receiver. It is repeated here for convenience:

$$P_r = \frac{P_t G_t G_r}{L_{path} L_{sys}}. \tag{2.115}$$

The range equation is expressed in terms of five variables that can be selected so that the power P_r is delivered to the receiver. Each of these variables may have design constraints that must be considered and traded off to optimize the link. For example, G_t is more than just a number representing the gain of an antenna. Its value has implications regarding the beamwidth, size and weight, and cost of an antenna. The system designer must work with subsystem and component designers to balance the performance and cost of each of the terms in implementing a solution. It is also important to note that for some applications the forward and reverse radio links may not be symmetric. For systems of interest to us the forward link, in the direction from a base station to a portable unit, looks much different from the reverse link. As such, a design must consider the requirements of each link as well as of the system as a whole. An examination of each term in the range equation will reveal how link design is performed.

First, consider the top-level link design parameters. One of the first terms to consider is actually one that is least obvious in Equation (2.115). Specifically, the operating frequency (wavelength) is woven into each of the variables of the range equation. For example, the free-space loss term L_{path} is obviously wavelength dependent, varying as the inverse of wavelength squared. The antenna gain terms have wavelength dependence as well. Even the transmitter power term may have some wavelength dependence associated with the power delivery system between the transmitter's final amplifier and the transmitting antenna. In addition, the use of a radio link operating frequency is regulated by the FCC and must also be chosen to produce manageable antenna dimensions. Typically, the choice of operating frequency has as much to do with spectrum availability as with technical desirability.

Once the operating frequency is chosen, the design may proceed. As stated earlier, our goal is to produce the minimum power P_r (as specified by the minimum required SNR plus margin) at the receiver input terminals. The process of taking this initial requirement and decomposing it into specific requirements allocated to each of the various subsystems of the link is called "requirement flow-down" or "allocation." A systems engineer must have the experience and knowledge of a broad range of technologies to know how to best flow down or allocate requirements among the various subsystems. It is always good practice, however, to engage the subsystem designers to ensure that the most informed choices are made and "buy-in" is obtained on the requirements from the design team.

There are a number of practical considerations surrounding the transmitter and determining its transmitted power P_t. Figure 2.31 is a block diagram of a simplified transmitter RF stage, which is helpful when discussing some of the design considerations. As its name implies, the

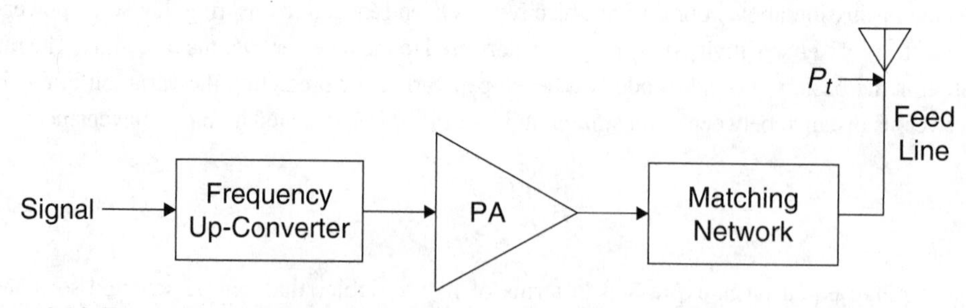

Figure 2.31 Transmitter RF Stage Block Diagram

power amplifier (PA) increases the power of the signal to be transmitted to some level in excess of the desired level P_t. A matching network may be required to match the amplifier output impedance to the antenna feed, which carries the signal to the antenna. If the transmitter is a base station, the feed may be a 150-foot or longer cable, so that power loss in the cable and connections may be appreciable. The power at the output of the PA should be large enough to account for the losses in the matching network and feed system so that the power delivered to the antenna terminals is P_t as required. Of course, due care in the design and selection of these components is required to avoid wasting power, especially in a mobile unit.

Amplifier design is important as well; an amplifier that wastes power (that is, one that has high quiescent power dissipation) is undesirable since it will drain the battery of a mobile unit. On the other hand, low quiescent power amplifiers tend to be nonlinear to some degree. In cell phones, low quiescent power amplifiers often introduce problems arising from nonlinearity of the power amplifier, which introduces unwanted frequencies in the transmitted spectrum. We will discuss this further in a later chapter on signaling. Although it is important to design matching networks and feed systems to minimize loss, wide-bandwidth, low-loss components are difficult to design and expensive to implement so that a prudent trade-off may be required between performance and cost. Therefore, a systems engineer should engage circuit designers to optimize the performance of these subsystems to ensure maximum power delivery to the antenna.

Example

Suppose a power amplifier is operated at its maximum linear power output point P_{1dB} = 40 dBm. The matching network and connectors introduce a loss of 0.5 dB and the feed cable adds an additional loss of 3 dB. The transmitting antenna gain in the desired direction is 10 dBd. Find the EIRP for this radio transmitter.

Solution

EIRP is defined as the radiated power produced by an isotropic radiator producing the same power as the actual antenna in the desired direction. This is equivalent to calculating P_tG_t as defined previously, but several details must be considered first.

First, write the equation that will determine the EIRP in decibels relative to 1 milliwatt (dBm):

$$EIRP\big|_{dB} = P_{PA}\big|_{dB} - L_{network}\big|_{dB} - L_{feed}\big|_{dB} + G_t\big|_{dB}. \tag{2.116}$$

The first term on the right-hand side is the output power level of the power amplifier, which is given as 40 dBm. The last term is the gain of the transmitting antenna. Antenna gains are normally given in terms of either dBd or dBi references. To convert dBd to dBi, simply add 2.15 dB. This accounts for the gain of a dipole relative to an isotropic radiator. In this example, the EIRP is calculated as

$$\begin{aligned} EIRP\big|_{dB} &= 40\text{ dBm} - 0.5\text{ dB} - 3\text{ dB} + (10\text{ dBd} + 2.15\text{ dB}) \\ &= 48.65\text{ dBm}. \end{aligned} \tag{2.117}$$

Thus, the EIRP produced is 48.65 dBm or 73.3 W. (Note that another measure of radiated power is often used, the **effective radiated power** [ERP], which is the radiated power above that of a dipole. The ERP in this case would be 48.65 dBm – 2.15 dB = 46.65 dBm or 46.2 W.) This is a reasonable example of a single-channel radio transmitter performance for the base station of a cellular phone link. Note that a mobile unit cannot support this power level from either a battery or a heat dissipation perspective.

The transmitting antenna gain term G_t is also more than a number. Its value depends on the antenna design, dimensions, and use. For the forward link, there is greater flexibility at the base station to meet design constraints associated with the terms P_t and G_t than at the mobile unit. This reduces the need for high-gain, low-loss antennas in the mobile unit.

Similar considerations apply for the receiving antenna gain term G_r. For the mobile unit, the same antenna is used for both transmitting and receiving, and a device known as a diplexer is used to separate the signals. At the base station, separate antennas are used for transmitting and receiving, which eliminates the need for a diplexer and its associated loss.

Based on this discussion, it should be apparent that there is greater flexibility at the base station for implementing system components, especially in terms of transmit power and antenna gain. Mobile units are severely constrained by the amount of power they can radiate, the size of the antennas they can support, the volume of circuitry they can house, and the amount of heat they can dissipate. This allows the base station to "take up the slack" by providing more transmitter power in the forward direction and higher-gain antennas in both the forward and reverse links. Furthermore, since larger, higher-gain antennas and better receiver electronics can be accommodated at the base station, the return link can be designed to provide better receiver sensitivity. As a result, the mobile unit design constraints can be made less demanding.

As a first consideration, an inexperienced systems engineer might attempt to balance the forward and reverse links by making the transmitter and receiver designs equally demanding on both the base station and mobile units. Size, weight, cost, and power constraints, however, are far more difficult to meet for a mobile unit than they are for a base station. An asymmetric approach allows the apportionment of antenna gain and transmitter power in a manner that reduces the difficulty in meeting mobile unit performance requirements by shifting the "burden" to the base station where the design constraints are not as restrictive.

An Example Link Budget

In this example, we will examine both the forward link (or downlink) from the base station to the mobile unit and the reverse link (or uplink) from the mobile unit to the base station. Although these links are asymmetric, there are elements in common—the link ranges will be identical and the handset antenna is the same for transmitting and receiving. The frequencies of the forward and reverse links may not be the same, but they are often very close.

For the forward link we use the transmitter and receiver architecture described previously. The reverse link will be specified in a similar manner, with the handset as the transmitter and the

Table 2.1 Transmitter Parameters for Example Link Budget

Parameter	Forward Link Tx (Base Station)	Reverse Link Tx (Handset)
PA P_{1dB}(dBm)	43	27
Connector losses (dB)	1	0.1
Feed losses (dB)	4	0.5
Transmitter antenna gain (dBd)	12	0

base station as the receiver, and with some modifications to the architecture as described. In this example we will find the maximum link range that can be achieved given the described hardware. We will assume an operating frequency of 1900 MHz.

The base station RF stage is described by the parameters listed in Table 2.1, corresponding to Figure 2.31. Note that the base station produces a higher EIRP than the handset, as would be expected from the considerations discussed above.

For the receiver analysis, we assume that the entire receiver has been analyzed as shown here and modeled as a single RF unit with a gain and noise figure, feeding a baseband detector as shown in Figure 2.32. The parameters for the receiver model for the forward and reverse link are given in Table 2.2. The receiver antenna gain is higher than the transmitter antenna gain for the base station owing to the common practice of using multiple receiving antennas. The base station and handset receivers are assumed to have identical gains and noise figures to simplify the example. The antenna noise temperatures for the base station and handset are different. Base station antennas are often tilted down slightly to limit interference with nearby cells. The base station antenna is thus "looking" at the Earth and sees Earth-temperature noise (290 K). The handset antenna, being small, is not very directional. It "sees" both Earth and sky and consequently will be assumed to have a lower antenna noise temperature.

We begin the analysis with the range equation, Equation (2.20), which we repeat here for convenience:

$$P_r\Big|_{dB} = P_t\Big|_{dB} + G_t\Big|_{dB} + G_r\Big|_{dB} - L_{path}\Big|_{dB} - L_{sys}\Big|_{dB} - L_{margin}\Big|_{dB}. \tag{2.118}$$

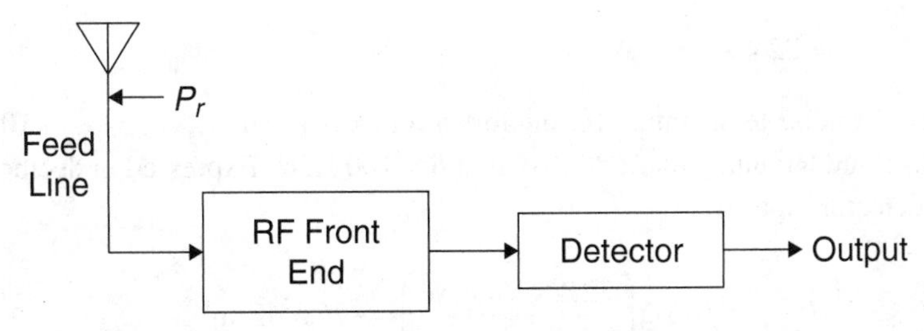

Figure 2.32 Simplified Model for Receivers

Table 2.2 Receiver Parameters for Example Link Budget

Parameter	Forward Link Rx (Handset)	Reverse Link Rx (Base Station)
Receiver antenna gain (dBd)	0	15
Feed losses (dB)	0.5	4
RF unit noise figure (dB)	8	8
RF unit gain (dB)	40	40
Detector required SNR (dB)	12	12
Antenna noise temperature (K)	60	290
Detector bandwidth (kHz)	100	100

Note that we have added an additional loss term to the range equation, to allow for design margin, as described in our discussion of receiver sensitivity. For illustrative purposes we choose a design margin of 3 dB for both links. We perform the analysis twice, once for the forward link and once for the reverse link. Using the parameters of Table 2.1, we find that $P_t|_{dB} = 43$ dBm, $G_t|_{dB}$ = 12 dBd or 14.15 dBi, $G_r|_{dB} = 0$ dBd or 2.15 dBi, and $L_{sys}|_{dB} = 1$ dB + 4 dB = 5 dB. If we can find the required signal power $P_r|_{dB}$ at the receiver input, we will be able to solve Equation (2.118) for the path loss. Then we can use Equation (2.22) to find the maximum range.

Determining the required signal power at the receiver input begins with a noise analysis. Since the antenna noise temperatures are not room temperature, we will use effective input-noise temperature to characterize the receivers rather than noise figure. Now a noise figure of $F|_{dB} = 8$ dB can be converted from decibels to a factor $F = 6.31$. The receiver's feedline loss of 0.5 dB implies a feedline noise figure of 0.5 dB, or $F_{feed} = 1.12$. Combining the feedline and receiver noise figures using Equation (2.101) gives $F_{total} = 1.12 + \frac{6.31-1}{1/1.12} = 7.08$. Then Equation (2.96) gives $T_e = (F_{total} - 1)\,T_0 = (7.08 - 1)290 = 1760$ K. The noise power at the detector input is given by Equation (2.89) as

$$\begin{aligned} P_{no} &= k(T + T_e)G_0 B \\ &= (1.38 \times 10^{-23})(60 + 1760)\left(\frac{1}{1.12} \times 10^4\right)(100 \times 10^3) \\ &= 22.4 \times 10^{-12} \text{ W}, \end{aligned} \tag{2.119}$$

where the receiver noise temperature for the forward link is $T = 60$ K, $G_0 = \frac{1}{1.12} \times 10^4$ (including the feedline loss and RF unit gain of 40 dB), and $B = 100$ KHz. Expressed in decibels, the noise power at the detector input is

$$P_{no}\Big|_{dB} = 10\log\left(\frac{22.4 \times 10^{-12} \text{ W}}{1 \text{ mW}}\right) = -76.5 \text{ dBm}. \tag{2.120}$$

Now the required signal-to-noise ratio at the detector input is 12 dB. This means that the signal power required at the detector input is

$$P_{ro}\Big|_{dB} = P_{no}\Big|_{dB} + 12\text{ dB} = -64.5\text{ dBm}. \tag{2.121}$$

Since the RF unit has a 40 dB gain, the signal power required at the receiver input is now determined to be

$$P_r\Big|_{dB} = -64.5\text{ dBm} - 40\text{ dB} = -104.5\text{ dBm}. \tag{2.122}$$

Given the required received signal power, we can substitute in the range equation and find

$$\begin{aligned} -104.5\text{ dBm} &= 43\text{ dBm} + 14.15\text{ dB} + 2.15\text{ dB} - L_{path}\Big|_{dB} - 5.5\text{ dB} - 3\text{ dB} \\ L_{path}\Big|_{dB} &= 155\text{ dB}. \end{aligned} \tag{2.123}$$

Once we have the maximum path loss, we are one step away from the maximum range. At a 1900 MHz frequency the wavelength is $\lambda = 0.158$ m. Then

$$\begin{aligned} L_{path}\Big|_{dB} &= 10\log\left(\left(\frac{4\pi d}{\lambda}\right)^2\right) \\ 155\text{ dB} &= 10\log\left(\left(\frac{4\pi d}{0.158}\right)^2\right) \\ d &= 737\text{ km}. \end{aligned} \tag{2.124}$$

Repeating this calculation for the reverse link, we start with the given items $P_t\Big|_{dB} = 27$ dBm, $G_t\Big|_{dB} = 0$ dBd or 2.15 dBi, $G_r\Big|_{dB} = 15$ dBd or 17.15 dBi, and $L_{sys}\Big|_{dB} = 0.1\text{ dB} + 0.5\text{ dB} = 0.6$ dB. Using an antenna noise temperature of $T = 290$ K, we find the required signal power at the receiver inpout to be $P_r\Big|_{dB} = -100.0$ dBm. The range equation then gives the maximum allowable path loss: $L_{path}\Big|_{dB} = 142.7$ dB. Finally, we can proceed as in Equation (2.124) to find the maximum range as $d = 171$ km.

We see from this analysis that the maximum range of the link is determined in this example by the reverse link and is $d = 171$ km. In practice it is usually considered desirable to adjust the forward and reverse link parameters so that the signal-to-noise ratios are the same in both cases. This gives users at both ends identical perceptions of the quality of service. The easiest way to "balance" the links is often to reduce the transmitted power in the forward link.

Conclusions

In this chapter the basic element of wireless communication, the radio link, was investigated. A combination of various kinds of expertise, including circuit design, antenna design, manufacturing, and basic link analysis, are utilized to "close the link" or guarantee that an adequate signal-to-noise ratio is delivered to the receiver detector. A systems-engineering approach allows for trading among the various system parameters to achieve a specific goal associated with the desired quality of service objectives.

The major topics covered in this chapter are each an introduction to a field of study. Students are encouraged to review current publications in each of the areas to obtain better understanding of each field as a prospect for specialization or to understand how recent advances can influence new designs. Suggested areas for study are

- Propagation of electromagnetic waves in free space (applied electromagnetics)
- Antenna patterns (antenna design)
- The range equation (link modeling)
- Thermal noise and noise analysis (systems analysis)
- System-level optimization of the radio link (communication system engineering)

Throughout the rest of this text, many of the topics we introduced in this chapter will be refined and expanded to provide a more accurate representation of a physical radio link, thereby providing the basis for better design and prediction of radio link performance. In Chapter 3 we investigate electromagnetic wave propagation in a terrestrial environment. Our discussion will reveal that measurement, analysis, and modeling are required to account for the phenomena that influence electromagnetic transmission in natural and man-made environments.

Problems

Problem 2.1

Consider a transmitting antenna of extent $L = 1$ m located at the origin and aligned with the z axis of a right-hand rectangular coordinate system in free space. Two isotropic receive antennas are located at a distance of 2 km from the origin and at $z = \pm 0.25$ km above and below the x–y plane. The antenna has a beamwidth factor of 66°. What is the highest frequency that can be transmitted so that the power at the receive antennas is no less than half the peak value?

Problem 2.2

An isotropic antenna is radiating a total power of P_0 W. A receive antenna is located 2 km from the radiator. It has a gain described by Equation (2.6) with loss factor $\eta = 0.9$ and an effective aperture of $A_{er1} = 600\ \text{cm}^2$. How far away can a second antenna with

$\eta = 0.8$ and effective aperture $A_{er2} = 1.0\ m^2$ be located so that both antennas receive the same power at their output terminals?

Problem 2.3

Compute the gains of the two antennas in the previous example for the case where the transmitter frequency is 800 MHz. What are the gains if the transmit frequency is 1.9 GHz? Express your answers in dBi.

Problem 2.4

Assume that the antennas in Problem 2.3 are dish antennas such that $A_{er} = \frac{\pi L^2}{4}$. If the beamwidth factor for either antenna is 51°, find the beamwidth of each antenna for each frequency.

Problem 2.5

A receiver is located 5 km from a transmitter in free space. The output at the terminals of the receive antenna is –87 dBm. The receive antenna has a gain of 2 dBd and there is a total of 2.8 dB of system loss. Calculate the transmitter EIRP in watts given a transmit frequency of 890 MHz. If the transmit antenna has a gain of 7.5 dBi, what is the transmit power in watts?

Problem 2.6

A transmitter radiates an EIRP of 1 W into free space. A receive antenna located at 1 km from the transmitter has a gain of 5 dBi. Assume that there are 2 dB of system loss.

A. If the system operates at 900 MHz, what is the power level at the output of the receive antenna terminals?
B. What is the power level at the output of the receive antenna terminals if the system operates at 1800 MHz?
C. How does the received signal level vary between cases A and B?

Explain what physically causes the received power to be different in these two cases, even though all other parameters do not change.

Problem 2.7

A transmitter in free space has an output power of 5 W and operates at 1.9 GHz. The transmit antenna has an effective aperture of 0.9 m^2 and the receive antenna has an effective aperture of 100 cm^2.

A. Calculate the free-space path loss at 10 km.
B. Calculate the transmit and receive antenna gains assuming $\eta = 1$.
C. What is the EIRP in decibels above a milliwatt?
D. Assuming 2 dB of system loss, compute the power at the output of the receive antenna in decibels above a milliwatt.

Problem 2.8

A 5 W transmitter is operating in free space at a frequency of 1800 MHz. The transmitter uses a circular aperture antenna with a 0.3 m diameter, loss factor $\eta = 1$, and beamwidth factor 55°. A receiver uses a dipole antenna and requires a minimum of –80 dBm to reliably detect the signal.

A. What is the maximum distance that the receiver can be from the transmitter to ensure reliable detection?
B. Two identical receivers are placed as far apart as possible while remaining within the 3 dB beamwidth of the antenna. What is the maximum distance that each receiver can be from the transmitter?
C. What is the maximum distance that the receivers described in part B can be from each other?

Problem 2.9

The resistors in the network of in Figure P 2.1 are in thermal equilibrium with their surroundings at room temperature. Find the RMS noise voltage produced by the network if the noise is measured in a 100 kHz bandwidth.

Problem 2.10

Find the lowest order lowpass Butterworth filter for which the difference between the 3 dB bandwidth and the noise bandwidth is less than 1%.

Problem 2.11

The RF amplifier for a certain receiver has a very simple two-pole bandpass filter with a 3 dB bandwidth of 2 MHz, centered at 10 MHz. The magnitude squared of the frequency response of the filter is

$$|H(f)|^2 = \frac{404 \times 10^{24}}{f^4 - 200 \times 10^{12} f^2 + 1.04 \times 10^{28}},$$

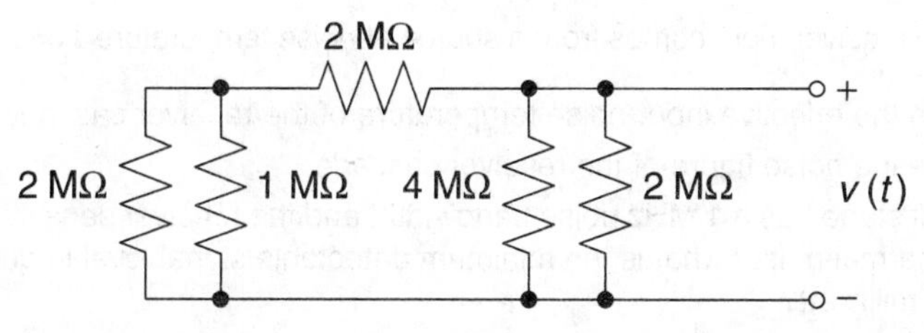

Figure P 2.1 A Resistor Network

where f is the frequency in hertz. The input to the amplifier is white noise with power spectrum $S_i(f) = N_0/2$. Calculate the power spectrum at the output of the amplifier.

Problem 2.12

Calculate the noise bandwidth of the amplifier in Problem 2.11.

Problem 2.13

A certain receiver has a noise figure of 9 dB, a gain of 50 dB, and a bandwidth of 1 MHz. A resistive noise source in thermal equilibrium at 290 K feeds the input to the receiver.

A. Calculate the output noise contributed by the amplifier.
B. What is the total output-noise power in decibels above a milliwatt?
C. What is the equivalent total noise power referenced to the receiver input?

Problem 2.14

In reference to the receiver of Problem 2.13, find the effective input-noise temperature in kelvin.

Problem 2.15

A receiver comprises four stages of amplification as follows:

Stage 1: $G_1\big|_{dB} = 10$ dB, $F_1\big|_{dB} = 5$ dB

Stage 2: $G_2\big|_{dB} = 20$ dB, $F_2\big|_{dB} = 10$ dB

Stage 3: $G_3\big|_{dB} = 40$ dB, $F_3\big|_{dB} = 12$ dB

Stage 4: $G_4\big|_{dB} = 40$ dB, $F_4\big|_{dB} = 12$ dB

Noise at the receiver input comes from a source at noise temperature $T_0 = 290$ K.

A. Compute the effective input-noise temperature of the receiver cascade.
B. Compute the noise figure of the receiver cascade.
C. If the last stage has a 1 MHz noise bandwidth, and the receiver sensitivity is 15 dB above thermal noise, what is the minimum detectable signal level in decibels above a milliwatt?
D. How does the minimum detectable signal change if the noise figure of the first stage is reduced to 2 dB?

Problem 2.16

For the system of Figure 2.30, assume that the IF amplifier has a 30 kHz bandwidth. Assume that the antenna noise temperature is 290 K and the receiver requires an SNR of at least 17 dB for reliable detection. What is the minimum required signal level?

Problem 2.17

Assume that the LNA in Figure 2.30 is moved to the antenna output terminals, just before the lossy transmission line. Under the conditions of Problem 2.16, what is the minimum required signal level? From the perspective of maximizing range, which arrangement is better and why?

Problem 2.18

A wireless link has the following parameters:

Effective isotropic radiated power, $EIRP|_{dB} = 50$ dBm

Path loss, $L_{path}|_{dB} = 145$ dB

System losses, $L_{sys}|_{dB} = 3$ dB

Rx antenna gain, $G_r|_{dB} = 3$ dBi

Rx noise bandwidth, $BW = 100$ KHz

Rx amplifier available gain, $G|_{dB} = 90$ dB

Rx noise figure, $F|_{dB} = 8$ dB

The noise temperature at the receiver antenna terminals is $T_0 = 290$ K.

A. 1. Find the signal power at the input to the receiver amplifier.
   2. Find the noise power spectrum at the input to the receiver amplifier.
   3. Find the signal-to-noise ratio at the receiver amplifier output.

B. A 300 ft feedline is added between the receiver's antenna and the amplifier input. The feedline has a loss of 1 dB/100 ft.
   1. Find the signal power at the input to the receiver amplifier.
   2. Find the noise power spectrum at the input to the receiver amplifier.
   3. Find the signal-to-noise ratio at the receiver amplifier output.

Problem 2.19

The downlink (base station to mobile unit) for an 800 MHz AMPS cellular system has the following parameters:

Transmitter power, $P_t|_{dB} = 35$ W

Loss between Tx and antenna, $L = 3$ dB

Tx antenna gain, $G_t|_{dB} = 10$ dBd

Rx antenna gain, $G_r|_{dB} = 2.9$ dBd

Rx noise bandwidth, $BW = 30$ KHz

Rx noise figure, $F|_{dB} = 9$ dB

The noise temperature at the receiver antenna terminals is $T_0 = 290$ K. The equipment and environment suggest that 6 dB of margin be used to assure performance under all conditions. For reliable voice communications, a received signal should be 17 dB above the noise. What is the maximum range between a mobile unit and base station to ensure good communications?

Problem 2.20

The uplink (mobile unit to base station) for an AMPS system has the following parameters:

Transmitter power, $P_t = 3$ W

Loss between Tx and antenna, $L|_{dB} = 3$ dB

Tx antenna gain, $G_t|_{dB} = 2.9$ dBd

Rx antenna gain, $G_r|_{dB} = 19$ dBd

Rx noise bandwidth, $BW = 30$ KHz

Rx noise figure, $F|_{dB} = 7$ dB

The noise temperature at the receiver antenna terminals is $T_0 = 290$ K.

The base station uses two antennas to obtain a 3 dB received signal improvement called a diversity gain. This is included in the given value for *Gr.* Assuming a 6 dB margin and a 17 dB receiver sensitivity, what is the maximum range at which a mobile unit can be located to ensure good communications?

CHAPTER 3

Channel Characteristics

Introduction

In the preceding chapter we developed the range equation assuming a free-space environment—that is, a perfect vacuum—with the nearest objects (other than the receiving antenna) being infinitely far away. Apart from applications such as communication between satellites, most wireless systems operate in a cluttered environment. We will refer to this environment as a "real-world" environment, since it represents the world in which we live. Unlike free space, a real-world environment includes an atmosphere through which radio waves must propagate and a variety of objects such as the Earth's surface, trees, and buildings.

The atmosphere and objects between and around the transmitter and receiver significantly affect a radio signal and may impair the overall system performance by attenuating or corrupting the signal. Objects in the propagation path may reflect, refract, and absorb some part of the signal. Signals may diffract around edges of buildings or other obstructions. Rough surfaces or groups of small reflecting surfaces such as the leaves on trees may scatter the signal in many directions. The Earth's surface can act as a reflector, producing a profound effect on path loss owing to destructive interference between a signal and its reflection. Over long propagation paths the atmosphere itself may absorb some of the electrical energy in a radio signal and dissipate that energy as heat. Variations in the density of the atmosphere may cause a signal to refract away from a preferred direction. The layered structure of the atmosphere may also cause a signal to be partially reflected.

Each of these real-world effects may be more or less significant depending on the specific application. For example, the effect of atmospheric absorption is much more troublesome for long-distance applications than for the shorter propagation distances normally associated with cellular telephone and personal communication applications. The varying density of the atmosphere allows radio waves to bend, leading to important applications in over-the-horizon communications and radar detection. Like atmospheric absorption, these refraction effects are of less importance for cellular telephone applications. For the applications that we will consider, atmospheric effects are negligible. This leaves reflection, diffraction, and scattering as the important short-distance real-world effects.

It is important to recognize that these real-world propagation effects are not entirely detrimental. In dense urban areas, and even in the downtown of a small city, buildings may obstruct the direct line-of-sight path between a mobile unit and any base station. In the absence of a direct propagation path to every location, one might expect that there would be many "shadowed" areas in which there would be no radio coverage. In practice, reflection, diffraction, and scattering "fill in the holes" and allow radio communication even when no direct path exists.

In the previous chapter we showed that the inevitable presence of noise in the receiver implies that a certain minimum power must be received to guarantee an acceptable quality of service. An important part of the design function performed by the systems engineer is to ensure that received signal power is adequate over the intended coverage area. In our development of the free-space range equation we showed that signal power decays with distance as d^{-2}, where d is the free-space distance between the transmitter and receiver. The range equation thus provides a propagation model that can be used to predict received signal power, assuming, of course, that the propagation environment is adequately represented as free space. We will find, however, that in a real-world environment the free-space range equation is only a first step in modeling propagation between the transmitter and receiver. For example, when reflection from the surface of the Earth is taken into account, we find that the signal power may decay as d^{-4} rather than as d^{-2}. Reflection, diffraction, and scattering caused by buildings and structures and by terrain features such as hills and ravines cause additional complexity in modeling propagation. It has been known for many years that at the carrier frequencies used for cellular and personal communication systems the received signal strength can vary significantly from point to point throughout the intended coverage area. These signal strength variations are so irregular, and so dependent on minute details of the physical environment, that it is practical to describe them only statistically. We will thus find that a systems engineer cannot as a practical matter guarantee adequate signal power over the entire coverage area. The best that can be done is to guarantee adequate signal power *with high probability* over *a high fraction* of the coverage area. Our aim in this chapter is to describe some of the phenomena that significantly alter signal propagation and to describe some of the models that can supplement the free-space range equation to enable systems engineers to design while taking these propagation effects into account.

Our chapter begins with an exploration of "macroscopic" or large-scale effects. We present a simple "two-ray" model that shows the consequences of reflection from the Earth's surface. A number of empirical models have been developed that relate average received signal power to propagation range, taking into account the heights of the transmitting and receiving antennas, the level of urban development, and features of the terrain. No single model has proved to be both simple enough to be usable and robust enough to be accurate under all circumstances. We will briefly describe the Okumura model and then present the Hata model and the Lee model in some detail. These models allow calculation of the median received signal power. Large-scale "shadowing" (including the effects of both reflection and diffraction) causes received signal strength to differ considerably from the median at different locations in the coverage area. We will introduce log-normal fading, a statistical model that gives a designer a basis for calculating the additional power needed to ensure a high probability of adequate coverage.

Next, the chapter moves on to "microscopic" or small-scale effects. The combination of reflection, diffraction, and scattering can cause multiple copies of the same signal to be received with different propagation delays. We again employ a simple two-ray model to show that this multipath reception can cause fading that, under certain circumstances, can be frequency selective. We follow our two-ray development with a more general model, the Rayleigh model, and introduce the concept of coherence bandwidth. The coherence bandwidth gives designers a basis for determining when frequency-selective fading will be an issue. Microscopic propagation effects also include fading caused by motion of a transmitter or receiver. We will find that this kind of fading is related to the Doppler spread of signals received over different propagation paths. We will once again demonstrate the effect by using a simple two-ray model, and then we introduce the concept of coherence time to provide a time reference for the speed with which fading occurs.

It is important to keep in mind that the motivation for developing the various propagation models is to give systems engineers the tools needed to design wireless systems that can meet quality of service objectives at an affordable cost. The chapter concludes with a continuation of the link analysis begun in Chapter 2 to show how a link budget can be developed that includes allowances for fading and coverage outages. In the subsequent chapter we will show how coverage can be extended to a large number of subscribers covering a virtually unlimited geographical area.

Macroscopic Models 1: Reflection from the Earth's Surface

We observed in the introduction to this chapter that reflection from the surface of the Earth can significantly change the way in which received signal power depends on propagation range. We have several objectives in presenting a derivation of this effect. First, we will introduce the two-ray model, which we will use to good effect several times in this chapter. Next, reflection from the Earth's surface is one example of the reflection phenomenon; it illustrates what can happen when radio waves reflect from any surface. Finally, in showing how the variation of signal power with propagation range changes from d^{-2} in free space to d^{-4} above a reflecting Earth, we introduce the concept of **path-loss exponent**. We will show in a later discussion that signal power varies with propagation range as $d^{-\nu}$ in a real-world environment, where the path-loss exponent ν typically takes values between 2 and 5.

Figure 3.1 is a representation of a uniform plane wave incident on the boundary between two different materials. In the most general case, some of the wave will be reflected from the boundary, and some of the wave will be transmitted. According to Snell's law, the angle of reflection from the boundary θ_r must equal the angle of incidence θ_i, where both angles are measured with respect to the normal to the boundary, as is shown in the figure. The angle of transmission into the second material θ_t will depend on the properties of the two materials. In general

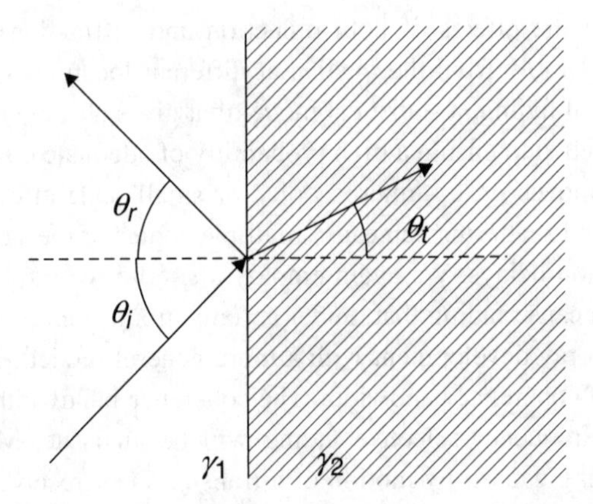

Figure 3.1 Plane Wave Incident upon a Planar Boundary

the incident, transmitted, and reflected signals will differ in amplitude and in phase, the relative amplitudes and phases depending on the properties of the materials and on the polarization of the incident wave. We will find it convenient to assume that the material labeled γ_1 in the figure is air, whose properties for the purposes of the present discussion do not differ significantly from those of free space. If the material labeled γ_2 in the figure is a perfect conductor, then there will be no transmitted signal, and the incident and reflected signals will have identical amplitudes. Alternatively, if the material γ_2 is a perfect dielectric, then there can be both reflected and transmitted signals. In this case the amplitudes of the reflected and transmitted signals will depend on the dielectric properties of the two materials. The significant factor in this case is that there is no energy loss in the reflection, and all of the power in the incident wave will be either reflected or transmitted into the second material. In considering reflection from the Earth, we find that the second material, the Earth, cannot be taken as either a perfect conductor or a perfect dielectric. It turns out that the Earth is well modeled as a "good" conductor. This means that the amplitude of the reflected wave is nearly equal to the amplitude of the incident wave, particularly when the angle of incidence θ_i is near 90°. Nonetheless, some of the power in the incident wave will be transmitted into the Earth. The transmitted wave will be rapidly attenuated, however, and all of the power transmitted into the Earth is turned into heat.

Figure 3.2 shows transmitting and receiving antennas located above a conducting plane representing the Earth. For the propagation ranges commonly encountered in cellular and personal communication applications, it is reasonable to treat the Earth's surface as essentially flat. In the figure, the transmitting antenna represents a base station, and the receiving antenna a mobile unit, as might make up a typical cellular data link. The transmitting antenna is at height h_t, the receiving antenna is at height h_r, and the distance along the Earth from transmitter to receiver is d. The figure shows two rays extending from the transmitter to the receiver: a direct ray of path length R_{dir} and an indirect ray that reflects from the Earth and has path length R_{ind}. Snell's law dictates that for the reflected ray $\psi_r = \psi_i$, a fact that can be used to find the location x of the reflection

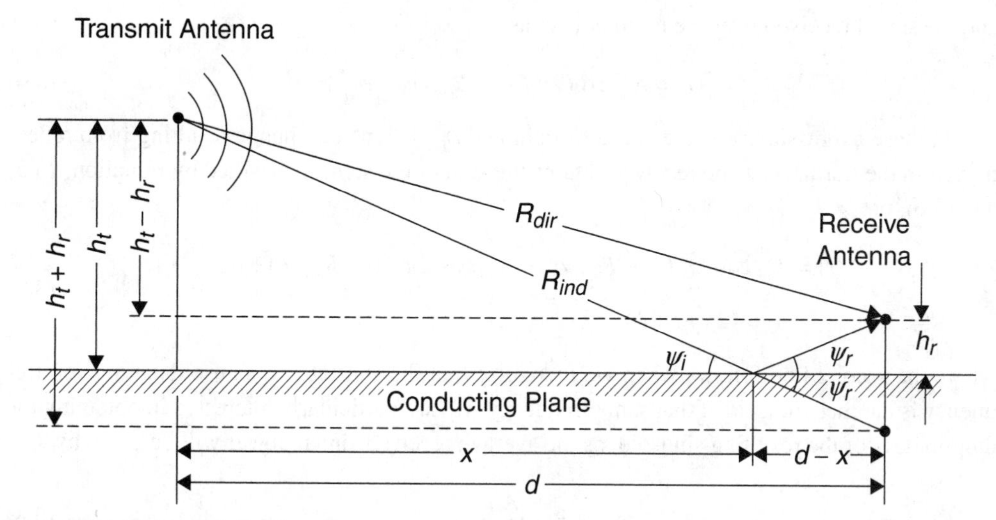

Figure 3.2 Ground Reflections

point. From the geometry of the figure, the length R_{dir} of the direct path is given by

$$R_{dir} = \sqrt{(h_t - h_r)^2 + d^2}. \tag{3.1}$$

For $d >> h_t + h_r$ we can use a Taylor series to obtain a useful approximation:

$$R_{dir} \cong d\left[1 + \frac{(h_t - h_r)^2}{2d^2}\right]. \tag{3.2}$$

The length of the indirect path is the sum of the path lengths of the incident and reflected parts. This length can be computed using straightforward geometry, but it is much simpler to recognize that the length of the indirect path can also be computed as the length of a path from the transmitting antenna to an "image" of the receiving antenna located h_r below the surface of the Earth. Thus we have

$$R_{ind} = \sqrt{(h_t + h_r)^2 + d^2}. \tag{3.3}$$

Once again a Taylor series expansion gives the approximation

$$R_{ind} \cong d\left[1 + \frac{(h_t + h_r)^2}{2d^2}\right]. \tag{3.4}$$

If the transmitted signal is a carrier wave of frequency f_c, we can write the signal received over the direct path as

$$r_{dir}(t) = A_{dir}\cos(2\pi f_c(t - R_{dir}/c)) \tag{3.5}$$

and the signal received over the indirect path as

$$r_{ind}(t) = A_{ind}\cos(2\pi f_c(t - R_{ind}/c) + \phi_{ind}). \tag{3.6}$$

In these expressions c is the speed of light and ϕ_{ind} is a phase change resulting from reflection from the Earth. Now the receiver obtains the sum of the signals specified by Equations (3.5) and (3.6), or

$$\begin{aligned} r(t) &= A_{dir}\cos(2\pi f_c(t - R_{dir}/c)) + A_{ind}\cos(2\pi f_c(t - R_{ind}/c) + \phi_{ind}) \\ &= A\cos(2\pi f_c t + \phi), \end{aligned} \tag{3.7}$$

where the second line follows when we recognize that the sum of two sinusoids at a given frequency is another sinusoid at that same frequency. We are particularly interested in obtaining the amplitude A of the resulting sinusoid, as the average received signal power will be given by

$$P_r = \frac{A^2}{2}. \tag{3.8}$$

Before combining the sinusoids, however, we observe that for reflection from a good conductor at a small grazing angle ψ_i, we have the approximation[1]

$$A_{ind} \approx A_{dir}, \tag{3.9}$$

and

$$\phi_{ind} \approx 180°. \tag{3.10}$$

The simplest way of combining the sinusoids in Equation (3.7) is to express each sinusoid as a phasor and to add the phasors. If $\mathbf{R}$ is the phasor representing the total received signal, then

$$\begin{aligned} \mathbf{R} &= A_{dir}e^{-j2\pi f_c R_{dir}/c} + A_{ind}e^{-j(2\pi f_c R_{ind}/c + \phi_{ind})} \\ &\cong A_{dir}e^{-j2\pi f_c R_{dir}/c} - A_{dir}e^{-j2\pi f_c R_{ind}/c} \\ &= A_{dir}e^{-j2\pi R_{dir}/\lambda} - A_{dir}e^{-j2\pi R_{ind}/\lambda}, \end{aligned} \tag{3.11}$$

where we made use of Equations (3.9) and (3.10) in the second line, and in the third line we substituted the wavelength λ for c/f_c. Next we make use of the identities

$$\begin{aligned} R_{dir} &= \frac{R_{dir} + R_{ind}}{2} + \frac{R_{dir} - R_{ind}}{2} \quad \text{and} \\ R_{ind} &= \frac{R_{dir} + R_{ind}}{2} - \frac{R_{dir} - R_{ind}}{2}. \end{aligned} \tag{3.12}$$

1. E. C. Jordan and K. G. Balmain, *Electromagnetic Waves and Radiating Systems,* 2nd ed. (Englewood Cliffs, NJ: Prentice Hall, 1968), 629.

These give

$$\begin{aligned}\mathbf{R} &= A_{dir} e^{-j2\pi(R_{dir}+R_{ind})/2\lambda}[e^{-j2\pi(R_{dir}-R_{ind})/2\lambda} - e^{j2\pi(R_{dir}-R_{ind})/2\lambda}] \\ &= j2A_{dir} e^{-j2\pi(R_{dir}+R_{ind})/2\lambda} \sin[2\pi(R_{ind} - R_{dir})/2\lambda].\end{aligned} \tag{3.13}$$

The magnitude of the phasor **R** is the amplitude A that we seek. From Equation (3.13) we obtain

$$A = |\mathbf{R}| = 2A_{dir} \, |\sin[2\pi(R_{ind} - R_{dir})/2\lambda]|. \tag{3.14}$$

Observe that depending on the path length difference $R_{ind} - R_{dir}$, the sine in Equation (3.14) can take any value between −1 and 1. When the sine takes the value 0, the direct and indirect rays arrive at the receiving antenna exactly out of phase and there is complete cancellation. When the sine takes the value ±1, the direct and indirect rays arrive exactly in phase, and there is complete reinforcement, leading to $A = 2A_{dir}$.

To understand the consequences of Equation (3.14), let us make use of the approximations given in Equations (3.2) and (3.4). We find

$$R_{ind} - R_{dir} \cong \frac{2h_t h_r}{d}, \tag{3.15}$$

so that Equation (3.14) becomes

$$A = 2A_{dir}\left|\sin\left(\frac{2\pi h_t h_r}{\lambda d}\right)\right|. \tag{3.16}$$

Recall that $\sin\theta \approx \theta$ for θ sufficiently small (for example, when $\theta = 0.5$ radians the error is less than 5%). Then for $d \gg \frac{2\pi h_t h_r}{\lambda}$ we may write

$$A \approx A_{dir}\frac{4\pi h_t h_r}{\lambda d}. \tag{3.17}$$

Substituting Equation (3.17) into Equation (3.8) gives

$$P_r \approx \frac{A_{dir}^2}{2}\left(\frac{4\pi h_t h_r}{\lambda d}\right)^2. \tag{3.18}$$

Now recall from Chapter 2 that for direct line-of-sight propagation the received signal power decreases as the square of the distance from the transmitter. In terms of amplitude, the value of A_{dir} decreases as the distance, that is,

$$A_{dir} = \frac{A_0}{d}, \tag{3.19}$$

where the constant A_0 may depend on transmitted power, antenna gains, and system losses, but not on distance. To be more explicit, if there were no reflection from the Earth, the received power would be given by

$$P_{r,dir} = \frac{A_{dir}^2}{2} = \frac{A_0^2}{2d^2}. \tag{3.20}$$

Using Equation (3.20), we can eliminate A_{dir} from Equation (3.18):

$$P_r \approx \frac{A_0^2}{2d^2}\left(\frac{4\pi h_t h_r}{\lambda d}\right)^2 = \frac{A_0^2}{2}\left(\frac{4\pi h_t h_r}{\lambda}\right)^2 \frac{1}{d^4}. \tag{3.21}$$

Now the range equation tells us that for line-of-sight propagation over the direct path alone,

$$P_{r,dir} = \frac{P_t G_t G_r \lambda^2}{(4\pi)^2 L_{sys}} \frac{1}{d^2}, \tag{3.22}$$

where we recognize that $R_{dir} \cong d$. Equating Equation (3.20) and Equation (3.22) gives us

$$A_0^2 = \frac{2P_t G_t G_r \lambda^2}{(4\pi)^2 L_{sys}}. \tag{3.23}$$

Substituting Equation (3.23) into Equation (3.21) gives

$$P_r \approx \frac{P_t G_t G_r \lambda^2}{(4\pi)^2 L_{sys}}\left(\frac{4\pi h_t h_r}{\lambda}\right)^2 \frac{1}{d^4} = \frac{P_t G_t G_r h_t^2 h_r^2}{L_{sys} d^4}. \tag{3.24}$$

Communications engineers would normally express Equation (3.24) in decibels, that is,

$$P_r\big|_{dB} = P_t\big|_{dB} + G_t\big|_{dB} + G_r\big|_{dB} + 20\log(h_t) + 20\log(h_r) - L_{sys}\big|_{dB} - 4\times 10\log(d). \tag{3.25}$$

The factor of 4 in front of the 10 log (d) in Equation (3.25) is the so-called path-loss exponent.

The result we have obtained demonstrates the rather dramatic impact of placing a simple reflecting plane in the presence of a transmitter and receiver. Although the result is an approximation that assumes low grazing angles and a conductive Earth, it does represent a reasonable model for unobstructed terrestrial propagation. In the next section we will recognize that the real world places additional obstructions between the transmitter and receiver in wireless systems, and we will begin to consider more robust, but also more complex, propagation models.

Example

Suppose a wireless base station transmits 10 W to a receiver 10 km away over a flat, conductive Earth. The transmitting antenna has a gain of 6 dB, and the receiving antenna has a gain of 2 dB. The base station antenna is at the top of a 30 m tower, and the receiving antenna is on top of a pickup truck at an elevation of 2 m. The transmitting frequency is 850 MHz, and there are 2 dB of system losses.

Find the received signal power using Equation (3.25), and compare with the received power obtained from the range equation without considering reflection from the Earth. Evaluate the accuracy of the approximation made in getting from Equation (3.16) to Equation (3.17).

Solution

The transmitted power can be expressed in decibels as

$$P_t\Big|_{dB} = 10\log\left(\frac{10\text{ W}}{10^{-3}\text{ W}}\right) = 40\text{ dBm}. \tag{3.26}$$

Equation (3.25) gives

$$\begin{aligned} P_r\Big|_{dB} &= 40\text{ dBm} + 6\text{ dB} + 2\text{ dB} + 20\log(30) + 20\log(2) \\ &\quad - 2\text{ dB} - 4\times 10\log(10\times 10^3) \\ &= -78.4\text{ dBm}. \end{aligned} \tag{3.27}$$

To compare with the result given by the range equation, we first calculate the wavelength as

$$\lambda = \frac{3\times 10^8}{850\times 10^6} = 0.353\text{ m}. \tag{3.28}$$

The free-space path loss is then

$$L_{path}\Big|_{dB} = 2\times 10\log\left(\frac{4\pi 10\times 10^3}{0.353}\right) = 111\text{ dB}. \tag{3.29}$$

The range equation gives

$$\begin{aligned} P_r\Big|_{dB} &= P_t\Big|_{dB} + G_t\Big|_{dB} + G_r\Big|_{dB} - L_{sys}\Big|_{dB} - L_{path}\Big|_{dB} \\ &= 40\text{ dBm} + 6\text{ dB} + 2\text{ dB} - 2\text{ dB} - 111\text{ dB} \\ &= -65.0\text{ dBm}. \end{aligned} \tag{3.30}$$

We see that ignoring reflection from the Earth gives an estimate of received power that is too optimistic by 13.4 dB (a factor of about 20).

To evaluate the approximation used to obtain Equation (3.17), we compute

$$\frac{2\pi h_t h_r}{\lambda d} = \frac{2\pi\times 30\times 2}{0.353\times(10\times 10^3)} = 0.107. \tag{3.31}$$

Then

$$\sin(0.107) = 0.107. \tag{3.32}$$

The approximation error is undetectable to three significant figures. Greater precision reveals an error of about 0.2%.

Macroscopic Models 2: Empirical Models

The purpose of macroscopic modeling is to provide a means for predicting path loss for a particular application environment. Any individual path-loss model has a limited range of applicability and will provide only an approximate characterization for a specific propagation environment. Nevertheless, modeling the communication path provides an extremely useful way to start a system design.

Modeling is useful for helping to understand how a proposed system will perform in a given environment, determining initial system layout, estimating the amount of equipment required to serve the application, and providing a means for predicting future expansion needs. The fact that predictions are only approximate, however, implies that results must be validated by measurement before financial investments are made to purchase equipment or obtain real estate. In the early days of cellular telephone development, base station equipment and purchase or lease of an antenna site could cost $1 million per cell. Clearly, confident predictions of how many cells are needed and their required locations and coverage areas can have a profound influence on the financial viability of a cellular system.

Although many models have been developed to cover the wide variety of environments encountered in wireless system design, we limit our discussion in this section to three outdoor models that are currently in use. These models are presented to demonstrate some of the current practice and to show how models can be used to perform radio system design in a real-world environment. All of the models we present are founded on the same set of underlying principles and the same approach to prediction. Although we present several models, our discussion is not meant to be a comprehensive treatment of all of the models available. To limit our discussion to a reasonable length we will not consider, for example, models that predict path loss in the indoor environment. These and other specialized outdoor models are well represented in the literature,[2] to which the reader is referred for further study.

The most widely used macroscopic propagation models are derived from measured data. Typically these models include the effects of terrain profile and some general characteristics of land use, such as whether the intended coverage area is urban or rural. In the previous section we found that a consequence of including reflection from the Earth in our propagation model was that the heights of the transmitting and receiving antennas became factors in determining the received signal power. The empirical models we will examine in this section also include the

2. A good starting point is T. S. Rappaport, *Wireless Communications: Principles and Practice*, 2nd ed. (Upper Saddle River, NJ: Prentice Hall, 2002), Chapter 4.

effects of antenna height, as well as the effects of carrier frequency. A typical formulation of an empirical model is to predict path loss based on measurements that use a specific reference system configuration, that is, a specific range of carrier frequencies, set of antenna heights, and terrain. Subsequently, correction factors are introduced to extrapolate the model to system configurations other than the reference configuration.

One of the most successful early attempts at providing a comprehensive model for the urban land mobile propagation environment was proposed by Okumura.[3] Using extensive measurements taken in Tokyo and the surrounding area, Okumura developed a set of curves that plot median signal attenuation relative to free space versus range for a specific system configuration. These curves span range values from 1 to 100 km and a frequency interval of 100 to 1920 MHz. Correction factors are provided to adjust the median path-loss value for differences in antenna height, diffraction, terrain profile, and land-use type. The Okumura model is entirely based on measured data and does not provide simple formulas for evaluation or extrapolation. Nevertheless, the Okumura model is considered simple to use and is widely employed for path-loss prediction in urban areas where it is most accurate. It is reported[4] to be accurate to around 10 dB to 14 dB compared with measurement.

The purely empirical nature of the Okumura model has made it difficult to incorporate into computer-based design tools. We will therefore focus our attention on two other models that are more analytic in nature and therefore more suitable for use in computer-aided analysis. These models are the Hata model and the Lee model.

The Hata Model

The Hata[5] model is an analytical model that is based on Okumura's data. It produces path-loss predictions that compare favorably with those produced by Okumura's model. Because the Hata model provides formulas as approximations to Okumura's curves, it is well suited for computer implementation. The model is founded on a standard equation for the median path loss of a system in an urban environment. Median path loss means that for a given set of measurements, half of the data shows greater path loss than the median, and half less. The Hata model states

$$\begin{aligned} L_{50}(urban)\big|_{dB} &= 69.55 + 26.16\log(f_0) - 13.82\log(h_{te}) \\ &\quad + (44.9 - 6.55\log(h_{te}))\log(d) - a(h_{re}), \end{aligned} \tag{3.33}$$

3. Y. Okumura, E. Ohmori, T. Kawano, and K. Fukuda, "Field Strength and Its Variability in VHF and UHF Land Mobile Service," *Review of the Electrical Communication Laboratory* 16, no. 9–10 (September–October 1968): 825–73.
4. Rappaport, *Wireless Communications,* 152.
5. M. Hata, "Empirical Formula for Propagation Loss in Land Mobile Radio Services," *IEEE Transactions on Vehicular Technology* VT-29, no. 3 (August 1980): 317–25.

where the terms are as follows:

$L_{50}\,(urban)|_{dB}$ is the median path loss in decibels for an urban environment.
f_0 is the carrier frequency in megahertz.
h_{te} is the effective transmitting (base station) antenna height in meters.
h_{re} is the effective receiving (mobile unit) antenna height in meters.
d is the range between transmitting and receiving antennas in kilometers.
$a(h_{re})$ is a correction factor for mobile unit antenna height.

The mobile unit antenna height correction factor turns out to depend on the coverage area. For a large city it is given by

$$a(h_{re}) = \begin{cases} 8.29[\log(1.54h_{re})]^2 - 1.1 \text{ dB} & \text{for } f_0 \le 200 \text{ MHz} \\ 3.2[\log(11.75h_{re})]^2 - 4.97 \text{ dB} & \text{for } f_0 \ge 400 \text{ MHz}, \end{cases} \tag{3.34}$$

and for a small to medium-sized city it is given by

$$a(h_{re}) = [1.1\log(f_0) - 0.7]h_{re} - [1.56\log(f_0) - 0.8] \text{ dB}. \tag{3.35}$$

The Hata model contains additional correction factors for land use. In a suburban area

$$L_{50}(suburban)|_{dB} = L_{50}(urban)\big|_{dB} - 2[\log(f_0/28)]^2 - 5.4, \tag{3.36}$$

and in open and rural areas

$$L_{50}(rural)\big|_{dB} = L_{50}(urban)\big|_{dB} - 4.78[\log(f_0)]^2 + 18.33\log(f_0) - 40.94. \tag{3.37}$$

The Hata model is applicable over the following range of parameters:

- $150 \text{ MHz} \le f_0 \le 1500 \text{ MHz}$
- $30 \text{ m} \le h_{te} \le 200 \text{ m}$
- $1 \text{ m} \le h_{re} \le 10 \text{ m}$
- $1 \text{ km} \le d \le 20 \text{ km}$

In all of these formulas, the logarithms are computed to the base 10. The "effective" antenna height is calculated by averaging the elevation of the terrain from 3 km to 15 km along the direction from the transmitting antenna to the receiving antenna. Antenna heights are measured as heights above this average.

The frequency range covered by the Hata model is not broad enough to include the PCS band at 1900 MHz. The European Cooperative for Scientific and Technical Research has proposed an extension to the Hata model that will extend its applicability to 2 GHz.

Example

A base station in a medium-sized city transmits from an antenna height of 30 m to a receiver 10 km away. The receiver antenna height is 2 m, and the carrier frequency is

850 MHz. Use the Hata model to determine the median path loss. Compare the median path loss with the path loss calculated using Equation (3.25) for propagation over a flat conductive Earth. How will the path loss predicted by the Hata model change if the receiving antenna is lowered to an effective height of 1 m?

Solution

For a medium-sized city, the mobile antenna correction factor is given by Equation (3.35), that is,

$$\begin{aligned} a(2) &= [1.1\log(850) - 0.7]\times 2 - [1.56\log(850) - 0.8] \\ &= 1.27 \text{ dB}. \end{aligned} \tag{3.38}$$

Then from Equation (3.33),

$$\begin{aligned} L_{50}(urban)\big|_{dB} &= 69.55 + 26.16\log(850) - 13.82\log(30) \\ &\quad + [44.9 - 6.55\log(30)]\log(10) - 1.27 \\ &= 160 \text{ dB}. \end{aligned} \tag{3.39}$$

For comparison we take only the path-loss terms from Equation (3.25):

$$\begin{aligned} L(Earth\ reflection)\big|_{dB} &= -20\log(h_t) - 20\log(h_r) + 40\log(d) \\ &= -20\log(30) - 20\log(2) + 40\log(10\times 10^3) \\ &= 124 \text{ dB}. \end{aligned} \tag{3.40}$$

Note that in Equation (3.25) all distances are in meters. The Earth reflection model does not agree well with the Hata model in this instance. This suggests that the Earth reflection model is not sophisticated enough to be used for path-loss prediction in an urban environment.

In the Hata model only the correction factor $a(h_{re})$ will change if the receiver antenna height is changed. For $h_{re} = 1$ m,

$$\begin{aligned} a(1) &= [1.1\log(850) - 0.7]\times 1 - [1.56\log(850) - 0.8] \\ &= -1.25 \text{ dB}. \end{aligned} \tag{3.41}$$

This produces a median path loss of

$$L_{50}(urban)\big|_{dB} = 162 \text{ dB}. \tag{3.42}$$

Lowering the height of the receiving antenna increases the path loss by about 2.5 dB. Recall that 3 dB is a factor of 2, and that twice the path loss means that the receiver will receive half of the former signal power.

Example

The base station in the previous example transmits 10 W through an antenna of 6 dB gain. The receiver, at a height of 2 m, has an antenna gain of 2 dB. There are 3 dB of system losses. Using the results of the previous example, find the median received signal power.

Solution

The transmitter power in decibels is

$$P_t\big|_{dB} = 10\log\left(\frac{10\ \text{W}}{0.001\ \text{W}}\right) = 40\ \text{dBm}. \tag{3.43}$$

Then the median received signal power is

$$\begin{aligned} P_{r,50}\big|_{dB} &= P_t\big|_{dB} + G_t\big|_{dB} + G_r\big|_{dB} - L_{sys}\big|_{dB} - L_{50}(urban)\big|_{dB} \\ &= 40\ \text{dBm} + 6\ \text{dB} + 2\ \text{dB} - 3\ \text{dB} - 160\ \text{dB} \\ &= -115\ \text{dBm}. \end{aligned} \tag{3.44}$$

The Lee Model

The Lee model[6] is based upon extensive measurements made in the Philadelphia, Pennsylvania–Camden, New Jersey, area, and in Denville, Newark, and Whippany, New Jersey, by members of the Bell Laboratories technical staff in the early 1970s. Like the Hata model, the Lee model is widely used in computer-automated tools for radio frequency design of personal communication systems. Like the Okumura and Hata models, the Lee model provides a method for calculating path loss for a specified set of "standard" conditions and provides correction factors to adjust for conditions different from the standard.

The Lee model treats the area within one mile of the base station transmitter differently from the remainder of the coverage area. This is done for several reasons. First, proximity to the transmitter often ensures adequate signal strength. Next, there is a high probability that propagation to a receiver near the base station will be line-of-sight. Finally, a receiver near the base of the transmitting tower may be below the main beam of the transmitting antenna, causing the antenna gain to be lower than normal. The reference distance of one mile is a somewhat arbitrary standard, but it ensures that the influence of the environment near the transmitting antenna does not significantly bias the path-loss prediction over the entire coverage area.

6. W. C. Y. Lee, *Mobile Communications Engineering* (New York: McGraw-Hill, 1998).

Example

As a preliminary to introducing the actual Lee model, consider a system for which line-of-sight conditions hold out to a reference range of d_0, after which there is a path-loss exponent of value v. Let us evaluate the received signal power at a range $d > d_0$.

The free-space range equation gives us, assuming free-space conditions out to range d,

$$\begin{aligned} P_r &= \frac{P_t G_t G_r}{L_{sys}\left(\frac{4\pi d}{\lambda}\right)^2} \\ &= \frac{P_t G_t G_r}{L_{sys}\left(\frac{4\pi d_0}{\lambda}\right)^2\left(\frac{d}{d_0}\right)^2}. \end{aligned} \tag{3.45}$$

If we replace the path-loss exponent of 2 with v for propagation from d_0 to d, then Equation (3.45) becomes

$$P_r = \frac{P_t G_t G_r}{L_{sys}\left(\frac{4\pi d_0}{\lambda}\right)^2\left(\frac{d}{d_0}\right)^v}. \tag{3.46}$$

Let us define a "reference" received power P_{r0} as the received power at distance d_0. Then Equation (3.46) becomes

$$P_r = \frac{P_{r0}}{\left(\frac{d}{d_0}\right)^v}. \tag{3.47}$$

Converting to decibels, we can write

$$P_r\big|_{dB} = P_{r0}\big|_{dB} - 10v\log\left(\frac{d}{d_0}\right). \tag{3.48}$$

The Lee model follows the form of Equation (3.48) using a reference distance d_0 of one mile. The reference power $P_{r0}|_{dB}$ and the path-loss exponent have been determined from measurements. The Lee model initially assumes a standard set of reference conditions and a flat terrain. Correction factors are added to account for differences from the reference conditions and to allow the path loss to be corrected on a point-by-point basis for the specific terrain profile. The standard conditions for the Lee model are as follows:

- Transmitter power is $P_0|_{dB} = 40$ dBm (10 W).
- Transmitting (base station) antenna height is $h_{t0} = 100$ ft.
- Transmitting (base station) antenna gain is $G_{t0}|_{dB} = 6$ dB with respect to the gain of a half-wave dipole (8.15 dB with respect to an isotropic radiator).

- Receiving (mobile unit) antenna height is $h_{r0} = 10$ ft. (The people on the Bell Labs technical staff must have been tall.)
- Receiving (mobile unit) antenna gain is $G_{r0}|_{dB} = 0$ dB with respect to the gain of a dipole antenna (that is, 2.15 dB with respect to an isotropic antenna).
- Carrier frequency is $f_0 = 850$ MHz.

The Lee model has been shown to be applicable for frequencies up to 2 GHz. It is most often expressed in terms of the median received power level as

$$P_{r,50}\Big|_{dB} = P_{1\text{-}mile}\Big|_{dB} - 10v\log\left(\frac{d}{d_0}\right) + \alpha_c, \tag{3.49}$$

where

$P_{r,50}\Big|_{dB}$ is the median received power level in decibels above a milliwatt
$P_{1-mile}\Big|_{dB}$ is the median received power level at the one-mile reference range in decibels above a milliwatt
v is the path-loss exponent for the specific type of area
d is the total range between the transmitter and receiver antennas
d_0 is the reference range of one mile
α_c is the correction factor in decibels for system parameters different from the standard

The correction factor α_c provides a means for converting median received power level from a prediction based on standard parameters to a prediction based on actual parameters. The correction factor is given by

$$\begin{aligned}\alpha_c = {} & 20\log\left(\frac{h_t}{h_{t0}}\right) + 20\log\left(\frac{h_r}{h_{r0}}\right) - n\log\left(\frac{f}{f_0}\right) \\ & + (P_t|_{dB} - P_0|_{dB}) + (G_t|_{dB} - G_{t0}|_{dB}) + (G_r|_{dB} - G_{r0}|_{dB}),\end{aligned} \tag{3.50}$$

where the parameters $P_t\Big|_{dB}, h_t, h_t, G_t\Big|_{dB}$ and $G_r\Big|_{dB}$ are the actual system parameters, and the parameter n is given by

$$n = \begin{cases} 20 & f < f_0 \\ 30 & f > f_0. \end{cases} \tag{3.51}$$

The correction factors for antenna height are consistent with Equation (3.25). The Lee model assumes that the mobile receiver's antenna height is not greater than 10 feet. The model

Table 3.1 Measured Reference Power and Path-Loss Exponent for Various Environments

Environment Type	1-Mile Reference Power (dBm) $P_{1-mile}\big\vert_{dB}$	Path-Loss Exponent ν
Free space	−45	2
Open areas	−49	4.35
Suburban areas	−61.7	3.84
Urban Newark	−64	4.31
Urban Philadelphia	−70	3.68
Urban San Francisco	−71	3.84
Urban Tokyo	−84	3.05
Dense urban New York	−77	4.8

may also include a correction term for diffraction that applies when the propagation path is blocked by an obstruction such as a hill.

Based on measurements, Lee provides values for the one-mile reference power and the path-loss exponent for various types of environment. These values are given in Table 3.1.

Lee's model as given by Equation (3.49) can be further corrected to include the effect of terrain that is not flat.[7] The additional correction factor is

$$20\log\left(\frac{h_{eff}}{h_t}\right), \tag{3.52}$$

where h_{eff} is an "effective" base station antenna height computed as shown in Figure 3.3. The two configurations shown in the figure are characterized by the property that the radio waves reflect from a slope rather than from a horizontal surface. In practice, terrain profiles can be determined using geographical databases that provide elevation versus geographic coordinates for the intended coverage area. Such databases are readily available for commercial applications in most areas worldwide. An array of "bins" can be defined along a radial line from the base station antenna. The average elevation and slope of each bin are taken as the elevation and slope at the center of the bin. An effective transmitter antenna height can then be calculated for the center of each bin. Once effective antenna height is determined, Equation (3.49) corrected by Equation (3.52)

7. W. C. Y. Lee, "Studies of Base-Station Antenna Height Effects on Mobile Radio," *IEEE Transactions on Vehicular Technology* VT-29, no. 2 (May 1980): 252–60.

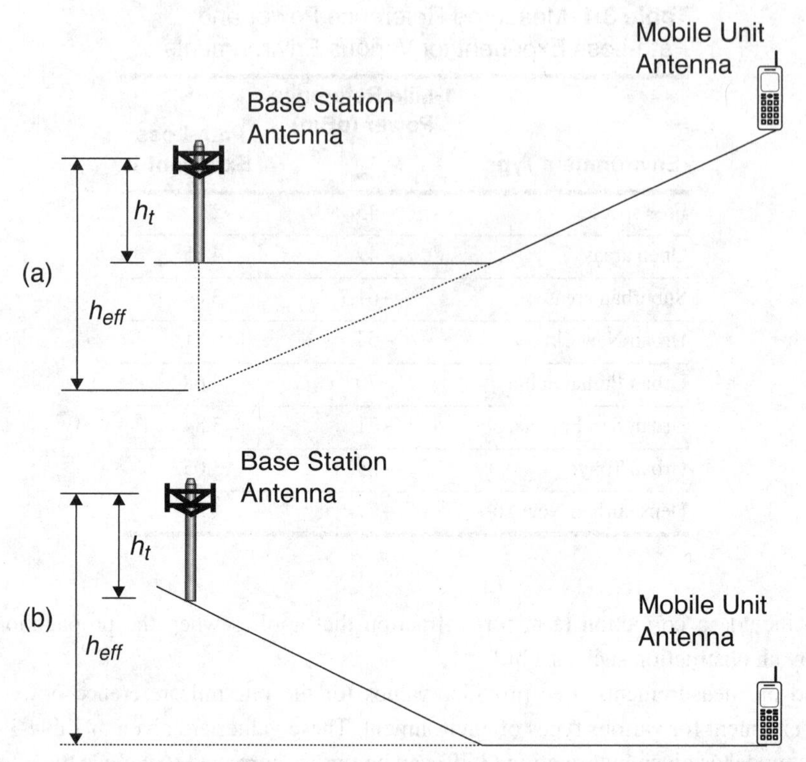

Figure 3.3 Computing the Effective Base Station Antenna Height: (a) Mobile Antenna on a Slope; (b) Base Station Antenna on a Slope

will give a prediction of median received signal power. Thus the local terrain profile can be used to modify the signal strength prediction.

Example

A base station in a medium-sized city transmits 43 dBm from an antenna height of 30 m to a receiver 10 km away over flat terrain. The receiver antenna height is 2 m, and the carrier frequency is 850 MHz. The base station antenna has a gain of 10 dBi, and the mobile unit antenna has a gain of 2 dBi. Use the Lee model to determine the median received signal power.

Compare the median received signal power with that obtained using the Hata model.

Solution

If we take urban Newark as a "medium-sized city" (compared to Philadelphia, Tokyo, or New York), then $P_{1-mile}\big|_{dB} = 64$ dBm and $v = 4.31$. Now 10 km is about 6.2 miles. Then

Equation (3.49) gives

$$P_{r,50}\Big|_{dB} = P_{1\text{-}mile}\Big|_{dB} - 10\nu\log\left(\frac{d}{d_0}\right) + \alpha_c$$

$$= -64 \text{ dBm} - 10 \times 4.31 \times \log\left(\frac{6.2}{1}\right) + \alpha_c \tag{3.53}$$

$$= -98.2 \text{ dBm} + \alpha_c.$$

To determine the correction factor, we observe that our 30 m base station antenna height is close to the 100 ft standard, and the 850 MHz carrier frequency also corresponds to the standard. The 2 m mobile unit antenna height is about 6.56 ft. We therefore have

$$\alpha_c = 20\log\left(\frac{6.56 \text{ ft}}{10 \text{ ft}}\right) + (43 \text{ dBm} - 40 \text{ dBm})$$
$$+ (10 \text{ dB} - 8.1 \text{ dB}) + (2 \text{ dB} - 2.1 \text{ dB}) \tag{3.54}$$
$$= 1.14 \text{ dB}.$$

As there is no correction needed for terrain, we obtain

$$P_{r,50}\Big|_{dB} = -98.2 \text{ dBm} + 1.14 \text{ dB} = -97.1 \text{ dBm}. \tag{3.55}$$

To compare with the Hata model, we begin with the path loss obtained in Equation (3.39), $L_{50}\,(urban)|_{dB} = 160$ dB. Then

$$P_{r,50}(Hata)\Big|_{dB} = 43 \text{ dBm} + 10 \text{ dB} + 2 \text{ dB} - 160 \text{ dB}$$
$$= -105 \text{ dBm}. \tag{3.56}$$

The Lee and Hata models differ in their predictions by about 8 dB. Neither model is in error; the difference represents the limited accuracy that can be obtained from simple models of real-world propagation. Clearly systems engineers must apply considerable engineering judgment based on experience and backed by field measurements before investing money based on the predictions of these models. The fact that cellular telephone systems work as well as they do is testimony to the fact that engineers are in fact able to apply the necessary judgment and use the models to produce successful designs.

Macroscopic Models 3: Log-Normal Shadowing

In the previous section we investigated a number of models that allow a systems engineer to predict median path loss as a function of the distance between transmitting and receiving antennas. The models also relate path loss to various system parameters such as antenna height and carrier frequency. The fact that these models predict median rather than absolute path loss suggests that measurements made at various locations at a given separation distance will produce a range of path-loss values. If we wish to design a communication system that will provide an acceptable quality of service not just "on the average" but at most locations in the coverage area, we will have to extend our path-loss models to enable us to predict not just median loss, but also the range of loss values we are likely to encounter at a given distance from the transmitter. One reason for the change in path loss with location at a given separation distance is that shadowing from various obstructions distributed over the signal path changes as the transmitter or receiver position is varied. It has been found that path-loss changes caused by shadowing are relatively gradual. This means that location must change by many wavelengths before a significant change in path loss will be observed. Another phenomenon, multipath propagation, is responsible for significant path-loss changes over distances of less than a wavelength. We will investigate these "microscopic" propagation effects in a later section of the current chapter.

In principle it should be possible to calculate the path loss for every location in the environment at a given distance between the transmitter and receiver. All that is needed is a detailed enough description of the physical environment between, and in the vicinity of, the transmitter and receiver, and powerful enough electromagnetics modeling software. In practice, of course, the detailed description of the environment is not available, and the calculations would be complicated. Consequently we will use a statistical model to account for the gradual variations in path loss produced by shadowing. In this model we will let the path loss $L_{path}\big|_{dB}$ be modeled as a random variable. If the mean of this random variable is $\bar{L}_{path}\big|_{dB}$, then we can write

$$L_{path}\big|_{dB} = \bar{L}_{path}\big|_{dB} + l_{path}\big|_{dB}, \tag{3.57}$$

where the term $l_{path}\big|_{dB}$ is a random variable with zero mean. The mean path loss $\bar{L}_{path}\big|_{dB}$ is reasonably well predicted by the median path loss obtained from the Okumura, Hata, or Lee model. It remains for us to characterize the random part $l_{path}\big|_{dB}$ of the path loss. Extensive measurements for a wide variety of environments have consistently demonstrated that the path loss, when expressed in decibels, follows a normal, or Gaussian, distribution. The probability density function $f_l(\xi)$ for $l_{path}\big|_{dB}$ is therefore

$$f_l(\xi) = \frac{1}{\sqrt{2\pi\sigma_{path}^2}} e^{-\frac{\xi^2}{2\sigma_{path}^2}}, \tag{3.58}$$

where σ_{path} is the standard deviation of the random variable $l_{path}\big|_{dB}$. We say that the path loss $L_{path}\big|_{dB}$, or equivalently its random part $l_{path}\big|_{dB}$, follows a **log-normal distribution**, where the logarithm comes from the fact that path loss is expressed in decibels. Note that the received signal power at a given separation distance can be written

$$\begin{aligned} P_r\big|_{dB} &= P_t\big|_{dB} + G_t\big|_{dB} + G_r\big|_{dB} - L_{sys}\big|_{dB} - L_{path}\big|_{dB} \\ &= P_t\big|_{dB} + G_t\big|_{dB} + G_r\big|_{dB} - L_{sys}\big|_{dB} - \bar{L}_{path}\big|_{dB} - l_{path}\big|_{dB} \\ &= \bar{P}_r\big|_{dB} - l_{path}\big|_{dB}, \end{aligned} \tag{3.59}$$

where $\bar{P}_r\big|_{dB}$ is the average received signal power at the given distance. Thus $P_r\big|_{dB}$ is also a log-normally distributed random variable with the probability density function $f_p(\rho)$ given by

$$f_P(\rho) = \frac{1}{\sqrt{2\pi\sigma_{path}^2}} e^{-\frac{(\rho - \bar{P}_r|_{dB})^2}{2\sigma_{path}^2}}. \tag{3.60}$$

Using the probability density function of Equation (3.60), we can express the probability that the received signal power $P_r\big|_{dB}$ is less than or equal to a given value $p_r\big|_{dB}$ as

$$\begin{aligned}\Pr[P_r\big|_{dB} \le p_r\big|_{dB}] &= \int_{-\infty}^{p_r|_{dB}} f_P(\rho)\,d\rho \\ &= \int_{-\infty}^{p_r|_{dB}} \frac{1}{\sqrt{2\pi\sigma_{path}^2}} e^{-\frac{(\rho - \bar{P}_r|_{dB})^2}{2\sigma_{path}^2}}\,d\rho.\end{aligned} \tag{3.61}$$

Similarly, the probability that the received signal power is greater than $p_r\big|_{dB}$ is

$$\Pr[P_r\big|_{dB} > p_r\big|_{dB}] = \int_{p_r|_{dB}}^{\infty} \frac{1}{\sqrt{2\pi\sigma_{path}^2}} e^{-\frac{(\rho - \bar{P}_r|_{dB})^2}{2\sigma_{path}^2}}\,d\rho. \tag{3.62}$$

The standard deviation σ_{path} is a parameter that must be chosen to calibrate the statistical model for a given environment. Measurements and empirical studies of extensive data reveal that the standard deviation of the log-normal distribution is approximately 8 dB for a wide range of urban and suburban environments. The standard deviation may be 1–2 dB higher for dense urban environments and 1–2 dB lower for open and rural areas.

The integrals encountered in Equations (3.61) and (3.62) cannot be evaluated in closed form. Owing to the importance of these integrals in a variety of statistical applications, however, they are widely tabulated. Usually tables provide standardized versions of these probability integrals. One standardized form, the Q-function, is defined as

$$Q(x) = \frac{1}{\sqrt{2\pi}} \int_{x}^{\infty} e^{-\frac{\xi^2}{2}}\,d\xi. \tag{3.63}$$

In terms of the Q-function, Equation (3.62) becomes

$$\Pr[P_r\big|_{dB} > p_r\big|_{dB}] = Q\left(\frac{p_r\big|_{dB} - \bar{P}_r\big|_{dB}}{\sigma_{path}}\right). \tag{3.64}$$

The Q-function has the useful property that

$$Q(-x) = 1 - Q(x) = \frac{1}{\sqrt{2\pi}} \int_{-\infty}^{x} e^{-\frac{\xi^2}{2}}\,d\xi. \tag{3.65}$$

This allows Equation (3.61) to be expressed as

$$\Pr[P_r|_{dB} \le p_r|_{dB}] = Q\left(\frac{\bar{P}_r|_{dB} - p_r|_{dB}}{\sigma_{path}}\right). \tag{3.66}$$

There is a table of Q-function values in Appendix A. Appendix A also presents an alternative standard form, the **error function**. The error function is tabulated in numerous mathematical tables and can also be evaluated using software applications such as MATLAB.

Example

A cellular base station transmits 20 W at 850 MHz from an antenna with 4 dBi of gain at a height of 40 m. A mobile receiver has an antenna gain of 0 dBi at a height of 1.5 m. There are 2 dB of system losses. The mobile unit is located at the perimeter of the cell, at a distance of 8 km from the transmitting antenna. The cell is in the suburbs of a small city.

Use the Hata model to find the median received signal power. Then find the probability that the received signal power will actually be 10 dB or more below the median. The standard deviation of received power is 8 dB.

Solution

Beginning with Equation (3.35), the correction factor for a small city is $\alpha(h_{re}) = 0.0136$ dB. Then Equation (3.33) gives $L_{50}(urban)|_{dB} = 155$ dB. Using Equation (3.36) to correct for land use, we obtain $L_{50}(suburban)|_{dB} = 145$ dB. Now we can use the median path loss to obtain the median received signal power:

$$\begin{aligned} P_{r,50}|_{dB} &= P_t|_{dB} + G_t|_{dB} + G_r|_{dB} - L_{sys}|_{dB} - L_{50}(suburban)|_{dB} \\ &= 43\text{ dBm} + 4\text{ dB} + 0\text{ dB} - 2\text{ dB} - 145\text{ dB} \\ &= -100\text{ dBm}. \end{aligned} \tag{3.67}$$

We will take this median received signal power as the average received signal power in our log-normal statistical model, that is,

$$\bar{P}_r|_{dB} = -100\text{ dBm}. \tag{3.68}$$

Now we wish to find $\Pr[P_r|_{dB} \le -110\text{ dBm}]$. Using Equation (3.66) with $\sigma_{path} = 8$ dB,

$$\begin{aligned} \Pr[P_r|_{dB} \le -110\text{ dBm}] &= Q\left(\frac{-100\text{ dBm} - (-110\text{ dBm})}{8\text{ dB}}\right) \\ &= Q(1.25) \\ &= 0.1056 \end{aligned} \tag{3.69}$$

using the Q-function table in Appendix A.

It is important to interpret the result given in Equation (3.69) correctly. The signal strength at any location near the cell boundary, whatever that strength happens to be, will not change, so long as the receiver and the surrounding environment remain motionless. If signal strength is measured at a number of locations near the cell boundary, however, the received signal power will be at or below –110 dB at about 10% of the locations. Our model tells us nothing about how far the receiver must be moved to go from a location with signal power above –110 dB to a location with power below –110 dB.

Example

A base station in a cellular system transmits at 890 MHz from an antenna of gain 6 dBi at a height of 35 m to a mobile unit with an antenna gain of 2 dBi at a height of 2 m. There are 1.5 dB of system losses. The cell is located in a large city. The mobile unit is at the edge of the cell, at a distance of 2 km from the base station. Suppose the receiver in the mobile unit requires a signal power of –90 dBm to provide the subscriber with an adequate quality of service (that is, the receiver sensitivity is –90 dBm). If $\sigma_{path} = 9$ dB, determine how much power the base station must transmit so that the received power will be adequate at 80% of locations at the cell limit.

Solution

We want to guarantee that $\Pr[P_r|_{dB} > -90\text{ dBm}] = 0.8$. Now Equation (3.64) gives us

$$\Pr\left[P_r|_{dB} > -90\text{ dBm}\right] = Q\left(\frac{-90\text{ dBm} - \bar{P}_r|_{dB}}{9\text{ dB}}\right) = 0.8. \tag{3.70}$$

Using Equation (3.65), this becomes

$$1 - Q\left(\frac{-90\text{ dBm} - \bar{P}_r|_{dB}}{9\text{ dB}}\right) = Q\left(\frac{90\text{ dBm} + \bar{P}_r|_{dB}}{9\text{ dB}}\right) = 1 - 0.8 = 0.2. \tag{3.71}$$

From the Q-function table in Appendix A we find

$$\frac{90\text{ dBm} + \bar{P}_r|_{dB}}{9\text{ dB}} = 0.84. \tag{3.72}$$

Solving gives the average receiver power needed at the cell limit:

$$\bar{P}_r|_{dB} = -82.4\text{ dBm}. \tag{3.73}$$

If we take the average receiver power as equivalent to the median power predicted, say, by the Hata model, we can write

$$-82.4\text{ dBm} = P_t|_{dB} + 6\text{ dB} + 2\text{ dB} - 1.5\text{ dB} - L_{50}(urban)|_{dB}. \tag{3.74}$$

Using the Hata model, we find that $L_{50}(urban)|_{dB} = 135$ dB , so that Equation (3.74) gives the required transmitter power, $P_t|_{dB} = 45.9$ dBm.

The difference between the average received signal power and the receiver sensitivity is called the **fade margin.** In this example a fade margin of 7.6 dB was needed to ensure adequate received power at 80% of locations 2 km from the transmitter.

Microscopic Models 1: Multipath Propagation and Fading

Introduction

In the previous section we developed a model to account for the variation of path loss with location, given a constant separation distance between the transmitting and receiving antennas. The primary physical phenomenon that was reflected in this model is shadowing by objects in the environment between the transmitter and receiver. The effect of shadowing is sometimes called **macroscopic** or **macro-scale fading**, or sometimes **large-scale fading**, as shadowing causes received signal strength to change as the receiver moves over distances that are long compared with a wavelength. The term *log-normal fading* is also used to describe this phenomenon, because the received signal power can be accurately modeled as a random variable with a log-normal probability distribution. In the current section we begin our examination of a different fading phenomenon. This fading is caused by **multipath propagation**, that is, reception of the same signal over two or more propagation paths. Multipath propagation causes the received signal power to change significantly as the receiver is moved over distances shorter than a wavelength and is consequently referred to as **microscopic** or **small-scale fading**.

Multipath propagation arises because of the presence of objects in the environment that can reflect or scatter the transmitted signal. Figure 3.4 suggests copies of a transmitted signal arriving

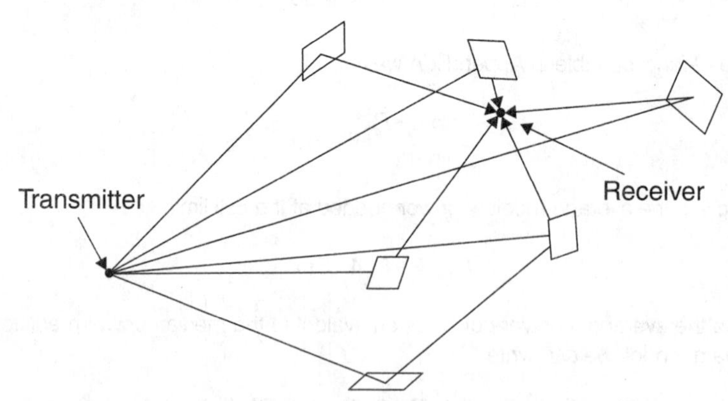

Figure 3.4 Multipath Propagation Caused by Reflection from Multiple Objects

at a receiver by multiple paths. Buildings and other man-made structures, hillsides, and the leaves of trees are examples of potential scatterers. In urban environments there may be many such scatterers, and copies of the transmitted signal may arrive at the receiver by a large number of simultaneous paths. When a signal reflects from a scattering surface, it may undergo a change in amplitude and a change in phase. Signals arriving at the receiver over multiple paths will then include these amplitude and phase changes. Further, each path will in general have a different length, so the received signals will arrive with slightly different delays. If the transmitter, the receiver, or any of the scatterers is in motion, signals will arrive at the receiver with a spread of Doppler shifts. This Doppler spread can cause the received signal strength to vary with time, even if the receiver's position does not change.

We will begin our study of micro-scale fading by considering a two-ray propagation model. This model, though simple, will allow us to discover and begin to quantify a number of important attributes of multipath propagation. The two-ray model will illustrate how constructive and destructive interference causes the received signal strength to be a function of receiver position. The simple model will also illustrate that the path loss of a multipath communication channel is frequency dependent, even when the receiver's location is fixed. This frequency dependence adds a new wrinkle to our propagation model. In some cases the loss is relatively constant over the bandwidth used by a single transmitter and receiver. In other cases the loss will vary within this bandwidth, leading to distortion of the received signal. We will find that the band of frequencies over which the loss remains approximately constant is dependent on the difference in delay times between the two arriving signals.

Once we have used the two-ray model to demonstrate the underlying causes of small-scale fading and frequency-dependent path loss, we will introduce a statistical model for the real-world case in which there are many received signals. The statistical model is called the Rayleigh model, after the probability distribution that plays a significant role in characterizing the amplitude of the received signal. The Rayleigh model will allow us to calculate the probability of a fade, much as the log-normal model allowed us to make that calculation for shadowing. To design a data link that will provide users with a high quality of service, allowance must be made for both macroscopic and microscopic fading. The statistical model will also give us a framework for defining a root-mean-square (RMS) delay spread, which we will use to obtain a quantitative estimate of the **coherence bandwidth**, that is, the bandwidth over which the loss remains approximately constant.

After establishing the Rayleigh model and frequency-dependent loss, we shall return to the two-ray model for insight into the case in which the receiver is in motion. We will be able to easily show how motion of the receiver produces a time-varying path loss. We will be able to characterize the timescale over which the loss varies in terms of the Doppler shifts in the carrier frequencies of the two received signals. The statistical model, which we will not treat in detail, yields a **coherence time**, defined as the time over which the path loss remains approximately constant. We will use the coherence time as a boundary between the cases of "slow" fading, in which the path loss remains nearly constant during the time required to receive a single pulse, and "fast" fading, in which the path loss changes while the pulse is in flight.

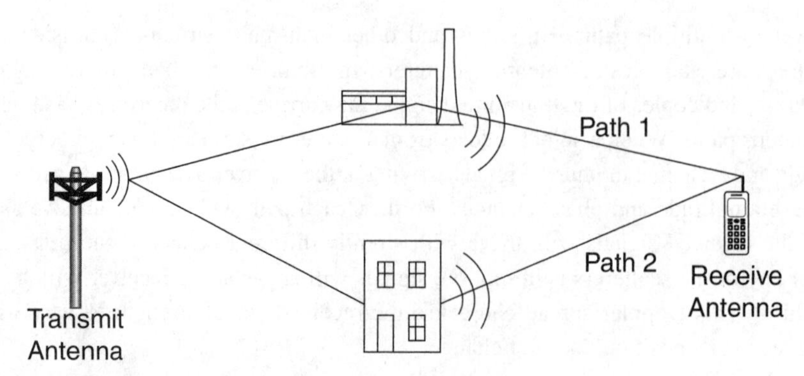

Figure 3.5 A Signal Received over Two Propagation Paths

A Two-Ray Model for Multipath Propagation: Stationary Receiver

Figure 3.5 shows a single transmitted signal traveling to a receiver over two propagation paths. To keep the model simple, we will assume that the transmitted signal is an unmodulated carrier. In practice, of course, the carrier will be modulated in some way with an information signal. One effect of modulation will be to give the transmitted signal a nonzero bandwidth. Although we do not show the modulation explicitly in the discussion to follow, we recognize that the transmitted signal's bandwidth makes the frequency response of the multipath channel a matter of concern.

The received signal can be written

$$r(t) = A_1 \cos(2\pi f(t - t_1) + \phi_1) + A_2 \cos(2\pi f(t - t_2) + \phi_2), \tag{3.75}$$

where

A_1 and ϕ_1 are respectively the amplitude and phase of the signal received over path 1
A_2 and ϕ_2 are the amplitude and phase of the signal received over path 2
t_1 is the propagation delay corresponding to path 1
t_2 is the propagation delay corresponding to path 2
f is the frequency

Following our approach in Equation (3.7), we will write the received signal as a single sinusoid:

$$r(t) = A\cos(2\pi ft + \phi). \tag{3.76}$$

We are particularly interested in the amplitude A of this combined received signal. Using phasors to combine the two sinusoids in Equation (3.75), we obtain

$$\mathbf{R} = A_1 e^{j(-2\pi ft_1 + \phi_1)} + A_2 e^{j(-2\pi ft_2 + \phi_2)}. \tag{3.77}$$

The magnitude of the resultant phasor **R** is a little easier to find if we factor out the first term. This gives

$$\mathbf{R} = A_1 e^{j(-2\pi f t_1+\phi_1)}\left(1+\frac{A_2}{A_1}e^{j[-2\pi f(t_2-t_1)+(\phi_2-\phi_1)]}\right). \tag{3.78}$$

Expanding the exponential in the parentheses using Euler's identity and combining real parts and imaginary parts gives

$$\begin{aligned}\mathbf{R} &= A_1 e^{j(-2\pi f t_1+\phi_1)}\left(1+\frac{A_2}{A_1}\cos[-2\pi f(t_2-t_1)+(\phi_2-\phi_1)]+j\frac{A_2}{A_1}\sin[-2\pi f(t_2-t_1)+(\phi_2-\phi_1)]\right)\\ &= A_1 e^{j(-2\pi f t_1+\phi_1)}\left\{\left(1+\frac{A_2}{A_1}\cos[-2\pi f(t_2-t_1)+(\phi_2-\phi_1)]\right)+j\left(\frac{A_2}{A_1}\sin[-2\pi f(t_2-t_1)+(\phi_2-\phi_1)]\right)\right\}.\end{aligned} \tag{3.79}$$

Now we can compute the magnitude as the square root of the square of the real part plus the square of the imaginary part:

$$\begin{aligned}A = |\mathbf{R}| &= A_1\sqrt{\left(1+\frac{A_2}{A_1}\cos[-2\pi f(t_2-t_1)+(\phi_2-\phi_1)]\right)^2+\left(\frac{A_2}{A_1}\sin[-2\pi f(t_2-t_1)+(\phi_2-\phi_1)]\right)^2}\\ &= A_1\sqrt{1+2\frac{A_2}{A_1}\cos[-2\pi f(t_2-t_1)+(\phi_2-\phi_1)]+\left(\frac{A_2}{A_1}\right)^2}\\ &= \sqrt{A_1^2+A_2^2+2A_1A_2\cos[-2\pi f(t_2-t_1)+(\phi_2-\phi_1)]},\end{aligned} \tag{3.80}$$

where we have simplified by combining cosine-squared and sine-squared terms. This result shows that the amplitude A of the received signal depends on the phase difference $\phi_2 - \phi_1$, on the delay difference $t_2 - t_1$, and, most interestingly, on the frequency f.

The amplitude of the combined signal given by Equation (3.80) can be larger or smaller than the amplitude of the signal received over either single path. If, for example, $\phi_2 - \phi_1 = 0$, and the path lengths differ by an integral number of wavelengths (that is, an even number of half-wavelengths), then the signals received over the two paths will combine constructively. In Equation (3.80) the cosine will take the value +1, giving $A = A_1 + A_2$. Alternatively, if $\phi_2 - \phi_1 = 0$ and the path lengths differ by an odd number of half-wavelengths, then the signals received over the two paths will combine destructively. In Equation (3.80) the cosine will take the value −1, and the amplitude of the combined signal will be $A = |A_1 - A_2|$. In general, the amplitude of the combined signal will take a value between these extremes. The amplitude of the received signal is not a function of time; this amplitude will remain constant as long as all of the parameters in Equation (3.80) remain constant. If the position of the receiver is changed, however, then the delay difference $t_2 - t_1$ will change, causing a change in the amplitude A. In an extreme case, with the two signals arriving from opposite directions, a change in receiver position of only a quarter-wavelength could cause one path to become a quarter-wavelength longer and the other path to become a quarter-wavelength shorter. The net effect could change the received signal strength from a maximum associated with constructive reinforcement to a minimum associated with destructive cancellation. This change in received signal strength with small changes in receiver position is often referred to as small-scale fading, since the distances over which the receiver must move to cause or end a fade are small compared with the wavelength. (At 850 MHz a wavelength is only 0.35 m.) It is important to note that the number of wavelengths by which paths 1 and 2 differ depends on the frequency f. This means that a small change in frequency can also result in a significant change in the amplitude of the combined received signal.

Example

Suppose the phase difference $\phi_2 - \phi_1$ is $\pi/3$, the amplitude difference is 2 dB, and path 2 is 150 m longer than path 1. Plot the received signal amplitude versus frequency, for frequencies from 850 to 855 MHz.

Solution

Let us suppose that path 1 produces the stronger signal. Then the 2 dB amplitude difference means that $20\log\frac{A_2}{A_1} = -2$ dB, so $A_2/A_1 = 0.794$. The delay difference $t_2 - t_1$ is given by $t_2 - t_1 = \frac{d_2 - d_1}{c} = \frac{150 \text{ m}}{3\times10^8 \text{ m/s}} = 0.5\ \mu s$. Equation (3.80) now gives

$$A = A_1\sqrt{1+(0.794)^2 + 2(0.794)\cos[-2\pi f(0.5\times10^{-6}) + \pi/3]}. \tag{3.81}$$

Figure 3.6 shows A/A_1 plotted against frequency. This figure is called a "standing wave pattern" in the electromagnetics literature. At 850.3 MHz the 0.5 μs difference in the propagation delays of the two paths represents an integral number of wavelengths, and the signals reinforce. At 851.3 MHz this same time difference corresponds to an odd number of half-wavelengths, and the signals partially cancel. We note that the ratio of maximum amplitude to minimum amplitude is $20\log(\frac{1.8}{0.2}) = 19.1$ dB.

In Figure 3.6 the relative amplitude is periodic in frequency with a period of 2 MHz. This means that a transmitted signal with a bandwidth much less than 2 MHz will propagate over a communication channel whose path loss is relatively constant over the signal bandwidth. A transmitted signal with a bandwidth comparable to 2 MHz, on the other hand, will experience a communication channel whose path loss varies significantly with frequency. The former case is referred to as "flat fading" and the latter case as "frequency-selective fading."

Example

Plot the relative amplitude described in Equation (3.81) over two 30 kHz frequency bands, one centered at 851.5 MHz and one centered at 852.5 MHz.

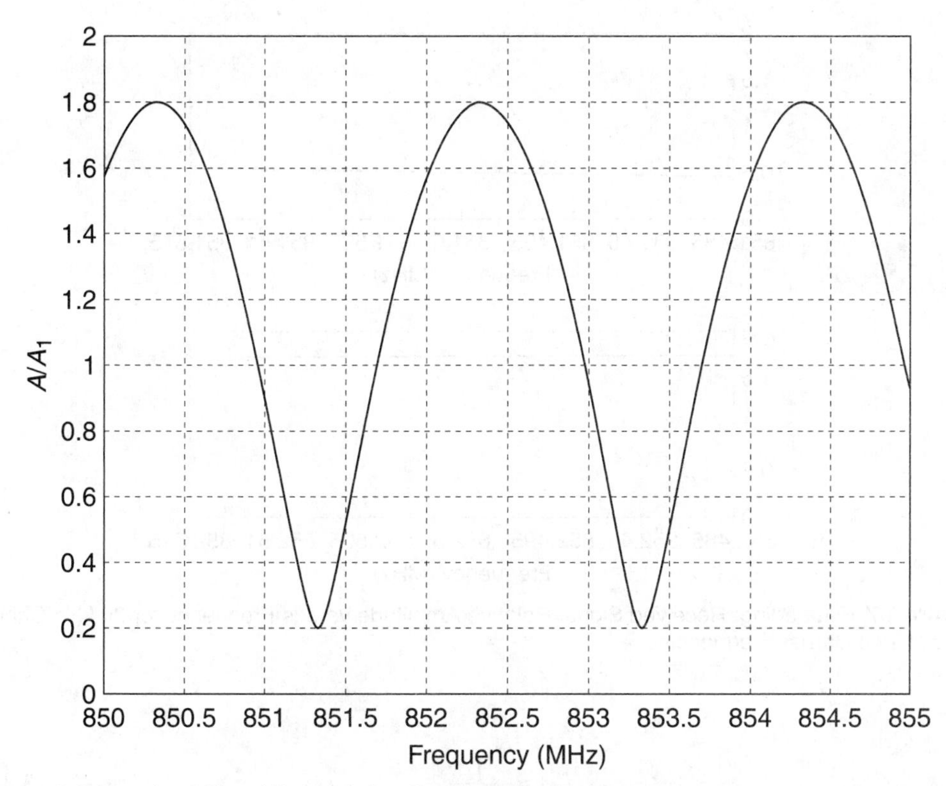

Figure 3.6 Received Signal Relative Amplitude versus Frequency

Solution

The required plots are shown in Figure 3.7. These graphs depict flat fading over two 30 kHz wide "channels." They show that in the given example, the received signal amplitude varies very little with frequency over a 30 kHz band, although the amplitudes are significantly different at the two carrier frequencies spaced 1 MHz apart.

In the example shown, the received signal amplitude varies with frequency in a periodic manner, with spectral nulls (fades) occurring every 2 MHz. Referring to Equation (3.80), we can see that in the general case the amplitude level will repeat whenever the frequency changes over an interval that causes the argument of the cosine to change by 2π radians. If f_{n-n} represents the null-to-null frequency interval, then

$$2\pi f_{n-n}(t_2 - t_1) = 2\pi \tag{3.82}$$

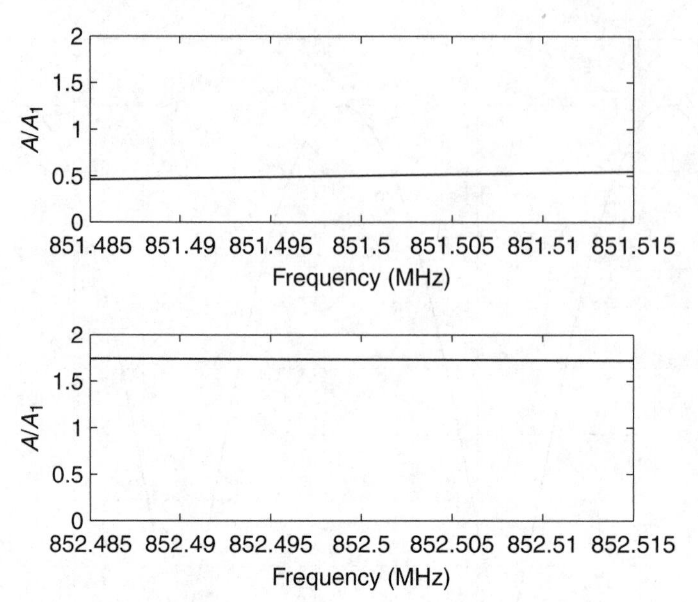

Figure 3.7 Flat Fading: Received Signal Relative Amplitude versus Frequency for 30 kHz Channels at Two Carrier Frequencies

or

$$f_{n-n} = \frac{1}{t_2 - t_1}. \tag{3.83}$$

We see that the null-to-null interval is inversely proportional to the "delay spread" $t_2 - t_1$. A modulated signal will experience flat fading if the signal bandwidth is much less than the reciprocal of the delay spread. Conversely, a modulated signal will experience frequency-selective fading if the signal bandwidth is comparable to or greater than the reciprocal of the delay spread.

Microscopic Models 2: Statistical Models for Multipath Propagation

Rayleigh Fading

It is common in dense urban environments for there to be no direct line-of-sight path from a transmitter to a receiver. In this case communication is made possible only by reflection and refraction from objects in the environment, particularly buildings. We will assume in what follows that the received signal is the sum of a large number M of waves all originating from the same transmitter, but all arriving from different directions, and each with its own amplitude, phase, and delay. Thus the received signal is

$$r(t) = \sum_{k=1}^{M} A_k \cos(2\pi ft + \theta_k), \tag{3.84}$$

where the angle θ_k is given by

$$\theta_k = -2\pi t_k + \phi_k,\ k = 1,\ldots,M. \tag{3.85}$$

Following our usual procedure, we use phasors to add the sinusoids in Equation (3.84),

$$\begin{aligned}\mathbf{R} &= \sum_{k=1}^{M} A_k e^{j\theta_k} \\ &= \sum_{k=1}^{M} A_k \cos\theta_k + j\sum_{k=1}^{M} A_k \sin\theta_k \\ &= X + jY,\end{aligned} \tag{3.86}$$

where $X = \sum_{k=1}^{M} A_k \cos\theta_k$, and $Y = \sum_{k=1}^{M} A_k \sin\theta_k$.

In a real-world environment we cannot measure the specific properties of all of the individual propagation paths that produce components of the received signal. We will therefore characterize the received signal statistically by representing $A_k, k = 1,\ldots,M$ and $\theta_k, k = 1,\ldots,M$ as random variables. A few reasonable assumptions will greatly simplify our characterization.

- The amplitudes $A_k, k = 1,\ldots,M$ are statistically independent of each other and are similarly distributed. The amplitudes are independent of the phases θ_k.
- The phases $\theta_k, k = 1,\ldots,M$ are statistically independent of each other and are uniformly distributed over the range 0 to 2π.

These assumptions, together with the additional assumption that the number of signal components M is large, allow us to invoke the central limit theorem and consequently to assert that the random variables X and Y defined in Equation (3.86) are statistically independent Gaussian random variables, with zero means and equal variances σ^2. The Gaussian model correlates well with measured data taken in representative environments.

The joint probability density function $f_{X,Y}(x, y)$ of X and Y is the familiar "bell curve," which can be written as

$$f_{X,Y}(x,y) = \frac{1}{2\pi\sigma^2} e^{-\frac{x^2+y^2}{2\sigma^2}}. \tag{3.87}$$

The cumulative probability distribution function $F_{X,Y}(x, y)$ is, for any real numbers x and y, the joint probability that $X \le x$ and $Y \le y$, that is,

$$\begin{aligned}F_{X,Y}(x,y) &= \Pr\{X \le x, Y \le y\} \\ &= \int_{-\infty}^{y}\int_{-\infty}^{x} f_{X,Y}(\xi,\eta)\,d\xi\,d\eta = \int_{-\infty}^{y}\int_{-\infty}^{x} \frac{1}{2\pi\sigma^2} e^{-\frac{\xi^2+\eta^2}{2\sigma^2}}\,d\xi\,d\eta.\end{aligned} \tag{3.88}$$

A little calculus reveals that

$$f_{X,Y}(x,y) = \frac{\partial^2}{\partial x\,\partial y} F_{X,Y}(x,y). \tag{3.89}$$

As in the two-ray model, we are interested in the amplitude A of the total received signal $r(t)$. Recall that A is the magnitude of the phasor $\mathbf{R}$, defined in Equation (3.86), that is,

$$A = |\mathbf{R}| = \sqrt{X^2 + Y^2}. \tag{3.90}$$

The amplitude A is a random variable, which means that we need to find its probability density function $f_A(a)$ rather than its value. Our approach will be to first find the cumulative probability distribution function $F_A(a)$ and then take the derivative as in Equation (3.89) (a single derivative in this case) to find $f_A(a)$. Now the phasor $\mathbf{R}$ can be written in polar form as

$$\mathbf{R} = Ae^{j\Theta}, \tag{3.91}$$

where A is given by Equation (3.90) and

$$\Theta = \tan^{-1}\frac{Y}{X}. \tag{3.92}$$

Using the dummy variables ξ, η, α, and θ, Equations (3.90) and (3.92) become

$$\begin{aligned} \alpha &= \sqrt{\xi^2 + \eta^2} \\ \theta &= \tan^{-1}\frac{\eta}{\xi}. \end{aligned} \tag{3.93}$$

We wish to find

$$\begin{aligned} F_A(a) &= \Pr\{A \le a\} \\ &= \iint_{\sqrt{\xi^2+\eta^2} \le a} f_{X,Y}(\xi,\eta)\,d\xi\,d\eta \\ &= \iint_{\sqrt{\xi^2+\eta^2} \le a} \frac{1}{2\pi\sigma^2} e^{-\frac{\xi^2+\eta^2}{2\sigma^2}}\,d\xi\,d\eta. \end{aligned} \tag{3.94}$$

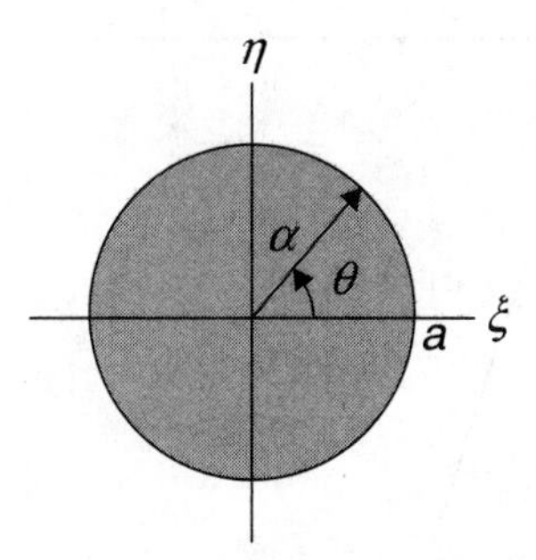

Figure 3.8 Region of Integration for Cumulative Distribution Function of Received Signal Amplitude

The region of integration is shown in Figure 3.8. Using Equation (3.93) to transform to polar coordinates gives us $\xi^2 + \eta^2 = \alpha^2$ and the area element $d\xi d\eta = \alpha d\theta d\alpha$.

Substituting in Equation (3.94) gives

$$\begin{aligned} F_A(a) &= \int_0^a \int_0^{2\pi} \frac{1}{2\pi\sigma^2} e^{-\frac{\alpha^2}{2\sigma^2}} \alpha \, d\theta \, d\alpha \\ &= \int_0^a \frac{1}{\sigma^2} e^{-\frac{\alpha^2}{2\sigma^2}} \alpha \, d\alpha. \end{aligned} \tag{3.95}$$

Finally,

$$\begin{aligned} f_A(a) &= \frac{d}{da} F_A(a) \\ &= \begin{cases} \frac{a}{\sigma^2} e^{-\frac{a^2}{2\sigma^2}}, & a \geq 0 \\ 0, & a < 0. \end{cases} \end{aligned} \tag{3.96}$$

The probability density function $f_A(a)$ for the amplitude of the received signal is known as a **Rayleigh density function**. It has a mean value m_A given by

$$m_A = \sigma\sqrt{\frac{\pi}{2}} \tag{3.97}$$

and a standard deviation σ_A given by

$$\sigma_A = \sigma\sqrt{2 - \frac{\pi}{2}}. \tag{3.98}$$

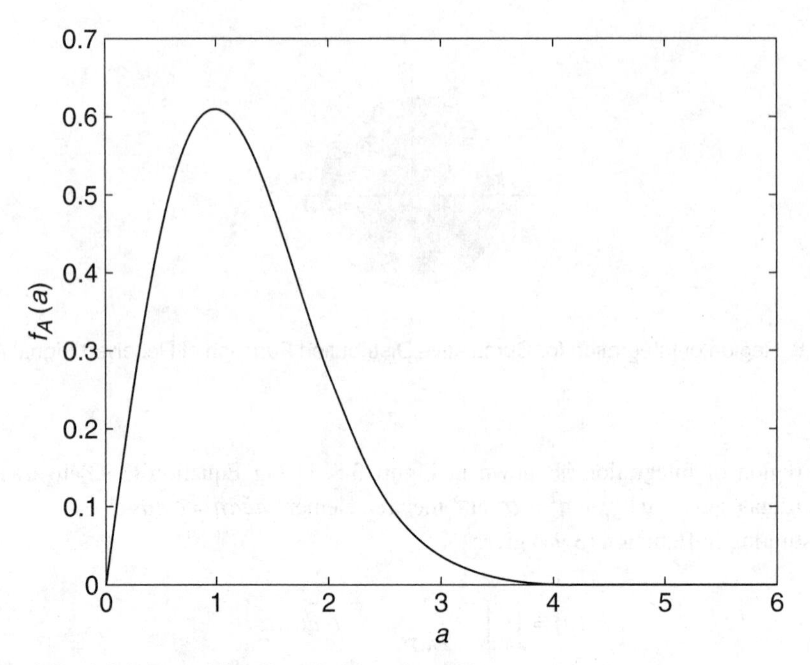

Figure 3.9 A Rayleigh Probability Density Function with $\sigma = 1$

Note that the standard deviation of the Rayleigh distribution is σ_A, given by Equation (3.98), and is not the parameter σ. A Rayleigh probability density function is illustrated in Figure 3.9 with $\sigma = 1$.

Figure 3.10 shows 100 samples of a Rayleigh-distributed random variable plotted in decibels. The horizontal axis is the sample number and does not represent time or position. This figure shows values of received signal amplitude that might be obtained from measurements taken in a multipath environment at positions a few wavelengths apart. The mean of the received signal amplitude has been set at zero decibels. Notice that a few of the samples show very deep fades.

Now that we have characterized the statistical behavior of the received signal amplitude, we can turn our attention to the received signal power. Average received power is related to amplitude by Equation (3.8), repeated here for convenience:

$$P_r = \frac{A^2}{2}. \tag{3.99}$$

Since the amplitude A is modeled as a random variable in the current analysis, so is the power P_r. We must then determine the probability density function for P_r. As we did for amplitude, we begin by seeking the cumulative probability distribution function $F_P(p) = \Pr\{P_r \leq p\}$. Now Equation (3.99) shows that $P_r \leq p$ whenever $A \leq \sqrt{2p}$ (recall that A cannot be negative).

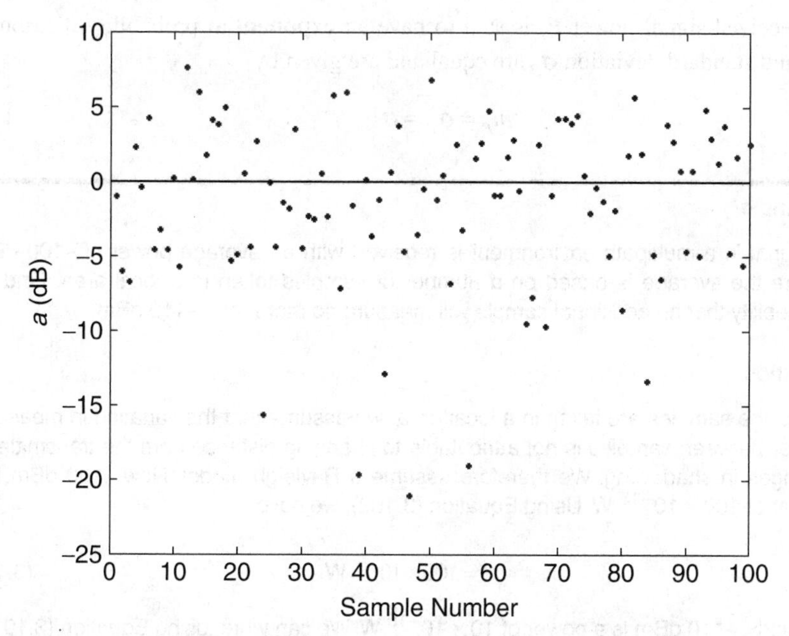

Figure 3.10 One Hundred Samples of a Rayleigh-Distributed Random Variable

We then have

$$\begin{aligned} F_P(p) &= \Pr\{P_r \leq p\} \\ &= \Pr\left\{A \leq \sqrt{2p}\right\} \\ &= \int_0^{\sqrt{2p}} f_A(\alpha)d\alpha \\ &= \int_0^{\sqrt{2p}} \frac{\alpha}{\sigma^2} e^{-\frac{\alpha^2}{2\sigma^2}} d\alpha. \end{aligned} \tag{3.100}$$

Following our customary practice, we obtain the probability distribution function for power by differentiating Equation (3.100):

$$\begin{aligned} f_P(p) &= \frac{d}{dp} F_P(p) \\ &= \frac{\sqrt{2p}}{\sigma^2} e^{-\frac{2p}{2\sigma^2}} \frac{1}{\sqrt{2p}} \\ &= \begin{cases} \frac{1}{\sigma^2} e^{-\frac{p}{\sigma^2}}, & p \geq 0 \\ 0, & p < 0. \end{cases} \end{aligned} \tag{3.101}$$

The received signal power P_r is said to have an **exponential** probability distribution. The mean m_P and standard deviation σ_P are equal and are given by

$$m_P = \sigma_P = \sigma^2. \tag{3.102}$$

Example

A signal in a multipath environment is received with an average power of –100 dBm, where the average is based on a number of samples taken in a local area. Find the probability that an additional sample will measure no more than –110 dBm.

Solution

Since the samples are taken in a local area, we assume that the variation in measured power between samples is not attributable to changing distance from the transmitter or changes in shadowing. We therefore assume a Rayleigh model. Now –100 dBm is a power of 100×10^{-15} W. Using Equation (3.102), we have

$$\sigma^2 = 100 \times 10^{-15} \text{ W}. \tag{3.103}$$

Similarly, –110 dBm is a power of 10×10^{-15} W. We can write, using Equation (3.101),

$$\begin{aligned}
\Pr\{P_r \le 10 \times 10^{-15} \text{ W}\} &= \int_0^{10\times10^{-15}} f_P(p)dp \\
&= \int_0^{10\times10^{-15}} \frac{1}{\sigma^2} e^{-\frac{p}{\sigma^2}} dp \\
&= -e^{-\frac{p}{\sigma^2}} \Bigg|_0^{10\times10^{-15}} \\
&= 1 - e^{-\frac{10\times10^{-15}}{\sigma^2}} \\
&= 1 - e^{-\frac{10\times10^{-15}}{100\times10^{-15}}} \\
&= 0.0952.
\end{aligned} \tag{3.104}$$

Example

A mobile receiver in a wireless system has a sensitivity of –95 dBm. As the receiver is moved in a local area, we wish to provide an 85% probability that the received signal will remain above the receiver sensitivity. What average received signal level is required?

Solution

First we recognize that –95 dBm is 316×10^{-15} W. Next, using Equation (3.101),

$$\begin{aligned}
\Pr\{P_r > 316\times10^{-15}\} &= \int_{316\times10^{-15}}^{\infty} \frac{1}{\sigma^2} e^{-\frac{p}{\sigma^2}} dp \\
&= -e^{-\frac{p}{\sigma^2}} \Big|_{316\times10^{-15}}^{\infty} \\
&= e^{-\frac{316\times10^{-15}}{\sigma^2}} \\
&= 0.85.
\end{aligned} \tag{3.105}$$

Solving for σ^2 gives $\sigma^2 = 1.95 \times 10^{-12}$ W. According to Equation (3.102), this is the average received signal power. Converting to decibels gives the answer as –87.1 dBm.

When one of the many signals that contribute to Equation (3.84) is much stronger than the others, the resulting composite signal can be modeled by a random variable having a Ricean probability distribution. This situation can occur, for example, when there is a direct line-of-sight path between the transmitting and receiving antennas in addition to multiple paths that involve scattering by reflection or diffraction. The Ricean distribution is named after S. O. Rice, who, in a classic paper,[8] used it to characterize a signal composed of a DC component plus thermal noise. The presence of a strong line-of-sight component implies that one of the components of Equation (3.86) will have a nonzero mean value. If we take the component with the nonzero mean as X, then Equation (3.87) must be modified to read

$$f_{X,Y}(x,y) = \frac{1}{2\pi\sigma^2} e^{-\frac{(x+m)^2+y^2}{2\sigma^2}}. \tag{3.106}$$

Following the steps in Equations (3.94) through (3.96) gives

$$f_A(a) = \begin{cases} \dfrac{a}{\sigma^2} e^{-\frac{(a^2+m^2)}{2\sigma^2}} \mathrm{I}_0\left(\dfrac{am}{\sigma^2}\right), & a \ge 0 \\ 0, & a < 0, \end{cases} \tag{3.107}$$

8. S. O. Rice, "Statistical Properties of a Sine Wave plus Random Noise," *Bell System Technical Journal* 27 (January 1948): 109–57.

where $I_0(\eta)$ is the zeroth-order modified Bessel function and is defined as the integral

$$I_0(\eta) = \frac{1}{2\pi}\int_0^{2\pi} e^{\eta\cos\theta}d\theta. \tag{3.108}$$

Bessel functions can be evaluated using mathematical tables or software applications such as MATLAB or Maple.

It is convenient to describe the ratio of power in the line-of-sight component of a signal $\frac{m^2}{2}$ to the power in the scattering components of the signal σ^2. Specifically, we define

$$K = \frac{m^2/2}{\sigma^2} \text{ or } K\big|_{dB} = 10\log\left(\frac{m^2}{2\sigma^2}\right). \tag{3.109}$$

Figure 3.11 is a plot of the Ricean probability density function for several values of $K\big|_{dB}$. Note that when the level of the line-of-sight wave is small (i.e., $K\big|_{dB} \to -\infty$), the density function approaches a Rayleigh density. As the line-of-sight component becomes large compared to the scattering components, the density function tends toward a Gaussian shape.

When a large line-of-sight component is present in a received signal, the effects of small-scale fading are reduced. This is because the peaks and nulls caused by constructive and destructive interference are less significant in relation to the average or steady component of the signal.

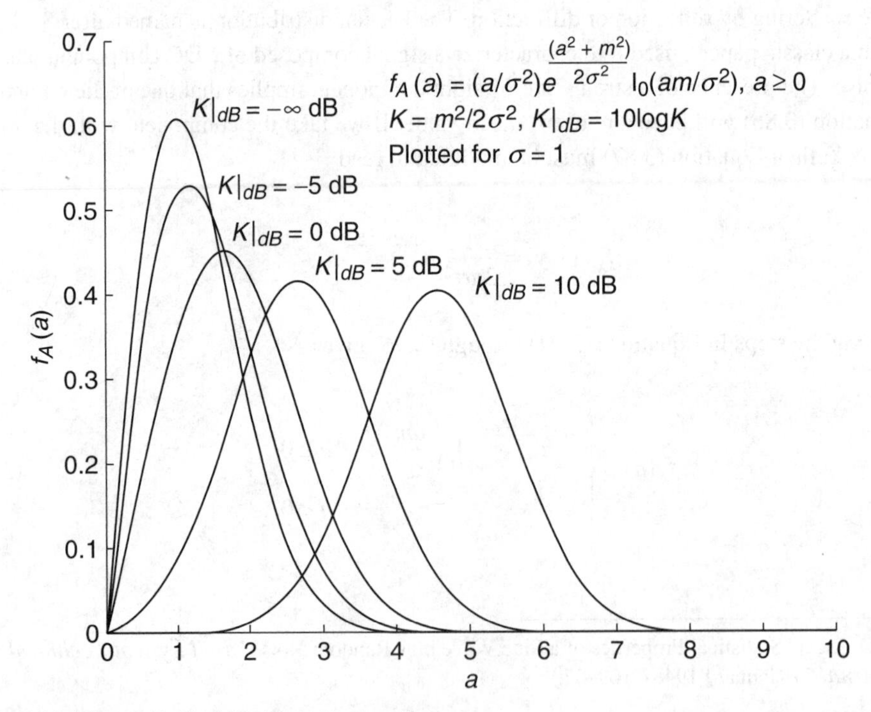

Figure 3.11 Ricean Probability Density Functions

Coherence Bandwidth

We learned from the two-ray model that in a multipath environment, fading is not only a consequence of changes in receiver position but is also caused by changes in frequency. We found that the frequency interval over which the received signal level remains approximately constant is inversely proportional to the delay spread, which in the two-ray model was $t_2 - t_1$. In the general multipath case there are many copies of the received signal arriving over different paths, each with a different delay. Figure 3.12 shows a typical standing wave pattern in which received signal amplitude is plotted versus frequency. The graph shows the same general effect as does Figure 3.6: There are frequencies at which the received signal components tend to add constructively, giving a large received amplitude; and there are frequencies at which the received signal components add destructively, leading to a small received amplitude. The plot of Figure 3.12 is more irregular than the two-ray result plotted in Figure 3.6. With reference to the standing wave pattern of Figure 3.12, we can define the coherence bandwidth B_{coh} as the frequency interval over which the received signal amplitude is relatively constant. The literature contains several definitions of coherence bandwidth, the difference depending on exactly what is meant by "relatively constant" signal amplitude. To get a quantitative sense of coherence bandwidth, we must examine the concept of delay spread a little more closely.

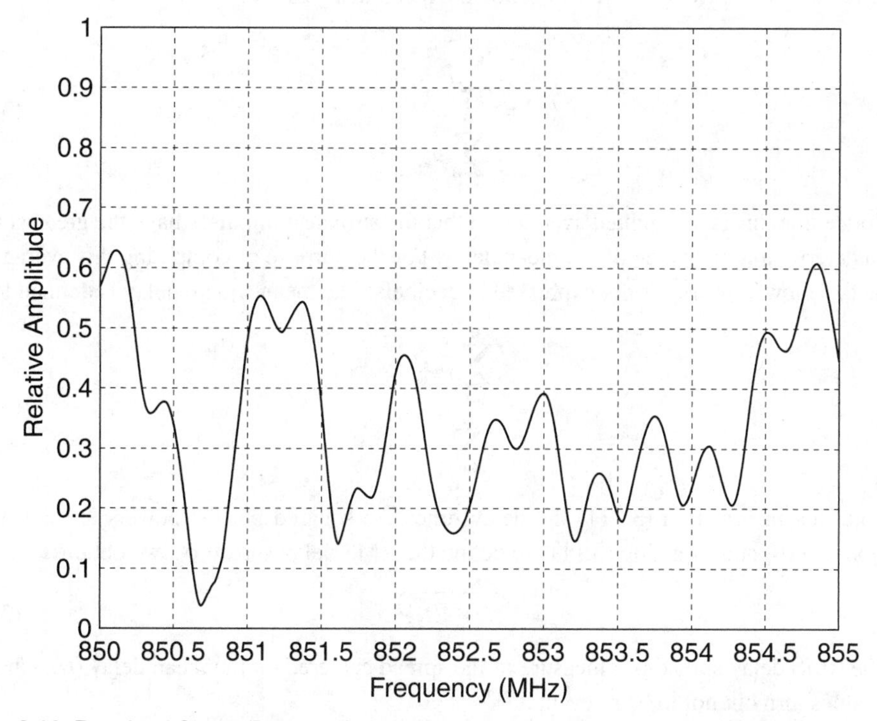

Figure 3.12 Received Signal Relative Amplitude versus Frequency for a Multipath Channel

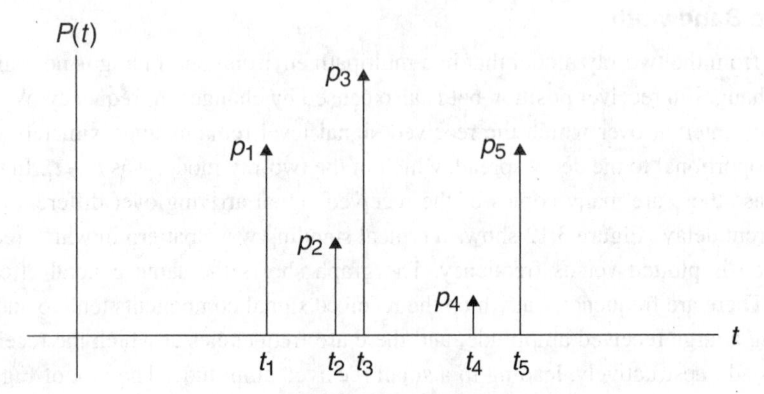

Figure 3.13 Power Delay Spread Function of a Multipath Communication Channel

Suppose the transmitted signal is an impulse. If there were a single path from transmitter to receiver, the received signal would also be an impulse, attenuated and arriving after a delay. In a multipath environment the total received signal will be a scattering of impulses, each arriving with a delay particular to the propagation path it traversed. Figure 3.13 shows a train of impulses having power levels $p_1,\ldots,p_M$ and arriving at times $t_1,\ldots,t_M$ respectively. In the figure, $M = 5$, and the plot is called the "power delay spread" function of the multipath communication channel. With reference to Figure 3.13, we can define the mean delay as

$$\langle t_k \rangle = \frac{\sum_{k=1}^{M} p_k t_k}{\sum_{k=1}^{M} p_k}. \tag{3.110}$$

Notice that this is a weighted average, so that the strongest impulses have the greatest influence in determining the value of the mean delay. For the purpose of computing this average and the one to follow, powers are not expressed in decibels. The mean square delay is defined to be

$$\langle t_k^2 \rangle = \frac{\sum_{k=1}^{M} p_k t_k^2}{\sum_{k=1}^{M} p_k}. \tag{3.111}$$

Note that in Equation (3.111) the delay times are squared but the powers are not. Using Equation (3.110) and Equation (3.111) to define the RMS delay spread σ_d, we obtain

$$\sigma_d = \sqrt{\langle t_k^2 \rangle - \langle t_k \rangle^2}. \tag{3.112}$$

The RMS delay spread is a measure of the spread centered on the mean delay $\langle t_k \rangle$. Absolute delay values turn out not to be relevant.

When the total number M of received components is large, it is convenient to model the power delay spread function of Figure 3.12 as a continuous function. We will represent this function by $p(t)$, where $p(t)$ has been normalized (by dividing by $\sum_{k=1}^{M} p_k$) so that $\int_0^\infty p(t)dt = 1$. There is some experimental evidence to show that a good model for $p(t)$ is exponential, that is,

$$p(t) = \begin{cases} \frac{1}{\sigma_d} e^{-\frac{t}{\sigma_d}}, & t \geq 0 \\ 0, & t < 0. \end{cases} \tag{3.113}$$

To formally treat the concept of "coherence,"[9] we consider the received signal at two neighboring frequencies f_1 and f_2. (We can arbitrarily choose f_2 to be the larger frequency.) The corresponding phasors $\mathbf{R}_1$ and $\mathbf{R}_2$ are each given by an equation like Equation (3.86). Now Equation (3.86) contains the angles θ_k, which depend on frequency and on path delay through Equation (3.85). The real and imaginary components X_1, Y_1, X_2, and Y_2 of the phasors $\mathbf{R}_1$ and $\mathbf{R}_2$ are four jointly Gaussian random variables. The joint probability density function for these random variables can be written out once the variances and covariances of the random variables are known. It turns out that all of the necessary variances and covariances can be obtained from Equation (3.86) by averaging over the time delays using Equation (3.113). Once we have a joint probability density function for X_1, Y_1, X_2, and Y_2, we can transform to polar coordinates as we did in developing Equation (3.96). The result is a joint probability density function for the received signal amplitudes A_1 at frequency f_1 and A_2 at frequency f_2. Finally, this joint density function can be used to calculate the correlation coefficient ρ between A_1 and A_2. The math is a little bit tricky, but the resulting correlation coefficient is

$$\rho = \frac{1}{1 + 2\pi(f_2 - f_1)\sigma_d}. \tag{3.114}$$

We can see from Equation (3.114) that the amplitudes of the signals at frequencies f_1 and f_2 will be highly correlated if $f_2 - f_1$ is small. The signal amplitudes will become less correlated as the frequencies move apart. Now we can define the coherence bandwidth as the frequency interval $f_2 - f_1$ for which the correlation coefficient ρ is at least 0.5. Solving Equation (3.114) with $\rho = 0.5$ gives $f_2 - f_1 = \frac{1}{\pi\sigma_d}$. Writing B_{coh} in place of $f_2 - f_1$, the coherence bandwidth is usually rounded off and written as

$$B_{coh} = \frac{1}{5\sigma_d}. \tag{3.115}$$

To appreciate the importance of coherence bandwidth, it is important to remember that the received signal is not a pure carrier (or an impulse) but a modulated carrier. As such the received signal will have some bandwidth of its own, B_{sig}. When $B_{sig} << B_{coh}$ we have flat fading. Although the received signal level may be enhanced by constructive interference or reduced by

9. W. C. Jakes, Jr., ed., *Microwave Mobile Communications* (New York: John Wiley & Sons, 1974), 50–52.

destructive interference, the signal level will remain nearly constant over the bandwidth B_{sig}. Alternatively, when $B_{sig} \gg B_{coh}$, we have frequency-selective fading. The signal level changes within the band B_{sig}, distorting the received signal.

It is interesting to compare the effects of flat and frequency-selective fading in the time domain. If the transmitted signals are digital, the duration of a pulse T_{sig} will be roughly $\frac{1}{B_{sig}}$, give or take a factor of 2. Now for flat fading,

$$
\begin{aligned}
B_{sig} &\ll B_{coh} \\
\frac{1}{T_{sig}} &\ll \frac{1}{5\sigma_d} \\
\sigma_d &\ll T_{sig}.
\end{aligned}
\tag{3.116}
$$

This means that the M copies of the received pulse all arrive within a time span that is short compared with the duration of a single pulse. As a result, when the M pulses are added at the receiver input, there is little distortion of the pulse shape. Now the time spread σ_d is short compared with the pulse duration, but it may not be short compared with the period of the carrier. Thus copies of the received pulse may arrive with carrier phases spanning all possible angles. The carriers may add constructively or destructively. This gives flat fading its characteristic property: There is little distortion to the shape of the received pulse, but its amplitude may be enhanced or decreased. For frequency-selective fading we have the opposite effect:

$$
\begin{aligned}
B_{sig} &\gg B_{coh} \\
\frac{1}{T_{sig}} &\gg \frac{1}{5\sigma_d} \\
\sigma_d &\gg T_{sig}.
\end{aligned}
\tag{3.117}
$$

In this case the received pulses are spread over a long enough time span that individual pulses can be resolved at the receiver. This means that when the transmitter produces a continuous train of pulses, each pulse will be spread by the multipath propagation so that there is intersymbol interference at the receiver. A multipath channel with a large RMS delay spread is sometimes described as **time dispersive**. Intersymbol interference must be dealt with at the receiver much as it is dealt with in a telephone modem. An equalization filter is used to flatten the channel frequency response and thereby undo the effect of the frequency-selective fading. Since the way the received signal level varies as a function of frequency will change when the receiver is moved, the equalizing filter must be adaptive. Adaptive equalization is usually implemented with digital signal processors. Design of adaptive equalizers is a fascinating topic that space does not allow us to treat in this text. We will, however, revisit this topic when we examine the ability of spread-spectrum modulation systems to give good performance in frequency-selective fading conditions.

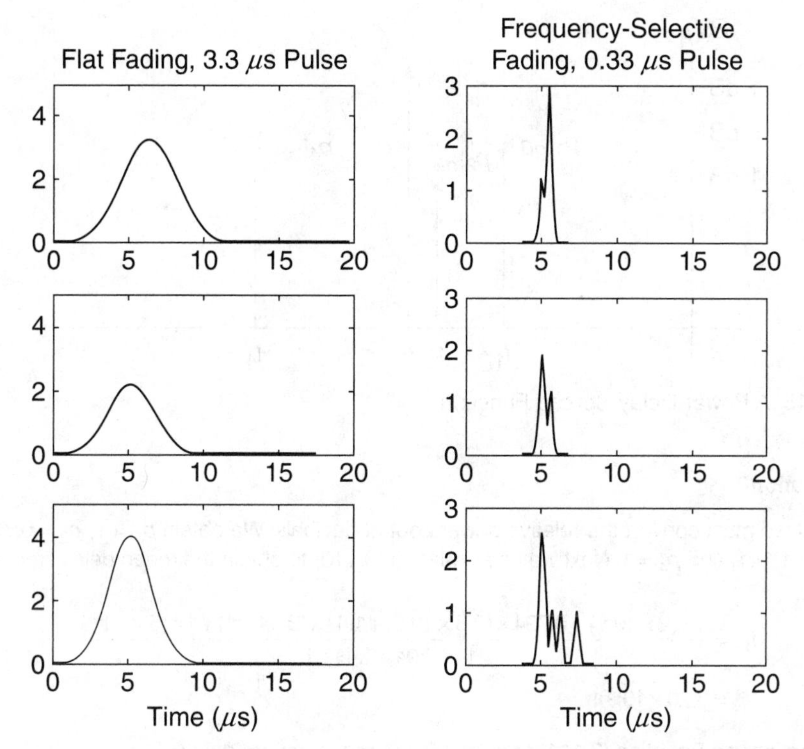

Figure 3.14 Flat and Frequency-Selective Fading in the Time Domain; the RMS Delay Spread Is 0.5 μs

Figure 3.14 shows pulses received over a simulated multipath channel. The RMS delay spread in the simulation was set to $\sigma_d = 0.5$ μs. The left-hand column shows pulses having a width of about $T_{sig} = 3.33$ μs. Since $T_{sig} >> \sigma_d$, this case represents flat fading. Notice that the three pulses shown all have nearly the same shape but different amplitudes. The right-hand column shows pulses having a width of about $T_{sig} = 0.33$ μs. This case shows frequency-selective fading. Observe that individual pulses that arrive over different paths can be resolved.

Example

Figure 3.15 shows the power delay spread for a particular multipath channel. The transmitter and receiver are both stationary. In the figure, the relative powers of the received impulses are $p_1|_{dB} = 0$ dB, $p_2|_{dB} = -1$ dB, $p_3|_{dB} = 1.5$ dB, and $p_4|_{dB} = 0$ dB. The corresponding arrival times are $t_1 = 2$ μs, $t_2 = 2.8$ μs, $t_3 = 3$ μs, and $t_4 = 5$ μs. Determine the RMS delay spread and the coherence bandwidth. If a communication signal using this channel has a bandwidth of $B_{sig} = 30$ kHz, will the signal experience flat fading or frequency- selective fading?

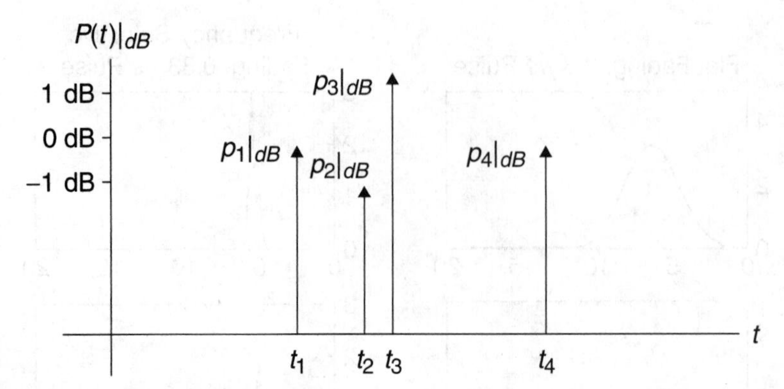

Figure 3.15 A Power Delay Spread Function

Solution

First we must convert the relative powers out of decibels. We obtain $p_1 = 1$, $p_2 = 0.794$, $p_3 = 1.141$, and $p_4 = 1$. Next we use Equation (3.110) to obtain the mean delay time:

$$\langle t_k \rangle = \frac{1\times(2\times10^{-6})+0.794\times(2.8\times10^{-6})+1.41\times(3\times10^{-6})+1\times(5\times10^{-6})}{1+0.794+1.41+1} \tag{3.118}$$
$$= 3.20\times10^{-6}\text{ s.}$$

Then we use Equation (3.111) to calculate the mean-square delay,

$$\langle t_k^2 \rangle = \frac{1\times(2\times10^{-6})^2+0.794\times(2.8\times10^{-6})^2+1.41\times(3\times10^{-6})^2+1\times(5\times10^{-6})^2}{1+0.794+1.41+1} \tag{3.119}$$
$$= 13.7\times10^{-12}\text{ s}^2,$$

and Equation (3.112) to give the RMS delay spread,

$$\sigma_d = \sqrt{13.7\times10^{-12}-(3.20\times10^{-6})^2} \tag{3.120}$$
$$= 1.86\times10^{-6}\text{ s.}$$

Now the coherence bandwidth is given by Equation (3.115):

$$B_{coh} = \frac{1}{5\times(1.86\times10^{-6})} \tag{3.121}$$
$$= 108\text{ kHz.}$$

Finally, for a communication signal with $B_{sig} = 30$ kHz, we have $B_{sig} \ll B_{coh}$, so the fading will be flat.

Microscopic Models 3: A Two-Ray Model with a Moving Receiver

In a real-world communication environment it is unusual for the mobile radio to remain motionless. People using cellular telephones commonly walk or ride in vehicles while talking on the phone. If the mobile unit is in motion during communications, the mobile antenna travels through the standing wave pattern caused by multipath propagation. The effect is the same whether the mobile unit is receiving or transmitting: The received signal level at the mobile unit or at the base station fluctuates as the location of the mobile unit's antenna changes. This is our first encounter with a **time-varying** communication channel. We are concerned with the rate at which the received signal level varies. If the signal level changes slowly with respect to the time that it takes to receive a single pulse, then individual pulses will not be distorted. If the received signal level remains above the receiver's sensitivity, an acceptable quality of service can be maintained regardless of the time-varying fades. On the other hand, if the received signal level changes significantly during reception of a single pulse, then the pulse amplitude will be distorted. One way of dealing with this kind of distortion is to modulate the carrier in such a way that no information is contained in the amplitude of the received signal. We will explore this approach further in the chapter on modulation. In the present and following section we assume that the base station is transmitting and the mobile unit is receiving. This assumption serves to simplify the explanations; the phenomenon is the same regardless of which end of the communication link is in motion.

A moving mobile receiver experiences a Doppler shift in the frequency of the received carrier. The amount of frequency shift depends on both the receiver's speed and the direction of the received wave with respect to the receiver's velocity. In a multipath environment, each propagation path will experience its own Doppler shift. We will see in the sequel that the interval of time over which the received signal remains approximately constant depends on the **Doppler spread**. This is in contrast to the results of the previous section, where we showed that the width of the frequency band over which the received signal level remains approximately constant depends on the delay spread. There is a time-frequency duality at work here. Delay spread (a time spread) is reciprocally related to coherence bandwidth. We will see that Doppler spread (a frequency spread) is reciprocally related to **coherence time**.

As we did in examining delay, we begin with a simple deterministic example. In fact, we begin by assuming a single propagation path (the one-ray case) and derive the expression for Doppler shift. Now Doppler shift is not in and of itself a form of distortion, especially since the frequency shifts encountered in terrestrial wireless systems are small. We include this derivation as a way of introducing notation and of reminding the reader of the way in which Doppler shift depends on receiver speed and direction. Once we have examined the one-ray case, we will move on to the two-ray case in which there is an actual "spread" between the frequencies of the two received signals. We will derive an expression showing how the received signal level varies with time, and we will demonstrate the reciprocal relation between the Doppler spread and the time over which the received signal level remains relatively constant. Having established these fundamental relationships in a simple case, we will proceed to the next section, in which we will use statistical techniques to model a more realistic multipath situation.

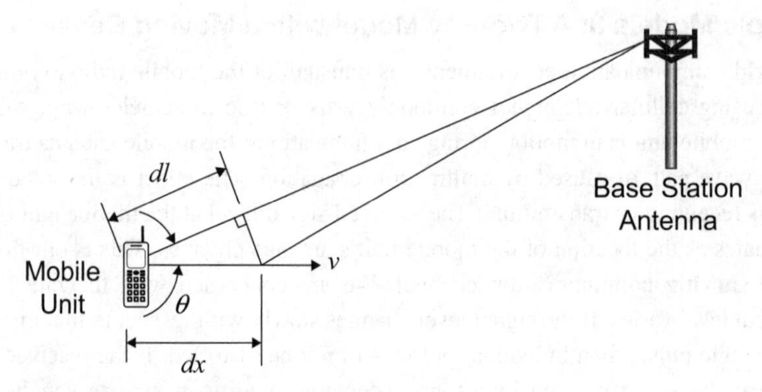

Figure 3.16 Geometry Illustrating the Doppler Shift

Figure 3.16 shows a mobile receiver moving to the right with velocity v. A radio wave is being received from a base station antenna at some distance from the mobile unit. The received wave makes an angle of θ with respect to the velocity of the mobile unit. Let us consider what happens during a very short interval of dt seconds. During this short interval the mobile moves to the right a distance $dx = v\,dt$, as shown in the figure. This change in position causes a shortening of the propagation path from the base station to the mobile. The change in path length is approximately $dl = dx\cos\theta = v\,dt\cos\theta$. Now this change in path length is short enough that the received signal level does not measurably change. Nevertheless, there will be a change $d\phi$ in the phase of the received carrier. This phase change can be calculated by the proportion

$$\frac{d\phi}{2\pi} = \frac{dl}{\lambda}, \tag{3.122}$$

where λ is the wavelength of the received carrier. Solving for $d\phi$, we obtain

$$\begin{aligned} d\phi &= \frac{2\pi}{\lambda} dl \\ &= \frac{2\pi}{\lambda} v\cos\theta\, dt. \end{aligned} \tag{3.123}$$

Even though the distance dx moved by the vehicle is short, as is the path length change dl, the phase change $d\phi$ might be significant. This is because the wavelength λ is also short at the frequencies used for wireless and personal communication systems. Now the rate of change of phase with time is frequency. Therefore the frequency shift ω_d experienced by the receiver is

$$\omega_d = \frac{d\phi}{dt} = 2\pi\frac{v}{\lambda}\cos\theta, \tag{3.124}$$

or, in hertz,

$$f_d = \frac{v}{\lambda}\cos\theta. \tag{3.125}$$

Example

Suppose in the preceding derivation the carrier frequency is f_c = 850 MHz, the receiver velocity is 70 mph, and the angle of the received wave is $\theta = 10°$. Find the Doppler shift f_d.

Solution

We can calculate the wavelength as

$$\lambda = \frac{3\times10^8}{f_c} = \frac{3\times10^8}{850\times10^6} = 0.353 \text{ m}. \tag{3.126}$$

The velocity of the receiver is $v = 31.3$ m/s. Equation (3.125) gives the Doppler shift as

$$f_d = \frac{v}{\lambda}\cos\theta = \frac{31.3}{0.353}\cos(10°) = 87.3 \text{ Hz}. \tag{3.127}$$

Now a frequency shift of 87.3 Hz is insignificant when compared with a carrier frequency of 850 MHz, or even when compared with a channel bandwidth of 30 kHz. We will see next, however, that the importance of the Doppler shift is greatly increased when there are multiple copies of the received signal.

Figure 3.17 shows a signal received over two propagation paths, with a receiver moving with velocity v as in Figure 3.16. We can write the received signal at time t_0 using Equation (3.75) as

$$r(t_0) = A_1\cos(2\pi f_c(t_0 - t_1) + \phi_1) + A_2\cos(2\pi f_c(t_0 - t_2) + \phi_2), \tag{3.128}$$

where

A_1 and ϕ_1 are respectively the amplitude and phase of the signal received over path 1
A_2 and ϕ_2 are the amplitude and phase of the signal received over path 2
t_1 is the propagation delay corresponding to path 1
t_2 is the propagation delay corresponding to path 2
f_c is the carrier frequency

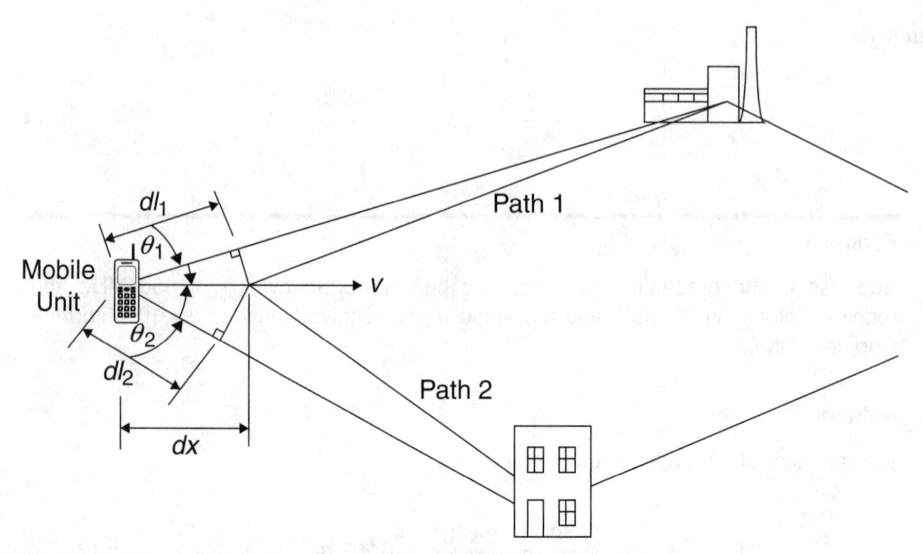

Figure 3.17 A Signal Received over Two Propagation Paths with the Receiver in Motion

If t is a short time interval, during which the receiver has moved through distance dx, then the received signal at time $t_0 + t$ is

$$\begin{aligned} r(t_0+t) &= A_1\cos(2\pi f_c(t_0+t-t_1)+\phi_1+d\phi_1) \\ &\quad + A_2\cos(2\pi f_c(t_0+t-t_2)+\phi_2+d\phi_2) \\ &= A_1\cos(2\pi f_c t+\phi_1'+d\phi_1)+A_2\cos(2\pi f_c t+\phi_2'+d\phi_2), \end{aligned} \tag{3.129}$$

where $\phi_k' = 2\pi f_c(t_0 - t_k) + \phi_k$, $k = 1,2$ are constant phase angles and

$$\begin{aligned} d\phi_1 &= 2\pi\frac{v}{\lambda}(\cos\theta_1)t \\ d\phi_2 &= 2\pi\frac{v}{\lambda}(\cos\theta_2)t, \end{aligned} \tag{3.130}$$

as in Equation (3.123). The phase changes $d\phi_1$ and $d\phi_2$ occur because the lengths of the propagation paths change as the receiver moves. Note that we are using t rather than dt to designate a "short" time increment. Although this change of notation will make the sequel easier to read, it does raise the issue of what we mean by "short." During the time interval t, the receiver moves a distance dx. We assume that dx is short enough that the receiver remains distant from the signal scatterers. We assume that the average loss over the path from the transmitter to the receiver does not change and that the receiver does not move into or out of shadows caused by buildings or other obstructions to propagation. We do not, however, require that dx be short compared with the wavelength λ. During the time interval t the receiver may move a number of wavelengths through the standing wave pattern, so that the received signal level may change significantly.

We now continue from Equation (3.129). As we have done previously, we wish to express the received signal as a single sinusoid

$$r(t) = A\cos(2\pi f_c t + \phi), \tag{3.131}$$

and, also as previously, we are particularly interested in the amplitude A. Using phasors to combine the two sinusoids in Equation (3.129) gives

$$\mathbf{R} = A_1 e^{j(\phi_1' + d\phi_1)} + A_2 e^{j(\phi_2' + d\phi_2)}. \tag{3.132}$$

As in Equation (3.78) we factor out the first term:

$$\mathbf{R} = A_1 e^{j(\phi_1' + d\phi_1)} \left(1 + \frac{A_2}{A_1} e^{j(\phi_2' - \phi_1' + d\phi_2 - d\phi_1)}\right). \tag{3.133}$$

Expanding the exponential in the parentheses using Euler's identity and combining real parts and imaginary parts gives

$$\begin{aligned}\mathbf{R} &= A_1 e^{j(\phi' + \phi_1)} \left(1 + \frac{A_2}{A_1}\cos(\phi_2' - \phi_1' + d\phi_2 - d\phi_1) + j\frac{A_2}{A_1}\sin(\phi_2' - \phi_1' + d\phi_2 - d\phi_1)\right) \\ &= A_1 e^{j(-2\pi f_1 + \phi_1)} \left\{\left(1 + \frac{A_2}{A_1}\cos(\phi_2' - \phi_1' + d\phi_2 - d\phi_1)\right) + j\left(\frac{A_2}{A_1}\sin(\phi_2' - \phi_1' + d\phi_2 - d\phi_1)\right)\right\}.\end{aligned} \tag{3.134}$$

As in Equation (3.80) the magnitude of the phasor is

$$\begin{aligned}A = |\mathbf{R}| &= A_1\sqrt{\left(1 + \frac{A_2}{A_1}\cos(\phi_2' - \phi_1' + d\phi_2 - d\phi_1)\right)^2 + \left(\frac{A_2}{A_1}\sin(\phi_2' - \phi_1' + d\phi_2 - d\phi_1)\right)^2} \\ &= A_1\sqrt{1 + 2\frac{A_2}{A_1}\cos(\phi_2' - \phi_1' + d\phi_2 - d\phi_1) + \left(\frac{A_2}{A_1}\right)^2} \\ &= \sqrt{A_1^2 + A_2^2 + 2A_1A_2\cos(\phi_2' - \phi_1' + d\phi_2 - d\phi_1)},\end{aligned} \tag{3.135}$$

where the phase difference $\phi_2' - \phi_1'$ is constant, but the phase difference $d\phi_2 - d\phi_1$ increases with time t. Using Equation (3.125), we can write

$$\begin{aligned}d\phi_1 &= 2\pi f_{d_1} t \\ d\phi_2 &= 2\pi f_{d_2} t,\end{aligned} \tag{3.136}$$

where f_{d_1} is the Doppler shift in the signal propagating over path 1 and f_{d_2} is the corresponding Doppler shift for path 2. Substituting in Equation (3.135) gives the amplitude of the received signal as

$$A = \sqrt{A_1^2 + A_2^2 + 2A_1A_2 \cos[\phi_2' - \phi_1' + 2\pi(f_{d_2} - f_{d_1})t]}. \tag{3.137}$$

Equation (3.137) should be compared with Equation (3.80). The two equations are nearly identical, except that the Doppler spread $f_{d_2} - f_{d_1}$ has replaced the delay spread $t_2 - t_1$, and time t has replaced frequency f. The dependence of Equation (3.137) on time shows that the received amplitude A will vary as the receiver moves.

Example

Suppose that two received signal components differ in amplitude by 2 dB. The carrier frequency is $f_c = 850$ MHz, and the phase difference is $\phi_2' - \phi_1' = \pi/6$. Further suppose that the receiver is moving at 70 mph, and the received signals arrive from 10° and 150° with respect to the receiver's direction of motion. Find and plot the relative amplitude A/A_1 as a function of time. Determine how long it takes for the received signal level to pass through one complete fading cycle.

Solution

Let us suppose that path 1 produces the stronger signal. Then the 2 dB amplitude difference means that $20\log\frac{A_2}{A_1} = -2$ dB, so $A_2/A_1 = 0.794$. The wavelength λ is given by

$$\lambda = \frac{3\times10^8}{850\times10^6} = 0.353 \text{ m}, \tag{3.138}$$

and the speed of the receiver is $v = 31.3$ m/s. The two Doppler frequencies f_{d_1} and f_{d_2} are given by

$$\begin{aligned} f_{d_1} &= \frac{v}{\lambda}\cos\theta_1 = \frac{31.3}{0.353}\cos(10^\circ) = 87.3 \text{ Hz} \\ f_{d_2} &= \frac{v}{\lambda}\cos\theta_2 = \frac{31.3}{0.353}\cos(150^\circ) = -76.8 \text{ Hz}. \end{aligned} \tag{3.139}$$

The Doppler spread is $f_{d_2} - f_{d_1} = -76.8 - 87.3 = -164$ Hz. Substituting in Equation (3.137),

$$A/A_1 = \sqrt{1 + 0.794^2 + 2\times0.794\cos(\pi/6 - 2\pi164t)}. \tag{3.140}$$

Figure 3.18 shows A/A_1 plotted against time. It is clear that the relative amplitude fades from a peak of 1.8 to a minimum of 0.2 and back as the receiver travels. Compare Figure 3.18 with Figure 3.6. Both of these figures show standing wave patterns but from

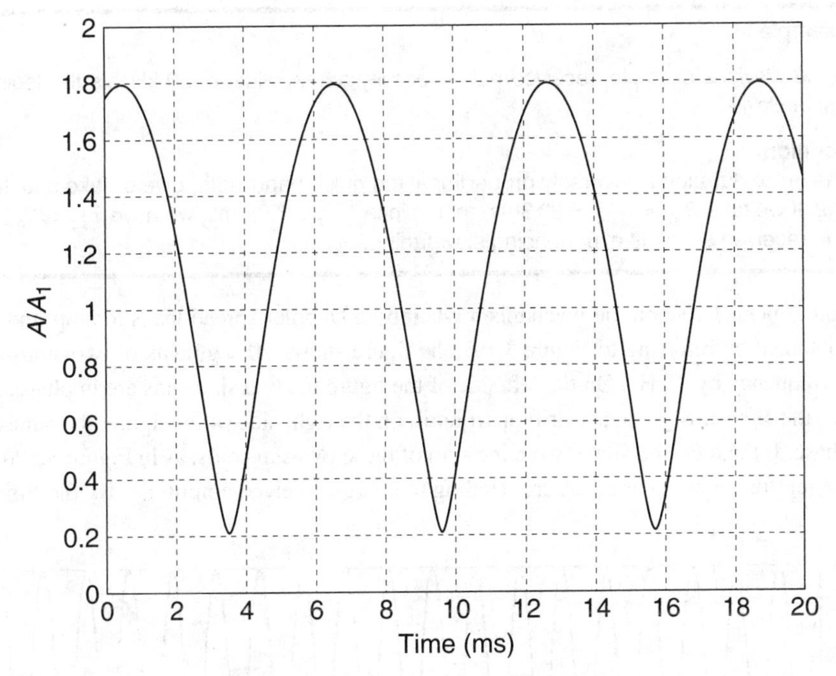

Figure 3.18 Received Signal Relative Amplitude versus Time

different perspectives. In Figure 3.18 the fading pattern is periodic in time. One complete fading cycle will occur when the argument of the cosine in Equation (3.140) varies through 2π radians. If we let t_{n-n} represent the null-to-null time interval, then

$$2\pi\left|f_{d_2}-f_{d_1}\right|t_{n-n}=2\pi$$

$$t_{n-n}=\frac{1}{\left|f_{d_2}-f_{d_1}\right|} \tag{3.141}$$

$$=\frac{1}{164}=6.09\text{ ms.}$$

Equation (3.141) shows that in the two-ray case, the received signal level is periodic in time with a period that is inversely proportional to the Doppler spread. The Doppler spread, in turn, depends on the speed of the receiver. In practice, two cases can be distinguished. Suppose the received signal is made up of pulses, each having a duration T_{sig}. If T_{sig} is very short compared with t_{n-n}, then the received signal level will not change significantly during the arrival of a single pulse. This situation is known as **slow fading**. On the other hand, if T_{sig} is comparable to or larger than t_{n-n}, then the received signal level will change during pulse arrival. This case is **fast fading**.

Example

In the previous example, received pulses occupy a bandwidth of 30 kHz. Is the fading fast or slow?

Solution

The pulse duration is inversely proportional the pulse bandwidth, give or take a factor of 2. If we take $T_{sig} \approx \frac{1}{30\times10^3} = 33.3\ \mu s$, then, since $t_{n-n} = 6.09$ ms, we have $T_{sig} << t_{n-n}$. The received signal is experiencing slow fading.

Another perspective on the mechanism whereby a Doppler spread leads to amplitude fading can be obtained by referring to Figure 3.19. The figure shows 20 segments of two sinusoids that differ in frequency by 25 Hz. On the left side of the figure the two sinusoids are in phase. As time increases, the sinusoids gradually drift apart until, on the right side of the figure, the sinusoids are out of phase. If the received signal were the sum of these two sinusoids, as in Figure 3.19(c), there would be reinforcement near time zero, leading to a large received amplitude. By the time 20 ms

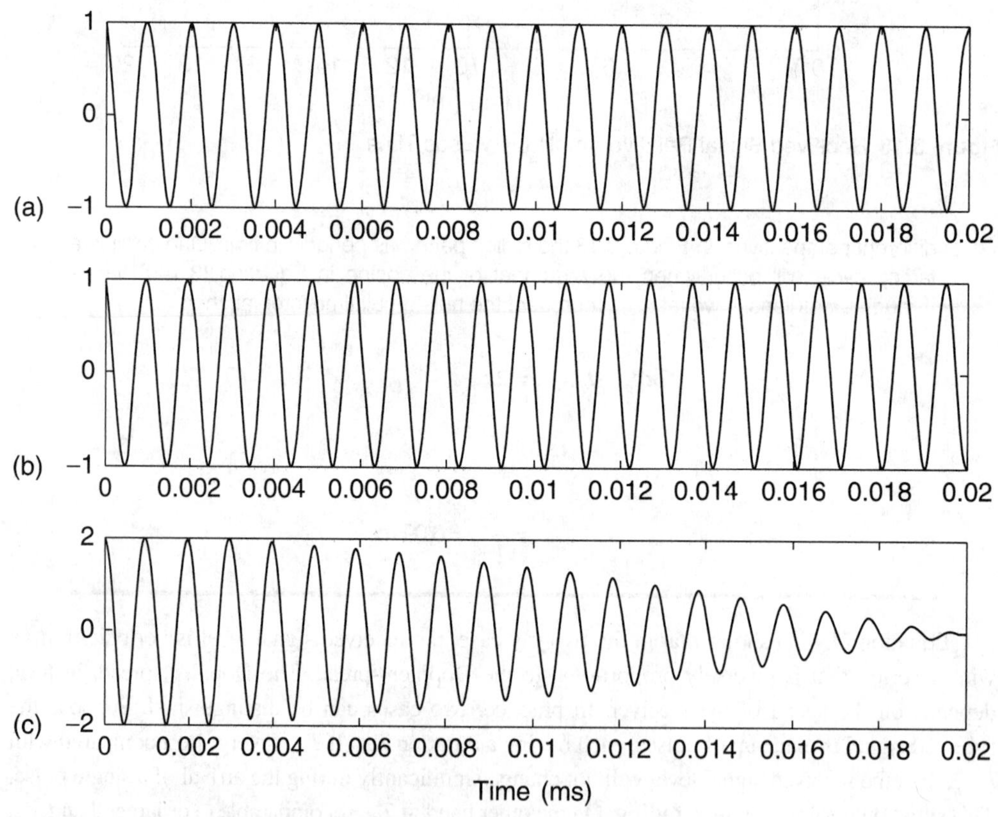

Figure 3.19 Phase Drift in Sinusoids of Slightly Different Frequencies: (a) 1000 Hz; (b) 1025 Hz; (c) the Sum of the Two Sinusoids

has elapsed, however, there would be cancellation, leading to a small received amplitude. This is the variation of amplitude with time that is expressed in Equation (3.137). The larger the frequency difference, the more rapidly the sinusoids will drift apart in phase, and the more rapidly reinforcement will be replaced by cancellation.

Microscopic Models 4: A Statistical Model with a Moving Receiver

A more realistic representation of the signal received in a multipath environment was given in Equation (3.84). In that case, as in the case we wish to consider here, there are many propagation paths, and the received signal is the sum of a large number of sinusoidal components. When the receiver is in motion, each component of the received signal will have its own Doppler shift. We can generalize Equation (3.129) and write

$$r(t) = \sum_{k=1}^{M} A_k \cos(2\pi f_c t + \phi_k' + d\phi_k), \tag{3.142}$$

where

$$d\phi_k = 2\pi f_{d_k} t = 2\pi \frac{v}{\lambda}(\cos\theta_k)t,\ k = 1, \ldots, M. \tag{3.143}$$

In phasor form Equation (3.142) becomes

$$\begin{aligned} \mathbf{R} &= \sum_{k=1}^{M} A_k e^{j(\phi_k' + d\phi_k)} \\ &= \sum_{k=1}^{M} A_k \cos(\phi_k' + d\phi_k) + j\sum_{k=1}^{M} A_k \sin(\phi_k' + d\phi_k) \\ &= X + jY, \end{aligned} \tag{3.144}$$

where $X = \sum_{k=1}^{M} A_k \cos(\phi_k' + d\phi_k)$ and $Y = \sum_{k=1}^{M} A_k \sin(\phi_k' + d\phi_k)$. When M is large, it is appropriate to model the term $d\phi_k$ as a random variable. The statistics of this random variable are determined, through Equation (3.143), by the statistics of θ_k, the signal arrival angle. In a dense urban environment, in which there are many reflections from buildings and no direct line of sight from the transmitting antenna to the receiving antenna, it is reasonable to model θ_k as having a uniform distribution, taking all angles from 0 to 2π with equal probability.

Let us now suppose that $\mathbf{R}_1$ and $\mathbf{R}_2$ represent phasors of the form of Equation (3.144), but measured a small interval of time, τ seconds, apart. (Since the receiver is moving, the two phasors will correspond to slightly different receiver positions.) When M is large, the four random variables X_1, Y_1, X_2, and Y_2 will have a jointly Gaussian probability distribution. This jointly Gaussian distribution is completely characterized by the variances and covariances of X_1, Y_1, X_2, and Y_2. These variances and covariances can be computed from Equation (3.144) by averaging over the arrival angle θ_k. Once we have a joint probability density function for X_1, Y_1, X_2, and Y_2, we can transform to polar coordinates as we did in developing Equation (3.96). The result is a joint probability density function for the received signal amplitudes A_1 at time t and A_2 at time $t + \tau$. Finally, this joint density function can be used to calculate the correlation coefficient ρ between A_1 and A_2. The math is a little bit tricky,[10] but the resulting correlation coefficient is

$$\rho = \mathrm{J}_0^2(2\pi f_d \tau), \tag{3.145}$$

where $\mathrm{J}_0(x)$ is the zero-order ordinary Bessel function of the first kind, and f_d is the maximum Doppler shift given by

$$f_d = \frac{v}{\lambda}. \tag{3.146}$$

The correlation coefficient ρ is plotted against $2\pi f_d \tau$ in Figure 3.20. We can see from the figure that the amplitudes of the signals at times t and $t + \tau$ will be highly correlated if τ is small. The signal amplitudes will become less correlated as τ becomes larger.

At this point we can formally define the coherence time T_{coh} as the interval τ at which the correlation coefficient ρ drops to 0.5. Solving Equation (3.145) with $\rho = 0.5$ gives $\tau \cong \frac{9}{16\pi f_d}$. Writing T_{coh} in place of τ, we have

$$\begin{aligned} T_{coh} &= \frac{9}{16\pi f_d} \\ &= \frac{9}{16\pi \frac{v}{\lambda}}. \end{aligned} \tag{3.147}$$

Over time intervals τ less than T_{coh}, the phase relationship between arriving signal components can be assumed to remain constant. Thus the amplitude of the resultant received signal will also remain approximately constant. Over time intervals τ greater than T_{coh}, the received signal components will have drifted in phase with respect to each other (remember Figure 3.19), and the amplitude of the resultant signal will not remain constant.

The significance of the coherence time is that it allows us to distinguish slow from fast fading for real-world, rather than for two-ray, channels. If the received signal consists of pulses of duration T_{sig}, then slow fading occurs when $T_{sig} << T_{coh}$ and fast fading occurs when $T_{sig} >> T_{coh}$.

---

10. Jakes, Jr., ed., *Microwave Mobile Communications*, 50–52.

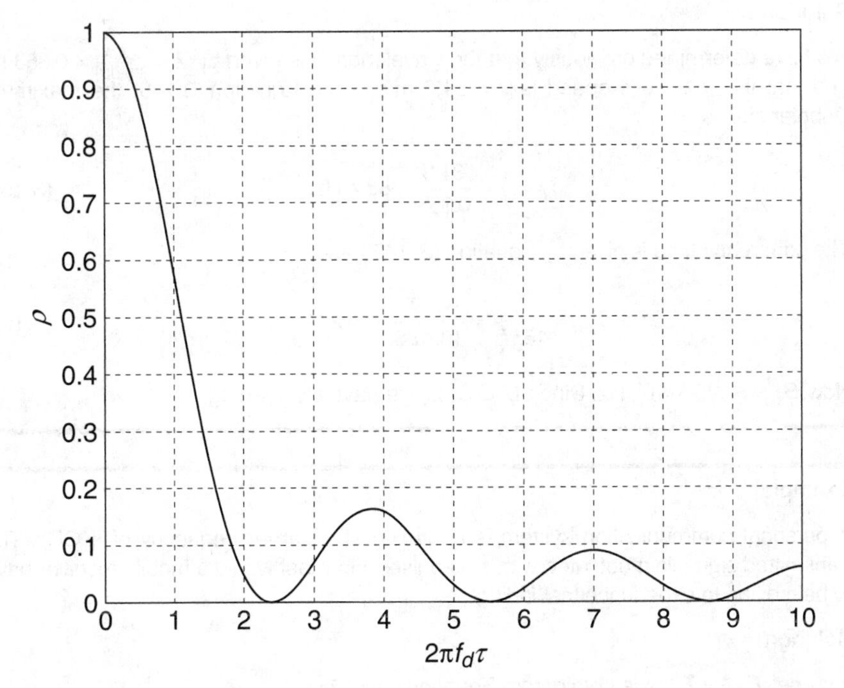

Figure 3.20 Correlation Coefficient for Carrier Amplitudes Separated in Time by τ Seconds

We can derive some additional insight by comparing slow and fast fading in the frequency domain. If we recognize that the modulated signal bandwidth is roughly the reciprocal of the pulse duration T_{sig}, then for slow fading

$$T_{sig} \ll T_{coh}$$

$$\frac{1}{B_{sig}} \ll \frac{9}{16\pi f_d} < \frac{1}{f_d}, \text{ so} \tag{3.148}$$

$$f_d \ll B_{sig}.$$

For fast fading the inequality is reversed:

$$f_d \gg B_{sig}. \tag{3.149}$$

Example

Suppose the carrier frequency of a cellular telephone system operating in an urban environment is 850 MHz and the signal bandwidth is 1.25 MHz. Suppose that the mobile receiver is moving at 70 mph. Find the coherence time. Determine whether the receiver is experiencing fast or slow fading.

Solution

We have determined previously that the wavelength λ is given by $\lambda = \frac{3\times10^8}{850\times10^6} = 0.353$ m, and that the receiver's speed is $v = 31.3$ m/s. From Equation (3.146) the maximum Doppler shift is

$$f_d = \frac{v}{\lambda} = \frac{31.3}{0.353} = 88.7 \text{ Hz}. \tag{3.150}$$

The coherence time is given by Equation (3.147) as

$$T_{coh} = \frac{9}{16\pi f_d} = \frac{9}{16\pi(88.7)} = 2.02 \text{ ms}. \tag{3.151}$$

Now $B_{sig} = 1.25 \times 10^6$ Hz. Since $B_{sig} >> f_d$, we have slow fading.

Example

A personal communication system is operating at a carrier frequency of 5 GHz. The transmitted signal is modulated with 1 μs pulses. How fast would a mobile receiver have to be moving to experience fast fading?

Solution

If we set $T_{coh} = T_{sig}$, we obtain from Equation (3.147)

$$1\times10^{-6} = \frac{9}{16\pi f_d}$$
$$f_d = 179 \text{ kHz}. \tag{3.152}$$

Then from Equation (3.146)

$$v = f_d\lambda = (179\times10^3)\frac{3\times10^8}{5\times10^9} = 10.7\times10^3 \text{ m/s}. \tag{3.153}$$

That is about 24,000 mph!

In concluding our discussion of multipath propagation, it is important to emphasize that coherence bandwidth and coherence time are independent concepts. A signal may experience flat or frequency-selective fading and may independently experience slow or fast fading. That means that there are four possible combinations, any of which may occur (if the mobile unit is moving fast enough).

Area Coverage

Now that we have examined a number of real-world propagation models, we turn our attention to our original goal, the design of a wireless communication system to provide an appropriate quality of service to a virtually unlimited number of users at an affordable cost. In designing a data link to serve a single user at a fixed location, the wireless system engineer must ensure an adequate received signal level at the location of the receiver's antenna. The macroscopic propagation models, such as those by Hata and Lee, allow prediction of the path loss between the transmitting and receiving antennas. The log-normal statistical model allows the designer to add a "fading margin" to the predicted path loss to compensate for uncertainties in the prediction. With estimates of path loss and fading margin, the designer is in a position to begin to specify tower heights, antenna gains, transmitter power, and other parameters that impact the received signal level. When designing a mobile wireless system, however, there is an additional consideration. For a system in which users may be located and moving anywhere within a given distance from a transmitting antenna, point-to-point path loss does not characterize the environment completely enough to support a reliable design. Adequate received signal level must be provided throughout the coverage area. Unfortunately, owing to the unpredictable presence of obstructions to propagation such as buildings and terrain features, it is impossible for the system designer to guarantee that a received signal will always be adequate everywhere within a given range of the transmitter. Therefore, in practice, systems are designed to provide an adequate received signal level over some fraction, say 90% or 95%, of the intended coverage area. Our task in this section is to explore how to predict this area coverage.

A signal is said to be adequate at a given location when the received signal power is above a required minimum level. This minimum level is called the "receiver sensitivity," and it depends on the receiver's noise figure and the minimum signal-to-noise ratio that the receiver requires to demodulate correctly. We will explore the relationship between signal-to-noise ratio and receiver performance in the chapters on modulation. For now we will assume that the receiver's sensitivity is a parameter that is given. Let us denote this receiver sensitivity by $p_{sens}\big|_{dB}$.

Suppose the desired coverage area is a circle of radius R centered on the transmitting antenna. In wireless telephone systems this coverage area is called a "cell." To guarantee adequate service quality at the cell boundary, we must ensure that the received signal power is greater than $p_{sens}\big|_{dB}$ at most locations on the boundary. If the received signal power at distance d from the transmitter is $P_r(d)\big|_{dB}$, we can define the **boundary coverage** as the probability $P_{p_{sens}}(R) = \Pr\{P_r(R)\big|_{dB} > p_{sens}\big|_{dB}\}$, where we have set the distance d to its value on the boundary R. Given a desired boundary coverage, we can use the macroscopic fading model, Equation (3.64), to determine the average received signal power $\overline{P}_r(R)\big|_{dB}$ required at the cell boundary.

By analogy with our definition of boundary coverage, let us define the **area coverage** F_u as the probability that the received signal power exceeds the receiver sensitivity throughout the area of the cell. Intuitively, F_u is the usable fraction of the cell area. In the following paragraphs we will show how the area coverage can be calculated from knowledge of the boundary coverage and the path-loss exponent. The system designer will make these calculations in the reverse order: First a desired area coverage will be determined. This will allow determination of the required boundary coverage. The boundary coverage, in turn, leads to determination of the required average received signal power.

As the receiver moves from the cell boundary toward the transmitting antenna, the received signal strength can be expected to increase. We can predict the average received signal power at an arbitrary distance d from the transmitting antenna by using Equation (3.48), that is,

$$\bar{P}_r(d)\big|_{dB} = \bar{P}_r(R)\big|_{dB} - 10\nu\log\left(\frac{d}{R}\right), \tag{3.154}$$

where, as previously, ν is the path-loss exponent. Note that inside the cell we have $d < R$. This makes the logarithm negative, and $\bar{P}_r(d)\big|_{dB} > \bar{P}_r(R)\big|_{dB}$. Now we can use Equation (3.64) to calculate the probability of obtaining an adequate received signal at any point a distance d from the transmitting antenna:

$$\begin{aligned} P_{p_{sens}}(d) &= \Pr[P_r(d)\big|_{dB} > p_{sens}\big|_{dB}] \\ &= Q\left(\frac{p_{sens}\big|_{dB} - \bar{P}_r(d)\big|_{dB}}{\sigma_{path}}\right) \\ &= Q\left(\frac{p_{sens}\big|_{dB} - \bar{P}_r(R)\big|_{dB} + 10\nu\log(d/R)}{\sigma_{path}}\right). \end{aligned} \tag{3.155}$$

Since the received signal power at any point is a random variable, let us say we have a "success" at a given point when the received signal power turns out to be above the receiver's sensitivity. Equation (3.155) gives the probability of "success" at any point a distance d from the transmitting antenna.

To develop some intuition about how area coverage can be calculated, let us perform a thought experiment. Suppose we identify a small area at distance d and azimuth θ from the transmitting antenna. We want this area to be small enough that the probability of success given by Equation (3.155) is the same at all points within this area. We want this area to be large enough, however, that we can obtain a large number of measurements of received signal strength uniformly spread over the area, such that these measurements will be independent. Let us denote this small area by $\Delta A(d, \theta)$. Now suppose we make N measurements of signal strength in our small area. The number N is proportional to the area, say, $N = \alpha\,\Delta A(d, \theta)$. Now some of these measurements will produce successes; the number of successes will be $P_{p_{sens}}(d)N = P_{p_{sens}}(d)\alpha\,\Delta A(d,\theta)$. We can find the area coverage for our small area by dividing the number of successes by the number of measurements:

$$\begin{aligned} F_u &= \frac{P_{p_{sens}}(d)N}{N} \\ &= \frac{P_{p_{sens}}(d)\alpha\,\Delta A(d,\theta)}{\alpha\,\Delta A(d,\theta)} \\ &= \frac{P_{p_{sens}}(d)\Delta A(d,\theta)}{\Delta A(d,\theta)}. \end{aligned} \tag{3.156}$$

Suppose we now identify two small areas, $\Delta A(d_1, \theta_1)$ at location (d_1, θ_1) and $\Delta A(d_2, \theta_2)$ at location (d_2, θ_2). We sample the received signal strength at N_1 locations in area $\Delta A(d_1, \theta_1)$ and at N_2 locations in area $\Delta A(d_2, \theta_2)$. Again, the number of samples is proportional to the area, that is, $N_k = \alpha\,\Delta A(d_k,\theta_k)$, $k = 1,2$. The total number of successes among our samples is now $P_{p_{sens}}(d_1)N_1 + P_{p_{sens}}(d_2)N_2 = \alpha\{P_{p_{sens}}(d_1)\Delta A(d_1,\theta_1) + P_{p_{sens}}(d_2)\Delta A(d_2,\theta_2)\}$. As previously, the area coverage is the total number of successes divided by the total number of samples, that is,

$$
\begin{aligned}
F_u &= \frac{\alpha\{P_{p_{sens}}(d_1)\Delta A(d_1,\theta_1) + P_{p_{sens}}(d_2)\Delta A(d_2,\theta_2)\}}{\alpha\{\Delta A(d_1,\theta_1) + \Delta A(d_2,\theta_2)\}} \\
&= \frac{P_{p_{sens}}(d_1)\Delta A(d_1,\theta_1) + P_{p_{sens}}(d_2)\Delta A(d_2,\theta_2)}{\Delta A(d_1,\theta_1) + \Delta A(d_2,\theta_2)}.
\end{aligned}
\tag{3.157}
$$

Equation (3.157) is easily generalized. Let us take the entire cell, a circle of radius R, and divide it up into small areas $dA = (dd')(d' \cdot d\theta')$, where we have written the distance from the transmitting antenna as d', generalizing from d_1 or d_2, and the azimuth as θ'. The area coverage is again the total number of successes divided by the total number of samples. Including the entire cell, this is

$$
\begin{aligned}
F_u &= \frac{\int_0^{2\pi}\int_0^R P_{p_{sens}}(d')\,dd'\,d'd\theta'}{\int_0^{2\pi}\int_0^R dd'\,d'd\theta'} \\
&= \frac{\int_0^{2\pi}\int_0^R P_{p_{sens}}(d')\,dd'\,d'd\theta'}{\pi R^2} \\
&= \frac{2}{R^2}\int_0^R P_{p_{sens}}(d')\,dd'\,d'd\theta'.
\end{aligned}
\tag{3.158}
$$

To evaluate Equation (3.158) we substitute Equation (3.155). This gives

$$
F_u = \frac{2}{R^2}\int_0^R Q\left(\frac{p_{sens}\big|_{dB} - \bar{P}_r(R)\big|_{dB} + 10\nu\log(d'/R)}{\sigma_{path}}\right) d'\,dd'. \tag{3.159}
$$

The integral in Equation (3.159) can be simplified![11] If we define

$$
a = \frac{p_{sens}\big|_{dB} - \bar{P}_r(R)\big|_{dB}}{\sigma_{path}} \tag{3.160}
$$

---

11. D. O. Reudink, "Properties of Mobile Radio Propagation above 400 MHz," *IEEE Transactions on Vehicular Technology* VT-23, no. 4 (November 1974): 143–59.

and

$$b = \frac{10 \nu \log e}{\sigma_{path}} \tag{3.161}$$

we can show that

$$F_u = Q(a) + e^{2\frac{1-ab}{b^2}} \left[1 - Q\left(\frac{ab-2}{b} \right) \right]. \tag{3.162}$$

Figure 3.21 shows Equation (3.162) plotted against σ_{path}/ν for various values of the boundary coverage $P_{P_{sens}}(R)$.

Example

Suppose the path-loss exponent is $\nu = 4$ and the path standard deviation is $\sigma_{path} = 8$ dB. If the coverage at the cell boundary is 90%, find the area coverage F_u.

Solution

We have $\sigma_{path}/\nu = 8/4 = 2$. If we follow the $P_{P_{sens}}(R) = 0.9$ curve, then Figure 3.21 gives $F_u = 0.97$.

Alternatively, given $P_{P_{sens}}(R) = 0.9$, we can use Equation (3.64) to determine $\frac{P_{sens}|_{dB} - \bar{P}_r(R)|_{dB}}{\sigma_{path}} = -1.28$, that is, $a = -1.28$ from Equation (3.160). Substituting $s_{path}/\nu = 2$ in Equation (3.161) gives $b = 2.17$. Finally, Equation (3.162) gives $F_u = 0.97$. This result means that the received signal power will exceed the receiver sensitivity over 97% of the cell area.

Example

In a certain cell the path-loss exponent is $\nu = 3.8$ and the path standard deviation is $\sigma_{path} =$ 9 dB. The system designer wishes to provide 90% area coverage; that is, the received signal power must exceed the receiver sensitivity over 90% of the cell area. Determine the coverage probability on the cell boundary. If the receiver sensitivity is −100 dBm, what must be the average received signal power at the cell boundary?

Solution

Two steps are required for the solution. In the first step we start with the given area coverage and calculate the needed boundary coverage. Using the parameters specified in the problem statement, $\sigma_{path}/\nu = 9/3.8 = 2.37$. Then Figure 3.21 with $\sigma_{path}/\nu = 2.37$ and $F_u = 0.9$ gives $P_{P_{sens}}(R) \cong 0.75$.

In the second step we start with the boundary coverage and calculate the required received signal power. Using Equation (3.64) with $P_{P_{sens}}(R) = \Pr[P_r(R)|_{dB} > P_{sens}|_{dB}] = 0.75$ gives $\frac{P_{sens}|_{dB} - \bar{P}_r(R)|_{dB}}{\sigma_{path}} = -0.674$, that is, $-0.674 = \frac{-100 \text{ dBm} - \bar{P}_r(R)|_{dB}}{9 \text{ dB}}$, or $\bar{P}_r(R)|_{dB} = -93.9$ dBm.

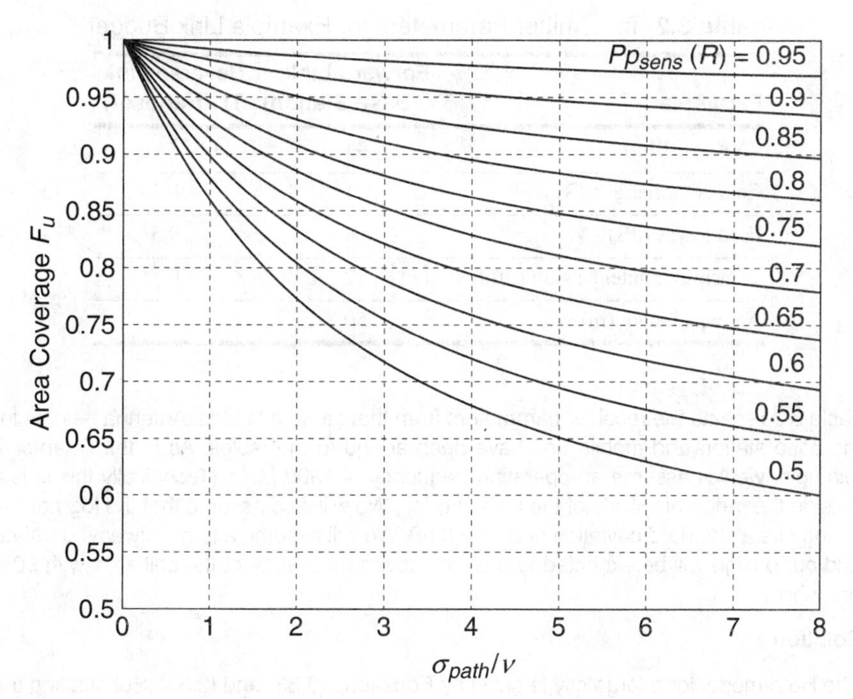

Figure 3.21 Reudink Curves: Area Coverage versus σ_{path}/v for Various Coverage Probabilities at the Cell Boundary

The solution can also be found analytically. With $F_u = 0.9$ and $b = \frac{10 \times 3.8 \times \log(e)}{9} = 1.83$, Equation (3.162) can be solved to give $a = -0.696$. (You can see why we might want to use Figure 3.21.) Then Equation (3.64) gives $P_{p_{sens}}(R) = Q = (-0.696) = 0.76$.

To find received signal power, we use Equation (3.160). This gives $-0.696 = \frac{P_{sens}|_{dB} - \bar{P}_r(R)|_{dB}}{\sigma_{path}} = \frac{-100 \text{ dBm} - \bar{P}_r(R)|_{dB}}{9 \text{ dB}}$, so that $\bar{P}_r(R)\Big|_{dB} = -93.7$ dBm.

The Link Budget

We now bring together some of the ideas developed in this chapter by revisiting the link budget example presented at the end of Chapter 2. In that example a set of transmitter and receiver parameters was given and the free-space path-loss model was used to calculate a maximum range for the data link. Here we will use the Hata model for path loss and include a margin for log-normal fading. We will again calculate the maximum range for the link; this range is the maximum cell radius.

Example

Table 3.2 below repeats the transmitter parameters assumed in the example of Chapter 2.

Table 3.2 Transmitter Parameters for Example Link Budget

Parameter	Forward Link Tx (Base Station)	Reverse Link Tx (Handset)
PA P_{1dB} (dBm)	43	27
Connector losses (dB)	1	0.1
Feed losses (dB)	4	0.5
Transmitter antenna gain (dBd)	12	0
Antenna height (m)	50	1.5

Table 3.3 repeats the receiver parameters from that same example. Antenna heights for the base station and mobile unit have been added to the tables. As in the Chapter 2 example, we will assume an operating frequency of 1900 MHz. (Technically this is just outside the range of validity of the Hata model.) We will also assume that the log-normal fading has a standard deviation of $\sigma_{path} = 8$ dB. We will assume a large city environment, and our design will be expected to produce coverage of 90% of the cell area with 90% probability.

Solution

The Hata model for a large city is given by Equations (3.33) and (3.34). Substituting the frequency and the antenna heights gives median path loss as a function of distance from the transmitter:

$$L_{50}\big|_{dB} = 132 + 33.8\log d. \tag{3.163}$$

Comparing Equation (3.163) with Equation (3.48) or Equation (3.154) shows that the path-loss exponent is $\nu = 3.38$.

Table 3.3 Receiver Parameters for Example Link Budget

Parameter	Forward Link Rx (Handset)	Reverse Link Rx (Base Station)
Receiver antenna gain (dBd)	0	15
Feed losses (dB)	0.5	4
RF unit noise figure (dB)	8	8
RF unit gain (dB)	40	40
Detector required SNR (dB)	12	12
Antenna noise temperature (K)	60	300
Detector bandwidth (kHz)	100	100
Antenna height (m)	1.5	50

Next we use the Reudink curves of Figure 3.21 to determine the required boundary coverage for an area coverage of 90%. Dividing the log-normal standard deviation by the path-loss exponent gives $\frac{\sigma_{path}}{\nu} = \frac{8}{3.38} = 2.37$. Entering Figure 3.21 with $\frac{\sigma_{path}}{\nu} = 2.37$ and $F_u = 0.9$ gives $P_{p_{sens}}(R) \cong 0.75$. Thus we need a 75% probability that the received signal power will exceed the receiver sensitivity at the cell boundary. Using this result in Equation (3.64) gives

$$\frac{p_r\big|_{dB} - \bar{P}_r\big|_{dB}}{\sigma_{path}} = -0.674, \tag{3.164}$$

where $P_r\big|_{dB} = p_{sens}\big|_{dB}$ is the receiver sensitivity and $\bar{P}_r\big|_{dB}$ is the average received signal power at the cell boundary.

The receiver sensitivity was shown in Chapter 2 to depend primarily on the required signal-to-noise ratio at the detector, the receiver noise figure, the signal bandwidth, and the antenna noise temperature. In the Chapter 2 example the values given in Table 3.3 led to $p_{sens}\big|_{dB} = -104.6$ dBm for the forward link and $p_{sens}\big|_{dB} = -104$ dBm for the reverse link. Equation (3.164) then gives $\bar{P}_r\big|_{dB} = -99.2$ dBm for the forward link and $\bar{P}_r\big|_{dB} = -98.6$ dBm for the reverse link.

Now we can put the pieces together by adding up the gains and losses, specifically,

$$\begin{aligned}\bar{P}_r\big|_{dB} = P_t\big|_{dB} + G_t\big|_{dB} - L_{t,connector}\big|_{dB} - L_{t,feed}\big|_{dB} \\ -L_{path}\big|_{dB} + G_r\big|_{dB} - L_{r,connector}\big|_{dB} - L_{r,feed}\big|_{dB} - L_{margin}\big|_{dB},\end{aligned} \tag{3.165}$$

where the antenna gains must be increased by 2.15 dB over the values given in Tables 3.2 and 3.3 to convert from dBd to dBi. As in the Chapter 2 example, we include a design margin of 3 dB. Substituting in Equation (3.165),

$$\begin{aligned}-99.2 \text{ dBm} = 43 \text{ dBm} + (12 \text{ dB} + 2.15 \text{ dB}) - 1 \text{ dB} - 4 \text{ dB} \\ -L_{path}\big|_{dB} + (0 \text{ dB} + 2.15 \text{ dB}) - 0 \text{ dB} - 0.5 \text{ dB} - 3 \text{ dB},\end{aligned} \tag{3.166}$$

which gives $L_{path}\big|_{dB} = 150$ dB for the forward link. Similarly, $L_{path}\big|_{dB} = 136$ dB for the reverse link.

Since the reverse link has the lower maximum path loss, the reverse link will determine the cell radius. Returning to the Hata model in Equation (3.163), we calculate the maximum cell size as 1.35 km. In Chapter 2 the maximum transmitter-receiver separation was calculated as 671 km. Including a real-world path-loss model and a fading margin has a significant effect on the predicted cell radius!

Conclusions

This chapter focused on propagation in the real world, and in particular on how this propagation can be modeled. Effective propagation models are necessary to provide system designers with the means to design reliable communication systems with a marketable quality of service at an affordable cost. In this section we present a brief review of the principal concepts introduced in this chapter.

- The chapter began with an examination of propagation over the surface of the Earth. We showed that the proximity of a conducting surface to the propagation path causes the path-loss exponent to increase from the free-space value of 2 to a value of 4. We also showed that the presence of the Earth's surface causes the transmitter and receiver antenna heights to become relevant. The model predicts that received signal power varies as 20 log (h), where h is the height of either antenna. This "correction" for antenna height shows up again in the Lee propagation model.
- We introduced two empirical models that predict median received signal power in urban, suburban, and rural environments. These models, by Hata and Lee, form the basis for commercial software packages that are used in cellular system design. It is important to point out that there are additional propagation models that space does not allow us to include in this text. In particular, there are specialized models for propagation in "microcells" that have cell radii much smaller than the Lee model limit of one mile. Microcells may be as small as a city block. There are also models that predict propagation in indoor environments, for design of systems to be deployed in shopping malls and offices.
- To account for variations from the median received signal power, particularly variations caused by shadowing of the radio signals, we introduced the log-normal statistical model. The log-normal model gives the designer the ability to specify a fading margin—that is, a level of received signal power above the receiver sensitivity—that will guarantee coverage with a specified reliability. The log-normal model is known as a large-scale or macro-scale model, reflecting the fact that shadowing effects are felt only over distances of many wavelengths.
- Multipath propagation also leads to variations in received signal power. This phenomenon is called small-scale or micro-scale fading, as the signal power can change significantly when the transmitter or receiver is moved over distances of less than a wavelength. We developed the Rayleigh statistical model to represent small-scale fading. This model allows us to calculate an additional fading margin. In practice, base stations protect themselves against small-scale fading by deploying multiple (often two) antennas separated by several feet. The chances that both antennas will experience a simultaneous fade are small. The use of multiple antennas to mitigate the effects of micro-scale fading is called "spatial diversity" reception.
- Multipath propagation also leads to a varying received signal strength as a function of frequency. We introduced coherence bandwidth as a measure of the frequency interval over which received signal strength remains constant. A modulated signal will experience flat fading or frequency-selective fading depending on whether its bandwidth is narrow or wide compared with the coherence bandwidth. Frequency-selective fading is an issue of increasing importance as data is carried over wireless systems at higher rates, leading to increased modulation bandwidths.
- Rayleigh fading and frequency-selective fading are phenomena that occur whether the transmitter and receiver are stationary or in motion. When the transmitter or receiver is in

motion, the received signal strength will vary with time. We introduced coherence time to measure the time interval over which the received signal power remains constant. Information signals whose pulses are short compared with the coherence time experience slow fading, and signals whose pulses are long compared with the coherence time experience fast fading. In cellular and personal communication systems the fading encountered is generally slow. This is because receiver velocities are limited to normal vehicle speeds. It is important to remember that flat/frequency-selective and slow/fast fading are independent phenomena. A receiver may experience any combination: flat and slow, flat and fast, frequency-selective and slow, and frequency-selective and fast fading.

- Area coverage is defined as the fraction of the intended coverage area in which the received signal power is above the receiver sensitivity. Shadowing, as represented by the log-normal fading model, is the principal reason that 100% coverage cannot be attained. We extended our statistical model so that area coverage can be related to the probability of receiving adequate signal power at the cell boundary. Often a system design begins with an area coverage specification. This specification then flows down to requirements on received signal power at a specified distance and subsequently to requirements on individual system components.
- We concluded the chapter by revisiting the link budget example introduced at the end of Chapter 2. We preserved the values of all of the parameters used in that example but added a realistic propagation model to replace the free-space model used originally. We showed that the predicted cell radius over which 90% of the cell area receives an adequate signal with a 90% probability is only 1.35 km in a dense urban environment. This example demonstrates the importance of effective propagation modeling, as the free-space model predicted a cell radius of over 600 km.

In Chapters 2 and 3 we concentrated on the models a system designer uses to ensure an adequate received signal power for a single receiver located somewhere within, or moving through, a coverage area. In Chapter 4 we begin to consider the design of systems intended to provide service to many users. In particular, we introduce the cellular concept, which is intended to allow a virtually unlimited number of users to access a communication system that is limited in transmitted power and total available bandwidth.

Problems

Problem 3.1

A wireless system base station transmits a 20 W signal from an antenna located 20 m above a flat Earth. The base station antenna has a gain of 6 dBi and the system losses are 2 dB. A receiver is located in a delivery truck with an antenna atop the truck at 3 m above the ground. The receiving antenna gain is 2 dBi and the system operates at 1.2 GHz.

A. Find the received signal power when the truck is 20 km from the base station. Express your answer in decibels above a milliwatt. (Hint: Use Equation (3.25).)
B. What would the received signal power be in the absence of ground reflections?
C. Calculate the accuracy of the approximation of Equation (3.17) in reference to Equation (3.16).

Problem 3.2

Assume that the approximations made in Equations (3.9) and (3.10) are valid only for grazing angles of 1° or less. Using the ground reflections diagram illustrated in Figure 3.2 and the base station and receiving antenna heights of Problem 3.1, determine the minimum distance d between the base station and receiver for which Equations (3.9) and (3.10) (and subsequently Equation (3.24)) are valid. Using this distance, find the error resulting from using the approximation of Equation (3.17) for Equation (3.16). Express your answer in decibels relative to the value obtained from Equation (3.16).

Problem 3.3

A wireless system base station operating at 850 MHz transmits a 30 W signal from an antenna located atop a 30 m tower. The base station antenna gain is 6 dBi and the system losses, including cable and connector losses between the transmitter and the antenna terminals, are 3 dB. A handheld receiver has an antenna gain of 3 dBi and is carried at an average height of 1.6 m. The receiver has a bandwidth of 30 kHz and a noise figure of 9 dB.

A. In order to provide an adequate quality of service (QoS) the signal level at the input terminals of the receiver must be no less than 17 dB above thermal noise. Assuming an antenna temperature of 300 K and a flat Earth, calculate the maximum operating range.
B. What would be the maximum operating range if the base station and handheld receiver were located in free space?

Problem 3.4

Assume that the base station of Problem 3.3 is deployed at position A. Suppose that an identical base station is to be deployed at position B, as close as possible to position A without causing undue cochannel (same-frequency) interference. Specifically, assume that a handheld receiver as described in Problem 3.3 is operating along the line joining A and B, and that at the maximum distance from A as determined in Problem 3.3, the signal-to-noise-and-interference ratio must be at least 16 dB (that is, 1 dB of interference is allowed). Assuming a flat Earth, find the minimum distance between A and B.

Problem 3.5

A wireless system base station operating at 1.2 GHz in an urban environment transmits 15 W from the top of a 30 m tower. The base station antenna's gain is 7 dBi and the system losses are 4 dB. A receiver operating in the vicinity of the base station is located at an average height of 1.6 m above a flat Earth and has an antenna gain of 2 dBi.

A. Calculate the median path loss when the receiver is 12 km from the base station using the Hata model.

B. Calculate the received signal power.

C. Calculate the received signal power using Equation (3.25) for the same range. Compare with your answer to part B and discuss the difference.

Problem 3.6

Assume that for the system of Problem 3.5 the receiver has a 30 kHz bandwidth and a 6 dB noise figure and operates in an environment that produces an antenna noise temperature of 290 K. The QoS objectives require at least 18 dB SNR.

A. Find the maximum operating range using the Hata model.

B. Find the maximum operating range using Equation (3.25) and discuss the difference.

Problem 3.7

Find the maximum operating range of the system described in Problem 3.5 and Problem 3.6 for suburban and rural environments.

Problem 3.8

Calculate the median path loss for the system of Problem 3.5 using the Lee model and the parameters describing New York City.

Problem 3.9

Find the maximum operating range for the system described in Problem 3.5 and Problem 3.6 using the Lee model for New York City. Compare the result with that of the Hata model and discuss the reasons for the difference.

Problem 3.10

The QoS for a certain system requires that the received signal power be greater than the receiver sensitivity level with probability P_Q. Express P_Q in terms of the fade margin $f_m|_{dB}$ and standard deviation of the path loss, σ_{path}.

Problem 3.11

A certain receiver has a thermal noise floor of –110 dBm. Suppose that the receiver requires a 15 dB SNR in order to adequately replicate the transmitted signal. The QoS objective requires that the received signal be above the receiver sensitivity level 85% of the time. Assuming σ_{path} = 7 dB, find the required fade margin in decibels. What is the minimum median signal level required at the receiver input to meet the QoS objective?

Problem 3.12

A base station operating at 1.8 GHz transmits a 44 dBm signal from an antenna located at 100 ft (effective height) above the pavement in urban Philadelphia. The transmitting antenna has a gain of 4 dBi and there are 2 dB of system losses between the transmitter and its antenna. The receiving antenna has a gain of 2 dBi and is located on the roof of a car 5 ft above the pavement.

A. Assuming that σ_{path} = 8 dB, calculate the probability that the received signal power is 10 dB above the median value at 10 mi. Use the Lee model.

B. Find the probability that the received signal power at 5 mi is greater than 10 dB above the median value at 10 mi.

Problem 3.13

Assume that the receiver of Problem 3.12 has a bandwidth of 200 kHz and a noise figure of 6 dB operating with an antenna noise temperature of 290 K. What is the maximum operating range for which there is a 90% probability that the received signal level is 15 dB above the receiver's thermal noise floor?

Problem 3.14

A public safety system has the following parameters:

Base station transmitter power = 35 W
Base station transmitting antenna gain = 6 dBi
Operating frequency = 1.2 GHz
System losses = 3 dB
Receiving antenna gain = 3 dBi
Receiving antenna height = 1.6 m
Receiver bandwidth = 30 kHz
Receiver noise figure = 4 dB
Antenna noise temperature = 290 K

The system is to be used in medium-sized city environments with σ_{path} = 6 dB. Adequate intelligibility of emergency communications requires that the received signal power be 12 dB above the receiver's noise floor (i.e., SNR = 12 dB).

A. Find the fade margin required to ensure that the received signal power is above the receiver sensitivity level with 95% probability.
B. Find the median signal level required to meet the 95% QoS objective.
C. What is the minimum effective height of the transmitting antenna required to achieve the QoS objective at 8 km? Use the Hata model.

Problem 3.15

Using Equation (3.80), show that the peaks of a two-ray model occur at

$$f = nf_0 + \frac{\Delta\phi}{2\pi} f_0$$

and the nulls occur at

$$f = nf_0 + \left(\frac{1}{2} + \frac{\Delta\phi}{2\pi}\right) f_0 ,$$

where f_0 is the frequency for which $\lambda_0 = d_2 - d_1$, the path difference, and $\Delta\phi = \phi_2 - \phi_1$.

A. What is the peak-to-peak (or null-to-null) frequency difference?

B. Verify that the expression for the peaks predicts the values observable in Figure 3.6.

Problem 3.16

What is the probability that a signal in a multipath environment exceeds its average power level?

Problem 3.17

By how many decibels must the average SNR exceed the minimum SNR to achieve a 95% probability that the received SNR will be above the minimum in a multipath environment?

Problem 3.18

A certain multipath channel has the power delay spread shown in Figure P 3.1, where $p_1|_{dB} = 0$ dB, $p_2|_{dB} = -3$ dB, $p_3|_{dB} = 3$ dB, and the corresponding times are $t_1 = 3\ \mu s$, $t_2 = 4\ \mu s$, and $t_3 = 6\ \mu s$. Approximate the maximum system data rate that can be supported without the need for equalization.

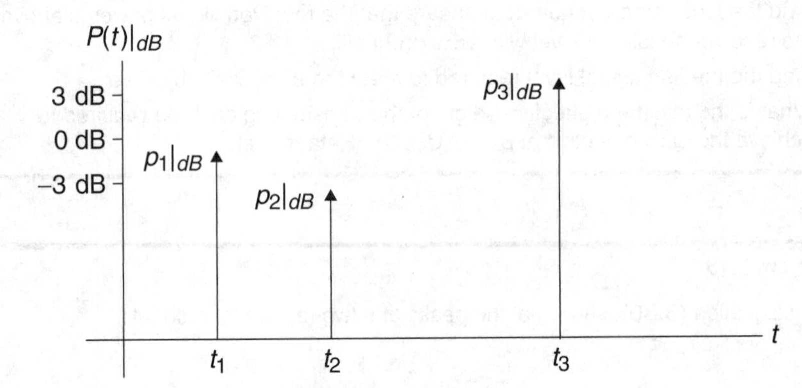

Figure P 3.1 A Power Delay Spread Function

Problem 3.19

A modern digital wireless communication system transmits at a center frequency of 1.9 GHz and has a bandwidth of 1.25 MHz and a data rate of 1.2288 Mbits/s. Assume that there are two signal paths that differ in amplitude by 1 dB and in phase by $\phi_2' - \phi_1' = \frac{\pi}{4}$. The receiver is moving at 60 mph and the signals arrive at –5° and 85° with respect to the direction of travel.

A. What is the Doppler spread?
B. What is the period of a fading cycle?
C. Is the fading fast or slow with respect to the signal? How fast would the receiver have to be moving to experience fast fading?

Problem 3.20

The sensitivity of a certain receiver is –104 dBm. The receiver is to be used in an urban environment where the path-loss exponent is 4.2 and σ_{path} = 9 dB. The system deployment criteria require that the received signal be above the receiver sensitivity over at least 90% of the coverage area. What is the required fade margin?

CHAPTER 4

Radio Frequency Coverage: Systems Engineering and Design

Motivation

The previous two chapters presented the fundamental concepts underlying the behavior of radio frequency propagation in a real-world environment. With these principles as a foundation, we return to the development of a solution to the general systems-engineering challenge posed in Chapter 1, that is, to design a wireless telecommunication system that will

- Support the communication of information from various sources, including speech, text, data, images, music, and video, in urban, suburban, and rural environments with quality approximating that of wired communications
- Be capable of expanding in geographic coverage
- Allow for virtually limitless growth in the number of users
- Support endpoints that are not geographically fixed and that may in fact be moving at vehicular speeds

The challenge as we have stated it is broad and devoid of any quantitative specification of the system's attributes. Nevertheless, the problem statement is not unlike one that a company might pose in the initial stages of developing a product or service.

Product or service development requires the investment of money, physical resources, and people. Sound business practice requires a clear assessment of the benefits to be gained by making the investment. The financial "return on investment" is often used to determine the viability of an investment in a new product or service. Based on an assessment of the return on investment, an enterprise (or, more accurately, the investors in the enterprise) can make informed decisions about the efficacy of proceeding with a new venture. This assessment may include quantifying the market opportunity in terms of the likely number of subscribers and the projected rate of growth in that number, the subscribers' willingness to pay for the service, the time to market (how long it will take to develop the service or product), the estimated cost of development, the projected profit margin, the product life cycle, and the competitive ability of companies offering similar products or services.

As discussed in Chapter 1, systems engineers play an important role, especially in the early process stages. As members of the product definition team, they represent the technical community, providing the insights necessary to ensure that the product concept is realistic and well defined. Systems engineers work with the business team to provide a complete, detailed, and quantitative definition of the product or service to be developed. This usually requires a strong interaction among the members of the technical community to ensure that the required technical competencies are available, the budget and schedule are realistic, and the technical risks are noted and realistically appraised.

Once a product or service has been defined in sufficient detail, key product requirements are assessed and analyzed to determine the high-level design or "system architecture" that best supports the project goals, within the constraints of resources, budget, and schedule. Simulations may be performed to estimate performance and the trade-off of performance against cost for various architectural alternatives under project constraints.

A system architecture may include a high-level technical description of the system and major subsystems along with their key functions and parameters. Development of a system architecture is often led by systems engineers working in collaboration with appropriate members of the technical team and the business team. In the next section we consider the overall system approach, or architecture, that might be used to implement the key characteristics of the system identified in our problem statement.

Requirements Assessment and System Architecture

As we proceed to investigate solutions to our stated problem, we will introduce some of the specific parameters and characteristics encountered in modern systems. We will also introduce realistic constraints as they are needed and when sufficient background has been presented to make them meaningful.

As a first consideration, we note that the allowable frequency range over which a proposed system may operate is usually fixed by government regulation and, in some cases, by international treaties. This implies that the operating frequency bands are finite and predetermined. In the United States, civilian communications policy is administered by the FCC, an agency created by Congress in 1934. Communications regulations are published in Volume 47 of the Code of Federal Regulations (CFR). The regulations pertaining to unlicensed radio services appear in Part 15 (designated 47CFR15). Each licensed radio service has its own part. Wireless services are assigned specific frequency bands to limit interference and ensure the reliability of communication services. Emission of radio frequency energy outside of the assigned frequency band must be limited in accordance with the rules governing the license. Licenses may also impose restrictions that vary by geographic area.

In most cases a wireless service is assigned a continuous range of frequencies. Let us designate this range as f_{sl} to f_{su}, where f_{sl} and f_{su} are the lower and upper limits, respectively, of the operating frequency range. Let the system bandwidth be denoted B_{sys}, where $B_{sys} = f_{su} - f_{sl}$. The fraction of this bandwidth to be used by an individual subscriber depends on the information source (speech, music, data, video, etc.) and on the quality of service to be supported. We suppose that the band f_{sl} to f_{su} is divided into N_{chan} subbands or "channels," each of bandwidth B_{chan}. We understand B_{chan} to be the minimum bandwidth required to convey the required information in both directions at the required quality of service (QoS) between two endpoints. Let the center frequency of each channel be given by

$$f_k = f_{sl} + \frac{B_{chan}}{2} + kB_{chan}, \quad k = 0,2,\ldots,N_{chan} - 1. \tag{4.1}$$

To simplify our discussions we also assume that each channel can support only one two-way radio link between a pair of endpoints at any instant of time. Therefore only N_{chan} simultaneous radio connections can be supported in the given spectrum B_{sys}. For systems of interest to us B_{chan} is much smaller than B_{sys}, so there are many available channels.

Example

Radio spectrum is a limited resource. Like land, "they aren't making any more of it." Although new technology allows the use of ever higher frequencies, the general pattern is that spectrum can be assigned to a new radio service only if it is taken away from an existing radio service. In 1983 the FCC allocated 40 MHz of spectrum for a new cellular telephone system. Six years later an additional 10 MHz was added. The allocated spectrum is in the bands 824–849 MHz and 869–894 MHz. This spectrum had previously been allocated to UHF television but became available when the spread of cable television reduced some of the demand for large numbers of broadcasting channels.

Each channel in the Advanced Mobile Phone Service (AMPS) consists of a 30 kHz "forward" link from the base station to the mobile unit and a 30 kHz "reverse" link from the mobile unit to the base station. All of the reverse links are contained in the band 824–849 MHz, and all of the forward links are in the band 869–894 MHz. In each two-way channel, the forward and reverse subchannels are separated by 45 MHz.

Systems that divide an allocated spectrum into channels that are to be shared among many users (as described previously) are classified as **frequency-division multiple access** (FDMA) or **frequency-division multiplexing** (FDM) systems. There are other ways to allocate channels among many users. For example, the entire spectrum allocation may be dedicated to a single user for a short period of time, with multiple users taking turns on a round-robin basis. In a later chapter we will discuss this approach and other alternatives for implementing multiple access. Our intention in this chapter is to focus on the cellular concept, a concept that is perhaps most easily understood in the context of an FDMA system.

Given a system bandwidth B_{sys} and a channel bandwidth B_{chan}, the number of distinct channels available is

$$N_{chan} = \frac{B_{sys}}{B_{chan}}. \tag{4.2}$$

It should be self-evident that the number of subscribers that a system can support is related to the number of available channels. The actual relationship between number of users and N_{chan}, however, is not so obvious. In a telephone system no user occupies a channel continuously, so a channel can support more than one user. In general, the number of users that can be supported is much greater than the number of channels. We will find that user behavior plays an important role in determining the ultimate capacity of an N_{chan}-channel system. In the early stages of development of the public telephone system, it was recognized that a direct connection between every pair of subscribers was impractical and unnecessary. The concept of "traffic engineering"—that is, planning the allocation of resources to allow these resources to be effectively shared—is central to the design of multiple-user systems. We will discuss the basics of traffic engineering later in this chapter, at which time we will answer the question of how many subscribers N_{chan} channels can support.

The simplest architecture for a wireless system might be to locate a single base station near the center of a coverage area and to adjust the transmitted power level and antenna height so that signal levels are sufficient to ensure the required quality of service within the coverage area. This is the common configuration for radio systems serving police and fire departments and for tow-truck and taxicab companies. Early mobile telephone systems were also designed along these lines. The Improved Mobile Telephone Service (IMTS), introduced in the 1950s and 1960s, is an example. Unfortunately by 1976 the IMTS system in New York City, operating with 12 channels, was able to serve only about 500 paying customers. There was a waiting list of over 3700 people, and service was poor owing to the high blocking (system busy) rate.

Although the number of subscribers (and the quality of service) could be increased by adding channels, systems of this design are limited in coverage area by the maximum practical transmitter power and the sensitivity of the receiver. This kind of system is referred to as being "noise limited." The performance within a given coverage area can be increased by adding repeater base stations, but doing so increases the cost of the service without increasing the number of customers who will share that cost. Increasing the coverage area may increase the number of potential customers, but without increasing the system's ability to provide them service. Without additional spectrum, adding more customers will degrade the system blocking rate. Clearly, neither of the goals of being *capable of expanding in geographic coverage* or allowing for *virtually limitless growth in the number of users* can be met with this approach.

Since spectrum is limited, the most promising avenue for increasing the number of users is to utilize the available spectrum more efficiently. Specifically, we wish to increase the number of subscribers who can simultaneously use a given channel within the specified coverage area. Normally, allowing multiple users to transmit on the same channel at the same time increases interference. The key, then, is to find ways to reduce, or eliminate, this so-called cochannel interference. One approach might be to geographically isolate the users of a given channel by physically separating them. This is the basic concept underlying development of the first cellular systems.

Suppose instead of a single high-powered, centrally located base station we envision many low-powered base stations, each serving only a portion of the desired coverage area. The smaller coverage area served by a single base station is called a **cell**. Since the area of a cell is small, each cell will contain only a limited number of subscribers. We can therefore distribute the available channels among the cells so that each cell has only a subset of the complete set of channels. Now if two cells are far enough apart, they can be assigned the same group of channels. This allows every channel to be reused, possibly many times, throughout the system's coverage area. If the same channel supports two different customers at the same time in different parts of the system, however, there is a potential for cochannel interference. We shall see that the system's ability to withstand cochannel interference and the size of the cells are the limiting factors in determining the number of subscribers that can be supported. A system with this configuration is said to be "interference limited." An important consequence of dividing the service area into cells is the need to transfer a call from one base station to another as a mobile unit moves through the service area. This transfer is called a "handoff" or "handover" and will be discussed in a subsequent section of this chapter. In the next section we will describe the cellular concept in some detail and demonstrate its ability to provide substantial, if not limitless, growth in both coverage and user capacity.

Cellular Concepts

Interference control is the key to allowing reuse of channels within a given geographic area. We therefore begin our discussion of cellular concepts by developing a simple model to predict the interference levels caused by geographically separated transmitters operating on a common channel.

Consider two cells whose base stations (BS_1 and BS_2 respectively) are separated by a distance D. Each base station has a circular coverage area of radius R as shown in Figure 4.1. Assume that the propagation environment is uniform within and completely surrounding the two cells and that the path-loss exponent is ν. Let the two base stations receive on the same channel, with center frequency f_1. Also shown in Figure 4.1 are two mobile units, MU_1 and MU_2, served by base stations BS_1 and BS_2 respectively. The mobile units are located at the boundaries of their respective cells, each at distance R from its base station. To keep the discussion simple, we consider transmission on the reverse channel, from mobile unit to base station, only.

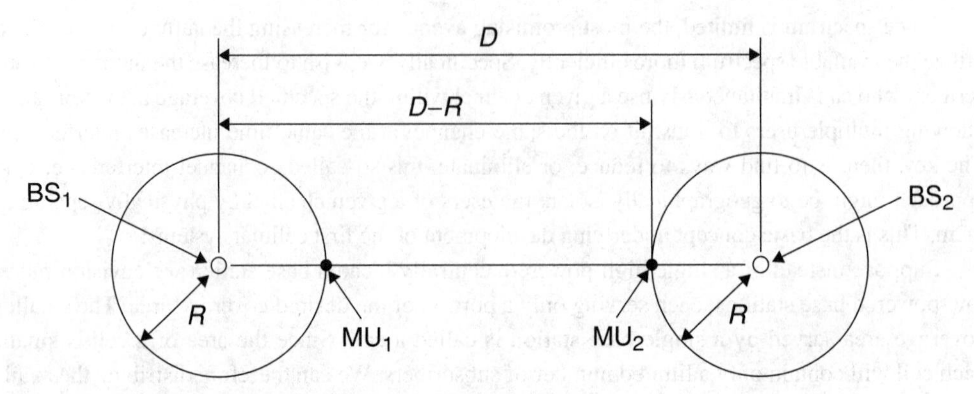

Figure 4.1 Cochannel Interference from a Single Source

From our work in Chapter 3, the level of the signal received at base station BS_1 from mobile unit MU_1 can be expressed as

$$P_1 = \frac{K_1}{R^\nu}, \tag{4.3}$$

where K_1 is a constant that incorporates the transmitter power, the antenna gains, the antenna heights, and so on. The received signal level at base station BS_1 from mobile unit MU_2 can be written

$$P_2 = \frac{K_2}{(D-R)^\nu} \cong \frac{K_2}{D^\nu}. \tag{4.4}$$

Now from the perspective of base station BS_1, P_1 represents "signal" and P_2 is cochannel interference. The signal-to-noise-and-interference ratio at base station BS_1 is

$$\frac{S}{I+N} = \frac{P_1}{P_2 + P_n} = \frac{\frac{K_1}{R^\nu}}{\frac{K_2}{D^\nu} + P_n}, \tag{4.5}$$

where P_n is thermal noise referred to the base station input. When the receiver noise is significantly greater than the interference—that is, when $P_n >> P_2$—we describe the system as "noise limited," as explained earlier. For this case,

$$\frac{S}{I+N} \cong \frac{P_1}{P_n} = \frac{K_1}{R^\nu P_n}. \tag{4.6}$$

When the received noise is significantly less than the interference—that is, when $P_n << P_2$—the system is "interference limited." Then we can write

$$\frac{S}{I+N} \cong \frac{P_1}{P_2} = \frac{K_1 D^\nu}{K_2 R^\nu}. \tag{4.7}$$

For identical mobile units transmitting at the same power level, Equation (4.7) becomes

$$\frac{S}{I} = \left(\frac{D}{R}\right)^{v}. \tag{4.8}$$

The locations of the mobile units in Figure 4.1 have been chosen to illustrate the worst case of **signal-to-interference ratio** (SIR). Mobile unit MU_1 is at the perimeter of its cell, as far as possible from the base station serving it. Mobile unit MU_2 is also at the perimeter of its cell, as close to base station BS_1 as it can get while remaining in the cell served by BS_2. It should be evident that for this geometry the signal-to-interference ratio would be the same if we calculated it at base station BS_2 or at either mobile unit.

In a realistic cellular system there are likely to be more than two cells, and cochannel interference may come from more than one source. If there are J cells surrounding base station BS_1, each containing a mobile unit, and all of these mobile units are transmitting on the same channel, then the signal-to-interference ratio at BS_1 can be written

$$\frac{S}{I} = \frac{P_1}{\sum_{j=2}^{J+1} P_j}, \tag{4.9}$$

where P_j, $j = 2,\ldots,J+1$ represent cochannel interference from mobile units in the surrounding cells. In the discussion that follows, we will see that in the most important case all of the interference sources are identical, and all are located at distances approximately D from the base station BS_1 receiver. In this case Equation (4.9) becomes

$$\frac{S}{I} = \frac{P_1}{\sum_{j=2}^{J+1} P_j} = \frac{P_1}{JP_2} = \frac{1}{J}\left(\frac{D}{R}\right)^{v}. \tag{4.10}$$

Equation (4.10) suggests that the ratio of separation distance D to cell radius R is the predominant factor in determining the signal-to-interference ratio $\frac{S}{I}$. The ratio $Q = \frac{D}{R}$ is usually termed the "frequency reuse ratio" or the "interference reduction factor." Quality of service objectives for the cellular system will usually dictate that some minimum signal-to-interference ratio, say, $S/I = \gamma_{min}$, be met throughout the area. This normally implies that D must exceed R by some factor greater than 1. Base stations BS_1 and BS_2 are then separated by enough distance that additional cells can be deployed between them. To avoid increasing the cochannel interference, these additional cells must operate on channel frequencies other than f_1. In this way an entire geographic area can receive contiguous coverage, while allowing frequency reuse at an acceptable level of cochannel interference.

When the number of available channels N_{chan} is sufficiently large, the channels may be organized into groups called "channel sets," F_i, $i = 1,\ldots,N$, and deployed among the cells as suggested by Figure 4.2. The assignment of sets of frequencies to each cell allows more than one

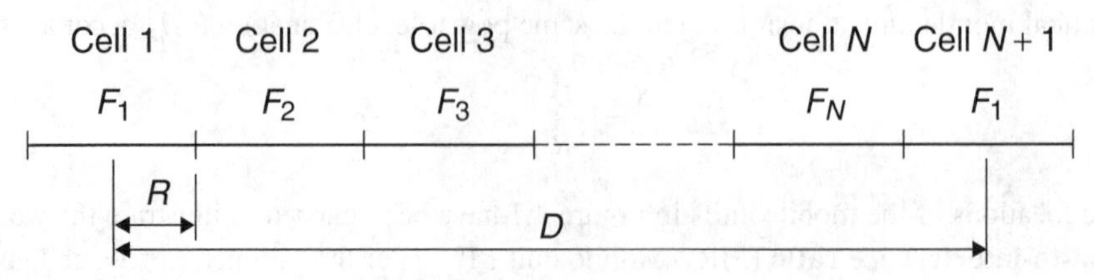

Figure 4.2 Deployment of Channel Sets in Neighboring Cells

subscriber to be serviced at a time in any given cell. The appropriate size of a channel set and the number of distinct channel sets then become design parameters.

In the single-base-station approach, popular before the introduction of frequency reuse, the location of users within the coverage area was not a matter of concern. In a cellular system the geographical distribution of users may matter. Since the allocated group of channels is broken up into channel sets, each cell can service only a limited number of customers. If all of the customers happen to congregate in a single cell, it is likely that there will not be enough channels available in that cell to provide adequate service. In real-world cellular systems the downtown areas of cities are more likely to experience a higher density of customers, at least during the working day, than are the peripheral areas. The higher customer density in some cells can be managed either by assigning more channels to these cells or by making these cells smaller so that they include fewer customers. Increasing the number of channels in a channel set causes a decrease in the number of channel sets that are available. The geometry of Figure 4.2 suggests that as the number of channel sets decreases, so does the separation distance D, leading to an increase in cochannel interference. On the other hand, decreasing the cell radius R requires more cells to cover the market area, leading to an increased investment in base station towers and transmitters. These considerations will be examined in detail later. For the present we will accept the simplest case and assume that subscribers are uniformly distributed throughout the market area.

Our design objectives require that the cellular system be capable of growth in both coverage area and user capacity. This implies that our system must have an ability to add cells to the periphery to expand the service area, and also an ability to add cells within the coverage area to increase the user capacity. It is certainly possible to lay out cells and assign channels in a way that is individually crafted for a given geographic area and customer distribution, but adding new cells for growth would then require custom reengineering. Such ad-hoc design could easily result in a need to redistribute cells and relocate base stations. Construction of a base station is costly and, in addition to the cost of erecting a tower, may involve the purchase of real estate and the approval of zoning boards. Consequently the capability to expand must be designed into the cellular system from the beginning. To facilitate expansion it is convenient to deploy cells that are uniform in shape and are located on a symmetric grid. We will now investigate the extent to which such a regular layout is possible and the principles on which the cell layout is based.

Given a specified service area, placing cells to cover that area requires that some assumption be made about the shape of the coverage regions of the individual base stations. The simplest

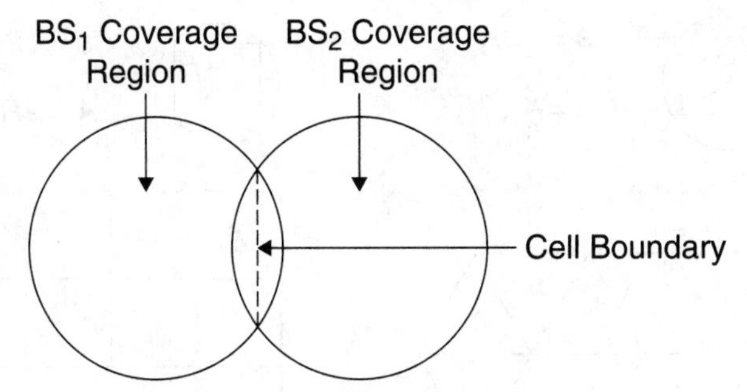

Figure 4.3 The Boundary between Two Circular Cells

assumption is that coverage regions will be circular, based on the principle that equal-level signal contours surrounding a transmitting antenna are circles with the base station at the origin. Given what we know about propagation, however, it is unlikely that real-world coverage areas will be circular in any but a statistical sense. In any specific instance, the shape of a coverage region is not circular but is rather somewhat amorphous and highly dependent on terrain and the location of surrounding obstacles. Keep in mind, however, that our objective in configuring cells is to provide a high likelihood that a mobile unit will be adequately served by its nearest base station. To accomplish this purpose, determining the shape of the actual coverage area is not as important as defining the boundaries over which the mobile unit will operate while communicating with the specific base station. Our approach, then, is to initially assume that the base station coverage areas are circular and to focus on developing an efficient method for locating and growing the cells. Historical experience has shown that once the cells are deployed, system parameters may be adjusted to provide adequate performance in the vast majority of circumstances without the need for extraordinary engineering efforts.

Although we will assume that the coverage regions are circular, a little reflection will show that it is not possible to uniformly cover a geographic area with circles without having gaps or overlaps in coverage. We naturally seek to determine a layout pattern that has no gaps. Also, to ensure that the number of cells deployed is minimal, we wish to minimize the overlaps. Figure 4.3 shows two overlapping circular coverage regions. In the region of overlap there is a straight-line boundary that divides the locations for which the signal from base station BS_1 is stronger than the signal from base station BS_2 and vice versa. Consequently, if we cover our market area with overlapping circular coverage regions arranged on some kind of a regular grid, the actual cells will be polygons. Now it turns out that there are only three regular polygonal shapes for which it is possible to completely "tile" an area without overlaps or gaps. These shapes are the equilateral triangle, the square, and the hexagon: regular polygons of three, four, or six sides, respectively. Figure 4.4 shows the tile pattern for each of these polygons and also shows how a single polygon can be inscribed in a circle of radius R, the cell (or coverage) radius. Observe that within the circle of radius R the area covered by the hexagonal pattern is greatest.

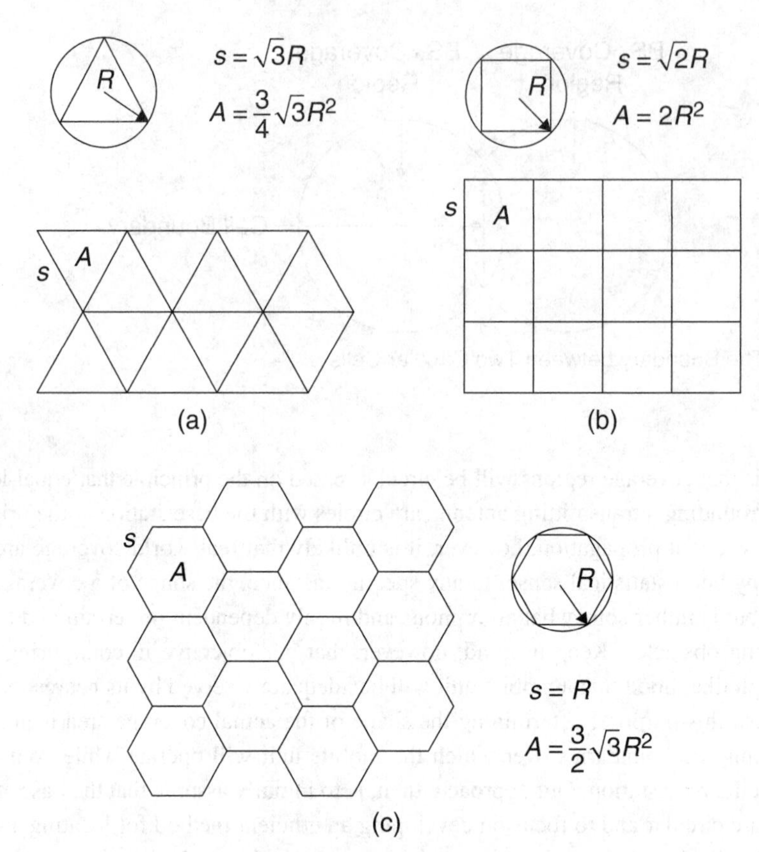

Figure 4.4 Covering a Plane Area with Regular Polygons: (a) Equilateral Triangles; (b) Squares; (c) Hexagons

Also observe that overlap is minimized when circular cells are deployed along a hexagonal grid as is shown in Figure 4.5. The hexagonal layout is evidently the most economically efficient one, as it requires the fewest cells to cover a given area. For these reasons, a hexagonal cell layout is chosen as the basis for designing cellular systems. In the discussion that follows, we will treat cells as having hexagonal shapes. Although this is never precisely true in fact, the assumption provides a means for developing concepts about frequency reuse and cell size that may be applied in practice. Further, the hexagonal layout does provide a starting point for real-world design. We therefore proceed to discuss the geometrical properties of a hexagonal grid as they relate to cell layout and frequency reuse.

Figure 4.6 shows a hexagonal cell and some of its neighbors. We assume that all of the cells have radius R and that the base stations are located in the centers of the cells. The hexagonal geometry dictates that adjacent cells are located at multiples of 60° surrounding any given cell, and the

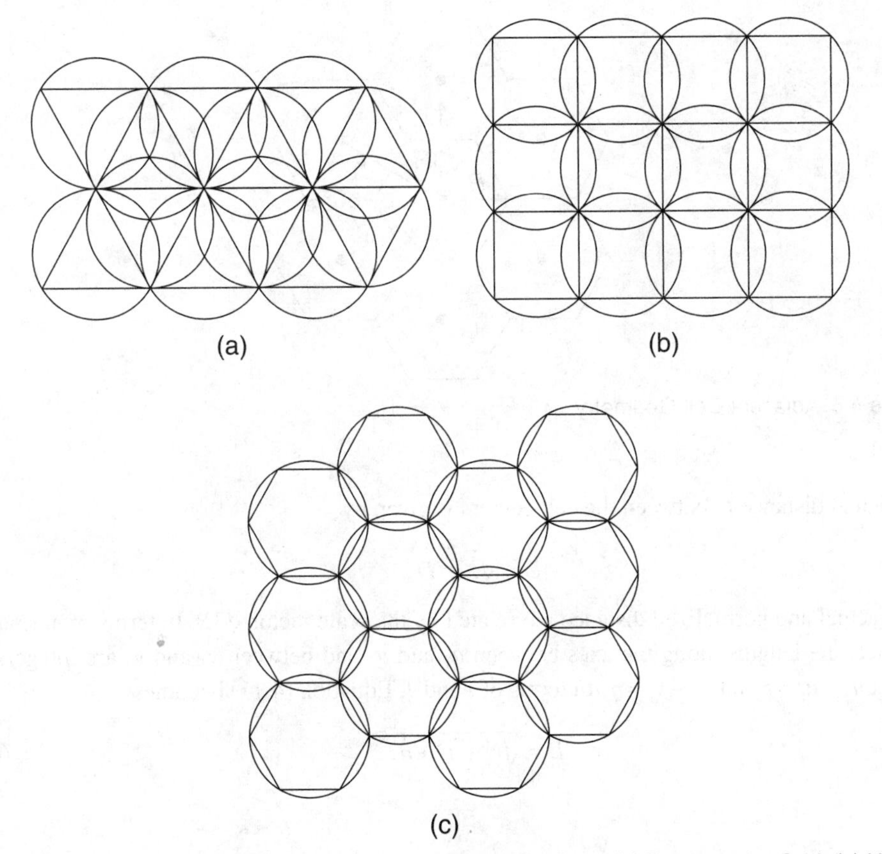

Figure 4.5 Overlap in Circular Cells Using a (a) Triangular Grid; (b) Rectangular Grid; (c) Hexagonal Grid

separation between adjacent cell centers is $\sqrt{3}R$. Following V. H. MacDonald[1], we locate position in the array of cells by choosing a coordinate system (u, v) such that the positive coordinate axes intersect at a 60° angle and the unit distance along either axis is equal to $\sqrt{3}R$, which is the separation between adjacent cells. Figure 4.7 shows these coordinate axes. The normalized distance $\hat{D}$ between any two arbitrary cells whose centers are located at the coordinates (u_1, v_1) and (u_2, v_2) can be found using the law of cosines as illustrated in Figure 4.8. We obtain

$$\begin{aligned}\hat{D} &= \sqrt{(u_2-u_1)^2+(v_2-v_1)^2-2(u_2-u_1)(v_2-v_1)\cos 120^\circ} \\ &= \sqrt{(u_2-u_1)^2+(v_2-v_1)^2+(u_2-u_1)(v_2-v_1)}.\end{aligned} \tag{4.11}$$

1. V. H. MacDonald, "The Cellular Concept," *Bell Systems Technical Journal* 58, no. 1 (January 1979): 15–41.

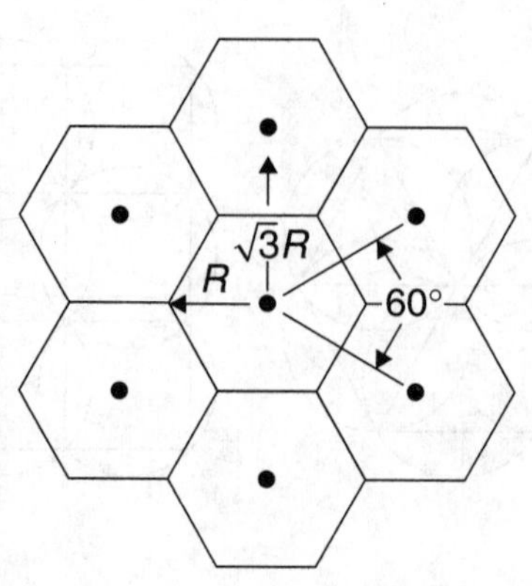

Figure 4.6 Adjacent Cell Geometry

The actual distance D between the cell centers is given by

$$D = \sqrt{3}R \cdot \hat{D}, \tag{4.12}$$

since actual and normalized distance are related by the scale factor $\sqrt{3}R$. In terms of normalized distance, the lengths along the axes between u_1 and u_2 and between v_1 and v_2 are integers. We can let $u_2 - u_1 = i$ and $v_2 - v_1 = j$. In terms of i and J, Equation (4.11) becomes

$$\hat{D} = \sqrt{i^2 + j^2 + ij}. \tag{4.13}$$

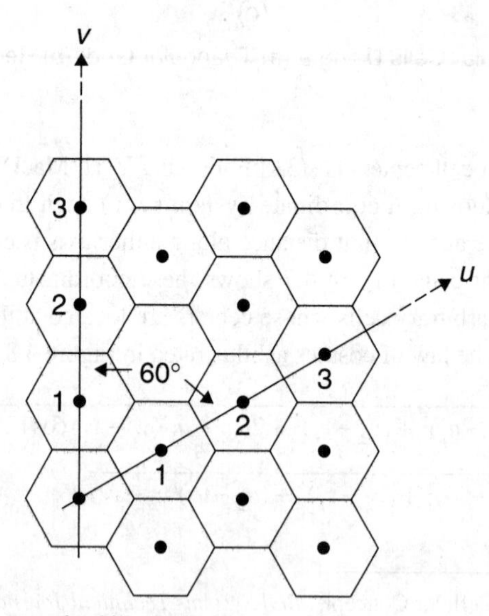

Figure 4.7 Coordinate System for Hexagonal Layout

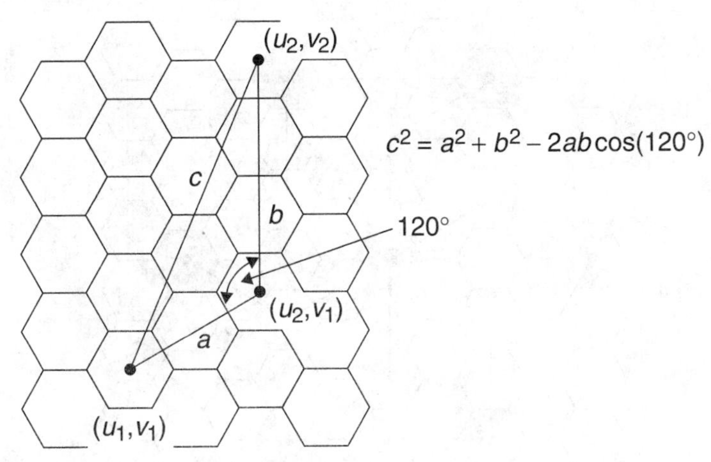

Figure 4.8 Computation of Separation Distance

If we choose the center of any cell as a reference location, then the distance to the center of any other cell can be expressed as a function $\hat{D}(i,j)$ of the integers i and j. The orientation of the coordinate system is arbitrary, so that for any i and j there will be six cells whose centers lie at distance $\hat{D}(i,j)$ from the reference. These cells will surround the reference at multiples of 60°.

Figure 4.9 illustrates this geometry for the case $i = 2, j = 2,\ \hat{D}(2,2) = \sqrt{2^2 + 2^2 + 2 \cdot 2} = 2\sqrt{3}$. Note that the centers of the six cells at distance $D(i, j)$ from the reference form the corners of a large hexagon whose radius and sides have lengths equal to the separation distance $\hat{D}(i,j)$. The location of any one of the six corners of the large hexagon can be found by starting at the reference point, moving i cells in a direction perpendicular to any side of the reference cell, turning counterclockwise 60°, and moving j cells in the new direction. (Similar results are obtained by turning clockwise instead of counterclockwise, or by moving j cells before moving i cells; however, the first method is usually adopted by convention.)

Recall from our discussion following Equation (4.10) that the signal-to-interference ratio S/I is determined by the frequency reuse ratio $Q = D/R$, where D is the distance between a pair of cells that use the same set of channels. In the hexagonal geometry, Equation (4.12) shows that the frequency reuse ratio (and therefore the signal-to-interference ratio) is actually determined by the normalized distance $\hat{D}$ between the cells, that is,

$$Q = \sqrt{3}\hat{D}. \tag{4.14}$$

If we choose a particular cell as our reference, we can use the required signal-to-interference ratio to determine a minimum separation distance $\hat{D}$ between cells that use the same channel set. Once $\hat{D}$ is established, we can use Equation (4.13) to find corresponding integers i and j and then use these integers to locate all of the neighboring cells that reuse the same set of channels. Table 4.1 lists some representative values of the location integers i and j, the corresponding

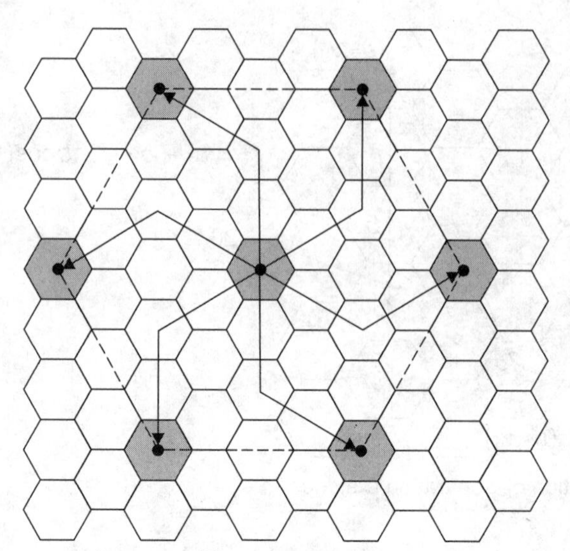

Figure 4.9 Cells at Distance $\hat{D}(2,2) = 2\sqrt{3}$ from a Reference Cell

values of $\hat{D}$ given by Equation (4.13), the values of $\hat{D}^2$ (we will need this later), and the frequency reuse ratios given by Equation (4.14). Notice that because i and j are integers, only certain distances $\hat{D}$ are physically possible. This means that only certain values of frequency reuse ratio Q are possible as well. If the signal-to-interference ratio requirement leads to a required value of Q that is not achievable, then the next higher value that is achievable must be used.

Table 4.1 Tabulation of Values of Normalized Distance and Frequency Reuse Ratio

i	j	$\hat{D}$	$\hat{D}^2$	Q
1	0	1	1	1.73
1	1	1.73	3	3
2	0	2	4	3.46
2	1	2.65	7	4.58
2	2	3.46	12	6
3	0	3	9	5.2
3	1	3.61	13	6.24
3	2	4.36	19	7.55
3	3	5.2	27	9
4	0	4	16	6.93

We observed earlier that the required distance $\hat{D}$ between cells that share the same channel set is usually large enough that there are cells in between that can use other channel sets. This suggests the following procedure for assigning channels to cells. Choose any cell and assign to it channel set F_1. Call this cell A. Using the procedure described previously, use the location integers i and j to identify all the cells that can use channel set F_1. Label all of these cell A. Now choose any unlabeled cell and assign to it channel set F_2. Label this cell B. Using the same i and j, locate all cells that can be assigned channel set F_2 and label them B. Continue this procedure until all cells have been assigned a channel set and labeled. A collection of contiguous cells, A, B, C, ..., that between them account for all of the channel sets is called a **cluster**. We will use the symbol $K_{cluster}$ to designate the number of cells in a cluster. This procedure is illustrated in Figure 4.10 for a reuse distance $\hat{D} = 2$, which corresponds to $i = 2, j = 0$. For the layout of the figure, we find that $K_{cluster} = 4$.

Since each cell in a cluster is assigned a unique channel set, and all of the channels sets are accounted for, there must be $K_{cluster}$ distinct channel sets. If the system has an allocation of N_{chan} channels, then the number of channels in a channel set—that is, the number of channels in a cell—is N_{cell} given by

$$N_{cell} = \frac{N_{chan}}{K_{cluster}}. \tag{4.15}$$

It is important to realize that each cluster has all N_{chan} channels available, and that these channels are reused in every cluster. If the entire market area is covered by M clusters of cells, then the total number N_{market} of channels available to subscribers is given by

$$N_{market} = MN_{chan}. \tag{4.16}$$

Two things should now be apparent: First, the number of subscribers that the entire system can accommodate can be vastly larger than the number N_{chan} of channels, since the same channel can be used simultaneously by different subscribers in different locations in the system. Second, to increase the geographic area covered by the cellular system, it is necessary only to add clusters of cells around the periphery. Since adding new clusters of cells increases M, the number N_{market} of available channels is also increased. Thus the added geographic area can bring in additional customers without increasing the usage burden on the channels.

Intuitively one might expect that the number $K_{cluster}$ of cells in a cluster is related to the reuse distance $\hat{D}$. A large distance $\hat{D}$ implies that many cells lie between "cochannel" cells that use the same channel set. This suggests that the cluster size should also be large. To make this intuitive picture more precise, we determine the relationship between $K_{cluster}$ and $\hat{D}$. First note that a hexagon of radius R has an area A_{cell} given by

$$A_{cell} = \frac{3\sqrt{3}}{2}R^2 . \tag{4.17}$$

Next, choose an arbitrary reference cell, call it cell A, and identify the six surrounding cells (also A) that share the same channel set. Recall that the centers of the six cochannel cells form a large

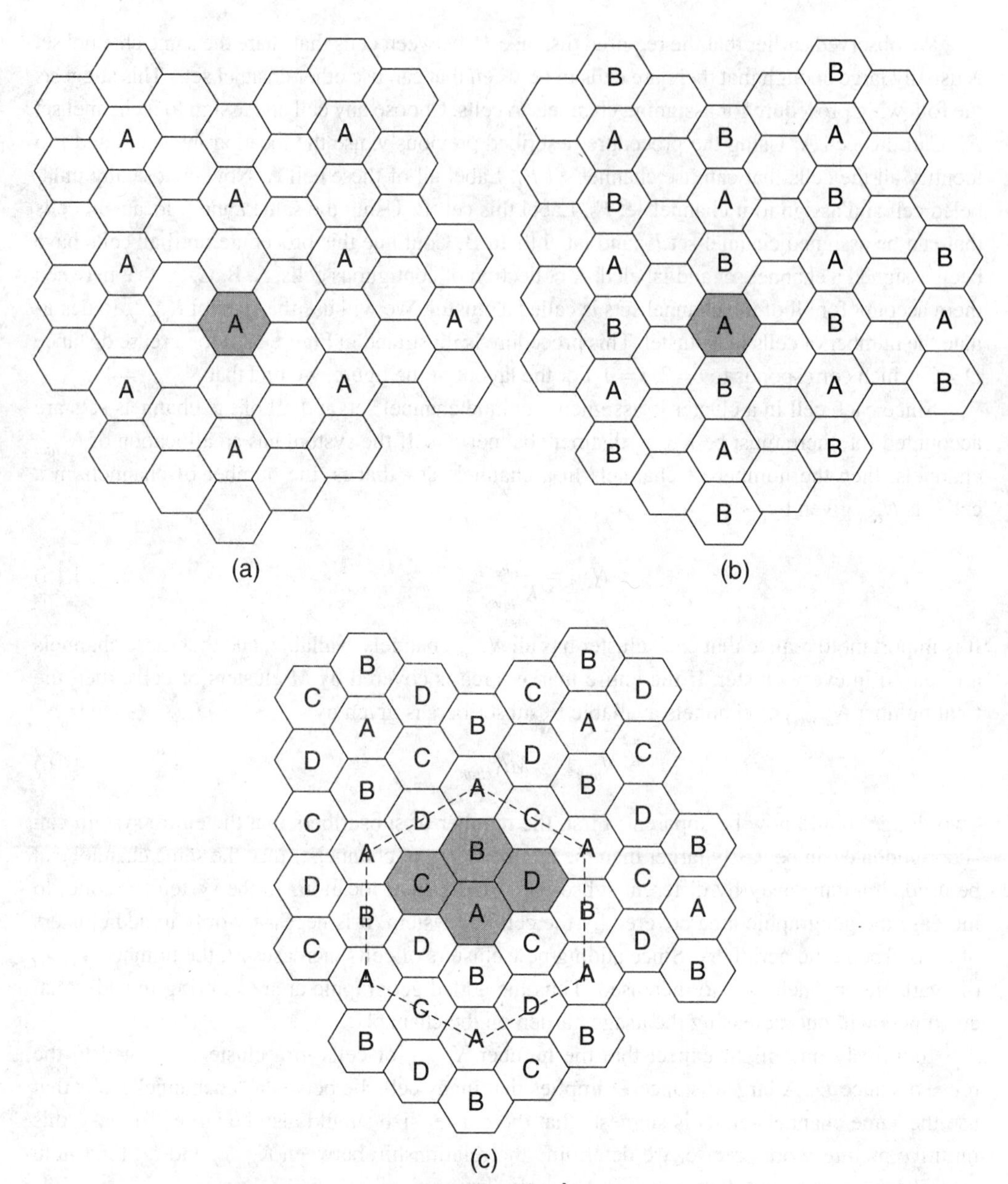

Figure 4.10 Cell Layout Process for Reuse Distance $\hat{D}(2,0) = 2$, Producing a Four-Cell Cluster

hexagon, with normalized radius $\hat{D}$ or actual radius D. The geometry is shown in Figure 4.9 and also in Figure 4.10(c). The large hexagon has an area A_{hex} given by

$$A_{hex} = \frac{3\sqrt{3}}{2}D^2. \tag{4.18}$$

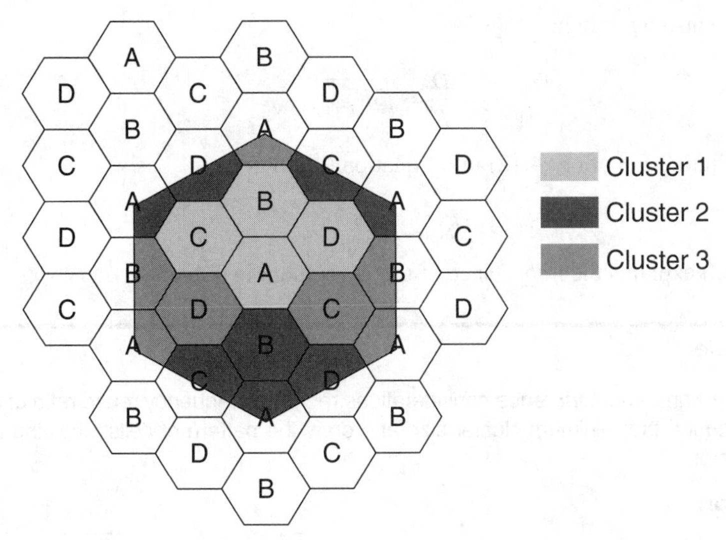

Figure 4.11 Every Large Hexagon Contains Three Clusters

Now we need to determine how many clusters are contained inside the large hexagon. We can associate a cluster with each cell A, but some of these clusters are only partially contained in the large hexagon. Figure 4.10(c) shows a typical situation. There will always be one complete cluster associated with the reference cell inside the large hexagon. The sides of the large hexagon meet at each vertex at an angle of 120°. Three of these vertices include an additional complete cluster. The remaining three vertices include a third complete cluster. Thus the large hexagon always includes three complete clusters regardless of the particular value of $\hat{D}$. Figure 4.11 shows the three clusters for the case $\hat{D} = 2$. The last step is to put the pieces together. The area of the large hexagon is given by Equation (4.18). Dividing by three gives the area of one cluster:

$$A_{cluster} = \frac{\sqrt{3}}{2} D^2. \tag{4.19}$$

Dividing the area of a cluster by the area of a cell gives the number of cells per cluster:

$$\begin{aligned} K_{cluster} &= \frac{A_{cluster}}{A_{cell}} \\ &= \frac{\frac{\sqrt{3}}{2} D^2}{\frac{3\sqrt{3}}{2} R^2} \\ &= \frac{1}{3}\left(\frac{D}{R}\right)^2. \end{aligned} \tag{4.20}$$

This result can also be written

$$Q = \frac{D}{R} = \sqrt{3K_{cluster}}. \tag{4.21}$$

Finally, substituting Equation (4.12) into Equation (4.20) gives

$$K_{cluster} = \hat{D}^2. \tag{4.22}$$

Equation (4.22) explains the importance of the $\hat{D}^2$ column in Table 4.1.

Example

Suppose signal-to-interference considerations require a frequency reuse ratio of at least $Q = 4.0$. Find the minimum cluster size and draw the pattern of cells, showing at least one cluster.

Solution

Equation (4.14) and Equation (4.13) give $Q = \sqrt{3}\hat{D}$ with $\hat{D} = \sqrt{i^2 + j^2 + ij}$. No combination of integer values for i and j will give a value for Q of 4.0. Rounding up to the nearest acceptable value gives $\hat{D} = \sqrt{2^2 + 1^2 + 2 \cdot 1} = \sqrt{7}$ and $Q = \sqrt{3 \cdot 7} = 4.58$, with $i = 2$ and $j = 1$. (See Table 4.1.) Then Equation (4.22) gives $K_{cluster} = \hat{D}^2 = 7$. Figure 4.12 shows a seven-cell cluster and some cochannel cells of channel set A.

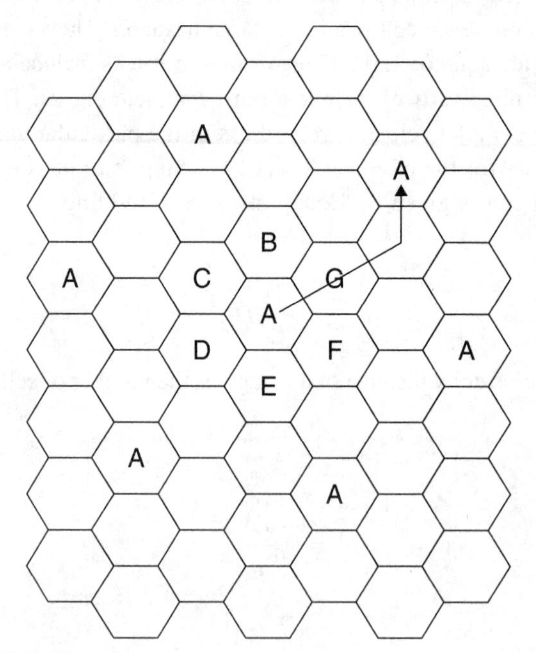

Figure 4.12 A Seven-Cell Cluster

Example

A market area for a cellular system is covered by 63 cells organized into 9-cell clusters. The system is allocated enough radio spectrum to support N_{chan} = 800 channels. Determine the percentage increase in the number of channels available to subscribers N_{market} if the cluster size is reduced to $K_{cluster}$ = 7. What might limit reducing the cluster size still further?

Solution

Using Equation (4.16) with $M = \frac{63 \text{ cells}}{9 \text{ cells/cluster}} = 7 \text{ clusters}$ gives $N_{market} = 7 \times 800 = 5600$ channels. If the cluster size is reduced to 7, then $M = \frac{63 \text{ cells}}{7 \text{ cells/cluster}} = 9 \text{ clusters}$, and $N_{market} = 9 \times 800 = 7200$ channels. The percentage increase is $\frac{7200-5600}{5600} \times 100\% = 28.6\%$.

Reducing the cluster size reduces the distance between cochannel cells, raising the level of cochannel interference. Cochannel interference provides the limit on how small the cluster size can be. Equation (4.21) shows how the cluster size is related to the frequency reuse ratio. In the next section we will establish the quantitative relationship between frequency reuse ratio and signal-to-interference ratio.

Estimation of Interference Levels

Cochannel Interference

The final example of the previous section showed that for a given cell size, the number of customers that a cellular system can support is maximized if the cluster size is minimum. Cochannel interference was seen to be the factor that limits the extent to which cluster size can be reduced, since reducing the cluster size has the effect of reducing the frequency reuse ratio $Q = D/R$. In this section we will explore the relation between the frequency reuse ratio and the signal-to-interference ratio. We will then introduce the concept of adjacent-channel interference, which also has an effect on efficient frequency reuse.

The first U.S. cellular system, AMPS, was designed with the goal of approximating "toll quality" speech communications using analog modulation. Toll quality refers to a subjective evaluation of the quality of long-distance wired communications in the public switched telephone network. The speech quality is measured by tests in which a number of people assess specific speech samples to develop a "mean opinion score" for each sample. To determine AMPS system parameters, systems engineers developed simulators that mimicked a typical signal as corrupted by fading, noise, and interference. These simulators were used to conduct tests to determine minimum signal-to-noise and signal-to-interference ratios that would allow the quality of service objectives to be met. The results of these early studies determined that quality objectives could be met under these conditions.

1. The signal-to-noise ratio is no less than 18 dB over 90% of the coverage area for cells limited by receiver noise.
2. The signal-to-interference ratio is no less than 17 dB over 90% of the coverage area for cells limited by interference.

Modern systems are predominantly digital and are capable of communicating many types of information in addition to speech. QoS metrics can be defined for each information type and used to determine the required signal-to-noise and signal-to-interference ratios.

We begin our analysis by considering the interference from the nearest cochannel base stations as shown in Figure 4.13. Although the diagram depicts a seven-cell cluster, the results apply to clusters of any size. We assume that the receiver noise is negligible compared to the interference. The reference base station is in the center of the diagram.

As a first approximation, consider that a mobile unit is located on the cell boundary at a distance equal to the cell radius R from the reference base station. This is the farthest that a mobile unit should be from its serving base station. The nearest cochannel sources are mobile units in the cochannel cells and are all approximately at the reuse distance D from the reference base station. The signal-to-interference ratio S/I is given by Equation (4.9) with $J = 6$ interference sources:

$$\frac{S}{I} = \frac{P_1}{\sum_{j=2}^{7} P_j}. \tag{4.23}$$

Now if all of the mobile units have the same parameters and the environment is uniform in all directions, then

$$P_2 = P_3 = \cdots = P_7. \tag{4.24}$$

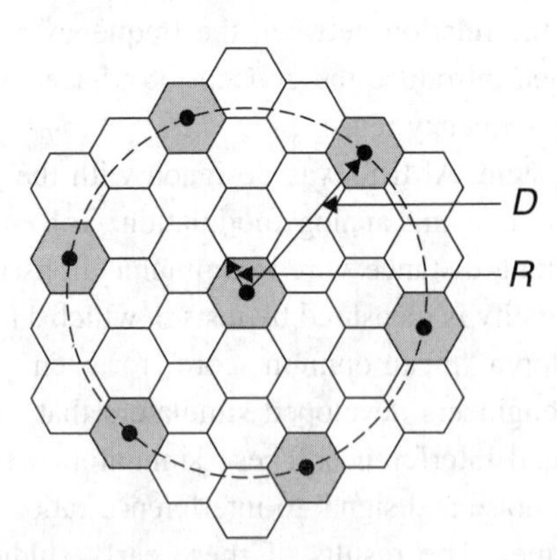

Figure 4.13 First-Tier Cochannel Interference Sources

If we further assume that the path-loss exponent is ν, then

$$\frac{S}{I} = \frac{\frac{K}{R^{\nu}}}{6\frac{K}{D^{\nu}}} = \frac{1}{6}\left(\frac{D}{R}\right)^{\nu} = \frac{1}{6}Q^{\nu}, \tag{4.25}$$

where Q is the frequency reuse factor. Using Equation (4.21),

$$\frac{S}{I} = \frac{\left(\sqrt{3K_{cluster}}\right)^{\nu}}{6}. \tag{4.26}$$

Example

Suppose, as in the AMPS system, that a signal-to-interference ratio of 18 dB is required. The path-loss exponent is $\nu = 4.0$. Considering only the nearest cochannel interference sources, find the minimum cluster size.

Solution

$\left.\frac{S}{I}\right|_{dB} = 18$ dB means $\frac{S}{I} = 63.1$. Using Equation (4.26), $63.1 = \frac{\left(\sqrt{3K_{cluster}}\right)^{4}}{6}$, so $K_{cluster} = 6.49$. A check of Table 4.1 shows that the nearest admissible cluster size larger than 6.49 is $K_{cluster} = 7$, achieved with $i = 2, j = 1$.

Table 4.2 repeats the information shown in Table 4.1 but includes a column of values of signal-to-interference ratio assuming that the path-loss exponent $\nu = 4$. Only first-tier interference is taken into account, but the first-tier interference predominates. In the presence of a deep fade, or when no first-tier sources are present, interference from second or further tiers may be noticeable.

Table 4.2 Approximate Signal-to-Interference Ratio for Several Reuse Ratios with $\nu = 4$

i	j	$\hat{D}$	$K_{cluster}$	Q	$\left.\frac{S}{I}\right|_{dB}$
1	0	1	1	1.73	1.8
1	1	1.73	3	3	11.3
2	0	2	4	3.46	13.8
2	1	2.65	7	4.58	18.7
2	2	3.46	12	6	23.3
3	0	3	9	5.2	20.8
3	1	3.61	13	6.24	24.0
3	2	4.36	19	7.55	27.3
3	3	5.2	27	9	30.4
4	0	4	16	6.93	25.8

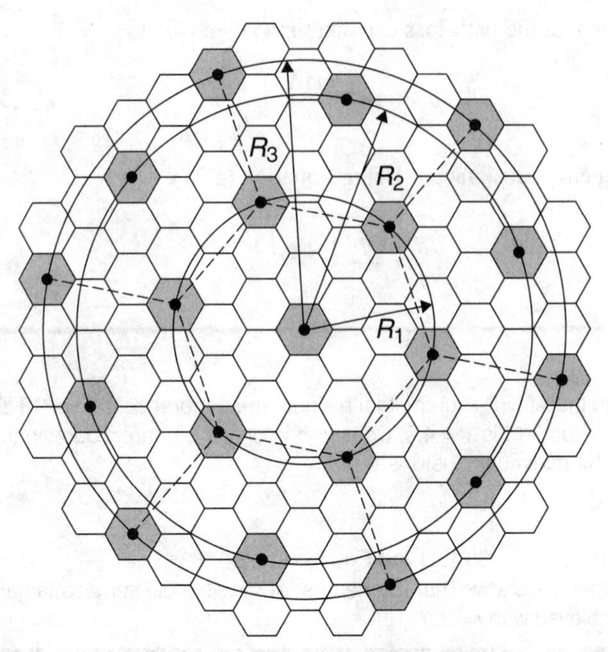

Figure 4.14 First-, Second-, and Third-Tier Cochannel Interference

Figure 4.14 shows the geometry of the second and third cochannel interference tiers. The second-tier radius is the center-to-center distance between large hexagons, that is,

$$R_2 = \sqrt{3}R_1 = \sqrt{3}D. \tag{4.27}$$

From the geometry, the third-tier radius is simply double the side of a large hexagon, or

$$R_3 = 2D. \tag{4.28}$$

For a path-loss exponent of $\nu = 4$, interference from the second tier is

$$\begin{aligned} \left.\frac{S}{I}\right|_{2,dB} &= 10\log\frac{1}{6}\left(\frac{\sqrt{3}D}{R}\right)^4 \\ &= 10\log\frac{1}{6}\left(\frac{D}{R}\right)^4 + 10\log(\sqrt{3})^4 \\ &= \left.\frac{S}{I}\right|_{dB} + 9.54 \text{ dB}, \end{aligned} \tag{4.29}$$

indicating that the interference from the second tier is 9.54 dB below the level of interference from the first tier. Similarly, the interference from the third tier is 12 dB below the first-tier level. It is rarely necessary to consider interference from tiers farther out, but it may be necessary to do so in situations where the fading is exceptionally severe, or where the propagation environment has a small or widely varying path-loss exponent. Should the need to consider additional tiers of interference arise, the method presented is directly applicable.

Adjacent-Channel Interference

Adjacent-channel interference is interference to a receiver listening on a given channel from a transmission occurring on an adjacent channel. A cellular receiver must be designed to receive on all channels in the cellular system band, as a particular telephone connection may be assigned to any of the possible channels. The receiver separates one channel from another by using a highly selective filter. The passband bandwidth of the filter is equal to the channel bandwidth. The filter must cut off sharply at the passband edges, so that signals in the adjacent channels will not be passed to the demodulator. Now a "brick wall" filter, which cuts off abruptly and completely at the passband edges, is a physical impossibility. It turns out, moreover, that sharp cutoff filters may be far too expensive for mass consumer markets. Analog filters require a large number of components to achieve high selectivity (sharp cutoff). Also, for both analog and digital implementations, the performance of a sharp cutoff filter is very sensitive to small errors in the component or coefficient values.

Figure 4.15 shows the frequency response of a tenth-order Chebyshev filter with a 30 kHz passband. The center frequency has been arbitrarily selected as 870.03 MHz. This frequency response is

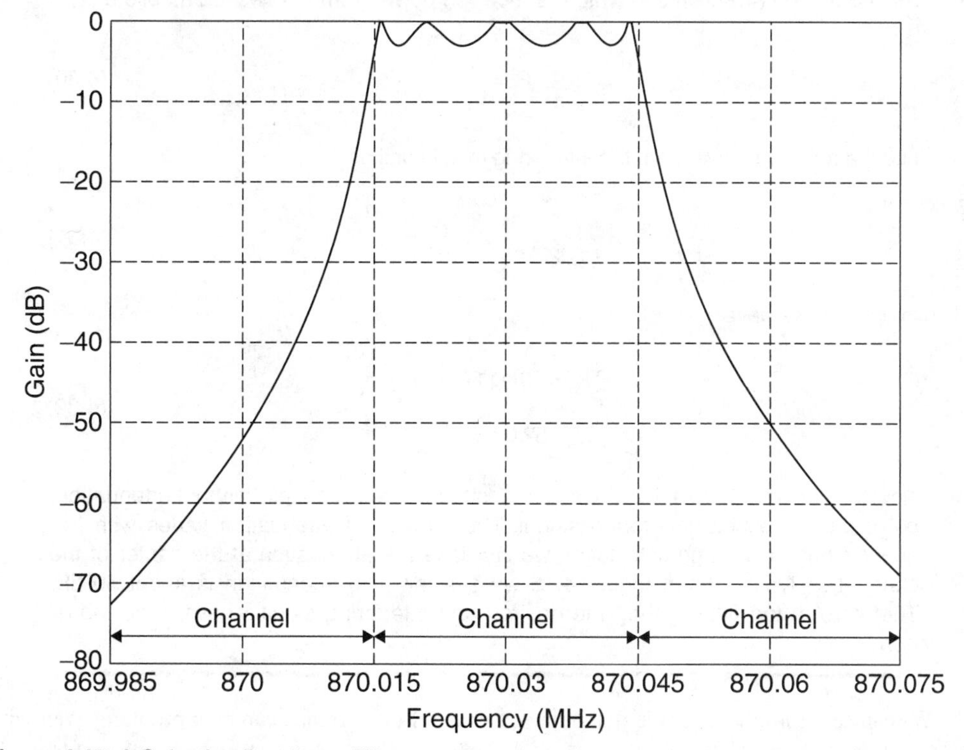

Figure 4.15 A Selective Bandpass Filter

not intended to represent any specific manufacturer's equipment but to suggest what a selective band-pass frequency response might look like. Notice from the figure that this filter produces an attenuation of about 50 dB at the center of the adjacent channels, 30 kHz above or below the filter's center frequency. One might imagine that 50 dB of attenuation in the adjacent channel might be enough to render interference from these channels negligible. The following example illustrates that the situation might be otherwise.

Example

A base station receiver in an urban environment is receiving two signals from mobile units, one desired and one in an adjacent channel. The mobile units use identical equipment and are transmitting at equal power levels; the path-loss exponent is $\nu = 4$. The distance from the desired mobile unit to the base station is d_1, and the distance from the interfering mobile unit to the base station is $d_1/20$. (This would be the situation, for example, if the desired mobile unit were one mile from the base station and the interfering mobile unit were one city block from the base station.) If the base station uses a filter like the one shown in Figure 4.15 to separate signals, find the relative power levels of the two received signals after filtering.

Solution

As in Equation (4.3), we can write the received power from the desired mobile unit as

$$P_1 = \frac{K}{d_1^4} \tag{4.30}$$

and the received power from the interfering mobile unit as

$$P_2 = \frac{K}{(d_1/20)^4} = 20^4 \frac{K}{d_1^4} = 20^4 P_1. \tag{4.31}$$

In decibels we have

$$\begin{aligned} P_2\big|_{dB} &= 10\log(20^4) + P_1\big|_{dB} \\ &= 52.0 \text{ dB} + P_1\big|_{dB}. \end{aligned} \tag{4.32}$$

Now the filter of Figure 4.15 passes the desired signal essentially without attenuation, but it attenuates the interfering signal. The amount of attenuation varies with frequency, but for a rough calculation we can take the attenuation at the center of the channel as typical. From Figure 4.15 we see that this attenuation is about 50 dB. Therefore at the filter output, the desired and interfering signals have about equal power.

We can see from the example that adjacent-channel interference can be a problem, even with highly selective channel filtering. Several strategies are available for dealing with this problem.

A common strategy in the broadcast services is to avoid using adjacent channels in the same market area. This strategy is used in both AM and FM broadcasting and in television. In cellular systems, however, the number of channels available translates directly into the number of customers who can be supported, which, in turn, translates directly into revenue. Channels are too valuable to be set aside for interference avoidance.

The power level of a mobile unit transmitter can be controlled dynamically, so that it transmits less power when it is nearer the base station than it does when it is at a cell edge. Power control has been in use in some form since the earliest cellular systems. In modern cellular systems a mobile unit's transmitted power is adjusted in 1 dB increments every few milliseconds, to keep the power level received at a base station constant as the mobile unit moves over the cell's coverage area.

Finally, channels can be partitioned so that adjacent channels are not assigned to the same cell or to cells that are immediate neighbors. This will guarantee that an interference source cannot get physically close to a base station receiver. When cluster sizes are small, however, the available channels will be divided up among a relatively small number of cells. In this case it may be difficult to avoid assigning adjacent channels to the same or nearby cells, and adjacent-channel interference may significantly limit how small the clusters can be made.

Cellular System Planning and Engineering

The Key Trade-offs

We have seen how the concept of frequency reuse leads to a cellular architecture that can allow for almost limitless expansion in the geographic area and the number of subscribers that the system can service. In configuring a cellular layout two parameters are of key importance; these are the cell radius R and the cluster size $K_{cluster}$. We begin this section by summarizing the design trade-offs in which these parameters are central. We will then focus on issues regarding system expansion. Although a cellular system can be expanded in geographic coverage simply by adding cells at the periphery, we have not addressed how a system can expand in user density. Since the history of cellular telephones has been characterized by rapid growth, the experience of all cellular start-ups has been a rapid expansion in subscribers. We will discuss two methods for dealing with an increasing subscriber density: sectoring and cell splitting.

The cell radius governs both the geographic area covered by a cell and also, for a given subscriber density, the number of subscribers that the cell must service. Simple economic considerations suggest that the cell size should be as large as possible. Since every cell requires an investment in a tower, land on which the tower is placed, and radio transmission equipment, a large cell size minimizes the cost per subscriber. The cell size is ultimately determined by the requirement that an adequate signal-to-noise ratio be maintained over the coverage area. As we saw in the previous chapter, a number of system parameters—transmitter power, receiver noise figure, antenna height, for example—are involved in determining the signal-to-noise ratio. Transmitter power is particularly limited in the reverse direction, as the mobile units are small and battery powered. A new cellular installation in a previously untapped market might start with a cell radius of several miles.

Given a cell radius R and a cluster size $K_{cluster}$, the geographic area covered by a cluster is

$$\begin{aligned} A_{cluster} &= K_{cluster} A_{cell} \\ &= K_{cluster} \frac{3\sqrt{3}}{2} R^2, \end{aligned} \tag{4.33}$$

using Equation (4.17). If the market has a geographic area of A_{market}, then the number of clusters M is given by

$$M = \frac{A_{market}}{A_{cluster}} = \frac{A_{market}}{K_{cluster} \frac{3\sqrt{3}}{2} R^2}. \tag{4.34}$$

Recall that all of the available channels, N_{chan}, are reused in every cluster. Hence, to make the maximum number of channels available to subscribers, the number of clusters M should be large, which by Equation (4.34) shows that the cell radius should be small. Ultimately cell radius is determined by a trade-off: R should be as large as possible to minimize the cost of the installation per subscriber, but R should be as small as possible to maximize the number of customers that the system can accommodate.

Equation (4.34) also gives us a lead on the second key layout parameter. If we consider the cell radius as fixed, then the number of clusters can be maximized by minimizing the number of cells in a cluster $K_{cluster}$. Combining Equation (4.14) with Equation (4.22) gives us

$$K_{cluster} = \frac{Q^2}{3}, \tag{4.35}$$

showing that the cluster size depends only on the frequency reuse ratio Q. Now in selecting a value of $K_{cluster}$ we once again encounter a trade-off. Equation (4.25) shows that for an environment with a uniform path-loss exponent ν, Q is the only parameter governing the signal-to-cochannel-interference ratio. We find that to maximize the number of customers that the system can handle, the cluster size should be made as small as possible, while maintaining an adequate signal-to-interference ratio.

Example

When the AMPS cellular system was first deployed, the aim of the system designers was to guarantee coverage. Initially the number of users was not significant. Consequently cells were configured with an eight-mile radius, and a 12-cell cluster size was chosen. The cell radius was chosen to guarantee a 17 dB signal-to-noise ratio over 90% of the coverage area, using what were then (early 1980s) practical values for antenna height, base and mobile power levels, and so forth. Although a 12-cell cluster size provided more than adequate cochannel separation to meet a requirement for a 17 dB signal-to-interference ratio in an interference-limited environment, it did not provide adequate frequency reuse to service an explosively growing customer base. The system planners reasoned that a subsequent shift to a 7-cell cluster size would provide an adequate number of channels. Now according to Table 4.2, a 7-cell cluster size should provide an adequate 18.7 dB signal-to-interference ratio. The margin, however, is slim, and the 17 dB signal-to-interference ratio requirement could not be met over 90% of the coverage area. In the paragraphs below we will discuss a technique called "sectoring," which can increase the signal-to-interference ratio without necessitating an increase in the cluster size.

Sectoring

Up until now we have assumed that a base station antenna is located in the center of a cell and is omnidirectional; that is, it radiates uniformly in all horizontal directions. In the following discussion we will show that directional antennas can be used to advantage in reducing cochannel interference. Figure 4.16 depicts a cell that has been divided into three 120° sectors. The base station feeds three 120° directional antennas, each of which radiates into one of the sectors. The channel set serving this cell has also been divided, so that each sector is assigned one-third of the available number N_{cell} of channels. Figure 4.17 shows a seven-cell-cluster layout with 120°-sectored cells. It can be seen in the figure that mobile units in sector a of the center cell will receive cochannel interference from only two of the first-tier cochannel base stations, rather than from all six. Likewise, the base station in the center cell will receive cochannel interference from mobile units in only two of the cochannel cells. The signal-to-interference ratio of Equation (4.26) must now be modified to

$$\frac{S}{I} = \frac{\left(\sqrt{3K_{cluster}}\right)^{\nu}}{2}, \tag{4.36}$$

where the denominator has been reduced from 6 to 2 to account for the reduced number of interference sources.

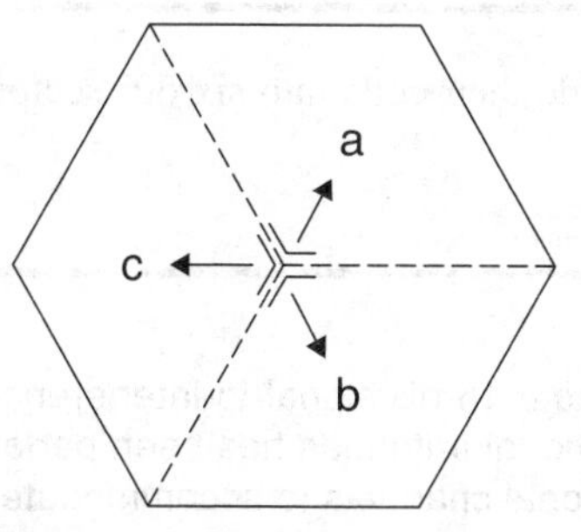

Figure 4.16 A Cell Divided into Three 120° Sectors

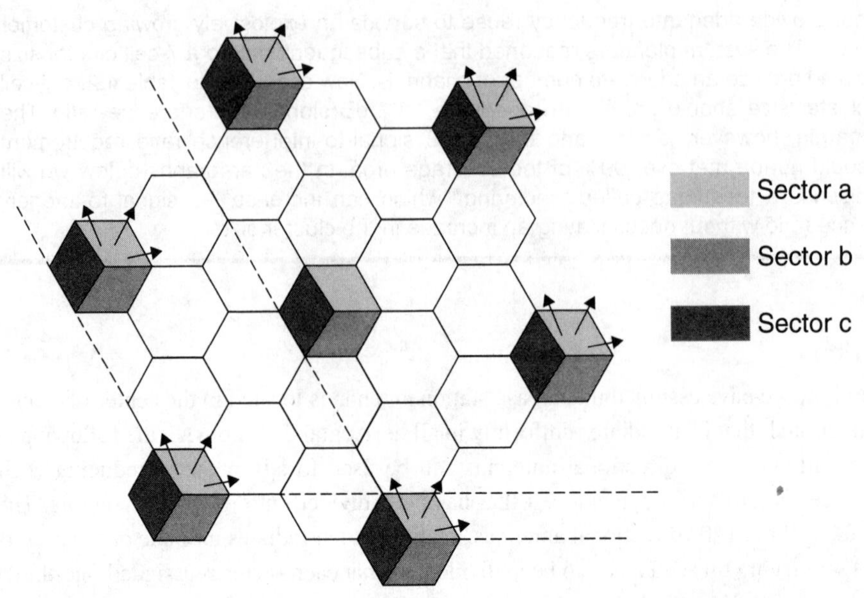

Figure 4.17 A Seven-Cell Cluster with 120° Sectors; the Arrows Suggest Base Station Radiation in Sector a

---

Example

Find the signal-to-interference ratio for a seven-cell-cluster layout with 120° sectors. Assume that the path-loss exponent is $\nu = 4$.

Solution

If $K_{cluster} = 7$, then Equation (4.36) gives $\frac{S}{I} = \frac{\left(\sqrt{3 \times 7}\right)^4}{2} = 221$, or $\left.\frac{S}{I}\right|_{dB} = 23.4$ dB.

Some cellular systems divide their cells into six 60° sectors. The analysis is similar to the 120°-sector case.

Example

A cellular system requires a 15 dB signal-to-interference ratio. A seven-cell-cluster layout with omnidirectional antennas has been performing adequately, but the system now needs additional channels to accommodate growth. By what percentage can the number N_{market} of channels be increased if 60° sectoring is introduced? The path-loss exponent is $\nu = 4$.

Solution

Figure 4.18(a) shows a seven-cell-cluster layout with 60° sectors. It can be seen that the shaded sector in the center receives cochannel interference from only one first-tier cell.

Then Equation (4.26), suitably modified, gives

$$\frac{S}{I} = \frac{\left(\sqrt{3K_{cluster}}\right)^{\nu}}{1} = \frac{\left(\sqrt{3\times 7}\right)^{4}}{1} = 441$$
$$\left.\frac{S}{I}\right|_{dB} = 10\log(441) = 26.4 \text{ dB}. \tag{4.37}$$

Since the signal-to-interference ratio exceeds the required 15 dB, we can try reducing the cluster size. Table 4.2 shows that the next smaller cluster size is $K_{cluster} = 4$. Figure 4.18(b) shows a four-cell-cluster layout. Evidently there is still only one source of cochannel interference. Then we have

$$\frac{S}{I} = \frac{\left(\sqrt{3K_{cluster}}\right)^{\nu}}{1} = \frac{\left(\sqrt{3\times 4}\right)^{4}}{1} = 144$$
$$\left.\frac{S}{I}\right|_{dB} = 10\log(144) = 21.6 \text{ dB}. \tag{4.38}$$

The signal-to-interference ratio is still above the requirement, so a further reduction in cluster size might be possible. Figure 4.18(c) shows a three-cell-cluster layout. For this case we have *two* interference sources, and

$$\frac{S}{I} = \frac{\left(\sqrt{3K_{cluster}}\right)^{\nu}}{2} = \frac{\left(\sqrt{3\times 2}\right)^{4}}{2} = 18.0$$
$$\left.\frac{S}{I}\right|_{dB} = 10\log(18.0) = 12.6 \text{ dB}. \tag{4.39}$$

Here the signal-to-interference ratio is too low, leading to the conclusion that the minimum acceptable cluster size is $K_{cluster} = 4$.

Recall that the number of clusters M varies inversely as the cluster size $K_{cluster}$. The number of channels N_{market}, on the other hand, is proportional to M, as all of the channel frequencies are reused in every cluster. We conclude that N_{market} varies inversely as cluster size. Then if the cluster size is reduced from seven to four, the total number of channels in the market area is increased by a factor of 7/4 = 1.75, which is a 75% increase.

The calculations in the preceding example are actually an idealization for several reasons. First, practical antennas have side lobes and cannot focus a transmitted beam into a perfect 120°

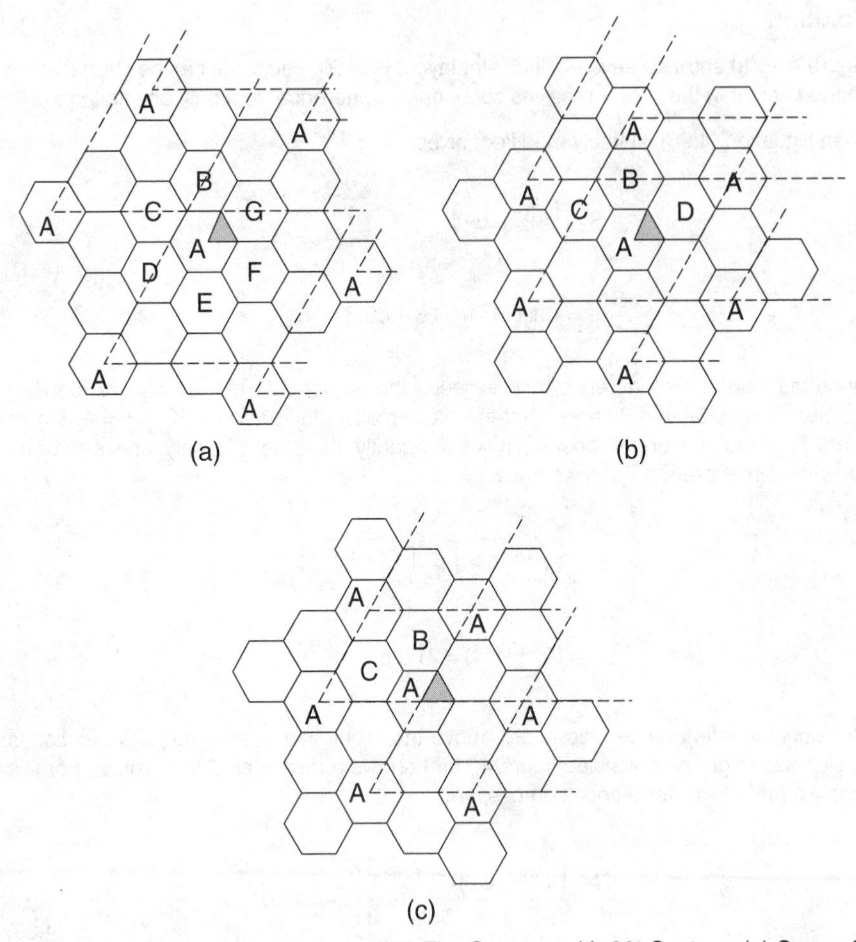

Figure 4.18 Cochannel Interference from First-Tier Sources with 60° Sectors: (a) Seven-Cell Clusters; (b) Four-Cell Clusters; (c) Three-Cell Clusters

or 60° sector. Some interference will be received in addition to that suggested by diagrams such as Figure 4.18. Next, it turns out that a given number of channels are not able to support as many subscribers when the pool of channels is divided into smaller groups. Thus a cell having N_{cell} channels will support fewer subscribers when the cell is divided into sectors. The reasons for this, and a quantitative assessment of how many subscribers a given number of channels can support, will be given later on in the present chapter. Finally, dividing a cell into sectors requires that a call in progress will have to be handed off (that is, assigned a new channel) when a mobile unit travels into a new sector. Although requiring handoffs between sectors as well as between cells does not directly reduce the number of customers that can be supported, it does increase the complexity of the system needed to support them.

Example

In the AMPS system channels are assigned to cells in such a way that both cochannel and adjacent-channel interference is minimized. This example describes how channels are assigned, assuming a seven-cell cluster size and 120° sectoring.

The AMPS system includes 50 MHz of spectrum divided into 25 MHz to be used for forward channels and 25 MHz to be used for reverse channels. With 30 kHz channels, there can be a total of 832 two-way channels. When the system was first deployed in the early 1980s, the FCC divided the 832 channels into two sets. In total, 416 adjacent "B-side" channels were allocated to a telephone carrier in each market, and the other 416 adjacent "A-side" channels were allocated to a nontelephone company, thereby ensuring competition in each market area. Of the 416 channels allocated to a single operating company, 21 channels were set aside for control purposes, leaving 395 two-way voice channels for each cellular system. Table 4.3 shows the first few dozen A-side channels listed in an array. The channel numbers 1, 2, 3, ..., refer to adjacent channels. The channels are listed in rows, where each row is 21 channels long. In a seven-cell cluster, the three columns labeled 1a, 1b, and 1c make up the channel set assigned to cell A. The three columns labeled 2a, 2b, and 2c make up the channel set assigned to cell B. This continues, with columns 7a, 7b, and 7c assigned to cell G. (Refer to Figure 4.12 or Figure 4.18(a).) Note that channels used within a cell are separated by at least seven channel spacings, or 210 kHz. Now within cell A, the channels in column 1a are assigned to sector a, the channels in column 1b are assigned to sector b, and the channels in column 1c are assigned to sector c. (Refer to Figure 4.17.) The pattern repeats in each of the cells. We see that within a sector, channels are separated by a spacing of at least 21 channels, or 630 kHz.

With a total of 395 voice channels and a table of 21 columns, a little arithmetic shows that the complete table will have 18.8 rows. In fact, some sectors will have 18 channels and some will have 19. Cells will have from 55 to 57 channels each.

Cell Splitting

The overall approach to laying out a cellular system is founded on ensuring a capability for systematic growth. When a new system is deployed, user demand is low, and users are assumed to be uniformly distributed over the area to be served. Initial system layout is designed to provide uniformly reliable coverage and uniform capacity over the entire service area. As new users subscribe to the cellular service, the demand for channels may begin to exceed the capacity of some base stations. As we mentioned earlier, this increased demand often first shows up in the downtown areas of cities, where the population is dense during the working day. In the previous section we showed how the number of channels available to customers, or equivalently, the density

Table 4.3 Part of the AMPS A-Side Channel Plan

1a	2a	3a	4a	5a	6a	7a	1b	2b	3b	4b	5b	6b	7b	1c	2c	3c	4c	5c	6c	7c
1	2	3	4	5	6	7	8	9	10	11	12	13	14	15	16	17	18	19	20	21
22	23	24	25	26	27	28	29	30	31	32	33	34	35	36	37	38	39	40	41	42
43	44	45	46	47	48	49	50	51	52	53	54	55	56	57	58	59	60	61	62	63
64	65	66	67	68	69	70	71	72	73	74	75	76	77	78	79	80	81	82	83	84
...	...	...	...	...	...	...	...	...	...	...	...	...	...	...	...	...	...	...	...	...
A	B		...			G	A	B		...			G	A	B		...			G

of channels per square kilometer, can be increased by decreasing the cluster size. Once a system has been initially deployed, however, a systemwide reduction in the cluster size may not be warranted, since user density does not grow at the same rate in all parts of the system. Cell splitting is a technique that provides the capability to add new smaller cells in specific areas of the system to support increased demand in those areas, while minimizing the need to modify the existing cell parameters. Cell splitting is based on cell radius reduction.

There are two challenges to increasing the system capacity by reducing the cell radius. Clearly, if cells are smaller there will have to be more of them, so additional base stations will be needed in the system. The challenge in this case is to reconfigure the system in such a way that existing base station towers do not have to be moved. The second challenge involves meeting an evolving and generally increasing demand that may vary dramatically between different geographic areas of a system. For example, a city center may have the highest density of users and therefore should be supported by cells with the smallest radius. The radii of cells in a typical system generally increase as one moves from urban to suburban to rural areas, since user density typically decreases as one moves away from a city center. The key challenge, then, is to add the minimum number of smaller cells wherever increased demand dictates the need for increased capacity. A gradual addition of new base stations and smaller cells implies that, at least for a time, the cellular system may have to operate with cells of more than one size.

Figure 4.19(a) shows a cellular layout with seven-cell clusters. Let us suppose that the cells in the center of the diagram are becoming congested, and cell A in the center is approaching user capacity. Figure 4.19(b) shows an overlay of smaller cells superimposed on the original layout. The new smaller cells have half the cell radius of the original cells. At half the radius, the new cells will have one-fourth of the area and will consequently need to support one-fourth the number of subscribers. Notice that one of the new smaller cells lies in the center of each of the larger cells. If we assume that base stations are located in the cell centers, this allows the original base stations to be maintained as the new cell pattern spreads outward from the center. Of course new base stations will have to be added for new cells that do not lie in the center of the larger cells.

Recall that the organization of cells into clusters is independent of the cell radius, so that the cluster size can be the same in the small-cell layout as it was in the large-cell layout. Recall that signal-to-interference ratio is determined by cluster size and not by cell radius. Consequently, if the cluster size is maintained, the signal-to-interference ratio will be the same after cell splitting as it was beforehand. If the entire system is replaced with new half-radius cells, and the cluster size is maintained, the number of channels per cell will be exactly as it was before, and the number of subscribers per cell will have been reduced. In large cities it is not uncommon for cellular systems to be configured into "microcells" whose radii are measured in hundreds of meters, rather than in kilometers.

When the cell radius is reduced by a factor, it is also possible, and desirable, to reduce the power in transmitted signals. The minimum required power level is determined by the need to maintain an adequate signal-to-noise ratio over a significant fraction of the cell area, which in turn requires a minimum signal-to-noise ratio at the cell radius. The following example illustrates the idea.

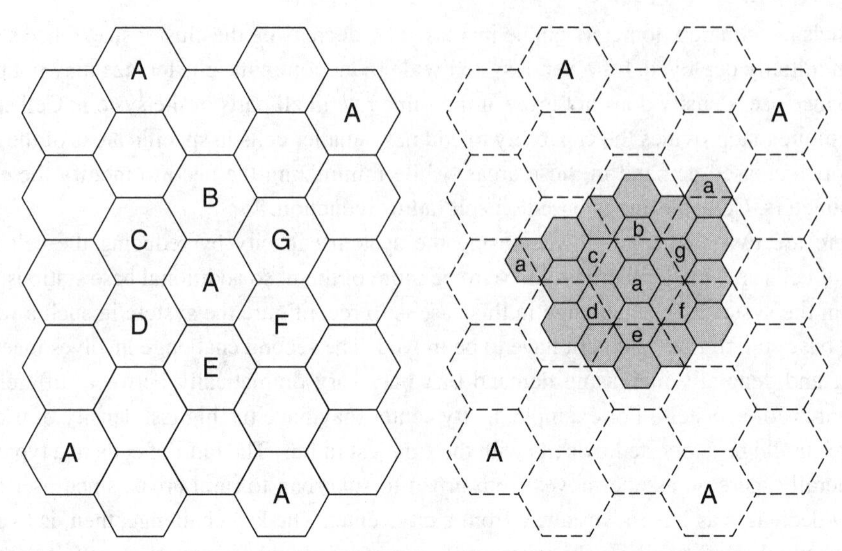

Figure 4.19 (a) Seven-Cell-Cluster Layout before Splitting; (b) Cell Splitting: Overlay of Half-Radius Cells

Example

In a certain cellular system the base stations radiate 15 W. Suppose the cells are split, and the new cells have half the radius that the original cells had. Find the power that the base stations in the new layout must transmit to maintain the signal-to-noise ratio at the cell boundaries. The path-loss exponent is $\nu = 4$.

Solution

Note that we are concerned with signal-to-noise ratio and not with signal-to-interference ratio. The former is determined by the received power and noise level, whereas the latter depends only on the cluster size. The noise level, however, is determined by the receiver noise figure, a parameter that is completely independent of the cell radius or layout. Therefore we can focus on the received signal power. Recall from Chapter 3 that the received signal power at distance d from the transmitting antenna is given by

$$P_r = \frac{P_{r0}}{\left(\frac{d}{d_0}\right)^{\nu}}, \tag{4.40}$$

where the reference power P_{r0} is the received power at some reference distance d_0. (See Equation (4.3).) The reference power P_{r0} is directly proportional to the transmitted power. Let P_{rOld} be the reference power measured at d_0 in the original cell configuration. For $\nu = 4$, the received power at the cell boundary $d = R_{old}$ is

$$P_r = \frac{P_{rOld}}{\left(\frac{R_{old}}{d_0}\right)^4}. \tag{4.41}$$

When the cells are split, the cell radius will become $R_{new} = R_{old}/2$. We want to maintain the received power at the cell radius at P_r. We have

$$P_r = \frac{P_{rOld}}{\left(\frac{R_{old}}{d_0}\right)^4} = \frac{P_{rNew}}{\left(\frac{R_{old}/2}{d_0}\right)^4}, \quad (4.42)$$

where P_{rNew} is the new reference power measured at d_0. Rearranging gives

$$P_{rNew} = P_{rOld}(1/2)^4. \quad (4.43)$$

The transmitted powers change in the same proportion as the reference powers. Therefore we have

$$P_{tNew} = 15(1/2)^4 = 0.938 \text{ W}. \quad (4.44)$$

In decibels the power reduction is $10\log(2^4) = 12$ dB, from $P_{tOld}\big|_{dB} = 41.8$ dBm to $P_{tNew}\big|_{dB} = 29.7$ dBm .

If a cellular layout is replaced in its entirety by a new layout with a smaller cell radius, the signal-to-interference ratio will not change, provided the cluster size does not change. Some special care must be taken, however, to avoid cochannel interference when both large and small cell radii coexist, as in the system of Figure 4.18. It turns out that the only way to avoid interference between the large-cell and small-cell systems is to assign entirely different sets of channels to the two systems. If a large-cell system becomes congested in the downtown, for example, channels can be taken away from the large-cell system to make up channel sets for the small-cell system. The capacity of the large-cell system will be reduced, but the large-cell system will now be used primarily in the suburbs, where the user density is low. The small-cell system also does not have a full complement of channels, but as the cell area is small, there may not be enough users per cell to demand a full channel set. As the small-cell system continues to spread, more and more channels can be reassigned from the large-cell to the small-cell system, until ultimately, the large-cell system is completely replaced.

Operational Considerations

Dividing a geographic market area into hexagonal cells and placing a base station in the center of each cell does not produce a cellular telephone system, although it is a necessary first step. In this section we describe certain operational issues that must be faced to allow the array of base stations to work together as a system. The treatment will be brief, as the intent is only to identify the issues and suggest some possible alternatives. It is not intended to provide a complete catalog of solutions. We start by describing, in general terms, how the base

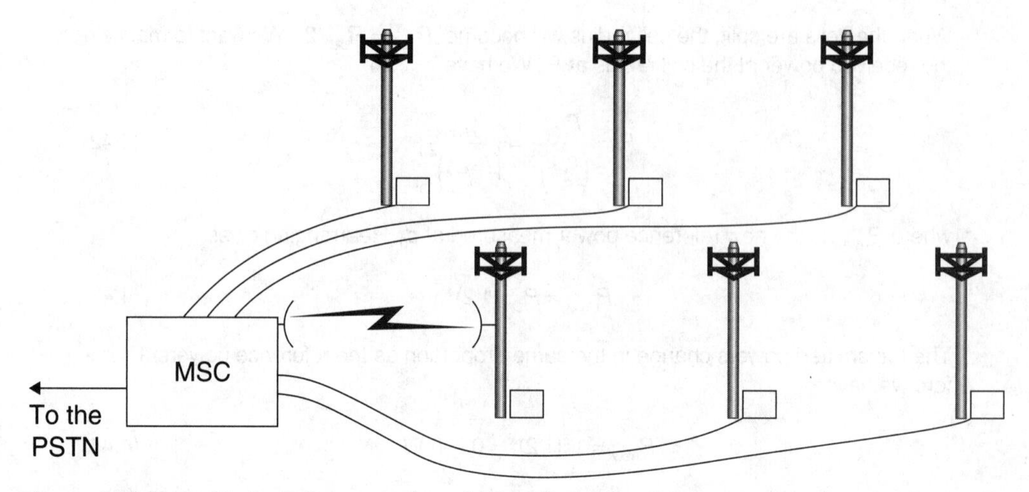

Figure 4.20 Base Stations Connected to a Mobile Switching Center

stations are connected to the telephone network. Next we discuss some options for assigning channels to cells, and finally, we describe how calls are handed off as a mobile unit travels between cells.

The Mobile Switching Center

The heart of a cellular system is the mobile switching center (MSC), also known as the mobile telephone switching office (MTSO). This is a large telephone switch that has dedicated connections to every base station and to the public switched telephone network (PSTN). Connections to the base stations can be by coaxial cable, by point-to-point microwave link, or by optical fiber. The MSC controls the frequencies on which the base station and mobile radios transmit and receive, and thus it is responsible for assigning channel sets to cells and for assigning channels to calls. The MSC orchestrates setup and disconnection of calls and manages handoffs between cells as will be described below. The MSC also maintains a list of subscribers and is able to authenticate callers and communicate with other MSCs to verify the identity of roamers from other cellular systems. All cellular calls, even those from a mobile unit to a mobile unit in the same system, are routed through the MSC. Rappaport describes an MSC as being able to handle 100,000 subscribers and up to 5,000 simultaneous calls.[2] Figure 4.20 suggests the arrangement.

2. T. S. Rappaport, *Wireless Communications: Principles and Practice,* 2nd ed. (Upper Saddle River, NJ: Prentice Hall, 2002), 14.

Dynamic Channel Assignment

In early cellular systems, the organization of channels into channels sets and the assignment of channel sets to cells were fixed. Once the channels were deployed, manual intervention was required to change a channel assignment. Thus if all of the channels in a given cell were in use, a subscriber wishing to place an additional call would be blocked and would receive a "busy" indication. Many modern systems allow changes in channel assignment under computer control. This provides the capability to dynamically adjust channel allocations based on varying user demands. Then if all channels in a given cell are in use, the MSC might be able to "borrow" a channel from a neighboring cell, if this can be done without causing an unacceptable level of cochannel interference. As a further extension of this idea, consider a market area consisting of an urban downtown surrounded by suburbs. During the business day, as people migrate toward the city center, demand in the downtown might increase as demand in the suburbs falls off. With dynamic channel assignment the MSC would be able to reassign channels from suburban cells to cells in the downtown to follow the demand. The algorithm for channel reassignment would have to take into account the cell locations, the propagation models, and the instantaneous traffic conditions to maintain an adequate separation difference between users operating on the same channel.

Handoff Concepts and Considerations

The cellular concept is fundamentally enabled by its ability to track the variations in a signal emanating from a mobile unit and to quickly transfer a connection among base stations as the mobile unit moves about a service area. Handoffs may occur between adjacent cells, between different sectors of the same cell, or between "large" and "small" cells of an overlaid pattern such as the one shown in 4.18(b). The purpose of a handoff is to ensure that a mobile unit is served by the base station that is assigned to serve the mobile unit's geographic location. When a mobile unit strays beyond the physical boundary of the cell or sector assigned to its serving base station, there is a significant potential for the mobile unit to cause or receive interference above the level required for adequate performance. Furthermore, when a mobile unit strays well beyond its serving cell boundary, the signal it receives may become so weak that the information needed to successfully complete a handoff cannot be communicated. In this case the connection may ultimately be dropped. A dramatic degradation in users' perception of the quality of service provided by a system may occur if a significant number of mobile units are being served beyond the cell boundaries of their serving base station or sector.

We have previously noted that the worst-case interference occurs when a mobile unit is located at a cell boundary. For this case the signal from the serving base station is weakest, and the signals from some of the interfering base stations may be strongest. Since this is the circumstance under which a handoff is imminent, we note that handoffs occur at precisely the times when the system is most susceptible to errors. It is therefore a good practice for the system to attempt to anticipate when a handoff might be needed and to initiate the handoff before the connection is broken owing to weak signals.

In most cellular systems a handoff involves a change from one channel to another. The decision to perform a handoff is made on the basis of signal strength measurements that attempt to determine whether the received signal level will be adequate to continue to meet quality of service requirements. The execution of a handoff involves a number of steps and may take a significant interval of time to complete. Given that a mobile unit may be moving with significant speed (in a vehicle, for example), a decision to perform a handoff is usually made in anticipation of the need. This may be accomplished by setting a threshold level for signal strength handoff somewhat above that needed to guarantee quality of service.

Handoff processes vary among systems, but in general the steps may involve measurement of signal strength; determination of the most appropriate channel to switch to; and coordination of the actions of the mobile unit, the base stations involved, and the switching functions by the MSC. The actual switching of a call from one base station to another should not be perceptible to a user during a connection. Since the switch generally involves tuning the mobile unit to a different channel, there is a possibility of losing a segment of the information being conveyed. The first AMPS system required that the handoff gap be no more than 100 ms to avoid the possibility of dropping a syllable of speech.

In the earliest cellular system designs, handoff control was centralized at the mobile switching center. The MSC monitored the signal strength received from each mobile unit. Signal strength could be averaged over a number of samples to avoid initiating a handoff prematurely in response to Rayleigh fading as the mobile unit moved. When average received signal strength dropped below the handoff threshold, the MSC would use extra receivers known as "location receivers" at the surrounding base stations to listen for the mobile unit. Based on the signal strength received from the mobile by each location radio, the MSC would make a determination about the cell or sector, if any, to which the call should be switched. In order to successfully carry out a handoff, a set of location receivers must be available, and there must be a free channel in the destination cell. It is easy to see that excessively frequent handoffs could overburden an MSC, resulting in calls being dropped before the handoff could be completed.

In some second-generation digital cellular systems, channels are shared among several calls in a time-multiplexed round-robin fashion. In such systems a mobile receiver is receiving data over its assigned channel only part of the time, so that the remaining time can be used to monitor base stations in adjacent cells. The mobile unit can then pass signal strength data on to its serving base station. In particular, the mobile unit can detect when signals from its own base station are becoming weaker, while signals from an adjacent base station are becoming stronger. The result is a "mobile assisted handoff" that requires less processing on the part of the MSC, resulting in greater reliability. Code-division multiple access (CDMA) systems offer a third handoff possibility. In these systems, to be discussed in detail in a later chapter, all of the users in a group of adjacent cells share a single channel. This opens the possibility that the signal from a given mobile unit may be received simultaneously by two base stations. The MSC can monitor the quality of both signals and at any given moment can select the better one. When one received signal clearly dominates, the MSC can complete the handoff without requiring any radios to actually switch frequencies. This procedure is called a "soft handoff," which greatly reduces the possibility of a call being dropped during a handoff.

In Chapter 3 we discussed the randomly varying nature of signals in a real-world environment. As a result of macro-fading and micro-fading phenomena, signal strength must be averaged to determine when a handoff is necessary. The accurate determination of the need for a handoff, and its successful completion, depends on a well-designed algorithm involving several variables, including the rate of change of the received signal level, the cell size, and the inferred mobile unit speed and direction. The penalty for a handoff failure is usually a dropped call, an event users consider most annoying. The use of resources to perform unnecessary or early handoffs may have a widespread degrading effect on overall system performance. It should be evident that well-designed handoff strategies are at the heart of a well-designed system. The design of handoff strategies is particularly challenging insofar as their effectiveness cannot be calculated by formula and usually requires large-scale simulation.

Traffic Engineering, Trunking, and Grade of Service

We have introduced the central concept of frequency reuse and discussed in detail the hexagonal cellular structure that follows from it. We have investigated the key design trade-offs involving cell radius and cluster size. We have explored avenues for expansion, including sectoring and cell splitting. Now we are ready to introduce the last piece of the puzzle. In this section we explore the connection between the number of radio channels a cell contains and the number of subscribers a cell can support.

Early in the development of the public switched telephone network it was recognized that having fixed connections between every possible pair of telephones is not only physically impractical and economically prohibitive but also (fortunately) unnecessary. The number of connections required to provide fixed connections between every possible pair of N telephones is given by

$$C_N = \sum_{k=1}^{N-1} k. \tag{4.45}$$

For example, 10 telephones require 45 connections, 100 telephones require 4950 connections, and 10,000 telephones require almost 50 million connections.

Since any given telephone needs to be connected to only one (or a relatively few) other telephones at any instant of time, and since the duration of a phone call, even a long phone call, is relatively short (typically measured in minutes), it is possible to deploy a pool of common resources that supports only requested connections for the duration of a call and allows resources to be returned to the pool when the call is ended. For example, consider a limited geographic area served by what is commonly called a telephone **central office**. Each of the telephones within the area is connected to the central office by a pair of wires. Within the central office any telephone can be connected to any other by joining the terminals of the wires. This is the function of the central office switch.

In a ten-digit telephone number the first three digits are the "area code," the second three digits are the "central office code," and the last four digits represent the "line number" within the office. Clearly there is a maximum of 10,000 phone numbers that can be associated with a given central office code. Therefore if we consider connections among telephones only within a single central office, a worst-case maximum of only 5000 connections would ever need to be made, compared with the 50 million identified in connection with Equation (4.45). In fact, since the duration of a call is relatively short, it would be rare indeed that more than a few hundred connections would be needed. The central office, of course, needs lines that connect to other central offices and to long-distance switches that allow connections to other area codes. Here again, resources are shared, so that the number of lines is much smaller than the number of possible connections. A line that connects switching offices and that is shared among users on an as-needed basis is called a **trunk**.

The fact that the number of trunks needed to make connections between offices is much smaller than the maximum number that could be used suggests that at times there might not be sufficient facilities to allow a call to be completed. A call that cannot be completed owing to a lack of resources is said to be **blocked**. The question then arises as to how to determine the quantity of equipment that is needed so that the event of a call being blocked is acceptably infrequent. It is interesting to note that similar questions about the quantity of shared resources arise in many situations we encounter daily, as in the following examples.

- How many bank tellers are needed to ensure an acceptable average wait during the noon hour on payday?
- How many trains and how many cars per train are needed to ensure that commuters from the suburbs will be served on the next arriving train?
- How many help desk agents should be on duty during the business day to support a help line with a waiting time of less than five minutes?

And finally, of most interest to us:

- How many channels per cell are needed in a cellular telephone system to ensure a reasonably low probability that a call will be blocked?

There are a number of common factors in all of these examples. First, there is a finite number of service providers. Next, there is a large base of customers, only a few of whom need service at any given moment; as a result, requests for service occur randomly in time. Next, the interval of time during which the service provider is occupied with a customer is random. Finally, the level of performance, or "goodness of service," can be quantified as a probability that service will be available when requested.

The frequency of service requests and the duration of service are not known in advance, but we consider them predictable in a statistical sense. These and similar service systems are studied in a branch of applied statistics commonly referred to as queueing or traffic theory. In the jargon of telecommunications, the term **trunking theory** is often used when referring to the application of queueing theory to determine the number of trunks required to support the connections between central offices. The term **traffic engineering** refers to the application of queueing theory to establishing the quantity of resources necessary to provide a given level of service. During the major growth years of the public switched telephone network, significant advancements were made in the field of queueing theory largely because of its particular relevance to telephone applications. Although the underlying theory is relevant to a wide range of situations that have enormous economic implications, our discussion here will be solely concerned with telephone traffic and its modern extension to cellular systems.

In the telephone system context the term **grade of service** (GOS) is used to mean the probability that a user's request for service will be blocked because a required facility, such as a trunk or a cellular channel, is not available. For example, a GOS of 2% implies that on the average a user might not be successful in placing a call on 2 out of every 100 attempts. In practice the blocking frequency varies with time. One would expect far more call attempts during business hours than during the middle of the night. Telephone operating companies maintain usage records and can identify a "busy hour," that is, the hour of the day during which there is the greatest demand for service. Typically, telephone systems are engineered to provide a specified grade of service during a specified busy hour. The busy hour chosen as a design target may be the busiest hour of the week, but it is not normally the busiest hour of the year. Facilities are rarely available to maintain the desired grade of service at 12:00 AM on New Year's Day or during the afternoon on Mother's Day.

User calling habits can be characterized statistically by two parameters: the average number of call requests per unit time λ_{user} and the average holding time $\bar{H}$. The parameter λ_{user} is also called the average arrival rate, referring to the rate at which calls from a single user arrive at the switch. The average holding time is the average duration of a call. The product $A_{user} = \lambda_{user}\bar{H}$—that is, the product of the average arrival rate and the average holding time—is called the **offered traffic intensity** or **offered load**. This quantity represents the average traffic that a user provides to the system. The term *offered* refers to the fact that a call request may or may not be honored, depending on the availability of a channel or trunk. Offered traffic intensity is a dimensionless quantity that is traditionally measured in erlangs, in honor of the pioneering efforts of the Danish mathematician A. K. Erlang in the early 1900s. One erlang is an arrival rate of one call per minute multiplied by a holding time of one minute. If, by fortuitous accident, a new call arrived just as the previous call terminated, and this continued to happen, then one erlang of traffic would continuously tie up one channel or trunk.

Example

Suppose during the busy hour a user makes an average of two calls per hour and holds each call an average of 15 min. Find the offered traffic intensity.

Solution

The average arrival rate is $\lambda_{user} = 2 \text{ calls/hour} = 2 \text{ calls}/60 \text{ min} = 1/30 \text{ calls/min}$. Then $A_{user} = \frac{1}{30} \text{ calls/min} \times 15 \text{ min} = 0.5 \text{ erlangs}$.

If the blocking probability or GOS is P_b, then $1-P_b$ represents the fraction of call requests that actually result in assignment of a channel. If we use $A_{c_{user}} = (1-P_b)\lambda_{user}$ instead of λ_{user} to calculate traffic intensity, we obtain the **carried traffic intensity** or **carried load** $A_{c_{user}}$. The carried load is a measure of the traffic actually carried by the system. In the computer industry this quantity is known as **throughput**.

Example

Suppose in the preceding example that the GOS during the busy hour is 10%. Find the carried load for an individual user.

Solution

The carried load is $(1-P_b)A_{user} = (1-0.1)0.5 = 0.45$ erlangs.

Call arrivals or requests for service are modeled as a Poisson random process. (See Appendix B for a detailed discussion of the model.) The Poisson model has been demonstrated through long experience to be very accurate. It is based on the assumption that there is a large pool of users who do not cooperate in deciding when to place calls. Holding times are very well predicted using an exponential probability distribution. This implies that calls of long duration are much less frequent than short calls. If the traffic intensity offered by a single user is A_{user}, then the traffic intensity offered by N users is $A = NA_{user}$. This follows directly from the fact that a call arrival rate of λ_{user} for a single user implies an aggregate call arrival rate of $\lambda = N\lambda_{user}$ for N users. The purpose of the statistical model is to relate the offered traffic intensity A, the grade of service P_b, and the number of channels or trunks K needed to maintain the desired grade of service.

Two models are widely used in traffic engineering to represent what happens when a call is blocked. The **blocked calls cleared** model assumes that when a channel or trunk is not available to service an arriving call, the call is cleared from the system. In the public switched telephone network, a call that cannot be serviced is switched to a "reorder" signal generator. (A reorder signal is similar to, but slightly more rapid than, the familiar busy signal that indicates that a called party is using the telephone.) The customer will eventually hang up and may try again later. The call arrival rate λ_{user} includes calls that are retries as well as new calls; the switch cannot distinguish one arriving call from another. The second model is known as **blocked calls delayed**. In this model a call that cannot be serviced is placed on a queue and will be serviced when a channel or trunk becomes available. This model represents the familiar "please hold for the next available operator" system that is used by many commercial organizations.

Use of the blocked-calls-cleared statistical model leads to the celebrated Erlang B formula that relates offered traffic intensity A, grade of service P_b, and number of channels K. A derivation is provided in Appendix B. The Erlang B formula is

$$P_b = \frac{\frac{(A)^K}{K!}}{\sum_{n=0}^{K} \frac{(A)^n}{n!}}. \tag{4.46}$$

Owing to the factorials, this formula can be a little tricky to evaluate using a calculator. Some tips for numerical evaluation using MATLAB are included in Appendix B. The Appendix also contains a table of values of offered load for various values of blocking probability and number of channels.

When the blocked-calls-delayed model is used, the "grade of service" refers to the probability that a call will be delayed. In this case the statistical model leads to the Erlang C formula,

$$P[delay] = \frac{\frac{(A)^K}{(K-A)(K-1)!}}{\frac{(A)^K}{(K-A)(K-1)!} + \sum_{n=0}^{K} \frac{(A)^n}{n!}}. \tag{4.47}$$

In the case that a call is delayed, there is interest in knowing how long the delay is likely to last. The probability that the delay will last more than t seconds is

$$P[delay > t] = P[delay] e^{-\frac{(K-A)t}{\bar{H}}}. \tag{4.48}$$

Example

In a certain cellular system an average subscriber places two calls per hour during the busy hour and holds calls for an average of 3 min. Each cell has 100 channels. If blocked calls are cleared, how many subscribers can be served by each cell at a 2% GOS?

Solution

Using the Erlang B table in Appendix B with $K = 100$ and $P_b = 2\%$, we find that $A = 87.972$ erlangs. Now an individual subscriber offers a load of A_{user} = 2calls/60 min × 3 min = 0.1 erlang. Thus the maximum number of subscribers is $N = \frac{A}{A_{user}} = \frac{87.972}{0.1} \cong 880$ subscribers.

Example

Suppose in the previous example that the channels have been divided into two groups of 50 channels each. Each subscriber is assigned to a group and can be served only by channels from that group. How many subscribers can be served by a two-group cell?

Solution

Using the Erlang B table with $K = 50$ and $P_b = 2\%$, we find that A_{group} = 40.255 erlangs. Then the maximum number of subscribers per group is $N_{group} = \frac{A_{group}}{A_{user}} = \frac{40.255}{0.1} \cong 403$. Counting both groups, the maximum number of subscribers is $N = 2N_{group} \cong 806$ subscribers.

Example

Continuing the example, suppose that the set of channels has been divided into four groups of 25 channels each. How many subscribers can be served by a cell now?

Solution

With $K = 25$ and $P_b = 2\%$, the Erlang B table gives $A_{group} = 17.505$ erlangs. Then $N_{group} = \frac{A_{group}}{A_{user}} = \frac{17.505}{0.1} \cong 175$. Counting all four groups gives $N = 4N_{group} \cong 700$ subscribers maximum.

An important concept to learn from this series of examples is that the number of subscribers that can be supported by a given number of channels decreases as the pool of channels is subdivided. We can express this in terms of the **trunking efficiency** ξ, defined as the carried load per channel, that is,

$$\xi = \frac{A_c}{K} = \frac{(1 - P_b)A}{K}. \tag{4.49}$$

Example

If the GOS is 2%, find the trunking efficiency for 100 channels offered as (a) a single group, (b) two groups of 50 channels each, and (c) four groups of 25 channels each.

Solution

Using the offered load data from the previous series of examples, we can fill in Table 4.4:

Table 4.4 Trunking Efficiency for 100 Channels

No. of Groups	A_{group} (erlangs)	A (erlangs)	A_c (erlangs)	ξ
1	87.972	87.972	86.21	0.86
2	40.255	80.50	78.89	0.79
4	17.505	70.02	68.62	0.69

With a single group of 100 channels, each channel carries 86% of the load it could carry if it were continuously in use. When the pool of channels is subdivided into four groups, then to maintain a 2% GOS, each channel can carry only 69% of a full load.

Example

An AMPS system supports 395 voice channels per service block in each market area. For a seven-cell-cluster size, a cell can be assigned a maximum of 57 channels. As a complete pool, the 57 channels can support an offered load of $A = 44.222$ erlangs at a 1% blocking probability. The carried load is $A_c = (1-0.01)44.222 = 43.78$ erlangs, and the trunking efficiency is $\xi = \frac{43.78}{57} = 0.77$, or 77%.

If 120°-sector antennas are used to improve the signal-to-interference ratio, then there will be 19 channels available in each sector. The offered load is now $A = 3\times 11.230 = 33.69$ erlangs (the offered load for one sector of 19 channels at 1% GOS is 11.230 erlangs, and there are three such sectors). The carried load is $A_c = (1-0.01)33.69 = 33.35$ erlangs, and the trunking efficiency is $\xi = \frac{33.35}{57} = 0.59$, or 59%. The signal-to-interference ratio may be higher, but the amount of traffic that can be carried at the specified grade of service is substantially decreased.

The next example pulls together several ideas. It explores how system growth through sectoring is impacted by trunking efficiency considerations.

Example

A cellular system with an allocation of $N_{chan} = 300$ channels requires a 15 dB signal-to-interference ratio. A seven-cell-cluster layout with omnidirectional antennas has been performing adequately, but the system now needs additional channels to accommodate growth. An average subscriber places two calls per hour during the busy hour and holds each call an average of 3 min. Blocked calls are cleared, and the system offers a GOS of 5%. By what percentage can the number of subscribers be increased if 60° sectoring is introduced? The path-loss exponent is $\nu = 4$.

Solution

Let us begin by estimating the number of subscribers that a cell can support before sectoring. Each subscriber produces an offered load of A_{user} = 2 calls/60 min × 3 min = 0.1 erlang. For a seven-cell-cluster layout, the number of channels per cell is $N_{cell} = \frac{N_{chan}}{K_{cluster}} = \frac{300}{7} = 42.9$. Using the Erlang B table in Appendix B with a GOS of 5% and 43 channels, we find that each cell can support a carried load of $A = 37.565$ erlangs. The number N of subscribers is then $N = \frac{A}{A_{user}} = \frac{37.565}{0.1} \cong 376$ subscribers per cell.

We have encountered this sectoring problem previously as an example in the section titled "Sectoring." In that section we showed that using 60° sectors, the 15 dB signal-to-interference ratio could be maintained even if the cluster size is reduced to $K_{cluster} = 4$. Now with a reduced cluster size, the number of channels per cell is increased to $N_{cell} = \frac{N_{chan}}{K_{cluster}} = \frac{300}{4} = 75$. Each cell is divided into six sectors, so the number of channels per sector is $N_{sector} = \frac{75}{6} = 12.5$. From the Erlang B table with 12 channels we obtain (conservatively) $A_{sector} = 7.95$ erlangs. Then for the entire cell, $A = 6\times 7.95 = 47.7$ erlangs. This gives a number of subscribers per cell of $N = \frac{A}{A_{user}} = \frac{47.7}{0.1} \cong 477$.

We now see that the number of subscribers has increased by $\frac{477}{376} = 1.27$, or 27%. This is a substantial increase, but recall that the example in the "Sectoring" section showed an increase in the number N_{market} of channels of 75%. Counting the channels does not give a true picture of the increase in capacity, since dividing the cells into sectors causes a loss of trunking efficiency.

Conclusions

In this chapter we have focused on ways to design a wireless telecommunication system that will support a virtually unlimited number of users and span a virtually unlimited geographic range. We have shown that the frequency reuse concept arises as a natural solution in an environment in which the available radio spectrum is limited. We showed how a cellular organization provides a methodology for implementing frequency reuse. The following are the important ideas to take away as you conclude this chapter.

- Transmitter frequencies (i.e., channels) can be reused if the transmitters are separated by a sufficient distance. This leads to a system architecture consisting of a geographic array of cells, each containing a base station that is assigned a specific set of channels.
- A hexagonal cell structure allows an unlimited geographical area to be covered by cells with no gaps in coverage and with minimum overlap. A cellular layout is characterized by a pair of integers *i* and *j*. Starting at any given cell, another cell that uses the same set of channels can be located by moving *i* cells in a direction parallel to one of the sides of the cell, then turning counterclockwise 60° and moving *j* cells in the new direction.
- A set of cells that together are assigned all of the channels is called a cluster. The size of a cluster is given by $K_{cluster} = i^2 + j^2 + ij$. Thus only certain cluster sizes are physically possible.
- The key design parameters for a cellular layout are the cell radius R and the cluster size $K_{cluster}$. The cluster size can be written in terms of the frequency reuse ratio $Q = D/R$, where D is the distance between cochannel cells. We have shown that $K_{cluster} = Q^2/3$.
- The cell radius is determined by a trade-off. The radius should be as large as possible to include as many subscribers as possible. This minimizes the cost per user of the base station tower and equipment. On the other hand, reducing the cell radius increases the number of cells that are needed to cover the market area, which in turn increases the number of times that each channel is reused. This increases the geographic density and total number of subscribers that a system can support.
- The cluster size is also determined by a trade-off. The cluster size should be made as small as possible to maximize the number of channels per cell and the number of clusters in the system. This maximizes the density of subscribers, and it also maximizes the number of times that each

channel is reused. Maximizing the frequency reuse has the effect of maximizing the total number of subscribers that the system can support. On the other hand, the signal-to-cochannel-interference ratio is entirely determined by the frequency reuse ratio Q and therefore by the cluster size. We have shown that $S/I = (\sqrt{3K_{cluster}})^{v}/6$, where v is the path-loss exponent. The cluster size must be large enough to ensure that the signal-to-interference ratio can produce an adequate quality of service.

- Adjacent-channel interference can also limit subscriber density. Channels closely spaced in frequency are usually not assigned to the same cell.
- Cellular systems must be designed from the outset to allow for an expanding number of subscribers. Two expansion routes are sectoring and cell splitting. Dividing each cell into sectors limits the number of sources of cochannel interference. This may allow the cluster size to be reduced, with a resulting increase in the subscriber density.
- Cell splitting allows an array of cells to be replaced with a similar array of cells having a smaller radius. The cluster size, and therefore the level of cochannel interference, is not changed. The increase in user density can be dramatic, but additional base stations must be added to the system. It is important to take advantage of the geometry so that existing base stations do not have to be moved when cells are split.
- In order for the system to operate effectively, a mechanism must be provided to hand calls off from one base station to another as a subscriber moves from one cell to another. Handoff strategies were briefly discussed, and the role of the mobile switching center in managing handoffs and assigning channels was introduced.
- Traffic engineering uses a statistical model to determine how many users can be supported by a given number of channels. An important parameter is the grade of service: the probability that a call will be blocked because no channel is available for it. The Erlang formulas relate the number of channels, the grade of service, and the number of subscribers. We showed that if a set of channels is subdivided into groups, the trunking efficiency goes down and fewer subscribers can be supported at a given grade of service. This is precisely what happens when cells are divided into sectors, and some of the expected capacity gain is not realized owing to a loss in trunking efficiency.

Chapters 2, 3, and 4 have all been concerned with designing a communication system that can convey an amount of power between endpoints that is adequate to allow service. In Chapter 2 we investigated a point-to-point link in free space. In Chapter 3 we introduced real-world propagation conditions but continued our focus on a point-to-point link. In the present chapter we have shown how a single mobile switching center can communicate with a large number of users spread over a large geographic area. Conveying power, however, is not the same thing as conveying information. Establishing a power link is only a necessary first step. In the next few chapters we center our attention on methods of modulation and then characterize the information sources; that is, we will be concerned with techniques for conveying information over the power link that we have established.

Problems

Problem 4.1

The first FCC spectrum allocation for cellular mobile telephone service consisted of two bands totaling 40 MHz. The upper band, 870–890 MHz, was reserved for forward channels or downlink transmissions, and the lower band, 825–845 MHz, was reserved for reverse channels or uplink transmissions. Forward channels are those used for transmissions from a base station to a mobile unit, and reverse channels are those used for transmissions from a mobile unit to a base station. The first cellular systems were designed for telephone (voice) services only and required two 30 kHz channels for a single two-way telephone call.

A. How many complete channel pairs were allocated?

B. The FCC specified that channels be numbered sequentially with reverse channel 1 located at 825.030 MHz and forward channel 1 located at 870.030 MHz. Write a simple expression that specifies the channel center frequency given the channel number for both forward and reverse channels.

C. In 1989 an additional allocation of 10 MHz (5 MHz for each direction) was made as follows: 824–825 MHz and 845–849 MHz for reverse channels, and 869–870 MHz and 890–894 MHz for forward channels. How many additional channel pairs did this allocation provide, and what is the total number of channel pairs available in the augmented cellular band?

D. The channel numbers for the 845–849 MHz (and 890–894 MHz) band are a continuation of those in the original allocation. To simplify digital logic computation, however, the channel numbers in the 824–825 MHz band are represented as the decimal equivalent of the 9-bit two's complement of negative channel numbers starting at channel 0. For example, channel 0 is represented as 111111111, which is 1023 in decimal, and channel 1 is 11111110, which is 1022 in decimal. What are the channel numbers for the channel pairs in the lower 1 MHz bands? Write expressions for the channel center frequencies of the 824–825 MHz and 869–870 MHz band channels.

Problem 4.2

Consider a linear arrangement of base stations as shown in Figure P 4.1. Each base station is separated from its nearest neighbor by a distance $2R$, where R is the radius of the circular coverage area (cell) for each base station. Apart from the transmission frequency, assume that all base stations are identical and that the environment is uniform with path loss of v. Assume that the mobile unit (MU) is located at the edge of the serving cell, that is, the area covered by BS_0. Assume that the nearest or first-tier cochannel neighbors are BS_{-k} and BS_k, and the reuse distance is denoted as D. (The figure shows the special case $D = 2$.)

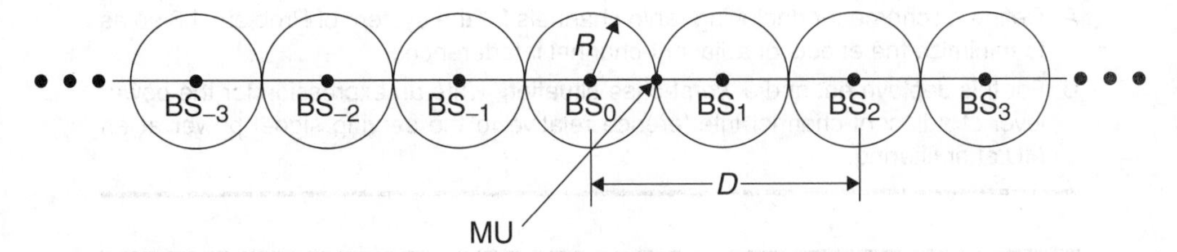

Figure P 4.1 Linear Arrangement of Cells

A. Write an expression for the frequency reuse ratio Q as a function of k.
B. Write a general expression for the S/I at the mobile unit from its first-tier cochannel neighbors (interference sources) as a function of k and v.

Problem 4.3

For the system in Problem 4.2, compute the S/I in decibels from the first-tier interference sources located at BS_{-3} and BS_3 for a path loss of $v = 3$. For this case, how does the S/I degrade if we include second-tier interference, that is, interference from BS_{-6} and BS_6?

Problem 4.4

For the initial 40 MHz spectrum allocation for cellular mobile telephones and the system of cells described in Problem 4.2, how many channel sets and how many complete channel pairs per set would be deployed for a reuse factor Q of 6?

Problem 4.5

For the FCC spectrum allocation described in Problem 4.1, channels are separated by 30 kHz. For any channel i, the channels $i+1$and $i-1$ are defined as the adjacent channels of channel i. In the absence of ideal or very sharp cutoff filters in the receiver, any channel i is subject to interference from its neighboring or adjacent channels. For example, a BS receiver operating on channel i will also receive signal energy from MUs operating on channels $i-1$ and $i+1$. The adjacent channel signal level, however, will be attenuated, depending on the out-of-band characteristics of the receiver filter. Assume that the adjacent-channel signal power is attenuated by a factor α after filtering and that all MUs transmit at the same power level. (Note: a similar situation exists for adjacent-channel BS transmissions.)

A. Define a scheme for deploying radio channels for the system of Problem 4.3 so as to minimize the effects of adjacent-channel interference.
B. For this deployment and a worst-case situation, write an expression for the power level of adjacent-channel interference relative to the serving signal power at an MU after filtering.

Problem 4.6

Write an expression for the worst-case signal-to-adjacent-channel-interference ratio at the output of the receiver filter at BS_0 on channel 4, when interference comes from MUs being served on channels 3 and 5 in BS_{-1} and BS_1 respectively.

Problem 4.7

Assuming that the coverage from any given BS is circular, calculate the percentage overlap when cell BSs are deployed on a triangular, rectangular, and hexagonal grid. See Figure 4.4 and Figure 4.5.

Problem 4.8

What is the smallest reuse distance that can be used to ensure a 20 dB S/I ratio at a cell boundary when the path-loss exponent is $\nu = 3.2$? How many cells are in a cluster for this reuse distance?

Problem 4.9

A certain metropolitan area covers 400 mi^2.

A. What is the minimum number of 1.5 mi cells required to provide radio coverage for the entire metropolitan area?
B. If the path-loss exponent ν is 3.5, what is the minimum cluster size that will ensure a 15 dB S/I at the cell boundary?
C. If there are 490 channels allocated to provide the service, how many channels are available in each cell and how many channels are available to serve users?

Problem 4.10

Extend Table 4.2 to provide values of $S/I\big|_{dB}$ for path-loss exponents $\nu = 3.5$ and $\nu = 3$, and plot $S/I\big|_{dB}$ versus Q, the reuse ratio, for each path-loss exponent, that is, 3, 3.5, and 4. Describe how path loss affects reuse, and discuss the ramifications to coverage and capacity. If you were designing a system for a given area, what would you need to know about the area and how would that information influence your cell parameters and layout? Assume that a large number of channels, N_{chan}, is available.

Problem 4.11

The radio resource density (or simply the radio density), ρ, is an interesting parameter that relates to the system's capacity to serve users simultaneously. The radio density is simply the number of radios per unit area. Develop an expression for ρ in terms of the reuse ratio Q, the cell area A_{cell}, and the number N_{chan} of channels.

Problem 4.12

The most expensive element in the deployment of a wireless system is the set of BSs and associated real estate. Assume that a metropolitan region of area A_{sys} is to be covered by a system allocated N_{chan} channels.

A. Given the path-loss exponent ν, develop an approximate relationship for the total cost CT of the BSs as a function of the radio density ρ, QoS objective in the form of the minimum S/I, and the cost per BS CBS. (Note: The approximation comes about in using the relationship for S/I.)

B. For a uniform deployment of cells, discuss how the total cost of BSs is affected by the spectrum allocation, radio density, and QoS.

C. How does the path-loss exponent ν affect the radio density?

Problem 4.13

For a path-loss exponent $\nu = 4$, compute $S/I|_{dB}$ for a reuse (cluster size) of seven including first- and second-tier interference sources. How much does the $S/I|_{dB}$ degrade when third-tier interference sources are included?

Problem 4.14

The approximation for S/I resulting from first-tier cochannel interferers, that is, $\frac{S}{I} \approx \frac{1}{6}\left(\frac{D}{R}\right)^{\nu} = \frac{1}{6}Q^{\nu}$, is poorest for small reuse ratios. Write an exact expression for the S/I resulting from first-tier cochannel interferers for a reuse of 1, and compute the $S/I|_{dB}$ for a path-loss exponent $\nu = 4$. Compare your answer with the result contained in Table 4.2.

Problem 4.15

For a certain deployment, suppose that adjacent channels are deployed only at adjacent BSs. Assuming that all BSs are identical, how much attenuation (in decibels) is required for an MU receiver filter to ensure a 40 dB margin against adjacent-channel interference in the worst case?

Problem 4.16

First-generation and most second-generation cellular systems relied on the geographic separation of cochannel radios to achieve a given *S/I* and therefore a given quality of service. Given the minimum required *S/I*, however, the reuse ratio and cluster size are determined. System capacity is related to reuse (cluster size) and is a maximum for a reuse of 1; however, the *S/I* ratio is minimized for a reuse of 1.

Suppose that a system could be devised so that the signal could be enhanced relative to the interference using signal processing. Indeed, such a system exists, and it uses a technique termed code-division multiple access (CDMA). If such a system were used, what is the processing gain it must provide in order to achieve an effective 17 dB *S/I* ratio for a reuse of 1 and uniform path-loss exponent $\nu = 4$? Consider both first- and second-tier cochannel interference sources, and use an exact formulation for the first tier.

Problem 4.17

In the ideal case, sectorization reduces cochannel interference by eliminating the interference from four of the six first-tier interferers. Realizable sectored antennas, however, do receive (and transmit) some energy outside the sector main lobe. Assume that the gain outside the main lobe is a uniform 30 dB down from the peak gain of the main lobe. Calculate the worst-case *S/I* ratio (first-tier interference solutions only) for $\nu = 4$ and a cluster size of seven. How does this compare to the cases of ideal sectored antennas and omnidirectional antennas?

Problem 4.18

Using the results of Problem 4.16, determine how 120° sectorization improves $S/I\big|_{dB}$ for a system of reuse of 1. Consider both first- and second-tier interference solutions.

Problem 4.19

Draw and label a 12-cell cluster. Assume that all cells in this cluster are to be split and overlaid with half-radius cells. Draw and label all the half-radius (split) cells necessary to completely cover the original 12-cell cluster. Assume that each of the original cells has a base station at its center and that these base stations will be reused after the split.

Problem 4.20

Assume that average users place two calls of 15 min duration per hour in a given system.

A. Given a seven-cell reuse (cluster size), compute the carried load per cell and trunking efficiency for a 2% GOS and 665-channel allocation.
B. Calculate the number of users that can be accommodated in each cell at the given GOS.
C. Compute the carried load and trunking efficiency if the reuse for the same seven cells were reduced to one.

CHAPTER 5

Digital Signaling Principles

Introduction

In Chapters 2 and 3 we learned that the first step in designing a communication system is to ensure adequate power transmission between locations. From engineering practice we developed some insight into how to optimize power transfer to obtain efficient communication links that can operate in the presence of noise and transmission impairments such as shadowing and multipath reception. In Chapter 4 we learned how frequency reuse enables the design of communication systems that serve multiple users over a growing geographical area. We are now ready to consider how information can be transferred across the links in such a communication system. We postpone consideration of what the information might be until Chapter 7. At this point we wish to create a "pipe" that can be used in a transparent way to convey information from a source to a destination.

Because most communication systems being designed today carry information in digital form, this chapter is focused on digital signaling principles. We begin with a discussion of baseband digital communications and then extend our view to include carrier-based digital communications. It is important to note in getting started that communication systems designed to carry information in digital form and communication systems designed to carry information in analog form are both analog systems. It is the information format that is digital or analog and not the signals transmitted over the wireless link. Conventionally, systems that carry information in digital format are known as "digital communication systems," and we will use this terminology in this chapter.

Digital communication systems differ from analog systems in the design goal of the physical communication link. Analog links are designed with the object of ensuring that the user at the receiving end of the link is provided with a faithful replica of the information waveform. For digital links, the goal is to provide the user at the receiving end with a faithful replica of the information. Preserving the waveform is not important, unless doing so contributes to this goal. The distinction between preserving waveform and preserving information will become clear as we investigate a simple baseband digital link. We will develop the basic indicator of performance quality, the bit error rate, and we will show how performance depends on signal design, as well as on other factors not entirely under the designer's control. We will investigate optimal receiver design for the additive white Gaussian noise (AWGN) channel.

Once we have presented the baseband digital link, we will add modulation. Modulation provides a number of advantages, not the least of which is that it enables radio transmission. To implement a data link based on a modulated carrier, a link designer must choose both symbol and modulation formats. The choice of format has implications for spectral efficiency, implementation complexity, power efficiency, adjacent-channel interference, and flexibility that the designer must take into account. In most cases, a systems engineer works within the constraints of a link budget, cost and complexity limits, spectrum regulations, and packaging and power concerns to select the proper signaling format.

This chapter presents several forms of digital modulation in common use. Special emphasis is placed on visualizing the time waveforms and power spectra for each format so that operation and performance can be understood. The section ends with a discussion of the various performance measures used to compare and facilitate selection of a modulation format.

Finally, spread-spectrum signaling is introduced. Spread spectrum is a modulation technique that broadens the bandwidth of the transmitted signal in a manner unrelated to the information to be transmitted. Spread spectrum was originally developed to provide concealment and security, but was subsequently found to be very effective in making signals resilient in the presence of interference and frequency-selective fading. We will encounter spread-spectrum techniques again in Chapter 6 as they also provide an effective multiple-access technique.

Baseband Digital Signaling

A baseband signal is a signal whose energy is concentrated around DC. A digital baseband signal often contains binary information represented as some kind of pulse train. Information is typically presented to a transmitter as a bit stream encoded in baseband form prior to modulation. In this section we introduce basic digital signaling architecture and techniques, performance analysis, and some typical improvement schemes.

Baseband Digital Communication Architecture

By definition a baseband digital communication system does not use modulation techniques to convey signals. As a consequence, it has the relatively simple architecture shown in Figure 5.1. Baseband digital communication systems or links may be used to convey signals between digital logic chips or between other wire-connected digital devices. The chief goal of the sending elements is to convert a stream of logical data into an analog waveform that can be sent along a tangible medium such as a pair of wires. The receiving elements examine the incoming analog waveforms, which may be noisy and distorted, and attempt to reconstruct the stream of logical data with a minimum of errors. The "logical data source" may produce signals that are inherently

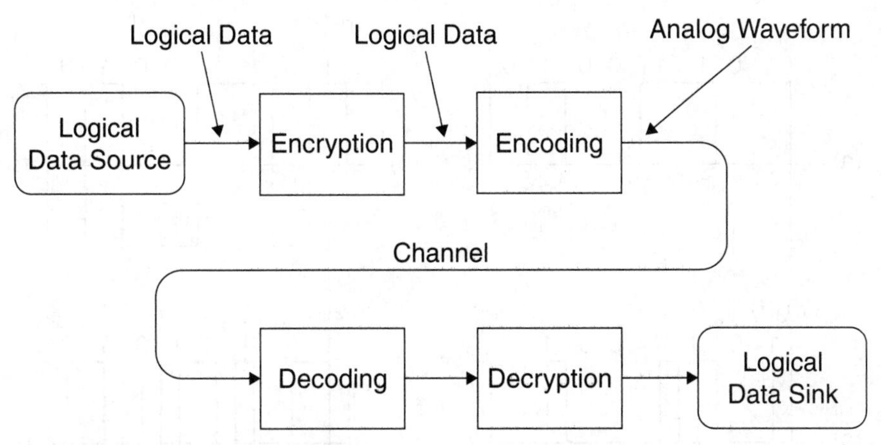

Figure 5.1 Diagram of a Baseband Digital Link

in a digital format, as in the case of a computer keyboard, or derived from an analog waveform that has been digitized. Techniques for digitizing waveforms, in particular speech waveforms, will be discussed in Chapter 7. The block in Figure 5.1 labeled "Encryption" has been included to suggest that the data from the source may be subject to digital processing prior to transmission. A variety of digital processing steps may be included, such as data compression, encryption to ensure privacy and authenticity, and coding for error control. Error control coding is also discussed in Chapter 7. Our focus in the current section is on the block labeled "Encoding," whose job is to map the logical symbols into a series of analog pulses that make up a waveform known as a "line code." At the receiving end of the link, the "Decoding" block reconstructs the stream of logical symbols. Next, digital processing steps such as error control decoding, decryption, and decompression are carried out before the data is delivered to its destination. We note in passing that each of the processing steps—digitization, compression, encryption, error control, and waveform generation—may be called "coding" in the literature. In the present chapter we confine the terminology to refer to the blocks labeled "Coding" and "Decoding" in Figure 5.1.

A **line code** is a pulse train or series of "symbols" that represents digital data as an analog waveform. Each symbol has a fixed duration called the **symbol period**. For systems in which one symbol represents one bit of information, the line code is classified as a binary line code. For a binary line code there are only two possible symbol waveforms that can be transmitted in each symbol period, one to represent a 1 and one to represent a 0. Some line codes utilize a set of more than two symbol waveforms that can represent multiple bits in a single symbol period.

There are many line codes in common use for representing digital data. Some typical forms are shown in Figure 5.2 where the bit sequence 10110010 is shown encoded four different ways. The upper left corner of the figure shows the familiar unipolar non-return-to-zero (NRZ) line code often used in transistor-transistor logic (TTL) circuits. The NRZ code gets its name from the fact that the voltage level does not return to zero during the symbol period. "Unipolar" refers to the fact that the two voltage levels used to represent the logical bits are zero and some nonzero (positive or negative)

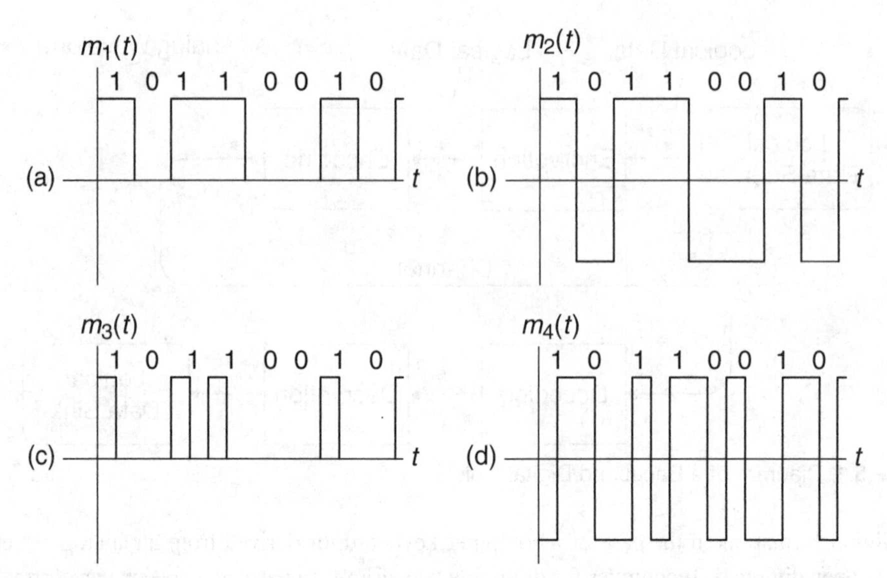

Figure 5.2 Examples of Line Codes Representing the Bit Stream 10110010: (a) Unipolar NRZ Signaling; (b) Polar NRZ Signaling; (c) Unipolar RZ Signaling; (d) Manchester Signaling

value. The upper right graph shows a polar NRZ line code. Polar codes use symmetric positive and negative voltage levels to encode 1s and 0s. The lower left plot shows a return-to-zero (RZ) code in which, as the name implies, the waveform returns to zero during the symbol period. This example shows an RZ code with a 50% duty cycle. In general, the duty cycle must be specified. The final plot, shown in the lower right corner of the figure, shows a Manchester code. This polar line code uses voltage transitions at the midsymbol period instead of voltage levels to carry information. For a symbol period of T_s, a 1 is represented as a transition from a negative to a positive voltage value at time $T_s/2$, and a 0 is represented as a transition from a positive to a negative voltage.

Each of these line codes has specific advantages and disadvantages. The unipolar NRZ line code is simple to implement and is easily extended to generate an "on-off keyed" modulated signal. The polar line code requires two voltage reference levels at the transmitter, but it produces signals with a zero DC level, which can be important when the channel is transformer coupled. Also, the symmetry of the polar line code makes it suitable for differential-mode transmission, which aids noise immunity. The polar line code is easily extended to generate "binary phase-shift keyed" modulation, which we will encounter later in the chapter. Both the unipolar RZ and the Manchester code guarantee that there is at least one edge in each pulse regardless of the data pattern. This can be useful in helping the receiver synchronize to the bit pattern. Although the Manchester code is more complex, it offers the advantage of a zero DC level. Both the unipolar RZ and Manchester line codes occupy a wider bandwidth than do the two NRZ line codes. In the figure, all of the symbols have rectangular shapes, but other shapes are often used. We will be especially interested in pulse shaping aimed at reducing the bandwidth of the transmitted signal.

Table 5.1 Symbol Encoding for the 2B1Q Line Code

Bit Pair	Voltage Magnitude	Current Symbol Polarity
00	1 V	Same as polarity of previous symbol
01	3 V	Same as polarity of previous symbol
10	1 V	Opposite of polarity of previous symbol
11	3 V	Opposite of polarity of previous symbol

Multiple bits can be encoded into a single symbol for more bandwidth-efficient transmission. For example, in the 2B1Q scheme pairs of bits are encoded into a single four-level symbol.[1] The algorithm for encoding is demonstrated in the following example.

Example

The 2B1Q line code, used in systems such as the Integrated Services Digital Network (ISDN), is a type of four-level code that converts two incoming bits into a symbol whose duration is twice the bit interval. The 2B1Q code is a "differential" code in which the voltage level of a given symbol depends on the incoming bit pair and also on the polarity of the previous symbol. Table 5.1 shows how the voltage levels are assigned. The encoder takes in a pair of bits. The first bit in the pair specifies whether the polarity is to be the same as that of the previous symbol or the opposite of that polarity. The second bit specifies whether the voltage magnitude is to be 1 or 3 V. As an example, the bit sequence 1010011101, assumed to arrive at a rate of 10 kbits/s, is shown encoded in Figure 5.3. Note that the symbol duration is twice the incoming bit duration. The bandwidth of the 2B1Q line code is thus half of the bandwidth that would be required for polar NRZ encoding.

Baseband Pulse Detection

In this section we address the problem of detecting symbol pulses in the presence of noise. We illustrate by assuming a binary line code received in additive white Gaussian noise.[2] The role of the receiving elements is to reliably determine whether a 1 or a 0 was sent for each symbol in the incoming waveform. We assume that the receiver makes its decisions one symbol at a time and that each decision is made independently. The analysis that follows is classic; details can be found in many communication systems textbooks.[3]

1. 2B1Q stands for "two binary, one quaternary" symbol.
2. The analysis can be easily extended to include multilevel line codes such as 2B1Q.
3. See, for example, R. E. Ziemer and R. L. Peterson, *Introduction to Digital Communication* (New York: Macmillan, 1992).

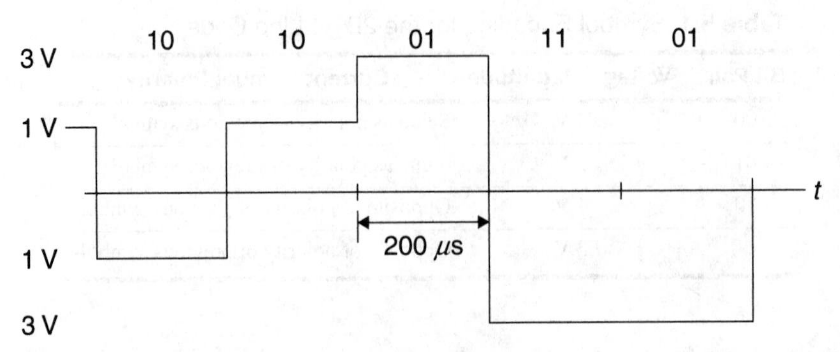

Figure 5.3 Encoding of 1010011101 as a 2B1Q Line Code

The **probability of error** is an important measure of digital receiver performance. This probability is calculated using an analytical model of a baseband system and is a basis for predicting the bit error rate (BER). The BER is an empirical performance measure given by the number of bits in error divided by the number of bits transferred during a test procedure. If the analytical model is accurate, and enough trials with enough samples per trial are performed during the test procedure, then we expect the probability of error to approximate the BER. We will use the two terms interchangeably, but keep in mind that they will be equal only if the analytical model is accurate.

A diagram of the system to be analyzed is shown in Figure 5.4. The diagram shows the signal and noise input to a baseband receiver. The signal is designated by $s_i(t)$, $0 \le t \le T_s$, where $i =$ 0 or 1 depending on whether the signal represents a 0 or a 1, respectively. The noise is written as $n(t)$, giving a received waveform of

$$r(t) = s_i(t) + n(t), \quad 0 \le t \le T_s, \quad i = 0,1. \tag{5.1}$$

The receiver consists of a filter, a sampler, and a threshold comparator. The filter has impulse response $h(t)$, to be discussed later. The sampler samples the filter output $y(t)$ at $t = kT_s$, $k = \ldots,0,1,2,\ldots$ to obtain the **decision statistic** $y(kT_s)$. The decision statistic is compared with a threshold V_T. The receiver will decide that a 1 was transmitted if $y(kT_s) \ge V_T$, and that a 0 was transmitted if $y(kT_s) < V_T$. In the following analysis we will set $k = 1$. Since decisions are made independently for each received pulse, the results will remain general.

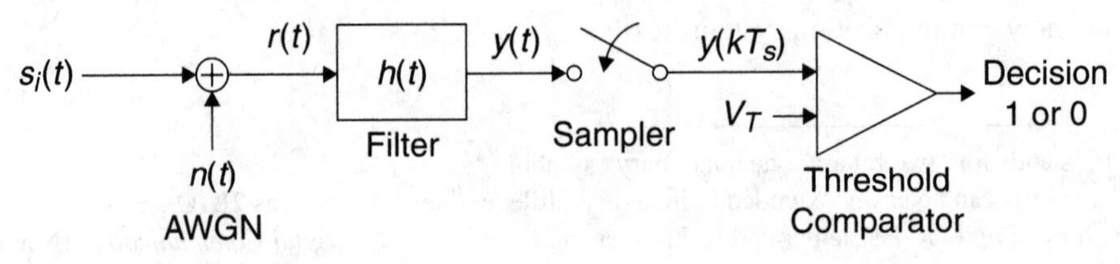

Figure 5.4 Baseband Pulse Receiver Architecture

Since the filter is linear and time-invariant, its output can be found by convolution. We can write

$$y(t) = r(t) * h(t)$$
$$= \int_{-\infty}^{\infty} r(\tau)h(t-\tau)d\tau \quad (5.2)$$
$$= \int_{0}^{T_s} r(\tau)h(t-\tau)d\tau,$$

where the limits on the last integral reflect the fact that the filter output is relevant only while a data pulse is being received. Substituting Equation (5.1) gives

$$y(t) = \int_{0}^{T_s} [s_i(\tau) + n(\tau)]h(t-\tau)d\tau$$
$$= \int_{0}^{T_s} s_i(\tau)h(t-\tau)d\tau + \int_{0}^{T_s} n(\tau)h(t-\tau)d\tau \quad (5.3)$$
$$= s_{oi}(t) + n_o(t), \quad i = 0,1.$$

Thus the filter output is the sum of a signal part and a noise part. Sampling at $t = T_s$ gives

$$y(T_s) = s_{oi}(T_s) + n_o(T_s), \quad i = 0,1. \quad (5.4)$$

The receiver's performance is characterized by its probability of error P_e. This is the probability that the receiver's decision is incorrect. Since there are two ways of making an incorrect decision, we can write

$$P_e = P[error \mid i = 0]P[i = 0] + P[error \mid i = 1]P[i = 1], \quad (5.5)$$

where $P[error \mid i = k]$, $k = 0$ or 1 is the conditional probability of error given that a value k was sent, and $P[i = k]$ is the probability that value k was sent. The probabilities $P[i = 0]$ and $P[i = 1]$ depend only on the data source. Statisticians call these probabilities ***a priori* probabilities**, because they describe what is known about the received bit *before* the receiver observes $r(t)$. We follow normal practice and assume $P[i = 0] = P[i = 1] = \frac{1}{2}$. The next step is to find $P[error \mid i = 0]$ and $P[error \mid i = 1]$. We will show the steps for the first of these in detail, since the calculation of the second term is similar.

Given that $i = 0$—that is, given that $s_0(t)$ is transmitted—the receiver makes an error if $y(T_s) \geq V_T$. We can then calculate $P[error \mid i = 0]$ by

$$P[error \mid i = 0] = P[y(T_s) \geq V_T \mid i = 0]$$
$$= P[s_{o0}(T_s) + n_o(T_s) \geq V_T]. \quad (5.6)$$

Now $s_{o0}(T_s)$ and V_T are constants, so the probability of error depends on the statistical behavior of the noise term $n_0(T_s)$. If we model the received noise $n(t)$ as Gaussian noise with zero mean and power spectrum $S_n(f) = \frac{N_0}{2}$, then the noise at the output of the filter will also be Gaussian with zero mean and power spectrum $S_{n_o}(f) = |H(f)|^2 S_n(f) = |H(f)|^2 \frac{N_0}{2}$, where $H(f)$ is the frequency response of the filter. The variance σ_n^2 of the noise $n_0(T_s)$ is equal to its average power, so we have

$$\begin{aligned}\sigma_n^2 = P_{n_o} &= \int_{-\infty}^{\infty} S_{n_o}(f)df \\ &= \frac{N_0}{2}\int_{-\infty}^{\infty} |H(f)|^2\, df.\end{aligned} \tag{5.7}$$

With this characterization of the noise, we can identify the decision statistic $y(T_s) = s_{o0}(T_s) + n_o(T_s)$ as a Gaussian random variable with mean $s_{o0}(T_s)$ and variance σ_n^2. We can then write the conditional probability of error of Equation (5.6) as

$$\begin{aligned}P[error\,|\,i=0] &= P[n_o(T_s) + s_{o0}(T_s) \geq +V_T] \\ &= \int_{V_T}^{\infty} \frac{1}{\sqrt{2\pi\sigma_n^2}} e^{-\frac{(\alpha - s_{o0}(T_s))^2}{2\sigma_n^2}} d\alpha \\ &= Q\left(\frac{V_T - s_{o0}(T_s)}{\sigma_n}\right),\end{aligned} \tag{5.8}$$

where $Q(x)$ is the complementary normal distribution function defined in Appendix A.

With the assumption that $i = 1$, a very similar calculation gives us

$$P[error\,|\,i=1] = Q\left(\frac{s_{o1}(T_s) - V_T}{\sigma_n}\right). \tag{5.9}$$

Putting Equations (5.8) and (5.9) together into Equation (5.5) gives the overall probability of error as

$$P_e = \frac{1}{2}Q\left(\frac{-s_{o0}(T_s) + V_T}{\sigma_n}\right) + \frac{1}{2}Q\left(\frac{s_{o1}(T_s) - V_T}{\sigma_n}\right). \tag{5.10}$$

The next step is to choose the threshold value V_T to minimize P_e. This is a matter of differentiating Equation (5.10) with respect to V_T and setting the derivative to zero. After some algebra, the optimum threshold value turns out to be

$$V_T = \frac{s_{o0}(T_s) + s_{o1}(T_s)}{2}. \tag{5.11}$$

Substituting for V_T in Equation (5.10) gives the probability of error

$$P_e = Q\left(\frac{s_{o1}(T_s) - s_{o0}(T_s)}{2\sigma_n}\right). \tag{5.12}$$

Example

Suppose polar keying (either polar NRZ or a Manchester code). In this case, $s_{o0}(T_s) = -s_{o1}(T_s)$. The probability of error is

$$P_e = Q\left(\frac{s_{o1}(T_s)}{\sigma_n}\right) = Q\left(\sqrt{\frac{s_{o1}^2(T_s)}{\sigma_n^2}}\right). \tag{5.13}$$

The second form of the expression, using the square root, will turn out to be convenient later on.

Now suppose unipolar keying is used. In this case $s_{o0}(T_s) = 0$. Then we have

$$P_e = Q\left(\frac{s_{o1}(T_s)}{2\sigma_n}\right) = Q\left(\frac{1}{2}\sqrt{\frac{s_{o1}^2(T_s)}{\sigma_n^2}}\right). \tag{5.14}$$

Equation (5.10) is illustrated in Figure 5.5. The figure shows the two Gaussian probability density functions for $y(T_s)$, one assuming $i = 0$ and the other assuming $i = 1$. The figure is drawn for the case of polar keying, with $s_{o0}(T_s) = -1$ and $s_{o1}(T_s) = 1$. The threshold is $V_T = 0$ as given by Equation (5.11). The conditional probabilities of error $P[error \mid i = 0]$ and $P[error \mid i = 1]$ are the shaded regions under the density functions.

Example

Data is transmitted using a polar NRZ line code. The noise at the output of the receiver filter is found to have an RMS value of 0.7 mV when no signal is received. When a 1 is transmitted, the signal at the receiver filter output has a value of 3 mV at the sample time. Find the probability of error.

Solution

Using Equation (5.13),

$$P_e = Q\left(\frac{3\text{ mV}}{0.7\text{ mV}}\right) = Q(4.29) = \int_{4.29}^{\infty} \frac{1}{\sqrt{2\pi}} e^{-\frac{\alpha^2}{2}} d\alpha = 0.9 \times 10^{-5}$$

This means that on the average 1 bit out of every 100,000 is expected to be in error.

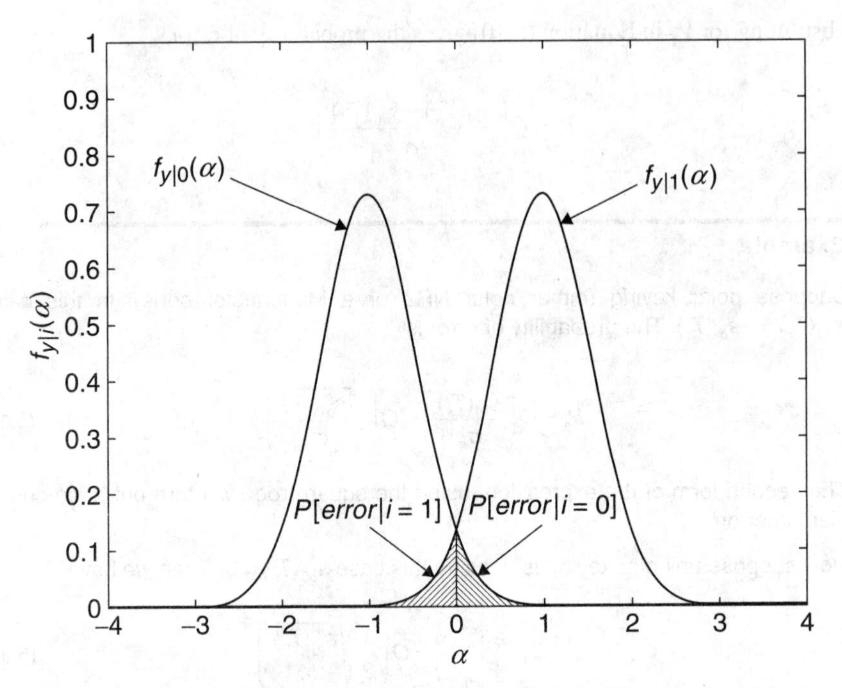

Figure 5.5 Probability Density Functions for the Decision Statistic Given $i = 0$ and $i = 1$

The Matched Filter

The choice of frequency response $H(f)$ for the filter of Figure 5.4 is not arbitrary. The filter has a significant effect on the performance of the receiver. If the filter bandwidth is chosen wide, then the line-code signals $s_i(t)$ will be passed with little distortion, but more noise than necessary will also pass through the filter. If the filter bandwidth is chosen narrow, then the line-code signals may be distorted to the point that correct decisions cannot be made.

Since our objective in designing the baseband pulse receiver is to minimize the probability of error, the appropriate objective in designing the receiver filter is also to minimize the probability of error. If we start with Equation (5.12), we can observe that the probability of error will be smallest when the argument $\frac{s_{o1}(T_s)-s_{o0}(T_s)}{\sigma_n}$ is as large as possible. Note that the numerator of this expression depends on the filter through

$$s_{o1}(T_s)-s_{o0}(T_s)=\int_0^{T_s}\left[s_1(\tau)-s_0(\tau)\right]h(T_s-\tau)d\tau \tag{5.15}$$

(see Equation (5.3).) The denominator depends on the filter through Equation (5.7). If we apply Parseval's theorem to Equation (5.7), we can write

$$\sigma_n^2=\frac{N_0}{2}\int_{-\infty}^{\infty}|H(f)|^2\,df=\frac{N_0}{2}\int_0^{T_s}h^2(t)dt. \tag{5.16}$$

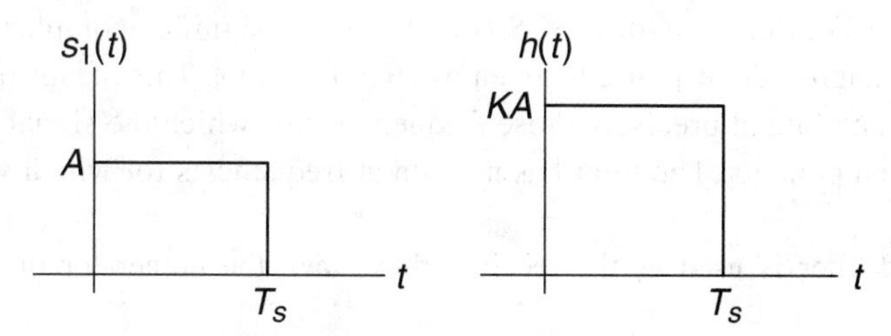

Figure 5.6 NRZ Line-Code Pulse and Matched Filter Impulse Response

The optimum filter, then, has an impulse response that maximizes the ratio

$$\frac{[s_{o1}(kT_s)-s_{o0}(kT_s)]^2}{\sigma_n^2}=\frac{\left[\int_0^{T_s}[s_1(\tau)-s_0(\tau)]h(T_s-\tau)d\tau\right]^2}{\frac{N_0}{2}\int_{-\infty}^{\infty}h^2(t)dt}. \tag{5.17}$$

It turns out[4] that the filter we seek is the one for which

$$h(t)=K[s_1(T_s-t)-s_0(T_s-t)]. \tag{5.18}$$

This filter is called a **matched filter** and is said to be matched to the difference signal $s_1(t)-s_0(t)$. The matched filter's impulse response is the difference signal running backward in time and delayed by the pulse duration T_s. The gain constant K is arbitrary, as it affects both the numerator and the denominator of Equation (5.17) in the same way. The delay T_s must be included to ensure that the receiver's filter is causal. This can be seen clearly in the following example.

Example

For polar keying, $s_0(t)=-s_1(t)$, so that $s_1(t)-s_0(t)=2s_1(t)$. For unipolar keying, $s_0(t)=0$, so that $s_1(t)-s_0(t)=s_1(t)$. Thus, for both polar keying and unipolar keying the impulse response of the matched filter is given by $h(t)=Ks_1(T_s-t)$. For NRZ pulses, the line-code pulse $s_1(t)$ and the impulse response $h(t)$ are shown in Figure 5.6.

Viewing the matched filter in the frequency domain helps us to understand its behavior. For a filter matched to $s(t)$, the impulse response is $h(t)=s(T_s-t)$ and the frequency response is

$$H(f)=S^*(f)e^{-j2\pi fT_s}, \tag{5.19}$$

4. Ziemer and Peterson, *Introduction to Digital Communication,* 155–58.

where $s(f)$ is the Fourier transform of $S(t)$ and the star signifies complex conjugate. We observe that the magnitude response is given by $|H(f)| = |S(f)|$. This means that the matched filter has significant gain at precisely those frequencies for which the signal $s(t)$ has significant frequency components. The filter has no gain at frequencies for which $s(t)$ has no components.

If a matched filter is used as the receiver filter, then the numerator of Equation (5.17) becomes

$$\begin{aligned}\int_0^{T_s}[s_1(\tau)-s_0(\tau)]h(T_s-\tau)d\tau &= \int_0^{T_s}[s_1(\tau)-s_0(\tau)]K[s_1(T_s-T_s+\tau)-s_0(T_s-T_s+\tau)]d\tau \\ &= K\int_0^{T_s}[s_1(\tau)-s_0(\tau)]^2 d\tau \\ &= KE_{diff},\end{aligned} \tag{5.20}$$

where E_{diff} is the energy in the difference signal $s_1(t)-s_0(t)$. The denominator becomes

$$\begin{aligned}\frac{N_0}{2}\int_0^{T_s}h^2(t)dt &= \frac{N_0}{2}\int_0^{T_s}K^2[s_1(T_s-t)-s_0(T_s-t)]^2 dt \\ &= \frac{N_0}{2}K^2E_{diff},\end{aligned} \tag{5.21}$$

since the energy in a signal remains the same if the signal is reversed in time. Now the argument of the Q-function is

$$\frac{[s_{o1}(kT_s)-s_{o0}(kT_s)]^2}{\sigma_n^2} = \frac{K^2E_{diff}^2}{\frac{N_0}{2}K^2E_{diff}} = 2\frac{E_{diff}}{N_0}. \tag{5.22}$$

Substituting in Equation (5.12), we have an expression for the minimum probability of error,

$$P_e = Q\left(\sqrt{\frac{E_{diff}}{2N_0}}\right). \tag{5.23}$$

It is interesting to note, and very important for applying these results, that the probability of error is independent of the shape of the transmitted pulses. The two factors that do affect the probability of error are the extent to which the two signals differ from each other, measured by E_{diff}, and the noise level, measured by N_0.

Example

Find the probability of error when a matched filter is used to receive unipolar NRZ signaling with 1 V pulses and a data rate of 100 kbits/s. Assume that the channel contributes additive white Gaussian noise with a power spectrum of 10^{-7} V²/Hz.

Solution

For unipolar signaling, $s_0(t) = 0$, so $E_{diff} = \int_0^{T_s} s_1^2(t)\,dt$. We are using 1 V rectangular pulses, so $E_{diff} = \int_0^{10^{-5}} 1^2\,dt = 10^{-5}$ V²s. The noise power spectrum is $\frac{N_0}{2} = 10^{-7}$, so $N_0 = 2 \times 10^{-7}$. The probability of error is then

$$P_e = Q\left(\sqrt{\frac{E_{diff}}{2N_0}}\right) = Q\left(\sqrt{\frac{10^{-5}}{4 \times 10^{-7}}}\right) = Q(5.0) \qquad (5.24)$$
$$= 2.9 \times 10^{-7}.$$

Example

For the unipolar NRZ line code of the previous example, find the waveform at the output of the matched filter. Ignore noise.

Solution

The signal pulse $s_1(t)$ and the matched filter impulse response $h(t)$ are as shown in Figure 5.6 with $A = 1$ V and $T_s = 10^{-5}$ s. The result of the convolution $y(t) = s_1(t) * h(t)$ is shown in Figure 5.7. At the sampling time $t = T_s$ the filter output is $y(T_s) = K \times 1^2 \times 10^{-5} = K \times 10^{-5}$ V²s. We note, first, that the output of the filter is maximum at the sampling time, and second, that the filter does not preserve the wave shape of the rectangular line-code pulse.

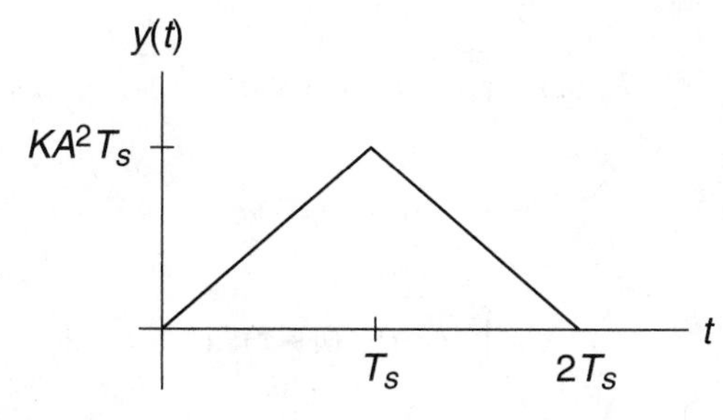

Figure 5.7 Matched Filter Response to a Rectangular NRZ Pulse

Correlation

A central function of most real-world wireless systems is to reliably detect and interpret signals that arrive at the receiver. Typically, a receiver sees a composite of desired signal, noise, and interference from other systems. In most communication applications the waveform of the information symbol is known, although the specific sequence of symbols that conveys the information is not. The receiver's task is to compare the received composite signal with a stored replica of the transmitted symbol and to detect that symbol's presence and polarity in the interference and noise. The matched filter in the receiver of Figure 5.4 contains a (time-reversed) copy of the transmitted symbol waveform as its impulse response. In this section we develop an alternative structure for the optimal receiver that uses a **correlator** for comparing signals.

A matched filter can be difficult to implement for some line-code signals and in some technologies. The correlator provides an alternative implementation that can be advantageous in some cases. Besides providing design engineers with an alternative implementation of the optimum receiver, the correlation concept contributes additional insight into the way an optimal receiver operates. To describe the correlator receiver we will introduce an operation on a pair of signals called **correlation**, which is a way of measuring the similarity between signals. Besides their use in optimal receivers, correlators can be found in applications such as synchronization circuits and radar where they are used to locate and identify specific waveforms in noise and interference. We will encounter correlators again when we study code-division multiple access systems in Chapter 6.

If $g_1(t)$ and $g_2(t)$ are energy signals, we define the **cross-correlation** $R_{12}(\tau)$ by

$$R_{12}(\tau) = \int_{-\infty}^{\infty} g_1(t) g_2(t+\tau) dt. \tag{5.25}$$

The parameter τ introduces a time shift in $g_2(t)$ before it is multiplied by $g_1(t)$ and integrated. This parameter is called the **lag**. The cross-correlation is independent of absolute time and takes a single numerical value at each value of lag. When $R_{12}(\tau)$ is plotted against τ it reveals the values of shift for which $g_1(t)$ and the shifted $g_2(t)$ are most similar.

Note that the order of the functions does matter in the cross-correlation operation, that is,

$$\begin{aligned} R_{21}(\tau) &= \int_{-\infty}^{\infty} g_2(t) g_1(t+\tau) dt \\ &= \int_{-\infty}^{\infty} g_1(t+\tau) g_2(t) dt \\ &= \int_{-\infty}^{\infty} g_1(u) g_2(u-\tau) du \\ &= R_{12}(-\tau). \end{aligned} \tag{5.26}$$

We see that cross-correlation is not commutative.

Although any two different signals may be compared using the cross-correlation function, it is often useful to compare a signal with a shifted version of itself. Such a comparison is called an **autocorrelation** function. Given an energy signal $g_1(t)$, the autocorrelation $R_{12}(\tau)$ is defined by

$$R_{11}(\tau) = \int_{-\infty}^{\infty} g_1(t)g_1(t+\tau)dt. \tag{5.27}$$

At zero lag, the autocorrelation function is the energy in the signal, that is,

$$R_{11}(0) = \int_{-\infty}^{\infty} g_1^2(t)dt = E_1. \tag{5.28}$$

It is straightforward to show that a signal correlates with itself at zero lag at least as well as it correlates with any shifted version of itself, specifically,

$$R_{11}(0) \geq |R_{11}(\tau)|. \tag{5.29}$$

From Equation (5.26) we see that the autocorrelation function has even symmetry, that is,

$$R_{11}(\tau) = R_{11}(-\tau). \tag{5.30}$$

Finally, it can be shown that the Fourier transform of the autocorrelation function is the energy spectrum of the signal. Formally, we have the Wiener-Khinchine theorem,

$$S_{11}(f) = |G(f)|^2 = \int_{-\infty}^{\infty} R_{11}(\tau)e^{-j2\pi f\tau}d\tau. \tag{5.31}$$

Periodic signals do not have finite energy, and so a modified definition of cross-correlation and autocorrelation function must be used. If $x_1(t)$ and $x_2(t)$ are periodic with common period T, then

$$R_{12}(\tau) = \frac{1}{T}\int_{-T/2}^{T/2} x_1(t)x_2(t+\tau)dt, \tag{5.32}$$

and

$$R_{11}(\tau) = \frac{1}{T}\int_{-T/2}^{T/2} x_1(t)x_1(t+\tau)dt. \tag{5.33}$$

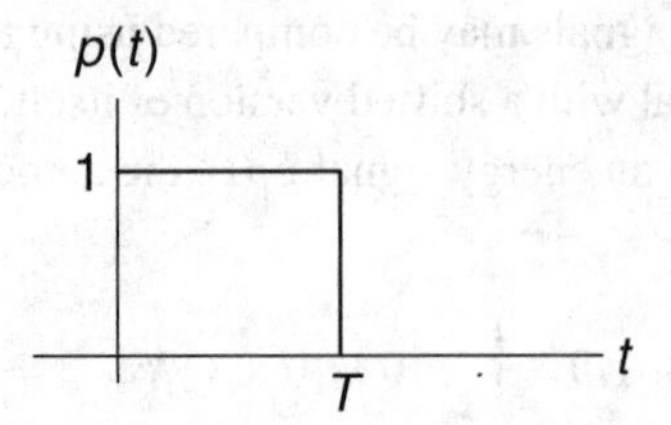

Figure 5.8 Pulse for Autocorrelation Example

The properties given by Equations (5.26), (5.29), and (5.30) apply to periodic signals as well as to energy signals. For periodic signals, the autocorrelation function at zero lag gives the average power rather than the energy, that is,

$$R_{11}(0) = \frac{1}{T}\int_{-T/2}^{T/2} x_1^2(t)dt = P_1. \tag{5.34}$$

Also, the Wiener-Khinchine theorem for periodic signals states that the Fourier transform of the autocorrelation function is the power spectrum and not the energy spectrum. Finally, we note that if $x_1(t)$ is periodic, then $R_{11}(\tau)$ will be periodic as well.

Example

Compute the autocorrelation function $R_{pp}(\tau)$ of the simple rectangular pulse $p(t)$ shown in Figure 5.8. Compute the energy spectrum of the pulse and find the total energy.

Solution

Figure 5.9 shows $p(t)$ and $p(t+\tau)$ plotted on common axes. The diagram is shown for values of lag in the range $0 \le \tau \le T$. In this range we have

$$R_{pp}(\tau) = \int_0^{-\tau+T} 1 \times 1dt = -\tau + T. \tag{5.35}$$

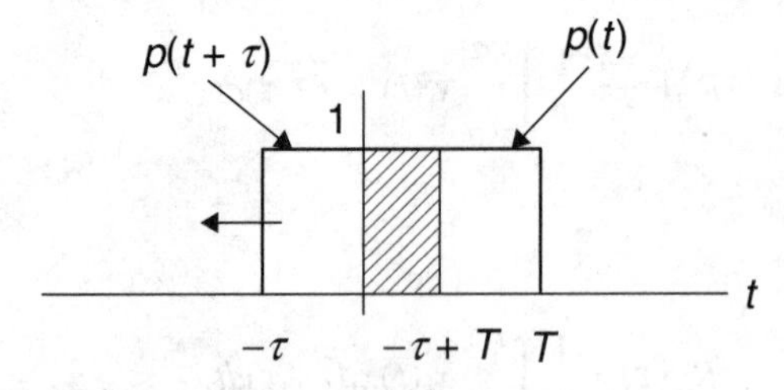

Figure 5.9 Depiction of Autocorrelation Calculation

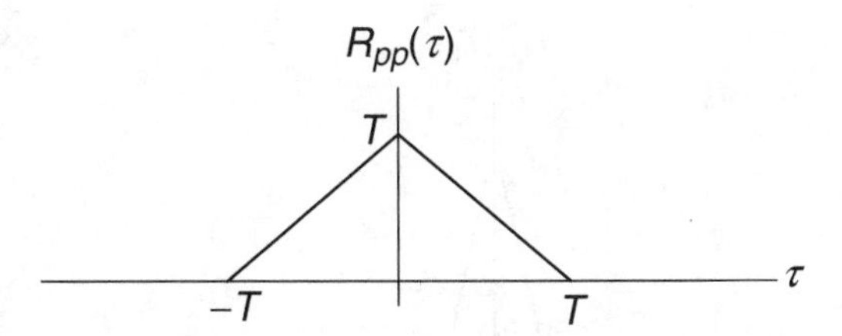

Figure 5.10 The Autocorrelation Function of a Rectangular Pulse

For $T < \tau$, the pulses $p(t)$ and $p(t+\tau)$ no longer overlap and $R_{pp}(\tau) = 0$. Once we have $R_{pp}(\tau)$ for $t \geq 0$, we can use the fact that $R_{pp}(-\tau) = R_{pp}(\tau)$ to complete the analysis. We have

$$R_{pp}(\tau) = \begin{cases} \tau + T, & -T \leq \tau < 0 \\ -\tau + T, & 0 \leq \tau \leq T \\ 0, & \text{otherwise.} \end{cases} \tag{5.36}$$

The result is plotted in Figure 5.10.

The energy spectrum of the pulse $p(t)$ is obtained by taking the Fourier transform of $R_{pp}(\tau)$. This gives

$$S_{pp}(f) = T^2 \operatorname{sinc}^2(Tf), \tag{5.37}$$

as expected. The pulse energy is $E_p = R_{pp}(0) = T$.

As should be apparent from the example, there is a close relationship between correlation and convolution. Correlation is essentially convolution without the "folding" operation.

Example

Consider the two-level waveform shown in Figure 5.11.

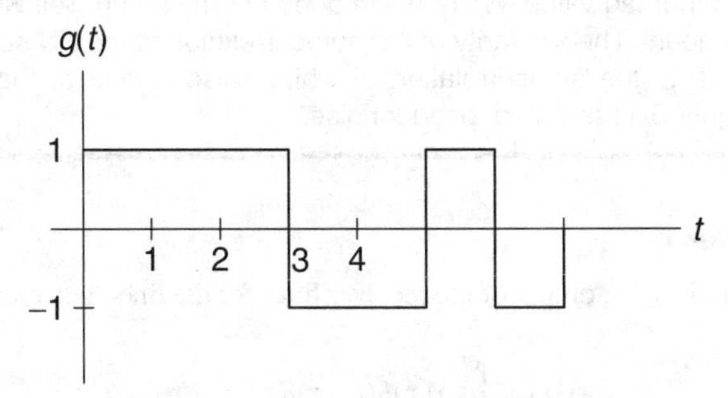

Figure 5.11 A Pseudonoise Binary Sequence Represented in a Polar NRZ Line Code

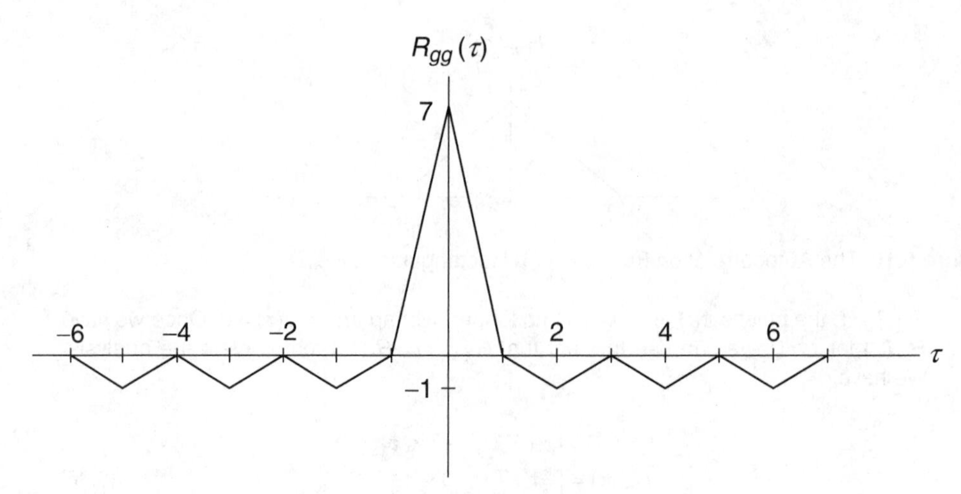

Figure 5.12 Autocorrelation of the Pseudonoise Sequence

This waveform is a pseudonoise (PN) sequence. Its properties give it wide application in communications, radar, and elsewhere. The autocorrelation of the PN sequence is shown in Figure 5.12.

Note that the autocorrelation function has a sharp peak at zero lag and is limited to small magnitudes at other values of lag. This suggests that the PN waveform might be useful as a synchronizing pulse, as it is not similar to shifted versions of itself. We will exploit the PN sequence further when we discuss code-division multiple access in Chapter 6.

Example

As a final example we show the autocorrelation function of a sample of a white Gaussian noise waveform. Figure 5.13(a) shows a sequence of 1000 noise samples. This is an energy signal because the length of the sequence is finite. The autocorrelation is shown in Figure 5.13(b).

Note that the autocorrelation at zero lag is at least ten times as great as the autocorrelation for any other lag value. White noise does not match up well with itself when shifted by any amount. The similarity of the autocorrelation of the PN sequence in the previous example to the autocorrelation of white noise is one of the reasons the sequence of Figure 5.11 is called "pseudonoise."

Correlation Receiver

According to Equation (5.2), the output of the receiver filter for the line-code receiver of Figure 5.4 is

$$y(t) = \int_0^{T_s} r(\tau)h(t-\tau)d\tau. \tag{5.38}$$

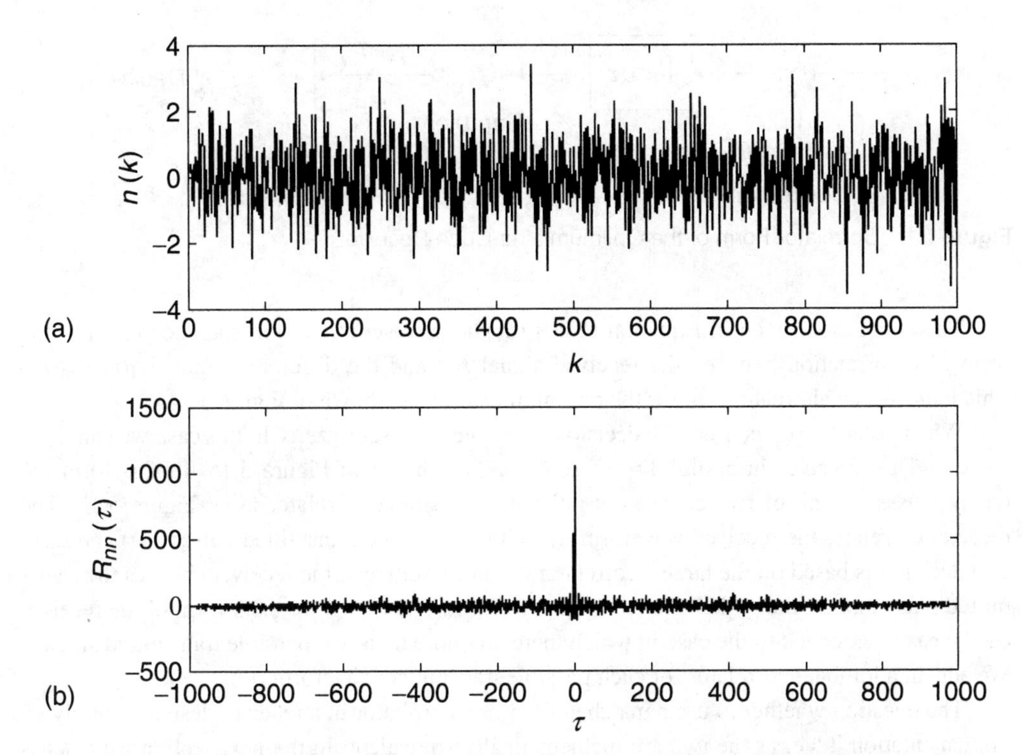

Figure 5.13 (a) Sample of White Gaussian Noise; (b) Autocorrelation of White Gaussian Noise Sample

Sampling at $t = t_s$ gives

$$y(T_s) = \int_0^{T_s} r(\tau)h(T_s - \tau)d\tau. \tag{5.39}$$

Now the optimum receiver uses a matched filter. Substituting the impulse response of the matched filter from Equation (5.18) gives

$$\begin{aligned} y(T_s) &= \int_0^{T_s} r(\tau)K\{s_1[T_s - (T_s - \tau)] - s_0[T_s - (T_s - \tau)]\}d\tau \\ &= K\int_0^{T_s} r(\tau)[s_1(\tau) - s_0(\tau)]d\tau. \end{aligned} \tag{5.40}$$

Recall that the value of the constant K does not affect the receiver performance. We will take $K =$ 1 to simplify the notation.

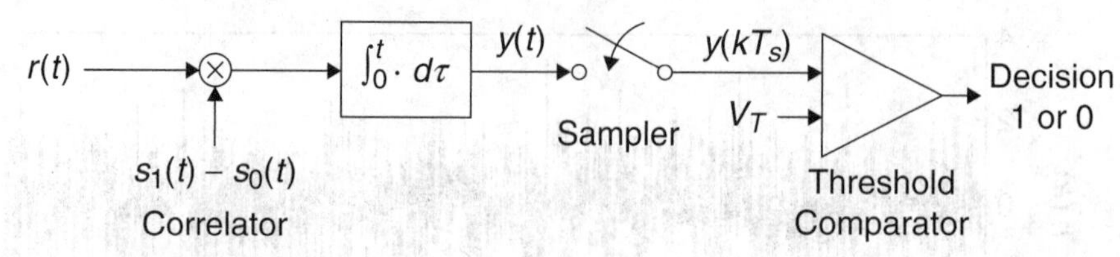

Figure 5.14 Correlator Form of the Optimum Line-Code Receiver

The expression given in Equation (5.40) for the receiver's decision statistic $y(T_s)$ has the form of a correlation between the received signal $r(t)$ and the difference signal $[s_1(t) - s_0(t)]$. This indicates an alternate form for the optimum receiver as shown in Figure 5.14.

When polar keying is used, the decision threshold V_T is set to zero. In this case we can draw the correlator receiver in a slightly different way, as shown in Figure 5.15. In this form the receiver uses a bank of two correlators rather than a single correlator as in Figure 5.14. The receiver correlates the received waveform $r(t)$ with each of the transmitted pulses $s_1(t)$ and $s_0(t)$. The decision is based on the largest correlator output. Intuitively, the receiver chooses the transmitted pulse that is "most like" the received waveform. The correlator bank form of the receiver can be easily extended to the case in which there are more than two possible transmitted signals. We add an additional correlator for each possible transmitted waveform.

The question whether to use a matched filter or a correlator in a receiver design is purely an implementation issue, as the two are mathematically equivalent. In the next section we discuss the performance of this optimum receiver architecture.

Receiver Performance

The performance of a receiver is judged primarily by its ability to reproduce a transmitted bit stream at a required level of accuracy, as measured by the bit error rate. Although it is never possible to have

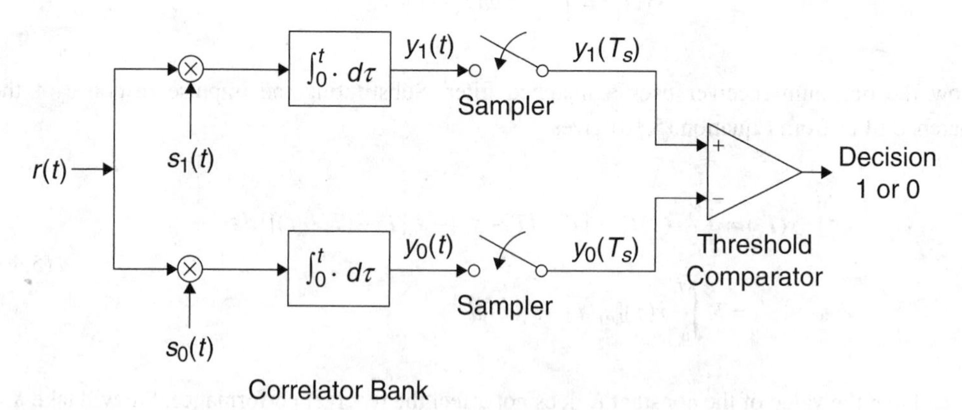

Figure 5.15 Optimum Receiver Based on a Correlator Bank

a bit error rate that is identically zero, it is possible to push the rate arbitrarily close to zero through the application of signal power and transmitter and receiver complexity (both increasing cost). For example, a link budget that allows more received signal power will improve the bit error rate.

The usual way of displaying the relationship between probability of error (as a measure of bit error rate) and received power is in a plot called a "waterfall" curve. This graph shows probability of error plotted logarithmically on the vertical axis versus an appropriate signal-to-noise ratio at the receiver plotted in decibels on the horizontal axis. The graph is specific to the choice of signaling and hardware implementation chosen for the communication system.

Before we can proceed to calculate waterfall curves, we must identify an appropriate signal-to-noise ratio that allows for easy comparison between signaling formats. For a given signaling scheme, let us define the energy per bit, E_b, as the average energy in a received bit. This is equivalent to the average received signal power divided by the data rate in bits per second. When the transmitted signal is a line code with pulses $s_1(t)$ and $s_0(t)$, we can identify E_1 as the energy in $s_1(t)$ and E_0 as the energy in $s_0(t)$. Then the average energy per bit is

$$\begin{aligned} E_b &= E_1 P[i=1] + E_0 P[i=0] \\ &= \frac{1}{2}E_1 + \frac{1}{2}E_0, \end{aligned} \tag{5.41}$$

assuming that 1s and 0s are *a priori* equally likely to be transmitted.

Example

Suppose we have a unipolar NRZ line code with A-volt pulses of T_s seconds duration. For a unipolar code, a pulse is transmitted for a 1 and no pulse is transmitted for a 0. We have $E_1 = A^2T_s$ and $E_2 = 0$. Then

$$E_b = \frac{1}{2}A^2T_s. \tag{5.42}$$

Now consider a polar NRZ line code with the same pulse parameters. In this case a positive pulse is transmitted for a 1 and a negative pulse for a 0. We have $E_1 = E_2 = A^2T_s$. For the polar line code,

$$E_b = A^2T_s. \tag{5.43}$$

In the analysis of the probability of error for an optimal line-code receiver, we showed in Equation (5.23) that $P_e = Q(\sqrt{\frac{E_{diff}}{2N_0}})$. The next step is to relate E_{diff} to E_b for each specific signaling scheme. This will give the probability of error in terms of the ratio E_b/N_0. The term E_b/N_0 (pronounced either "ee bee over en zero" or "ebno") is the traditional parameter used in comparing signaling schemes. This is the parameter plotted on the horizontal axis in the waterfall curve.

Dimensionally, E_b is measured in joules, or watt-seconds, and N_0 is measured in watts per hertz, or watt-seconds, so that the ratio is dimensionless. For any specific signaling scheme the signal-to-noise ratio can be related to E_b/N_0.

Example

For unipolar NRZ signaling, E_b is given by Equation (5.42). The difference energy E_{diff} is the energy in $s_1(t) - s_0(t) = s_1(t)$, which is A^2T_s. We have

$$E_{diff} = 2E_b, \tag{5.44}$$

so

$$P_e = Q\left(\sqrt{\frac{E_b}{N_0}}\right). \tag{5.45}$$

For polar NRZ signaling, E_b is given by Equation (5.43). The difference energy is

$$E_{diff} = 4A^2T_s = 4E_b. \tag{5.46}$$

Then we have

$$P_e = Q\left(\sqrt{\frac{2E_b}{N_0}}\right). \tag{5.47}$$

Table 5.2 compares the average energy per bit, the difference energy, and the probability of error for the line codes of Figure 5.2.

Table 5.2 Average Bit Energy, Difference Energy, and Probability of Error for Example Line Codes

Line Code	E_b	E_{diff}	P_e
Unipolar NRZ	$\frac{1}{2}A^2T_s$	A^2T_s	$Q\left(\sqrt{\frac{E_b}{N_0}}\right)$
Polar	A^2T_s	$4A^2T_s$	$Q\left(\sqrt{\frac{2E_b}{N_0}}\right)$
Unipolar RZ (50% duty cycle)	$\frac{1}{4}A^2T_s$	$\frac{1}{2}A^2T_s$	$Q\left(\sqrt{\frac{E_b}{N_0}}\right)$
Manchester	A^2T_s	$4A^2T_s$	$Q\left(\sqrt{\frac{2E_b}{N_0}}\right)$

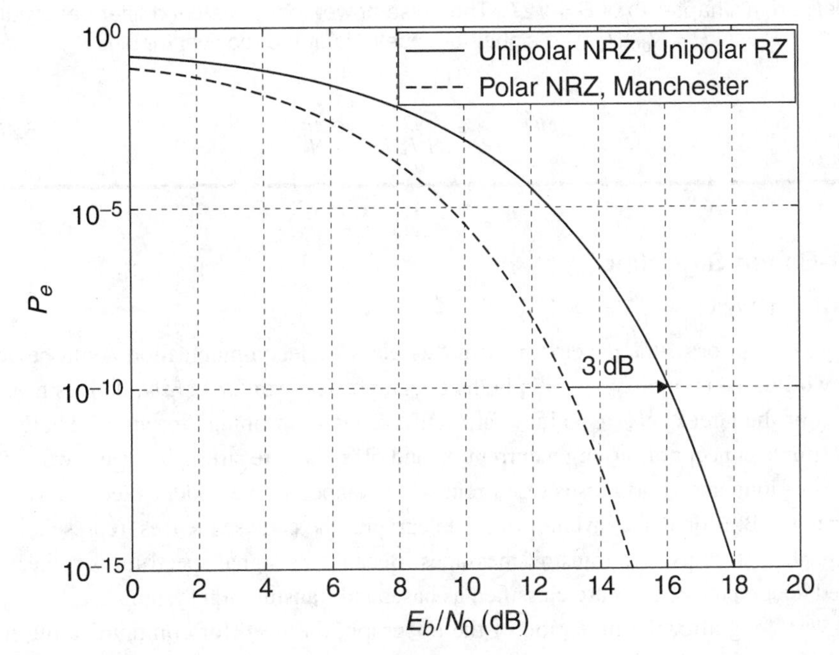

Figure 5.16 Waterfall Curves for Line Codes of Figure 5.2

Figure 5.16 shows waterfall curves for the line codes of Figure 5.2. It is easy to see from the curves that the polar line codes offer a performance advantage over the unipolar line codes. To achieve a given probability of error, a polar line code requires 3 dB less E_b/N_0 than does a unipolar line code. This is because the polar line codes have a larger difference energy for a given energy per bit. Polar line codes, however, can be more complicated to generate.

The waterfall curves show that probability of error can be a very sensitive function of E_b/N_0. Well into the curve a 3 dB change in E_b/N_0 can cause a very appreciable change in P_e. In practice, a graph of measured bit error rate versus E_b/N_0 will often plot out with the same form as the theoretical curve, but shifted to the right by 1 or 2 dB. This shift is called the **implementation loss** of the receiver and is caused by excess dissipation, distortion, and other design impediments in the receiver implementation.

In any given receiver implementation E_b/N_0 can be related to the signal-to-noise ratio. The following example shows the relationship in a specific case.

Example

Suppose a system uses polar NRZ signaling with T_s-second pulses. The average signal power at the receiver input is $P_s = E_b/T_s$. The noise power spectrum at the receiver input is $S_n(f) = N_0/2$. To find the noise power, we need to measure the noise through an appropriate bandwidth. The matched filter has a noise bandwidth (defined in Chapter 2) of $B = 1/2T_s$. The noise power in the matched filter bandwidth is then $P_n = N_0 B = N_0/2T_s$. The signal-to-noise ratio at the receiver input is

$$SNR = \frac{P_s}{P_n} = \frac{E_b/T_s}{N_0/2T_s} = 2\frac{E_b}{N_0}. \tag{5.48}$$

Carrier-Based Signaling

Modulation Overview

For many years prior to the development of "wireless," telecommunication could be achieved only via wire-based telegraphy, or simply the telegraph. The word *telegraph* derives from a concatenation of the Greek *tele,* meaning "far," and *graphien,* meaning "to write." These systems used the simple concept of turning a current on and off on a wire strung between two endpoints. Sequences of long and short pulses of current were mapped into a code, called the Morse code after Samuel F. B. Morse, the inventor of the telegraph. The code sequences, representing letters of the alphabet, were used to transmit messages. Because its signal spectrum is centered about DC, wired telegraphy systems are classified as baseband transmission systems.

Not very long after the invention of the telegraph, the need for communicating in situations where it was impractical to string wires became apparent. In particular, once a ship left visual contact with land it was virtually isolated and could not communicate, for example, the need for assistance. The pressing need to provide a means for communicating with vessels at sea was a major driver for the development of early "wireless" communication systems. The use of electromagnetic radiation to provide a link between endpoints is the common basis for these early "wireless telegraph" systems, as well as for the systems of today. Today, however, the use of wireless communications goes far beyond the needs of the early twentieth century and provides the freedom of untethered communication for anyone, at any time, and anywhere, a concept that was probably unfathomable little more than a century ago. The earliest wireless telegraph systems were based on technology inspired by Heinrich Hertz and developed by Guglielmo Marconi and others. Communication depended on the radiation of noise emitted by a spark gap transmitter and filtered by a tuned antenna system. The wireless telegraph is classified as a bandpass transmission system, because the transmitters produced bursts of noise in a relatively narrow bandwidth to represent the "on" and "off" signal elements.

Modulation is a bandpass process whereby message information is encoded onto a high-frequency carrier waveform. If we assume that the carrier is sinusoidal, we can imagine several means of "modulating," or modifying, it to carry messages across a link. In particular, the amplitude, frequency, or phase of the voltage or current waveform can be modified, as well as the radiated power or the wave polarization, since the propagating wave is electromagnetic.

The term *radio* became associated with carrier-based modulation systems only after the use of modulated carrier technology for voice transmission became common. This terminology persisted

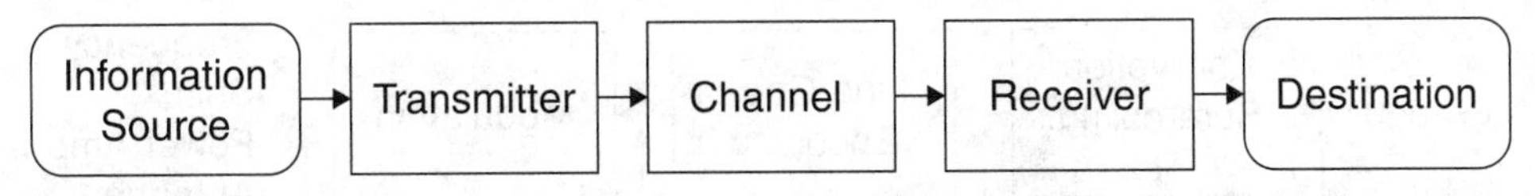

Figure 5.17 A High-Level Communication System Model

well into the digital modulation age. Interestingly, radio technology is now called "wireless" again, only this time it means wireless telephony rather than wireless telegraphy.

Modulated Carrier Communication Architecture

The architecture of a modulated carrier communication system is actually quite simple from a high-level perspective. The system consists of three basic blocks: a transmitter, a channel, and a receiver. A block diagram of the architecture is shown in Figure 5.17, derived from the classical work of Claude Shannon.[5] The transmitter receives input information (the message) and prepares it for transmission through the channel. For information communication to be reliable, the design of a transmitter must carefully consider the nature of the channel and the receiver to be used. In a carrier-based system, a transmitter includes a modulator to encode the information onto the carrier waveform.

For our discussions, a channel is the medium through which the modulated carrier is transmitted.[6] A channel may delay or distort a signal and add interference and noise before the signal arrives at the receiver. Channel models were discussed at length in Chapters 2 and 3. The modulation type is chosen, at least in part, to combat the signal-degrading effects of the channel.

A receiver has the difficult role of recovering the information encoded on the received carrier. At a minimum, a receiver must undo the work of the transmitter. Thus, a demodulator must be included in the receiver's functional components. The challenge resides in the fact that the channel may have altered or degraded the signal to the extent that information recovery is extremely difficult.

An important qualification is necessary at this point. When we speak of the receiver "recovering information," we mean that the receiver attempts to reproduce a stream of message bits (in a digital system) or a message waveform (in an analog system). A radio link—that is, the transmitter, channel, and receiver—has no knowledge of the content of the message it conveys. It may be possible to replace missing or corrupted parts of a message by inference from its context, but we consider for our discussions that such inference is performed outside the physical communication link.

More detailed block diagrams of the transmitter and receiver are shown in Figures 5.18 and 5.19, respectively. Note that the functions are paired; each operation performed in the transmitter has a corresponding operation performed in the receiver.

---

5. C. E. Shannon, "A Mathematical Theory of Communication," *Bell System Technical Journal* 27 (July 1948): 379–423.
6. Engineers use the term *channel* to refer to the part of the system that they cannot control.

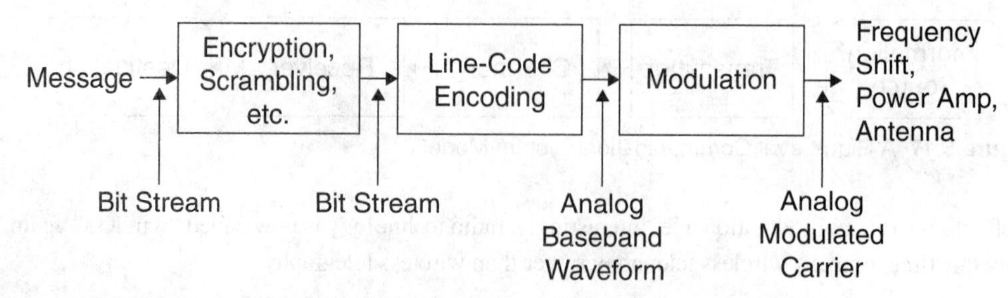

Figure 5.18 Transmitter High-Level Architecture

The transmitter comprises four major functional blocks. The first block, labeled "Encryption, Scrambling, etc.," includes a variety of operations, all of which are performed on the message data stream. These operations may include error control coding, which adds bits to the message stream to allow detection, and possibly correction, of bit errors. (Error control coding is discussed in Chapter 7.) This first block may also include scrambling and/or encryption to reduce the possibility of eavesdropping. The second major function is line-code encoding, in which the incoming logical bits are represented as a baseband analog waveform. The third major block is the modulator, the subject of this section. As we shall see, the choice of modulation scheme can have a major impact on system performance. The fourth major functional block, not shown, prepares the modulated signal for transmission. This block includes frequency shifters, filters, power amplifiers, and the antenna subsystem.

The receiver contains functional blocks that mirror those of the transmitter. Prior to the first block are a low-noise amplifier, filters, other amplifiers, and frequency converters that prepare the received signal for processing. The first block for our consideration is the demodulator. Its purpose is to recover the message line code from the modulated waveform. The second block is the line-code decoder, an element that recovers the bit stream from a possibly noisy and distorted line code. Subsequently, error control decoding, descrambling, and decryption are performed on the logical bit stream to yield an estimate of the original message bit stream.

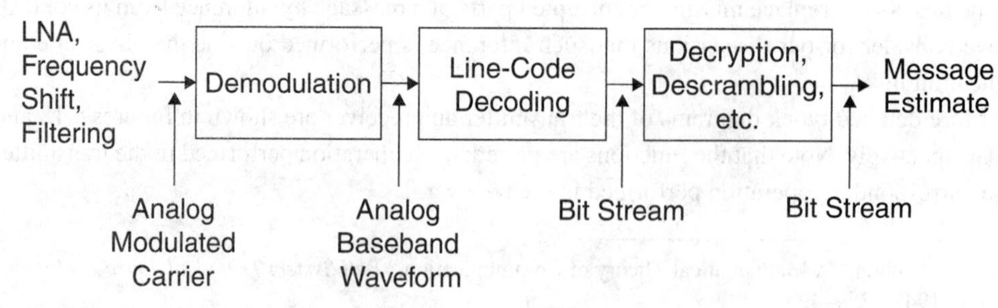

Figure 5.19 Receiver High-Level Architecture

Digital Modulation Principles

Over the years in which wireless systems have been in use, a wide variety of modulation methods have been developed, each possessing particular advantages and disadvantages. In this and subsequent sections, we introduce several methods that have been successfully used to implement second- and third-generation cellular telephone systems. Our focus in presenting these modulation techniques is to identify the particular features that are of importance in cellular telephone applications and to illustrate design trade-offs that a systems engineer must make when choosing a modulation method. The modulation methods that we present can be divided into two classes. The first group, sometimes referred to as "linear" modulation methods, includes several variations on **phase-shift keying** (PSK). (The term *keying* means the same thing as signaling and is a colorful holdover from telegraph days.) The second, "nonlinear" group includes **frequency-shift keying** (FSK) and its variations. The division into classes is primarily an editorial convenience. One of the modulation methods we will introduce is minimum-shift keying, which can be viewed either as a variety of phase-shift keying or as a variety of frequency-shift keying, depending on how one rearranges the defining equations.

Before we work our way through our brief catalog of modulation techniques, it is important to identify the particular properties of modulation methods that we find important in the cellular telephone application. For each type of modulation we develop time and frequency domain expressions for the transmitted signals. The frequency domain expressions lead to an identification of the bandwidth occupied by the signals. As we have seen in Chapter 4, adjacent-channel interference is a major concern in cellular applications. We therefore examine how the signal energy decreases outside the channel bandwidth. In addition to time and frequency domain considerations, we compare modulation methods on the basis of the E_b/N_0 needed to achieve a given probability of error and by their spectral efficiencies. Our discussions also encompass concerns about implementation issues. In particular, we describe why some modulation schemes work particularly well with inexpensive transmitter power amplifiers.

Bandwidth

Bandwidth is an important property of both baseband and modulated signals. Because the concept is sometimes a source of confusion, we take a moment to clarify the sense in which this term will be used in the following discussion.

It is difficult to precisely define "bandwidth" for two reasons. First, the signals used in communication systems are rarely bandlimited in any absolute sense. Although the frequency components of practical signals usually become smaller with increasing frequency difference from the carrier, they are never identically zero. Second, what is meant by a "significant" component varies with the application. Consequently, any definition of bandwidth requires a choice about which frequency components to include and which to ignore. There are several definitions in common use that reflect different choices and are useful in different applications. For example, the "3 dB" bandwidth is widely used in audio applications. We introduced "noise" bandwidth in the channel modeling discussion in Chapter 2.

When comparing alternative modulation methods there are two concepts of bandwidth that are particularly useful. The first reflects the fact that a modulated signal must pass through a bandlimited channel. The bandwidth of the channel is established by standards and by regulation and is physically set by filters in the final transmitter stages and the receiver front end. Normally we wish to design modulated signals that will pass through these bandlimiting filters with a minimum of distortion. (These filters do not include the receiver's matched filter, which, as we have seen, does not preserve the shape of its input waveform.) We can define the "bandwidth" of a signal as the bandwidth of an ideal brick-wall filter that will pass the signal without unacceptable distortion.

Example

Consider a line code that uses rectangular pulses $g(t)$. A unit-amplitude rectangular pulse of width T_s has spectrum given by

$$G(f) = T_s \operatorname{sinc}(T_s f). \tag{5.49}$$

This spectrum is plotted on a linear scale in Figure 5.20(a). The first null of the spectrum is located at frequency $1/T_s$. The energy in the pulse at frequencies from DC to

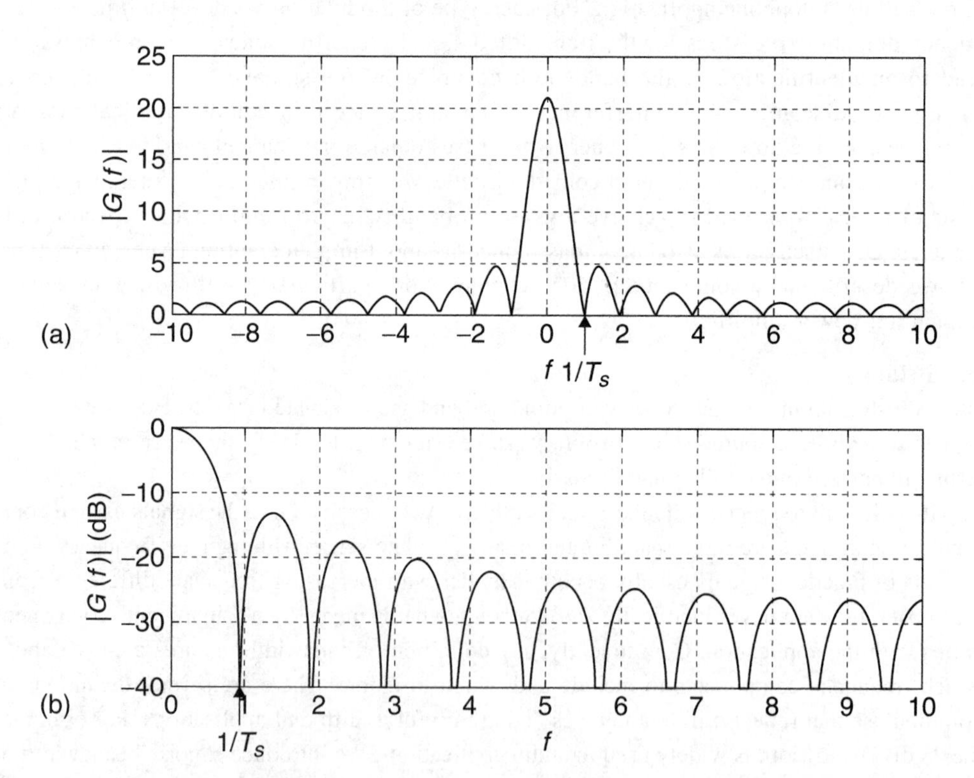

Figure 5.20 Spectrum of a Rectangular Pulse: (a) Linear Scale; (b) Decibel Scale

$1/T_s$—that is, $\int_{-1/T_s}^{1/T_s} |G(f)|^2\, df$—can be shown by direct calculation to be more than 90% of the total energy in the pulse. Therefore, the **first-null bandwidth**,

$$B = \frac{1}{T_s}, \tag{5.50}$$

is often taken as the bandwidth of an ideal filter that will pass the pulse without unacceptable distortion.

It is interesting in this context to ask what is the minimum bandwidth needed to transmit a given data sequence. Consider a line code given by

$$s(t) = \sum_{i=-\infty}^{\infty} a_i g(t - iT_s), \tag{5.51}$$

where $\{a_i\},\ \ i = \ldots, 0, 1, 2, \ldots$ represents the information, $g(t)$ represents a line-code pulse, and T_s represents the signaling interval. We will allow the $\{a_i\}$ to take arbitrary values; however, for polar signaling each a_i takes values +1 or −1, whereas for unipolar signaling each a_i takes values +1 or 0. The line-code pulse $g(t)$ may be arbitrary, with the understanding that we must be able to recover the values of the $\{a_i\}$ coefficients when $s(t)$ is sampled; that is, we require

$$s(kT_s) = a_k, \quad k = \ldots, -1, 0, 1, 2, \ldots. \tag{5.52}$$

The constraint of Equation (5.52) implies that the line-code pulses must satisfy

$$g(t) = \begin{cases} 1, & t = 0 \\ 0, & t = kT_s. \end{cases} \tag{5.53}$$

To see the frequency-domain effect of Equation (5.53), we sample $g(t)$ and then compute the Fourier transform. The sampled $g(t)$ can be written

$$g(t) \sum_{k=-\infty}^{\infty} \delta(t - kT_s) = \sum_{k=-\infty}^{\infty} g(kT_s)\delta(t - kT_s). \tag{5.54}$$

Equation (5.54) has Fourier transform $G_{eq}(f)$ given by

$$G_{eq}(f) = G(f) * \frac{1}{T_s} \sum_{k=-\infty}^{\infty} \delta\left(f - \frac{k}{T_s}\right) = \frac{1}{T_s} \sum_{k=-\infty}^{\infty} G\left(f - \frac{k}{T_s}\right). \tag{5.55}$$

The spectrum $G_{eq}(f)$ is called the **Nyquist equivalent spectrum** of $g(t)$.[7] The Fourier transform relationship between Equations (5.54) and (5.55) can be written explicitly as

$$G_{eq}(f)=\int_{-\infty}^{\infty}\left[\sum_{k=-\infty}^{\infty}g(kT_s)\delta(t-kT_s)\right]e^{-j2\pi ft}dt. \tag{5.56}$$

Substituting for $g(kT_s)$ from Equation (5.53) gives

$$G_{eq}(f)=\int_{-\infty}^{\infty}\delta(t)e^{-j2\pi ft}dt=1. \tag{5.57}$$

We see that to faithfully reproduce the information sequence $\{a_i\}, i=\ldots,0,1,2,\ldots$ we may use any convenient line-code pulse shape, provided that we choose a pulse whose Nyquist equivalent spectrum is a constant.

The right-hand term in Equation (5.55) reveals an important property of the Nyquist equivalent spectrum. Since this spectrum is constructed by summing copies of $G(f)$ repeated at intervals of $1/T_s$ Hz, we see that the Nyquist equivalent spectrum is periodic, with period $1/T_s$. Thus, to satisfy Equation (5.57), we need only ensure that the Nyquist equivalent spectrum is constant over one period, that is,

$$G_{eq}(f)=1, \quad -\frac{1}{2T_s}\le f\le\frac{1}{2T_s}. \tag{5.58}$$

Figure 5.21 shows three examples of line-code pulse spectra $G(f)$ and the corresponding Nyquist equivalent spectrum $G_{eq}(f)$. In Figure 5.21(a) the absolute bandwidth B of the pulse is less than $1/2T_s$. In this case, the Nyquist equivalent $G_{eq}(f)$ has gaps in the neighborhood of $n/2T_s$, where n is any odd integer. It is apparent that Equation (5.57) cannot be satisfied in this case, so the information coefficients $\{a_i\}$ cannot be recovered by sampling $s(t)$. In Figure 5.21(b) the absolute bandwidth of the pulse is identically equal to $1/2T_s$. In this case, Equation (5.58) can be satisfied, but only by the pulse spectrum

$$G(f)=\begin{cases}1, & -\frac{1}{2T_s}\le f\le\frac{1}{2T_s}\\ 0, & \text{otherwise.}\end{cases} \tag{5.59}$$

7. Some authors define the Nyquist equivalent spectrum as $T_sG_{eq}(f)$.

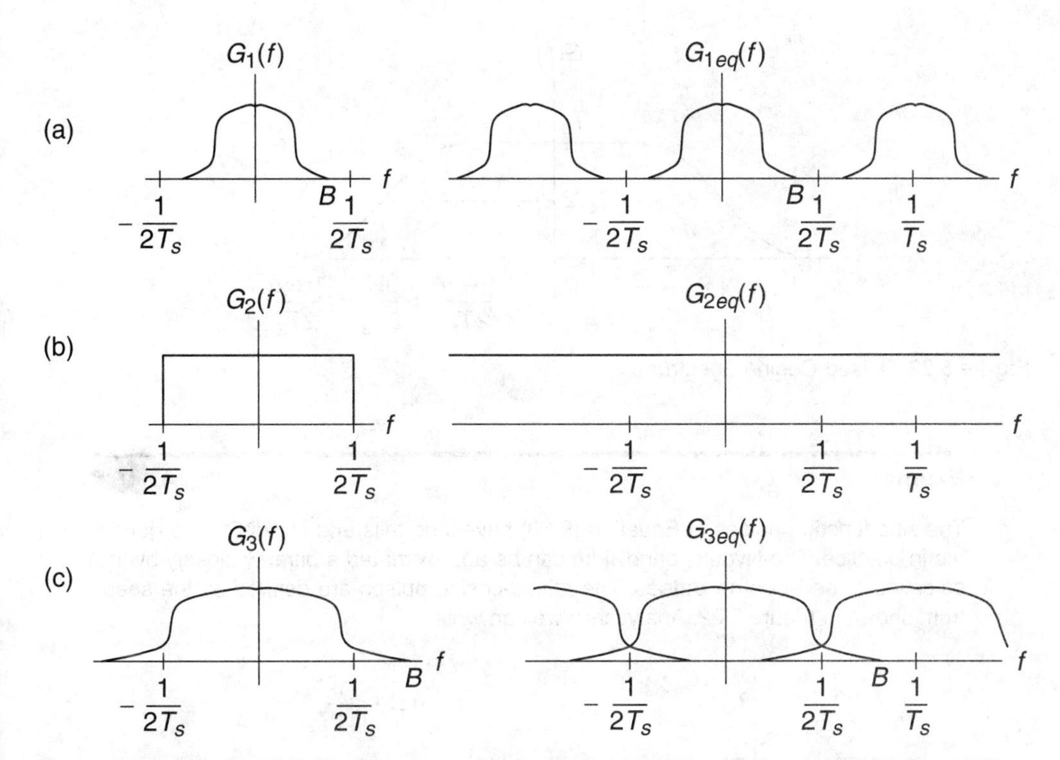

Figure 5.21 Baseband Pulse Spectra and Nyquist Equivalents: (a) $B < 1/2T_s$; (b) $B = 1/2T_s$; (c) $B > 1/2T_s$

The line-code pulse that has the spectrum given by Equation (5.59) is

$$g(t) = \text{sinc}(t/T_s). \tag{5.60}$$

In Figure 5.21(c) the absolute bandwidth B of the pulse $g(t)$ is greater than $1/2T_s$. In the Nyquist equivalent spectrum the copies of $G(f)$ overlap at frequencies that are odd multiples of $1/2T_s$. This overlap allows some choice in the shape of the pulse spectrum $G(f)$. Any pulse spectrum that is symmetrical about $1/2T_s$ in such a way that Equation (5.58) is satisfied will serve our needs.

The examples of Figure 5.21 lead to the conclusion that the minimum bandwidth line-code pulse that will interpolate a continuous signal between sample values spaced by T_s seconds is $1/2T_s$ Hz. This bandwidth is called the **Nyquist bandwidth**. The Nyquist bandwidth can only be achieved using the pulse of Equation (5.60). A wider choice of pulses is available if the bandwidth is allowed to be larger than the minimum.

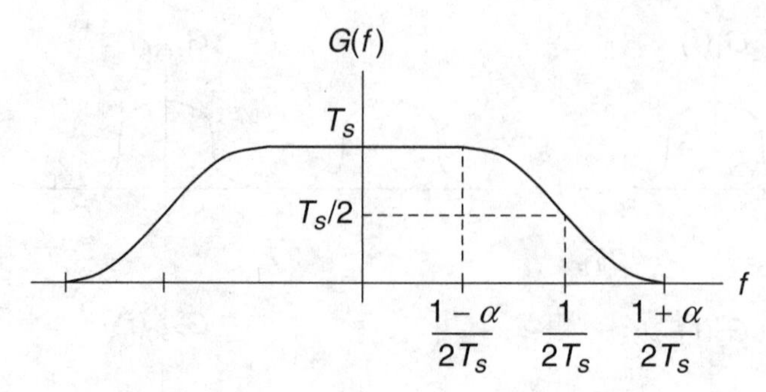

Figure 5.22 Raised-Cosine Spectrum

Example

The sinc-function pulses of Equation (5.60) have long tails and are difficult to generate in practice. The Nyquist bandwidth can be approximated arbitrarily closely by the so-called **raised-cosine pulses**. The raised-cosine pulses are defined by the spectrum shown in Figure 5.22. Analytically we can write

$$G(f)=\begin{cases} T_s, & 0\le |f| < \dfrac{1-\alpha}{2T_s} \\ \dfrac{T_s}{2}\left\{1-\sin\left[\dfrac{\pi T_s}{\alpha}\left(|f|-\dfrac{1}{2T_s}\right)\right]\right\}, & \dfrac{1-\alpha}{2T_s}\le |f| < \dfrac{1+\alpha}{2T_s} \\ 0, & \dfrac{1+\alpha}{2T_s}\le |f|, \end{cases} \tag{5.61}$$

where the parameter α satisfies $0 \le \alpha \le 1$ and is called the **rolloff parameter**. In the time domain the pulses are given by

$$g(t)=\operatorname{sinc}\left(\frac{t}{T_s}\right)\frac{\cos\left(\pi\alpha\dfrac{t}{T_s}\right)}{1-4\alpha^2\left(\dfrac{t}{T_s}\right)^2}. \tag{5.62}$$

When $\alpha = 0$, Equation (5.62) reduces to Equation (5.60). For $\alpha > 0$, Equation (5.62) falls off with time as $1/t^3$. This rapid decay makes these pulses easier to approximate with finite-duration pulses. The absolute bandwidth of the raised-cosine pulses is

$$B=\frac{1+\alpha}{2T_s} \tag{5.63}$$

which approximates the Nyquist bandwidth for small values of α. It is easy to see from Equation (5.62) that the raised-cosine pulses satisfy the condition of Equation (5.53).

The discussion leading to the definition of Nyquist bandwidth establishes a "complementary" sampling theorem. In the classical sampling theorem, we are given a continuous signal of absolute bandwidth B that we wish to sample in such a way that the signal can be reconstructed from the sequence of samples. In the present discussion, we start with a given sequence of samples and wish to interpolate a bandlimited waveform $s(t)$ through those samples in such a way that the sample sequence can be reconstructed by sampling $s(t)$. In the classical sampling theorem, we must sample at a rate at least $1/T_s = 2B$ samples/s to ensure that the original waveform can be reconstructed from the sample sequence. In the "complementary" theorem, we find that the "interpolating" line-code pulses $g(t)$ must have bandwidth at least $B = 1/2T_s$ to ensure that the samples can be recovered from the continuous signal. In the classical sampling theorem, too slow a sampling rate leads to "aliasing," which prevents the original signal from being accurately reconstructed. In the complementary theorem, a wider bandwidth B leads to aliasing (Figure 5.21(c)) that provides the designer with a choice of interpolating pulses. It is interesting to note the important role that the Nyquist bandwidth $B = 1/2T_s$ plays in both versions of the sampling theorem.

The second concept of bandwidth focuses on frequency components that appear outside the allocated channel. Channelized applications such as cellular telephone systems are very sensitive to adjacent-channel interference. An example in Chapter 4 showed that even adjacent-channel signal levels that appear very low could cause significant interference under the right conditions.

Typical standards on out-of-band frequency components of modulated signals specify that spectral levels be reduced below the carrier by 30 to 60 dB. Typically, the spectrum of a modulated signal will decrease as the difference from the carrier frequency increases. The rate of spectrum rolloff will be our measure of out-of-band spectrum.

It can be shown as a property of the Fourier transform that a time function that contains discontinuities produces a power spectrum that falls off with frequency as $1/f^2$. Expressed in decibels, this is a rolloff rate of 20 dB per decade. A continuous time function whose derivatives contain discontinuities has a power spectrum that falls off with frequency as $1/f^4$, or 40 dB per decade. A time function with a continuous derivative whose second derivative is discontinuous has a power spectrum that falls off with frequency as $1/f^6$, or 60 dB per decade, and so forth. Figure 5.20(b) shows a decibel plot of the spectrum of a rectangular pulse. This is the classic example of the spectrum of a time function that contains discontinuities. A rolloff rate of 20 dB per decade is very slow and not suitable for most channelized applications. As a result, rectangular pulses are seldom used as modulating signals. We will see as we discuss particular modulation methods how steps are taken to smooth the transmitted signals to obtain a faster spectrum rolloff.

Binary Phase-Shift Keying (BPSK)

The first modulation method that we will examine is **binary phase-shift keying** (BPSK). It is simple to implement and is very widely used for transmission of digital messages. It will serve us as a baseline against which other modulation methods are compared.

Suppose a message is given by the line-code signal

$$m(t) = \sum_{i=-\infty}^{\infty} a_i g(t - iT_s), \tag{5.64}$$

where the $\{a_i\}$ coefficients are ±1 reflecting polar keying and $g(t)$ represents the symbol pulse shape. The modulated signal is

$$x(t) = Am(t)\cos(2\pi f_c t), \tag{5.65}$$

where f_c is the carrier frequency and $\cos(2\pi f_c t)$ is the carrier. A block diagram of the transmitter is shown in Figure 5.23. There are two possible transmitted signals. On any given transmission the transmitter output is either $Ag(t)\cos(2\pi f_c t)$ or $-Ag(t)\cos(2\pi f_c t)$. This pair of signals can also be written $Ag(t)\cos(2\pi f_c t)$ and $Ag(t)\cos(2\pi f_c t + \pi)$, justifying the name "phase-shift keying."

If the line-code pulse $g(t)$ has power spectrum $G(f)$, and if we model the incoming bit stream as a sequence of independent random bits, each equally likely to be 1 or 0, then the transmitted signal has power spectrum $S_x(f)$ given by

$$S_x(f) = \frac{A^2}{4T_s}|G(f - f_c)|^2 + \frac{A^2}{4T_s}|G(f + f_c)|^2. \tag{5.66}$$

Therefore, if $G(f)$ is a baseband spectrum, then $S_x(f)$ will be a bandpass spectrum, with power concentrated near the carrier frequency. The bandwidth of the transmitted signal $x(t)$ will depend on the bandwidth of the line code $g(t)$.

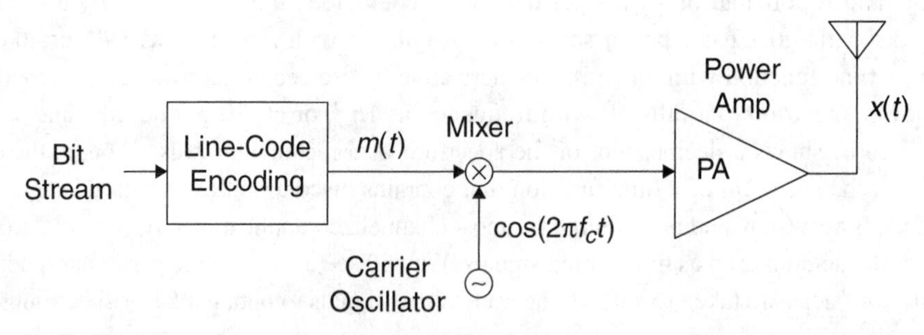

Figure 5.23 Transmitter for Binary Phase-Shift Keying

Because rectangular pulses have such a poor spectrum rolloff, the line-code pulses $g(t)$ used in channel-based applications are usually shaped to be smooth pulses. The raised-cosine pulse of Equation (5.62) is an attractive choice. The transmitted pulse $g(t)$ is often called a "root-raised-cosine" pulse because its spectrum is described by the square root of Equation (5.61). When the received pulse is passed through a matched filter, the output is a raised-cosine pulse. This arrangement produces smooth, bandlimited transmitted pulses that do not introduce intersymbol interference at the receiver output.

The Nyquist bandwidth of the transmitted signal is

$$B = 2\frac{1}{2T_s} = \frac{1}{T_s}. \tag{5.67}$$

The factor of 2 is needed because the modulated signal contains both upper and lower sidebands. Raised-cosine and root-raised-cosine pulses have a bandwidth that is absolutely limited, so there would appear to be no adjacent-channel interference for this modulation method. This conclusion turns out not to be quite true in practice, for reasons that we will explore when we discuss quadrature phase-shift keying.

Spectral efficiency is a parameter that measures the data rate normalized by the bandwidth. As we will be comparing spectral efficiencies of various modulation methods, it is important that we be precise about the bandwidth that we use as a normalizing factor. For BPSK modulation, one bit is transmitted per T_s-second pulse. The transmitted signal occupies a Nyquist bandwidth given by Equation (5.67). The spectral efficiency is then

$$\eta = \frac{1 \text{ bit}/T_s \text{ seconds}}{1/T_s \text{ Hz}} = 1 \text{ bit/s/Hz}. \tag{5.68}$$

Note that if we were using raised-cosine pulses with 100% rolloff ($\alpha = 1$), then the absolute bandwidth would be twice the Nyquist bandwidth, or $2/T_s$. The spectral efficiency computed using this bandwidth would be $\eta = 1/2$ bit/s/Hz, a figure that appears in many textbooks.

A receiver for a binary phase-shift keyed signal is shown in Figure 5.24. The "front end" contains a low-noise amplifier (LNA) and bandpass filter (BPF) to eliminate adjacent-channel interference and limit noise power. The front end may also contain circuits to shift the carrier frequency to a value more convenient for demodulation. The receiver front end is generic; all radio receivers begin with similar circuits, regardless of the modulation system used. The "line-code receiver" is the receiver of Figure 5.4. Its purpose is to decode the demodulated line code. The figure shows a matched filter receiver, but a correlator would do equally well. Additional processing on the digital bit stream may follow the line-code receiver. The "demodulator" is a new addition to the receiver, as compared to receivers for baseband systems. Strictly speaking, the demodulator should include a lowpass filter (LPF) to remove harmonics of the carrier frequency generated in mixing. The matched filter has a lowpass characteristic, however, and usually performs this function sufficiently.

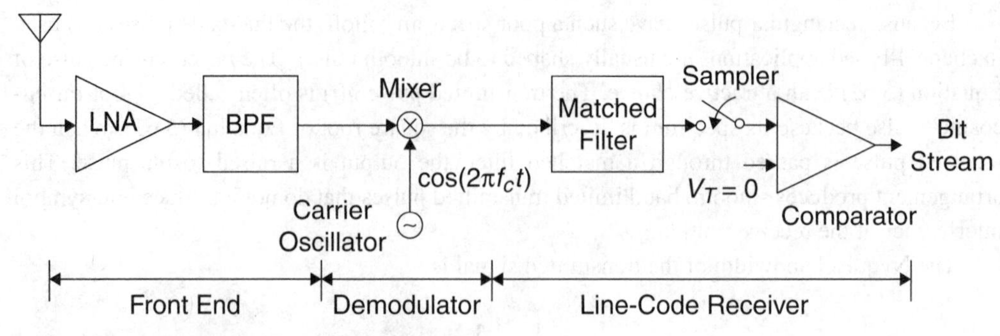

Figure 5.24 Receiver for Binary Phase-Shift Keying

To demodulate properly, the oscillator in the demodulator of Figure 5.24 must produce a sinusoid that is in phase with the received carrier. Although not shown in the figure, there will be a synchronizing circuit, usually a phase-locked loop, included for that purpose. A modulation system that includes a phase-locked receiver is said to be "coherent." We show in the next section that there is a "noncoherent" alternative that does not require synchronizing circuitry, but has slightly poorer performance.

As noted earlier, BPSK modulation uses polar signaling, as the two possible transmitted pulses differ only in sign. The coherent receiver shown in Figure 5.24 is an optimal receiver and achieves a probability of error given by Equation (5.47). The average energy per bit, E_b, is the energy in $Ag(t)\cos(2\pi f_c t)$, where we take the amplitude A as the amplitude at the receiver input, that is,

$$E_b = \frac{A^2}{2} E_g, \tag{5.69}$$

where E_g is the energy in $g(t)$. The factor E_g equals 1 if $g(t)$ is a root-raised-cosine pulse with spectrum given by the square root of Equation (5.61).

It is often convenient to write a modulated signal in the "quadrature" form

$$x(t) = x_I(t)\cos(2\pi f_c t) - x_Q(t)\sin(2\pi f_c t), \tag{5.70}$$

where $x_I(t)$ and $x_Q(t)$ are called the **in-phase** and **quadrature** components, respectively. Comparing Equation (5.70) with Equation (5.65), we see that for BPSK,

$$\begin{aligned} x_I(t) &= Am(t) \\ x_Q(t) &= 0. \end{aligned} \tag{5.71}$$

The quadrature form of Equation (5.70) is customarily plotted as a **signal constellation**. A signal constellation is a plot in the complex plane where sample values of the in-phase and

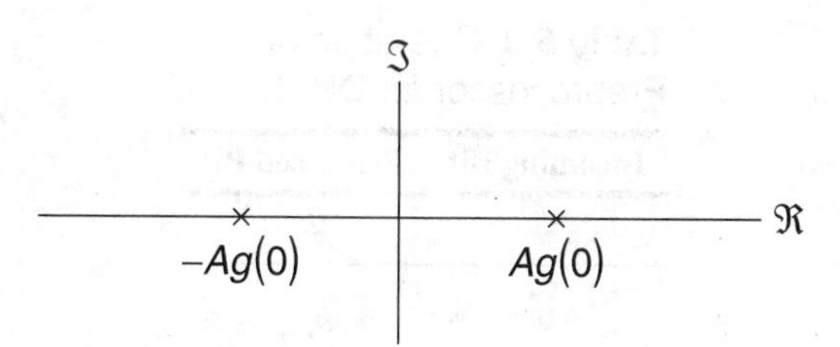

Figure 5.25 Signal Constellation for Binary Phase-Shift Keying

quadrature terms are plotted along the real axis and imaginary axes, respectively. For BPSK there will be two signal points on the real axis equidistant from the origin, representing the fact that $m(t)$ is a polar keyed waveform and possesses only two possible equal and opposite values. There is no component along the imaginary axis, because $x_Q(t) = 0$. The signal constellation for BPSK is shown in Figure 5.25. The two signal points are marked along the real axis at $\pm Ag(0)$.

Binary phase-shift keying is a widely used modulation technique because it is simple to implement and its E_b/N_0 is relatively low for a given probability of error. It has three relative weaknesses, however. The first, as we have already noted, is that it requires synchronization circuitry to produce coherent demodulation, thereby adding to implementation complexity. We introduce a noncoherent version in the next section. The second weakness is its relatively low spectral efficiency of 1 bit/s/Hz. We will discuss quadrature phase-shift keying, having twice the spectral efficiency, later in this chapter. In fact, there are other modulation schemes that have significantly higher spectral efficiencies than even quadrature phase-shift keying. Finally, BPSK signals do not have a constant envelope. Because of this, an associated transmitter requires a linear power amplifier, which can be inefficient and more expensive than alternative nonlinear power amplifiers. We will discuss this issue at greater length when we discuss quadrature phase-shift keying, as it also suffers from this disadvantage.

Differential Binary Phase-Shift Keying (DPSK)

Differential binary phase-shift keying, through the addition of a simple preprocessor at the transmitter, eliminates the necessity of a receiver synchronizing circuit. There is a small performance penalty, however, the nature and size of which we discuss below.

The preprocessor of Figure 5.26 is placed at the transmitter, ahead of the line-code encoder. The information bits arriving from the source are designated $\{d_k\}$. The processed bits sent to the

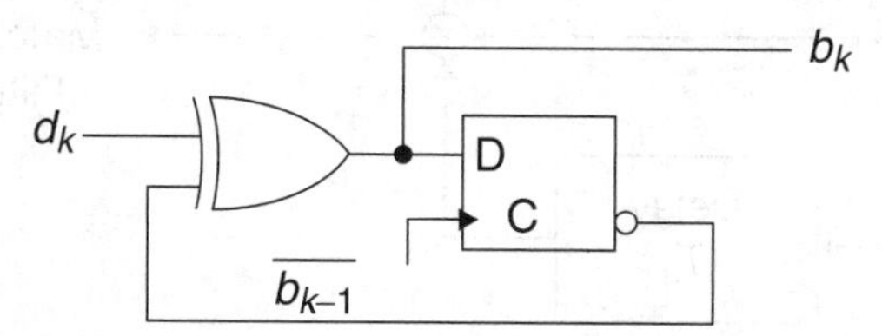

Figure 5.26 Preprocessor for DPSK

Table 5.3 Operation of Preprocessor for DPSK

Incoming Bit	Processed Bit
d_k	b_k
0	$\overline{b_{k-1}}$
1	b_{k-1}

line-code encoder are $\{b_k\}$. The preprocessor is a small state machine that implements the difference equation

$$b_k = \overline{b_{k-1}} \oplus d_k, \tag{5.72}$$

where "$\oplus$" designates the exclusive-or operation, and the overbar denotes complement. The preprocessor works as follows. If the incoming data bit d_k is a 0, then $b_k = \overline{b_{k-1}}$; that is, the bit sent to the transmitter will have a complementary value to the previous bit, and the transmitted pulse will have the opposite polarity from the previous transmitted pulse. On the other hand, if the incoming data bit d_k is a 1, then $b_k = b_{k-1}$. In this case, the bit sent to the transmitter will be the same as the previous bit, and the transmitted pulse will have the same polarity as the previous transmitted pulse. The operation of the preprocessor is summarized in Table 5.3.

A demodulator for differential binary phase-shift keying is very simple, as is illustrated in Figure 5.27. The oscillator and its synchronizing circuit are gone, and the inputs to the mixer are the current and previous received pulses.

Suppose the current and previous received pulses have opposite polarity. Then the inputs to the mixer are $a_k g(t)\cos 2\pi f_c t$ and $a_{k-1} g(t)\cos 2\pi f_c t$, where $a_k a_{k-1} = -1$, and we assume for simplicity that there is an integral number of cycles of carrier in T_s seconds. The output of the mixer is $-g^2(t)\cos^2 2\pi f t = -\frac{1}{2}g^2(t) - \frac{1}{2}g^2(t)\cos 2\pi 2 f_c t$. The term at twice the carrier frequency is eliminated by the lowpass effect of the matched filter. Only the $-\frac{1}{2}g^2(t)$ term is significant. The negative polarity ensures that (in the absence of noise) the line-code decoder will recover the data bit $d_k = 0$.

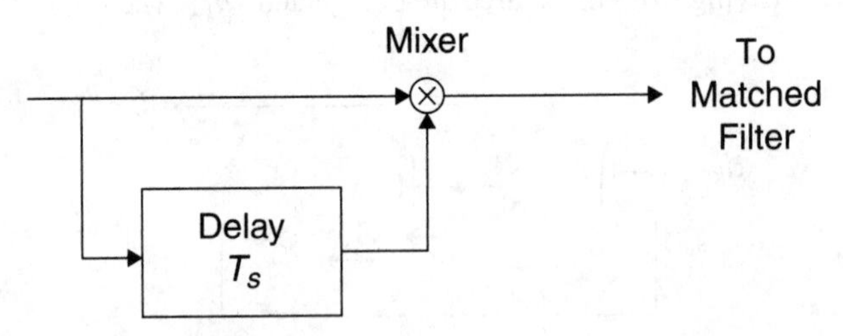

Figure 5.27 Demodulator for DPSK

Suppose instead that the previous and present received pulses have the same polarity. In this case, the inputs to the mixer are just as before, only now $a_k a_{k-1} = 1$. The significant part of the mixer output is $\frac{1}{2}g^2(t)$. The positive polarity ensures (again neglecting noise) that the line-code decoder will recover the data bit $d_k = 1$.

Example

Suppose the preprocessor has an initial state of $b_{-1} = 0$. Let the data from the information source be

$$d_k = 0,1,1,0,0,0,1,\ldots.$$

The output of the preprocessor will be

$$b_k = (0),1,1,1,0,1,0,0,\ldots,$$

where the initial state is shown in parentheses. The transmitted pulses have polarities given by

$$a_k = (-1),1,1,1,-1,1,-1,-1,\ldots$$

matching $\{b_k\}$. The receiver computes

$$a_{-1}a_0 = -1 \cdot 1 = -1$$
$$a_0 a_1 = 1 \cdot 1 = 1$$
$$a_1 a_2 = 1 \cdot 1 = 1$$
$$a_2 a_3 = 1 \cdot -1 = -1$$
$$a_3 a_4 = -1 \cdot 1 = -1$$
$$a_4 a_5 = 1 \cdot -1 = -1$$
$$a_5 a_6 = -1 \cdot -1 = 1$$
$$\ldots$$

leading to the output sequence

$$\widehat{d_k} = 0,1,1,0,0,0,1,\ldots,$$

which is identical to the bit sequence from the information source.

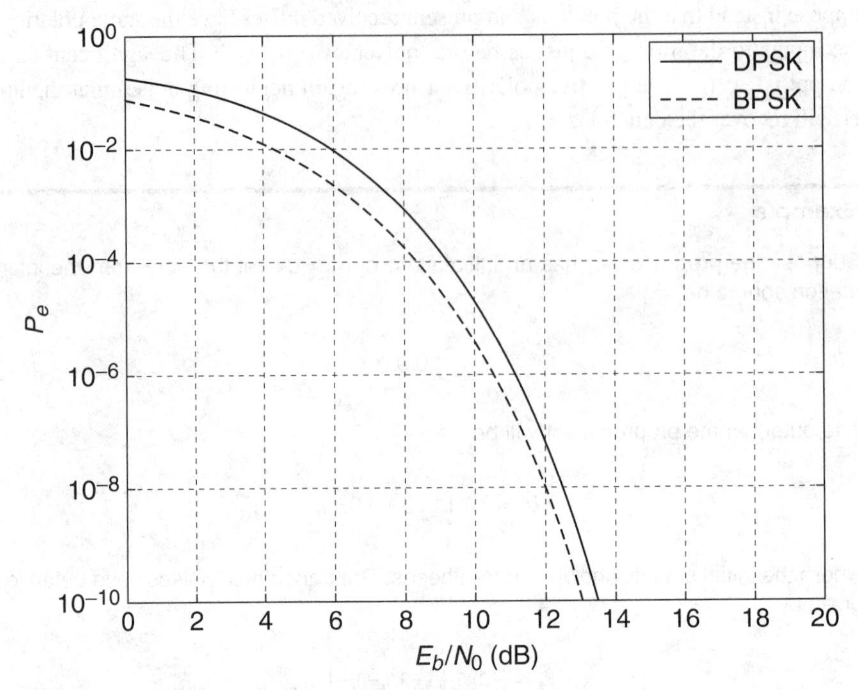

Figure 5.28 Probability of Error versus E_b/N_0 for DPSK and BPSK

In the BPSK receiver of Figure 5.24, only one of the inputs to the mixer is noisy since the mixer input from the oscillator is noise-free. In contrast, the DPSK demodulator of Figure 5.27 has two noisy mixer inputs. As a result, the performance of a DPSK receiver is poorer than that of a BPSK receiver. It turns out that for differential binary phase-shift keying, the probability of error is given by

$$P_e = \frac{1}{2} e^{-\frac{E_b}{N_0}}. \tag{5.73}$$

A waterfall curve plotting P_e versus E_b/N_0 for coherent BPSK and noncoherent DPSK is given in Figure 5.28. For a given probability of error, the required E_b/N_0 is about 1 dB higher for DPSK than it is for BPSK. This is a small difference in many applications, and DPSK is an attractive alternative to BPSK when receiver cost must be minimized.

We conclude by noting that the transmitted signals for DPSK and for BPSK are identical. Thus, both systems have the same bandwidth, the same adjacent-channel rolloff, and the same spectral efficiency. Both techniques produce transmitted signals whose amplitudes must be reproduced accurately at the receiver and hence require linear transmitter power amplifiers. The two methods differ slightly in performance and significantly in receiver complexity.

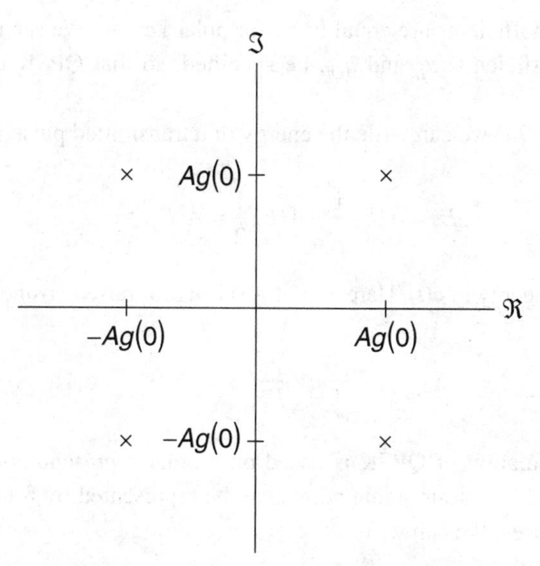

Figure 5.29 Signal Constellation for Quadrature Phase-Shift Keying

Quadrature Phase-Shift Keying (QPSK)

Quadrature phase-shift keying is a straightforward extension of binary phase-shift keying that doubles the spectral efficiency. A signal constellation for QPSK is shown in Figure 5.29. The distance of any of the four signal points from the origin is $\sqrt{2}Ag(0)$.

Two equivalent analytic formulations for QPSK can be derived from the signal constellation. The first is formulated on the rectangular-coordinate representation of the constellation. There are two possible values for the in-phase component, with equal magnitudes and opposite polarities, and two possible values for the quadrature component, also with equal magnitudes and opposite polarities. Thus, both the in-phase and quadrature components are polar keyed. Taking Equation (5.70) as a starting point and comparing with Equation (5.65), we can write

$$\begin{aligned} x(t) &= x_I(t)\cos(2\pi f_c t) - x_Q(t)\sin(2\pi f_c t) \\ &= Am_I(t)\cos(2\pi f_c t) - Am_Q(t)\sin(2\pi f_c t), \end{aligned} \tag{5.74}$$

where

$$\begin{aligned} m_I(t) &= \sum_{i=-\infty}^{\infty} a_{Ii} g(t - iT_s) \\ m_Q(t) &= \sum_{i=-\infty}^{\infty} a_{Qi} g(t - iT_s). \end{aligned} \tag{5.75}$$

The $\{a_{Ii}\}$ and $\{a_{Qi}\}$ coefficients are equal to ±1 for polar keying. We see that each transmission requires that two coefficients, a_{Ii} and a_{Qi}, be specified, so that QPSK transmits two bits per pulse.

From Equation (5.74) we can write the energy of a transmitted pulse as

$$E_s = A^2 E_g \frac{1}{2} + A^2 E_g \frac{1}{2} = A^2 E_g, \tag{5.76}$$

where again E_g is the energy in $g(t)$. Here, too, if $g(t)$ is a root-raised-cosine pulse, then $E_g = 1$ and we can write

$$A = \sqrt{E_s}. \tag{5.77}$$

The second formulation of QPSK is based on a polar representation of the constellation. Note that in Figure 5.29 the four signal points can be represented by four equal-amplitude signals with different phases. We can write

$$x(t) = \sqrt{2E_s} \sum_{i=-\infty}^{\infty} g(t - iT_s)\cos(2\pi f_c t + \theta_i), \tag{5.78}$$

where $\theta_i = \pi/4, 3\pi/4, -\pi/4, -3\pi/4$. (We are continuing to assume that the carrier has an integral number of cycles in T_s seconds.) The term *phase-shift* keying derives from this formulation. A transmitter for QPSK is shown in Figure 5.30. This block diagram shows a direct implementation of Equation (5.74). It is essentially two BPSK transmitters running in parallel, with carriers differing in phase by $\pi/2$.

The power spectrum of the transmitted signal is given by Equation (5.66). From this equation we see that the bandwidth of a QPSK signal is identical to the bandwidth of a BPSK signal that uses the same pulse interval T_s and the same pulse shape $g(t)$. The Nyquist bandwidth of $x(t)$

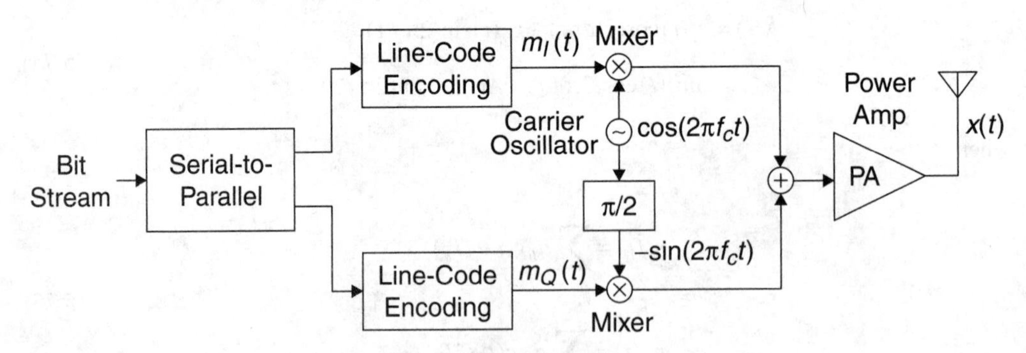

Figure 5.30 Transmitter for Quadrature Phase-Shift Keying

is $B = 1/T_s$, as in Equation (5.67). As was the case for BPSK, the pulse shape $g(t)$ must be chosen to give a rapid spectral rolloff in adjacent channels. Again, root-raised-cosine pulses are a common choice.

At two bits per transmitted pulse, the spectral efficiency of the QPSK signal is

$$\eta = \frac{2\text{ bit}/T_s\text{ seconds}}{1/T_s\text{ Hz}} = 2\text{ bit/s/Hz}, \tag{5.79}$$

normalized to the Nyquist bandwidth. We see that this is twice the spectral efficiency of BPSK. A segment of QPSK signal using 100% rolloff ($\alpha = 1$) root-raised-cosine pulses is shown in Figure 5.31. In this figure the pulses occur every $T_s = 1$ ms and the carrier frequency is $f_c = 8000$ Hz. It is clear in the figure that the envelope of the signal $x(t)$ is not constant, and in fact there are moments at which the envelope passes through zero. If the shape of the envelope is to be maintained when $x(t)$ is transmitted, then the power amplifier in the transmitter must be a linear amplifier. In reality, it is difficult to make a power amplifier both linear and efficient. Linear power amplifiers tend to convert a significant fraction of the power they draw from their power supplies into heat. For base station operation, the efficiency of the transmitter's power amplifier may not be a significant issue, but for handset operation this inefficiency is an extra drain on the battery. Efficient power amplifiers are power amplifiers that convert nearly all of the power

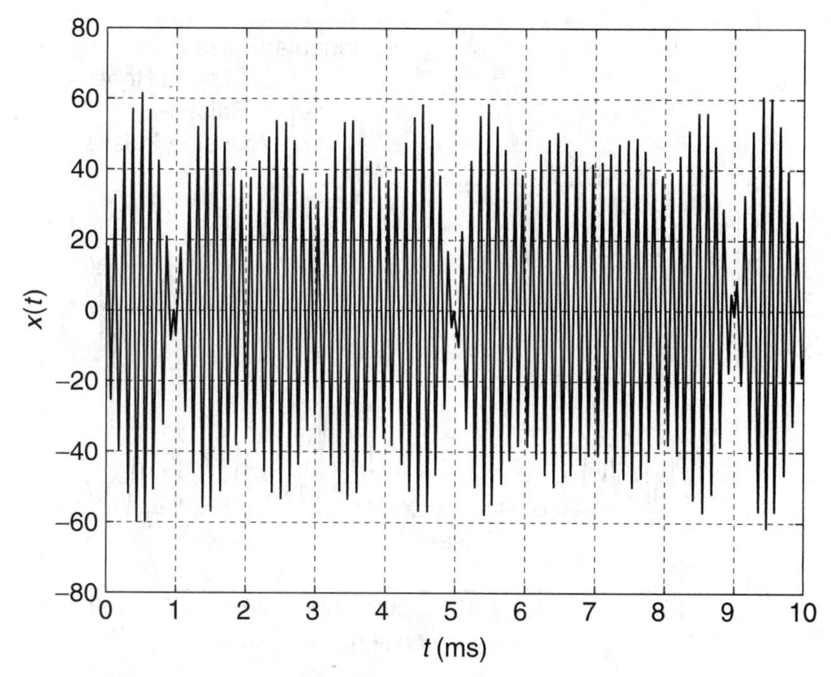

Figure 5.31 A QPSK Signal with Root-Raised-Cosine Pulses

drawn from the supply to radio frequency power. They are designed to drive the transistors in their output stages between saturation and cutoff. Thus, they tend to compress the envelopes of the signals they amplify. One consequence of this distortion is to create out-of-band frequency components that are a source of adjacent-channel interference. In the extreme case, severe compression of the envelope creates "edges" at the instants when the envelope passes through zero. Recall that the power spectrum of such a discontinuous signal falls off with frequency as $1/f^2$, essentially as a sinc function does.

The requirement for linear power amplification is the most significant drawback to using QPSK in cellular telephone applications. This is particularly true for the reverse link, where the transmitter is in the mobile handset. A mobile handset transmitter must be inexpensive to manufacture and must conserve the small handset battery power. Following our next discussion on error probability for QPSK, we return to this issue and discuss how QPSK can be modified to reduce its sensitivity to distortion caused by nonlinear amplification.

Example

A QPSK signal was simulated for 2000 bits of random input data. The pulse shape is 100% rolloff root-raised-cosine. The pulse duration is $T_s = 10\ \mu s$ and the carrier frequency is $f_c = 400$ kHz. An estimate of the power spectrum is shown in Figure 5.32. The QPSK signal was clipped by passing it through a hard limiter and then bandpass

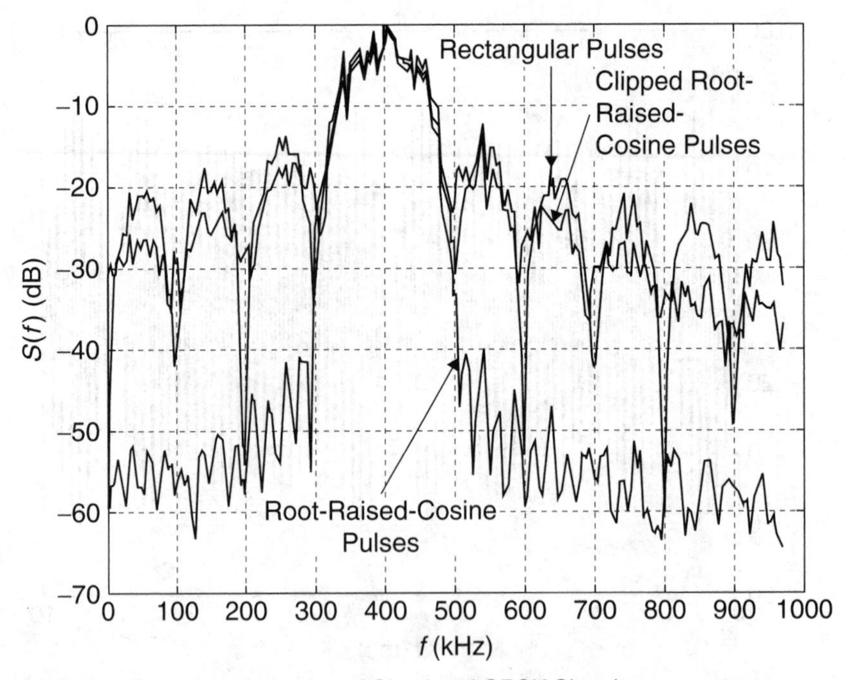

Figure 5.32 Power Spectrum Estimates of Simulated QPSK Signals

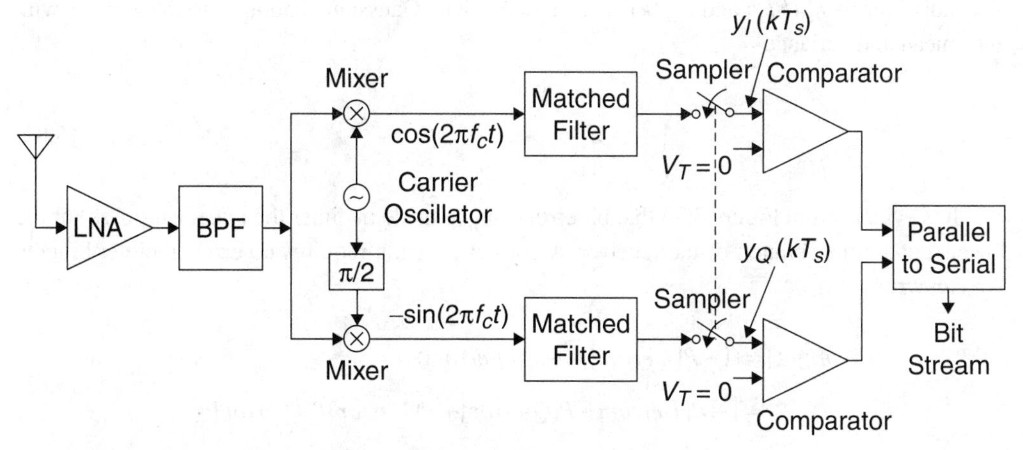

Figure 5.33 Receiver for Quadrature Phase-Shift Keying

filtered to remove harmonics of the carrier frequency. The spectrum of the clipped signal is also shown in the figure. For comparison, the power spectrum of a QPSK signal using rectangular pulses is also shown.

The bandwidth of the root-raised-cosine QPSK signal is 200 kHz. It is clear that clipping the signal raises the level of the power spectrum outside the desired signaling channel. The spectrum of the clipped signal shows side lobes that are nearly as high as those for a rectangular pulse shape.

A receiver for a QPSK signal is shown in Figure 5.33. The receiver has the structure of two BPSK receivers operating in parallel, with oscillators that differ in phase by $\pi/2$. This is a synchronous receiver, and the carrier oscillator that generates $\cos(2\pi f_c t)$ must be kept in phase with the received carrier. The synchronizing circuit is not shown in the figure.

At every sampling time $t = kT_s$, each branch produces a decision statistic; the statistic produced by the upper branch in the figure is labeled $y_I(kT_s)$ and the statistic produced by the lower branch is labeled $y_Q(kT_s)$. It can be shown, by tracing signals through the receiver, that

$$\begin{aligned} y_I(kT_s) &= s_{Io}(kT_s) + n_{Io}(kT_s) \\ y_Q(kT_s) &= s_{Qo}(kT_s) + n_{Qo}(kT_s), \end{aligned} \tag{5.80}$$

where $s_{Io}(kT_s)$ and $s_{Qo}(kT_s)$ are proportional to the in-phase (real-axis) and quadrature (imaginary-axis) components of the signals shown in the signal constellation of Figure 5.29, that is,

$$\begin{aligned} s_{Io}(kT_s) &= \pm\frac{A}{2}E_g \\ s_{Qo}(kT_s) &= \pm\frac{A}{2}E_g. \end{aligned} \tag{5.81}$$

The noise terms $n_{Io}(kT_s)$ and $n_{Qo}(kT_s)$ are independent Gaussian random variables, each with zero mean and variance

$$\sigma^2 = \frac{N_0 E_g}{4}. \tag{5.82}$$

It is evident from Figure 5.33 that bit errors can be made in either the upper (in-phase) or the lower (quadrature) branch of the receiver. A correct decision requires no error in either branch. We can write

$$\begin{aligned} P[correct] &= (1 - P[I\text{-}error])(1 - P[Q\text{-}error]) \\ &= 1 - P[I\text{-}error] - P[Q\text{-}error] + P[I\text{-}error]P[Q\text{-}error]. \end{aligned} \tag{5.83}$$

Now even if $P[I\text{-}error]$ and $P[Q\text{-}error]$ are as large as 10^{-2}, the product $P[I\text{-}error]$ $P[Q\text{-}error]$ will be on the order of 10^{-4}, which is negligible by comparison. We can therefore write

$$P[correct] \cong 1 - P[I\text{-}error] - P[Q\text{-}error], \tag{5.84}$$

which gives

$$P[error] = P_{e,s} \cong P[I\text{-}error] + P[Q\text{-}error]. \tag{5.85}$$

The s in the subscript of $P_{e,s}$ stands for "symbol," to distinguish the probability of error given by Equation (5.85) from the probability of a bit error, to be calculated below. Since the in-phase branch and the quadrature branch of the receiver are each matched filter binary receivers, we can find $P[I\text{-}error]$ and $P[Q\text{-}error]$ by following the steps leading to Equation (5.23). We omit the details and state the result as

$$P[I\text{-}error] = P[Q\text{-}error] = Q\left(\sqrt{\frac{A^2 E_g}{N_0}}\right) = Q\left(\sqrt{\frac{E_s}{N_0}}\right), \tag{5.86}$$

where we used Equation (5.76) in the last step. Substituting Equation (5.86) into Equation (5.85) gives the probability of symbol error as

$$P_{e,s} \cong 2Q\left(\sqrt{\frac{E_s}{N_0}}\right). \tag{5.87}$$

To allow us to compare QPSK with other modulation systems, we need to obtain an expression for the bit error probability, $P_{e,b}$, as a function of E_b/N_0, where E_b is the energy per bit.

Given that E_s is the energy per pulse ("symbol") and that QPSK transmits two bits per pulse, we can write

$$E_b = \frac{E_s}{2}. \tag{5.88}$$

Then Equation (5.87) becomes

$$P_{e,s} \cong 2Q\left(\sqrt{2\frac{E_b}{N_0}}\right). \tag{5.89}$$

To find the probability of a bit error we need to know how many bit errors occur on the average when a symbol error occurs. To know that, we need to know how the bits are assigned to the symbols. Let us assume that an incoming bit pair is assigned to the coefficient pair (a_{Ii}, a_{Qi}) as shown in Figure 5.34. With this assignment, an in-phase error causes an error in the first bit, and a quadrature error causes an error in the second bit. Errors in both bits can be caused only by simultaneous in-phase and quadrature errors, an event whose probability we have already shown to be negligible. Thus, for the given bit assignment, each symbol error nearly always results in a single bit error.

Suppose a large number N of bits is generated by the source. To transmit these bits, the QPSK transmitter generates $N/2$ pulses. If the probability of symbol error is $P_{e,s}$, then the number of symbols in error is $\frac{N}{2}P_{e,s}$. Each symbol error causes one bit error, so the number of bits in

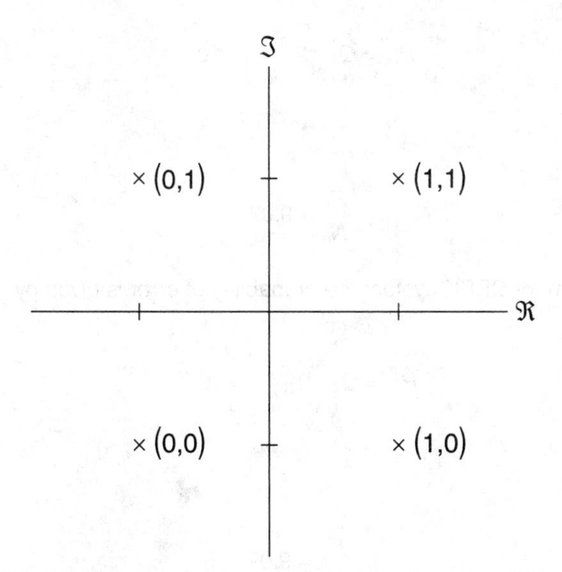

Figure 5.34 QPSK Signal Constellation Showing Bit Assignment

error is also $\frac{N}{2}P_{e,s}$. Now the probability of bit error is the number of bits in error divided by the number of bits received. We have

$$\begin{aligned} P_{e,b} &= \frac{(N/2)P_{e,s}}{N} = \frac{P_{e,s}}{2} \\ &= Q\left(\sqrt{2\frac{E_b}{N_0}}\right), \end{aligned} \tag{5.90}$$

where the last line comes from Equation (5.89). Comparing Equation (5.90) with Equation (5.47), we see that, somewhat counter to intuition, the probability of bit error is given by the same formula for BPSK and for QPSK.

Example

A data source generates a bit stream at 100 kbits/s. The bits are transmitted using BPSK over an additive white Gaussian noise channel, achieving a probability of bit error of 10^{-5}. Suppose we wish to replace the modulator and demodulator with a QPSK system. How much will the receiver sensitivity change if we maintain the probability of bit error at 10^{-5}?

Solution

For the original BPSK system the probability of error is given by

$$P_{e,b} = Q\left(\sqrt{2\frac{E_b}{N_0}}\right) = 10^{-5}. \tag{5.91}$$

Solving gives

$$\frac{E_b}{N_0} = 9.09. \tag{5.92}$$

For the replacement QPSK system the probability of error is given by

$$\hat{P}_{e,b} = Q\left(\sqrt{2\frac{\hat{E}_b}{N_0}}\right) = 10^{-5}, \tag{5.93}$$

so

$$\frac{\hat{E}_b}{N_0} = 9.09 \tag{5.94}$$

as well. The "hat" symbol is used to distinguish parameters of the QPSK system from parameters of the BPSK system. We assume that the noise level is independent of the modulation system used. For BPSK, one pulse is transmitted every $T_s = 10^{-5}$ seconds. The average received power needed to achieve the required probability of error (that is, the receiver sensitivity) is

$$P_s = \frac{E_b}{T_s}. \tag{5.95}$$

For QPSK, one pulse is transmitted every $\hat{T}_s = 2T_s$ seconds. The energy per pulse is $\hat{E}_s = 2\hat{E}_b$. The receiver sensitivity is

$$\hat{P}_s = \frac{\hat{E}_s}{\hat{T}_s} = \frac{2\hat{E}_b}{2T_s} = \frac{E_b}{T_s} = P_s, \tag{5.96}$$

using Equations (5.92), (5.94), and (5.95). Thus changing modulation methods has not changed the receiver sensitivity at all.

What is the advantage to switching to QPSK in this case? The channel bandwidth for BPSK, estimated using the Nyquist bandwidth, is $B = \frac{1}{T_s}$. For QPSK, however, the bandwidth is $\hat{B} = \frac{1}{\hat{T}_s} = \frac{1}{2T_s} = \frac{B}{2}$. Thus, by switching to QPSK, we are able to reduce the bandwidth in half, with no performance penalty.

Offset QPSK (OQPSK)

Suppose in the transmitter of Figure 5.30 we dėlay the output of the quadrature channel line-code generator by $T_s/2$ seconds. An example of the two line-code signals for QPSK and OQPSK is shown in Figure 5.35. Rectangular pulses are shown to make the diagram easier to read. The line-code offset affects the phase changes that are possible in the OQPSK signal. The possibilities are most easily seen with reference to the signal constellation shown in Figure 5.34. When $m_I(t)$ changes polarity, the selected signal point moves between the right half plane and the left half plane. When $m_Q(t)$ changes polarity, the signal point changes between the upper half plane and the lower half plane. Because $m_I(t)$ and $m_Q(t)$ do not change at the same time, changes never occur between diagonally opposed signal points. In terms of bit pattern, there can never be a $(0,0) \rightleftarrows (1,1)$ or $(0,1) \rightleftarrows (1,0)$ transition, since each input bit affects one line code and the two line codes do not change simultaneously.

The fact that signal changes never occur between diagonally opposed signal points is significant, because it implies that the amplitude of the modulated signal never passes through zero during a transition. An example of an OQPSK signal is shown in Figure 5.36 using the same parameters as those used in Figure 5.31. By comparing the two figures, you can see clearly how the OQPSK signal envelope remains relatively constant and does not dip to zero as the QPSK envelope occasionally does.

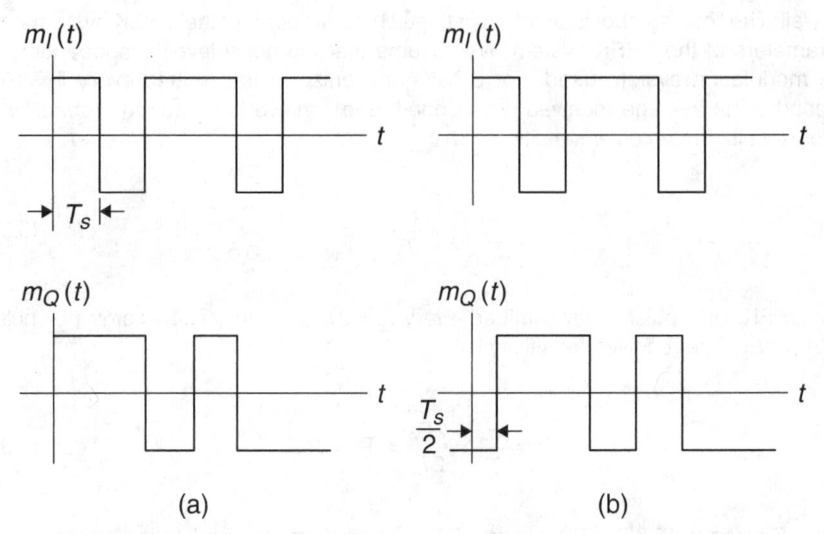

Figure 5.35 In-Phase and Quadrature Line-Code Signals: (a) QPSK; (b) OQPSK

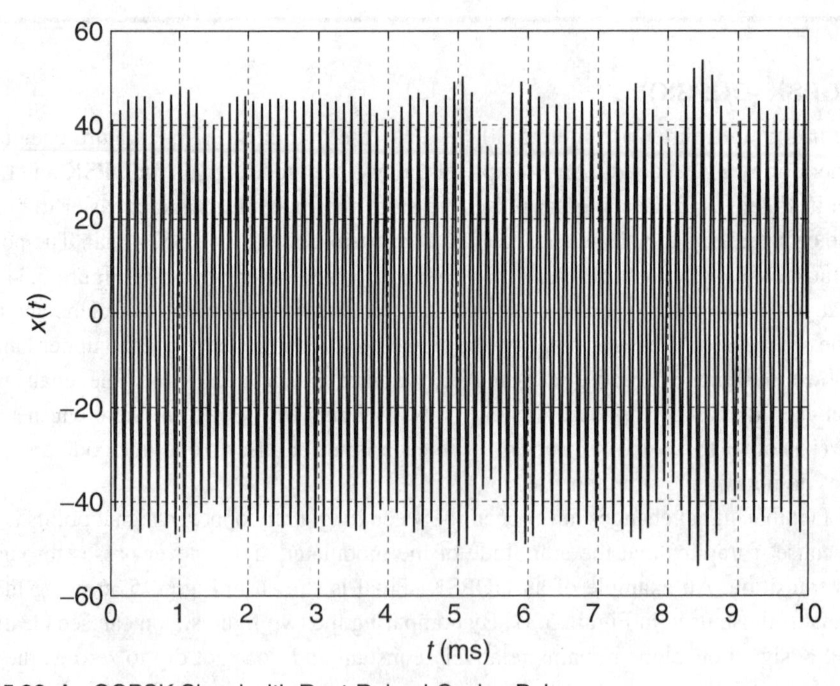

Figure 5.36 An OQPSK Signal with Root-Raised-Cosine Pulses

Variations in the amplitude of the transmitted signal are of no consequence if the transmitter's amplifiers and filters are linear. We have discussed how amplifier efficiency can be improved, however, if some clipping of the transmitted signal is allowed. The consequences of clipping a QPSK signal are that the spectrum of the signal develops side lobes that can interfere with signals in adjacent channels. The advantage to OQPSK is that clipping the amplitude has a significantly smaller effect. Since the signal amplitude does not dip to zero, the clipped signal does not develop the "near discontinuities" that cause the spectrum to broaden.

Example

QPSK and OQPSK signals were simulated for 2000 bits of random input data. The pulse shape is 100% rolloff root-raised-cosine. The pulse duration is $T_s = 10\ \mu s$, and the carrier frequency is $f_c = 400$ kHz. Estimates of power spectra are shown in Figure 5.37. Both the QPSK and OQPSK signals were clipped by passing them through hard limiters and then bandpass filtered to remove harmonics of the carrier frequency. The spectra of the original QPSK signal and of both clipped signals are shown in the figure.

Both the QPSK and OQPSK signals suffer from increased out-of-band spectral "growth" when clipped. The side lobes for the clipped OQPSK signal are about 10 dB lower than the side lobes for the clipped QPSK signal, however.

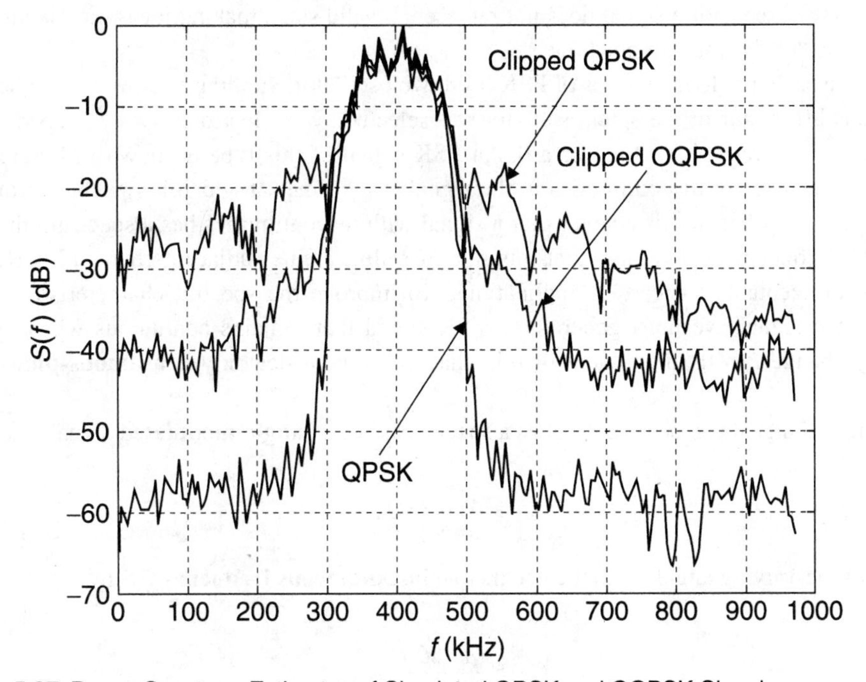

Figure 5.37 Power Spectrum Estimates of Simulated QPSK and OQPSK Signals

Offset QPSK, when it is not clipped or distorted, has the same power spectrum as QPSK, and hence the same bandwidth. It has the same spectral efficiency and the same relation between $P_{e,b}$ and E_b/N_0 as QPSK. The two modulation methods differ only in the timing of the bit transitions and in the consequent shape of the transmitted waveform.

Example

QPSK and OQPSK are used in the cdmaOne and cdma2000 cellular telephone systems. Both these systems use QPSK in the forward link and OQPSK in the reverse link, where the handset transmitter requires an efficient power amplifier.

Frequency-Shift Keying (FSK)

Binary frequency-shift keying (BFSK) is a modulation method in which a constant-amplitude sinusoid is transmitted. The sinusoid switches back and forth between two possible transmission frequencies. One frequency is used to represent a 1 and the other to represent a 0. It is possible to make use of more than two frequencies. Such "m-ary" FSK signals, where m signifies the number of symbols or frequencies used, are of practical importance, but space does not permit discussing them in the current chapter. Frequency-shift keying is a venerable modulation method that has been in use at least since the 1920s. FSK was attractive for wired and wireless telegraph because the absence of a signal does not represent a valid state, making it easy to distinguish a telegraphic "off" from a dropped line.

Some early implementations of FSK used two oscillators running at different frequencies. The signal from one or the other oscillator was selected by an electronic switch, depending on whether the current bit was a 1 or a 0. An FSK signal of this type is shown in Figure 5.38. Notice that there are discontinuities in the signal at the times at which it switches frequency. We observed earlier in this chapter that a signal with discontinuities has a spectrum that falls off with frequency as $1/f^2$, and that this rate of rolloff creates adjacent-channel interference that is unacceptable for certain applications. To improve the spectral characteristics of the transmitted signal, we must generate an FSK signal that remains continuous when the frequency changes. In the remainder of this chapter we consider only **continuous-phase** FSK signals.

An FSK signal can be thought of as a special case of an **angle-modulated** signal

$$x(t) = A\cos\theta(t). \tag{5.97}$$

For the time-varying angle $\theta(t)$, let us define the **instantaneous frequency** $f(t)$ by

$$f(t) = \frac{1}{2\pi}\frac{d\theta(t)}{dt}. \tag{5.98}$$

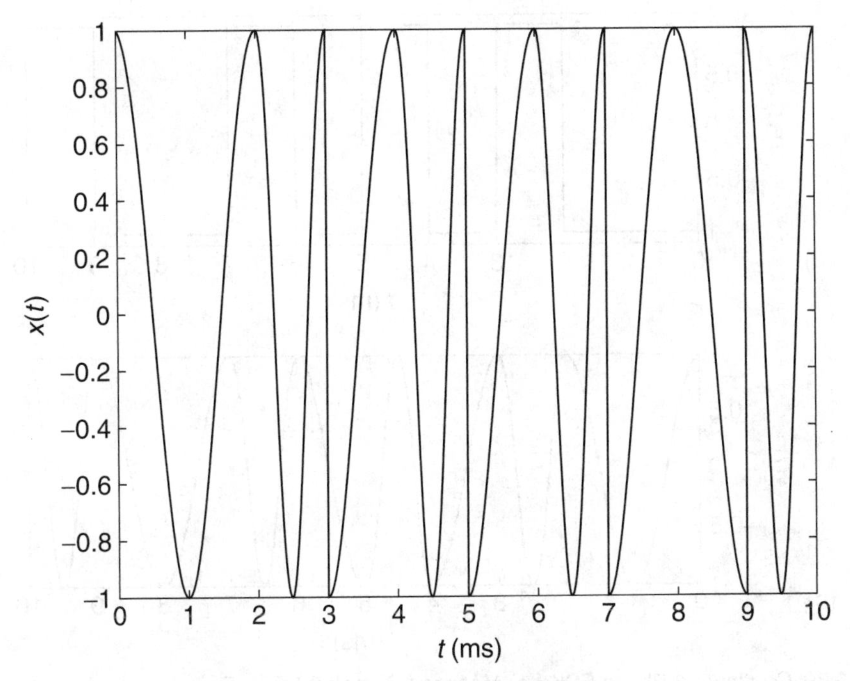

Figure 5.38 Discontinuous-Phase Frequency-Shift Keyed Signal

We can invert Equation (5.98) and write the angle $\theta(t)$ in terms of the instantaneous frequency as

$$\theta(t) = \theta_0 + 2\pi \int_0^t f(\tau)d\tau, \tag{5.99}$$

where $\theta_0 = \theta(0)$.

To specialize the general angle-modulated formulation to frequency modulation, let us specify that the instantaneous frequency $f(t)$ is given as

$$f(t) = f_c + k_f m(t), \tag{5.100}$$

where f_c is the carrier frequency, k_f is a system parameter called the **frequency sensitivity**, and $m(t)$ is the message. We can now substitute Equation (5.100) into Equation (5.99) and then into Equation (5.97) to obtain a general expression for a frequency-modulated signal:

$$x(t) = A\cos\left[2\pi f_c t + 2\pi k_f \int_0^t m(\tau)d\tau + \theta_0\right]. \tag{5.101}$$

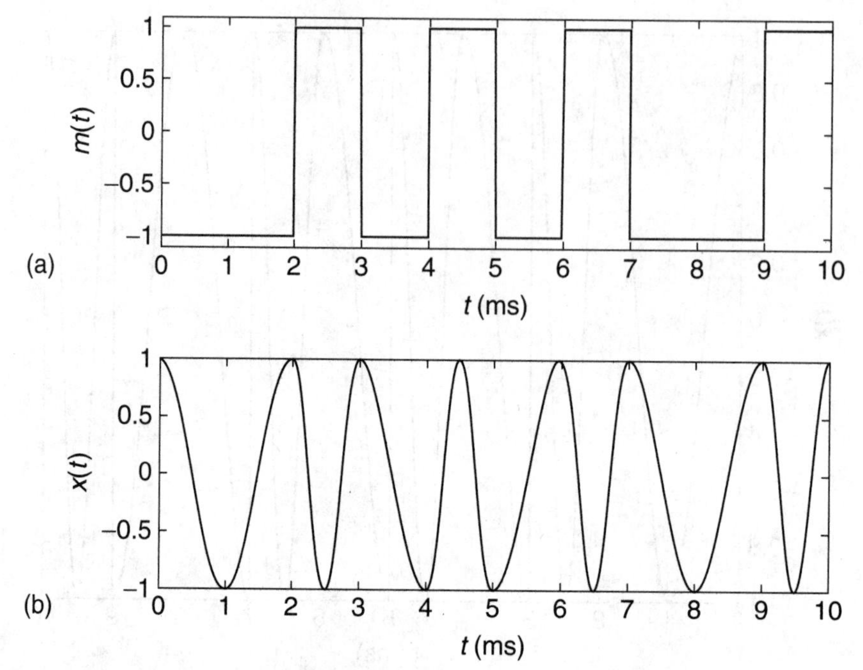

Figure 5.39 Continuous-Phase FSK: (a) Message Signal; (b) FSK Signal

To further specialize Equation (5.101) to FSK, let the message $m(t)$ be a polar NRZ line code, as shown in Figure 5.39(a). We denote the pulse width as T_s and the peak pulse value as m_p. The resulting FSK signal is shown in Figure 5.39(b). With a rectangular message signal, Equation (5.100) gives the instantaneous frequency as

$$f(t) = f_c \pm k_f m_p. \tag{5.102}$$

Each of the two possible values of instantaneous frequency differ from the carrier frequency by

$$\Delta f = k_f m_p, \tag{5.103}$$

a quantity called the **peak frequency deviation**. It is customary to designate an FSK signal by the pulse interval T_s and by a **modulation index** h, defined as the ratio of the peak frequency deviation to the message Nyquist bandwidth, that is,

$$h = \frac{\Delta f}{1/2T_s} = 2\Delta f T_s. \tag{5.104}$$

To find the bandwidth of the FSK signal, it is necessary to calculate the Fourier transform of Equation (5.101). This challenging feat was accomplished for a periodic message by Carson in a landmark 1922 paper.[8] We will not reproduce the details of Carson's work, but the result can be summarized in a rule of thumb for bandwidth known as "Carson's rule." Carson's rule states that the bandwidth B_{FSK} of the FSK signal is approximately given in terms of the peak frequency deviation Δf and the message bandwidth B_m as

$$B_{FSK} \approx 2(\Delta f + B_m). \tag{5.105}$$

The presence of Δf in the bandwidth expression is intuitively reasonable, but the additional bandwidth attributed to the message was a surprising result when first published. For analog FM, it is normal practice to make the frequency deviation term dominate, as performance is better for larger Δf. For binary FSK, the performance does not improve with deviation, provided Δf exceeds a minimum value (to be discussed below). Therefore, it is common practice to keep the peak frequency deviation small to limit the bandwidth.

Example

An FSK signal was simulated for 20,000 bits of random input data. The message is an NRZ polar waveform. The pulse duration is T_s = 1/15,000 seconds, the carrier frequency is f_c = 55 kHz, and the peak frequency deviation is Δf = 5 kHz. The modulation index is therefore h = 2/3. An estimate of the power spectrum is shown in Figure 5.40.

If we use the first-null bandwidth for B_m, then Carson's rule estimates the FSK bandwidth B_{FSK} as

$$B_{FSK} \approx 2(\Delta f + B_m) = 2(5000 + 15000) = 20 \text{ kHz}. \tag{5.106}$$

The Carson's rule bandwidth is marked on the figure. It can be seen that Carson's rule gives a somewhat generous estimate, corresponding in this case to the "second-null" bandwidth of the FSK signal.

The spectral efficiency of a binary FSK signal is given by

$$\eta = \frac{1/T_s \text{ bits/s}}{B_{FSK} \text{ Hz}} \approx \frac{1}{2T_s(\Delta f + B_m)}. \tag{5.107}$$

8. J. R. Carson, "Notes on the Theory of Modulation," *Proceedings of the Institute of Radio Engineers* 10 (1922): 57–64.

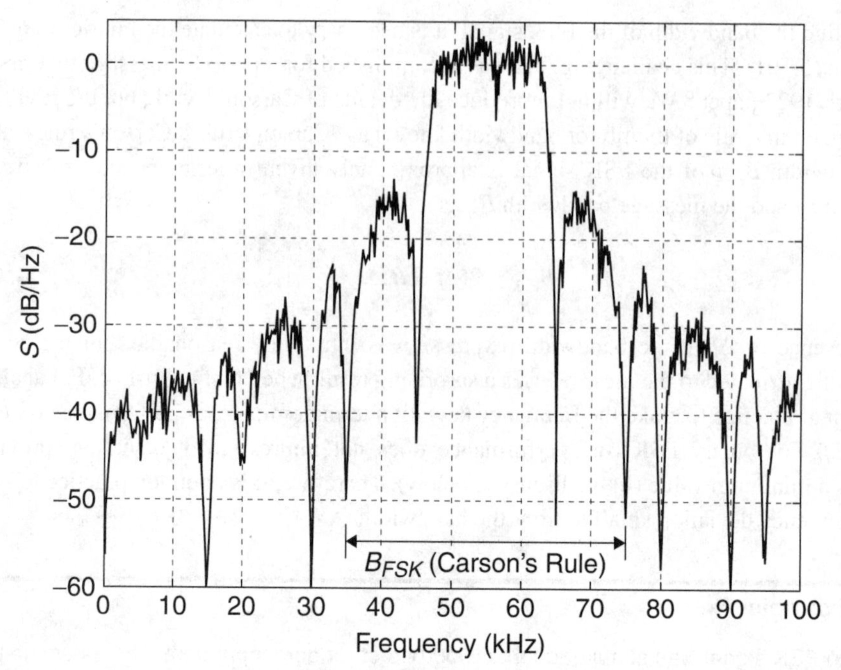

Figure 5.40 Power Spectrum Estimate of a Simulated FSK Signal

If we take the Nyquist bandwidth $\frac{1}{2T_s}$ for B_m (optimistically, if the message is a polar NRZ signal), we obtain

$$\eta \approx \frac{1}{2\Delta f T_s + 1} = \frac{1}{h+1} < 1 \text{ bits/s/Hz}. \tag{5.108}$$

Thus, we see that the spectral efficiency of a binary FSK signal is poorer than that of a BPSK signal.

The rate at which the spectrum of the FSK signal rolls off with frequency depends on the "smoothness" of the signal $x(t)$. When the message $m(t)$ has the appearance of Figure 5.39(a), the phase function $2\pi k_f \int_0^t m(\tau)d\tau$ is a sequence of straight lines, as shown in Figure 5.41.

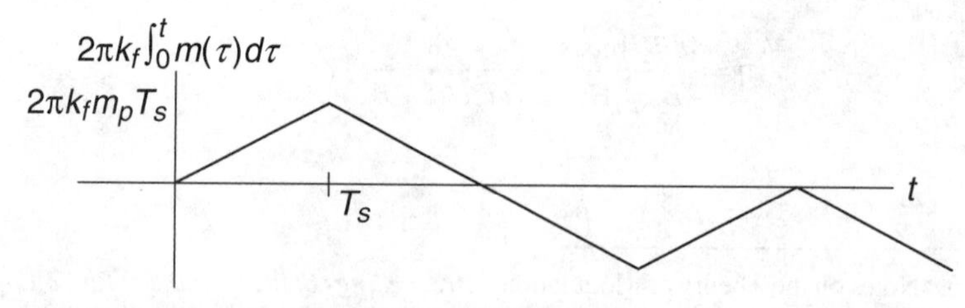

Figure 5.41 FSK Phase Function for NRZ Polar Message

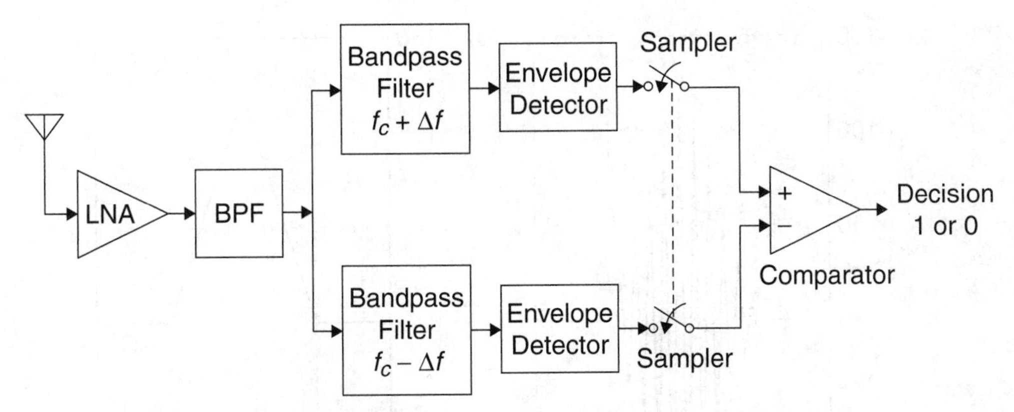

Figure 5.42 Noncoherent FSK Receiver

This is a continuous signal, even though $m(t)$ is discontinuous. It follows that $x(t)$ is a continuous signal, although the corners in the phase function imply that $x(t)$ will not have a continuous derivative. A continuous signal with a discontinuous derivative has a power spectrum that falls off with frequency as $1/f^4$. This is an adequate rolloff rate to avoid adjacent-channel interference for many applications. We will discuss below a simple technique for improving the rolloff rate for demanding applications, such as cellular telephone systems.

Frequency-shift keyed signals can be demodulated coherently or noncoherently. Coherent demodulation offers a small performance benefit, but it can be a challenge to implement, given that there are two transmitted sinusoids at different frequencies. Noncoherent demodulation is commonly used and is the technique discussed here.

A receiver for an FSK signal is shown in Figure 5.42. The received signal is applied to two bandpass filters, one tuned to each of the possible transmitted frequencies. On any given transmission, one of the bandpass filters will produce a large output signal and the other will produce an output signal that is filtered noise. These signals are applied to envelope detectors, and the outputs of the envelope detectors are sampled and compared. The larger envelope will normally indicate which frequency was transmitted.

Performance of the receiver can be improved if the bandpass filters are matched to the transmitted pulses. The response $y(t)$ of a filter matched to a rectangular burst of sinusoid is shown in Figure 5.43(a), and the envelope $e(t)$ is shown in Figure 5.43(b). Because the noncoherent receiver is not synchronized to the phase of the transmitted sinusoid, there is no guarantee that the waveform $y(t)$ peaks at the sample time $t = 0$. The envelope always peaks at $t = 0$, however. Thus, the envelope detector is needed to ensure that the sample value will always be a large positive number when a sinusoid of the appropriate frequency is present.

Performance of the coherent FSK receiver is given by Equation (5.23). To find E_{diff}, consider the difference signal

$$s_1(t) - s_0(t) = A\cos(2\pi f_1 t + \phi_1) - A\cos(2\pi f_0 t + \phi_0), \quad 0 \le t \le T_s, \tag{5.109}$$

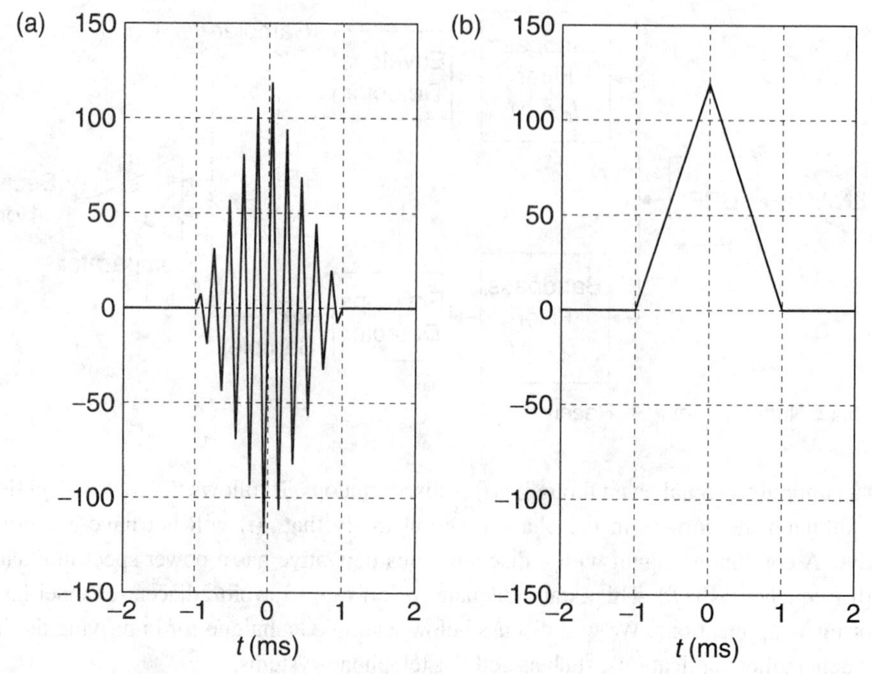

Figure 5.43 (a) Response of a Matched Filter to a Rectangular Burst of Sinusoid; (b) Envelope of Matched Filter Response

where $f_1 = f_c + \Delta f$, $f_0 = f_c - \Delta f$, and the phase angles ϕ_1 and ϕ_0 have whatever values are necessary to maintain a continuous-phase signal. The energy E_{diff} is given by

$$\begin{aligned} E_{diff} &= \int_0^{T_s} [s_1(t) - s_0(t)]^2 dt \\ &\cong \frac{A^2 T_s}{2} - 2A^2 \int_0^{T_s} \cos(2\pi f_1 t + \phi_1)\cos(2\pi f_0 t + \phi_0)dt + \frac{A^2 T_s}{2}, \end{aligned} \tag{5.110}$$

where the approximation is good if there is an integral number of carrier cycles in T_s seconds for both frequencies f_1 and f_0, or if there are so many carrier cycles in T_s seconds that the energy contributed by a partial cycle at the end of the interval can be ignored. It is a common practice to choose the peak frequency deviation Δf so that the integral in the middle term in Equation (5.110) is zero. This means that the cross-correlation between the two possible transmitted pulses $s_1(t)$ and $s_0(t)$ is zero. Two signals whose cross-correlation is zero are said to be **orthogonal**. In Equation (5.110), orthogonality can be guaranteed through careful choice of f_1 and f_0. Alternatively, we can obtain approximate orthogonality by choosing these frequencies high enough that there are many cycles in T_s seconds at either frequency. This approach can be easily verified by evaluating the integral of the product of cosines in Equation (5.110) and comparing the result with $A^2T_s/2$.

For orthogonal signals $s_1(t)$ and $s_0(t)$, the difference energy E_{diff} is given by the sum of the energy in $s_1(t)$ and the energy in $s_0(t)$. For FSK, both these signals have energy $A^2T_s/2$, which is also the energy per bit, E_b. Thus, for orthogonal equal-energy signals we have

$$E_{diff} = 2E_b, \tag{5.111}$$

which then implies

$$P_e = Q\left(\sqrt{\frac{2E_b}{2N_0}}\right) = Q\left(\sqrt{\frac{E_b}{N_0}}\right). \tag{5.112}$$

Comparison of Equation (5.112) with Equation (5.47) shows that even with coherent detection, FSK requires 3 dB more E_b/N_0 than either BPSK or QPSK for the same probability of bit error.

For noncoherent detection of FSK, there is a slight performance penalty, just as there is for noncoherent detection of BPSK. We omit the derivation, but the result is that the probability of error is given by

$$P_e = \frac{1}{2}e^{-\frac{E_b}{2N_0}}. \tag{5.113}$$

Comparing Equation (5.113) with Equation (5.73) leads us to the interesting conclusion that noncoherent FSK suffers a 3 dB penalty compared with noncoherent BPSK (that is, with DPSK), just as coherent FSK suffers a 3 dB penalty compared with coherent BPSK.

Given that FSK has a wider bandwidth than BPSK, a poorer spectral efficiency, a poorer out-of-band spectral rolloff, and a greater E_b/N_0 needed to produce a given probability of error, we might well ask the reason for this modulation method's continued popularity. There are three primary reasons. First, we have already mentioned the "always on" feature that was attractive for wireless telegraph. Second, noncoherent FSK transmitters and receivers have been simple to implement in a wide variety of technologies. The third reason is especially relevant for cellular systems or any systems requiring an inexpensive mobile terminal. As can be seen from Equation (5.101), the amplitude of a transmitted FSK signal is always constant, regardless of the message $m(t)$, and the value of the amplitude carries no information. Thus, there is no loss of information or broadening of the spectrum of the transmitted signal if the transmitter's power amplifier clips the modulated waveform. Frequency-shift keying is often the modulation method of choice when transmitter cost and power requirements must be kept to a minimum.

Gaussian Frequency-Shift Keying (GFSK)

The out-of-band spectral rolloff of an FSK signal can be improved if the transmitted signal $x(t)$ is made smoother. We have seen that when the message $m(t)$ is represented as a polar NRZ signal, the phase function $2\pi k_f \int_0^t m(\tau)d\tau$ will have "corners," and this will lead to "corners" in $x(t)$ at the instants when the frequency changes. To further reduce the out-of-band emissions, we can apply further smoothing to the message signal $m(t)$. The simplest approach is to filter the polar NRZ line code before sending $m(t)$ to the FSK transmitter. For example, a lowpass filter will eliminate the discontinuities in $m(t)$, which in turn will eliminate the corners in the phase function. The characteristics of the lowpass filter, however, can have significant impact on other performance characteristics. For example, lowpass filtering a polar NRZ signal can lead to a significant increase in intersymbol interference if the filter parameters are not chosen appropriately. Furthermore, Equation (5.102) shows that overshoot in the filter response can increase the peak frequency deviation, which, according to Equation (5.105), will increase the bandwidth.

A popular filter for smoothing the message $m(t)$ in FSK systems is the **Gaussian filter**. It has an impulse response

$$h(t) = \frac{1}{\sqrt{2\pi\sigma^2}} e^{-\frac{t^2}{2\sigma^2}}, \tag{5.114}$$

where σ is related to the bandwidth B of the filter by

$$\sigma = \frac{\sqrt{\ln 2}}{2\pi B}. \tag{5.115}$$

This filter produces no overshoot in response to a step input. The Gaussian filter has a frequency response $H(f)$ that also has a Gaussian shape:

$$H(f) = e^{-\frac{\ln 2}{2}\left(\frac{f}{B}\right)^2}. \tag{5.116}$$

The degree of smoothing, and the corresponding degree of intersymbol interference, is determined by the bandwidth of the filter relative to the bandwidth of an NRZ pulse. We can identify a "smoothing parameter" as the ratio $\frac{B}{1/T_s} = BT_s$. If $BT_s \gg 1$, there is very little smoothing and very little intersymbol interference. On the other hand, if $BT_s \ll 1$, there is a great deal of smoothing and significant intersymbol interference. A popular compromise is to set the smoothing parameter at $BT_s = 0.5$. Figure 5.44 shows a single 1 μs Gaussian-filtered pulse for $BT_s = 0.05$, $BT_s = 0.5$, and $BT_s = 10$. When $BT_s = 0.05$, the pulse is very smooth but stretched out over about 16 μs. When a train of such pulses is transmitted, each pulse will interfere with several of its neighbors. When $BT_s = 10$, the pulse is hardly filtered at all and remains nearly rectangular. The remaining case, $BT_s = 0.5$, is clearly a compromise between these extremes. Figure 5.45(a) shows a polar NRZ waveform and Figure 5.45(b) shows the result of passing that waveform through a Gaussian filter with $BT_s = 0.5$. It is apparent that the filter has

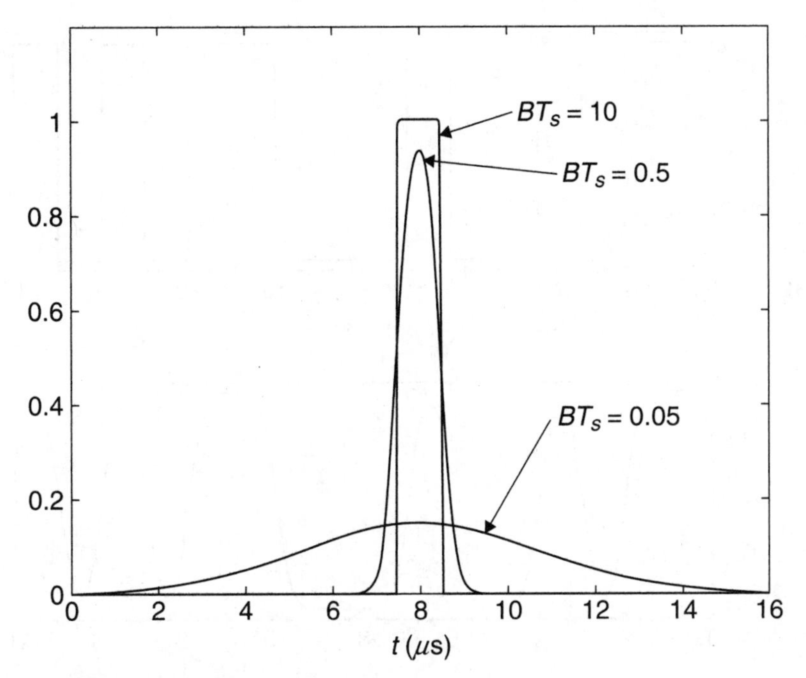

Figure 5.44 Gaussian-Filtered Pulses

smoothed the vertical edges of the NRZ signal. It is also apparent that the pulse amplitudes are not all equal but seem to depend on the bit pattern. This is a consequence of intersymbol interference between the pulses. When used as a modulating signal, this filtered message will produce a peak frequency deviation that differs slightly from bit to bit.

Because a Gaussian-filtered message signal produces an FSK signal $x(t)$ with continuous derivatives of all orders, the FSK spectrum falls off very rapidly with frequency. The fact that all derivatives are continuous eliminates the possibility of spectrum spreading due to time discontinuities. Gaussian FSK is used in the GSM cellular telephone system and also in the 802.11 and Bluetooth wireless systems.

Example

An FSK signal was simulated for 20,000 bits of random input data. The message is an NRZ polar waveform smoothed with a Gaussian filter having $BT_s = 0.5$. The pulse duration is $T_s = 1/15{,}000$ seconds, the carrier frequency is $f_c = 55$ kHz, and the peak frequency deviation is $\Delta f = 5$ kHz. An estimate of the power spectrum is shown in Figure 5.46. Also shown is the power spectrum from Figure 5.40 for the FSK signal without Gaussian filtering.

The figure shows clearly how rapidly the Gaussian FSK spectrum rolls off outside of the channel bandwidth.

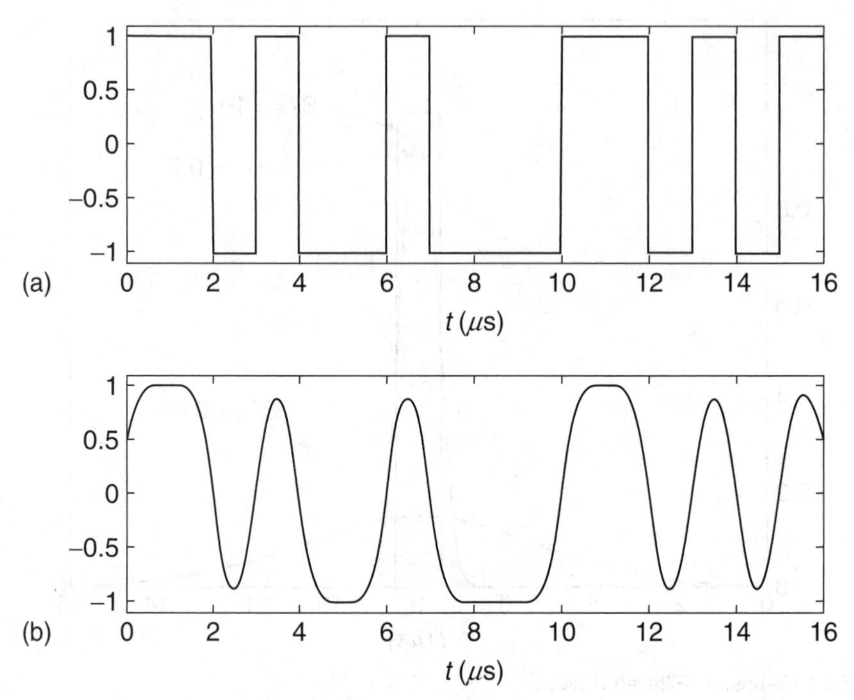

Figure 5.45 (a) Polar NRZ Message Signal; (b) Gaussian-Filtered Message Signal

Minimum-Shift Keying (MSK)

Carson's rule, Equation (5.105), suggests that the bandwidth of an FSK signal can be reduced if the peak frequency deviation Δf is made small. If Δf is made too small, however, the probability of error for a given E_b/N_0 increases from the values given by Equation (5.112) or Equation (5.113). Intuitively, if Δf is made too small, the passbands of the bandpass filters in Figure 5.42 will overlap, and both filters will respond when an input sinusoid at either $f_1 = f_c + \Delta f$ or $f_0 = f_c - \Delta f$ is present. This will make it difficult to reliably distinguish between the two transmitted frequencies.

For reliable FSK reception, it is desirable that the signals $\cos(2\pi f_1 t)$ and $\cos(2\pi f_0 t)$ be orthogonal, or nearly so. Orthogonality is assumed in deriving Equations (5.112) and (5.113). It can be verified by direct calculation that the smallest peak frequency deviation that satisfies the orthogonality requirement is

$$\Delta f = \frac{1}{4T_b}, \tag{5.117}$$

where T_b is used to represent *bit* duration, so as not to cause confusion in the discussion below. When Δf takes the minimum value given by Equation (5.117), the modulation index is

$$h = 2\Delta f T_b = \frac{1}{2}. \tag{5.118}$$

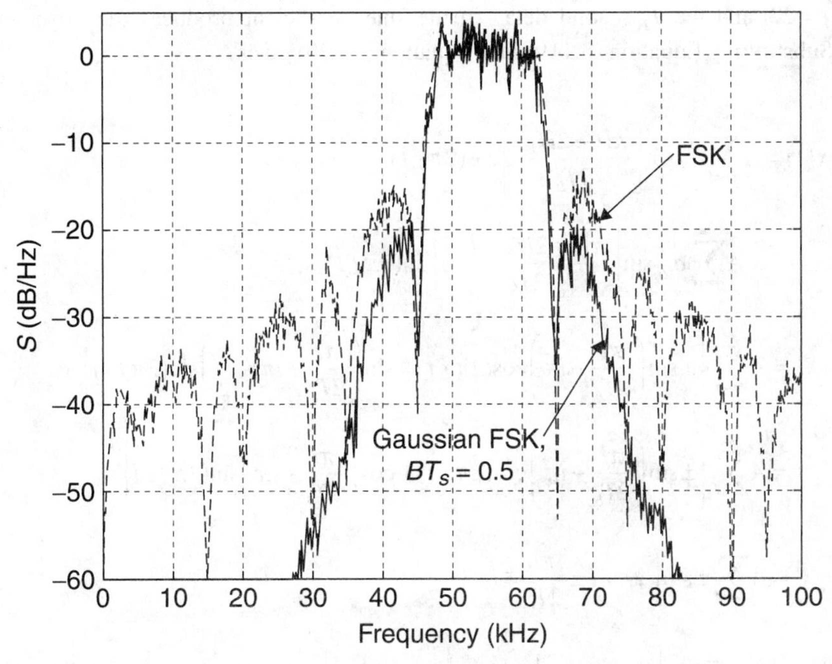

Figure 5.46 Estimated Power Spectra of Simulated FSK and Gaussian FSK Signals

Frequency-shift keying with Δf given by Equation (5.117) or h given by Equation (5.118) is known as **minimum-shift keying** (MSK).

An MSK signal can be considered as a special case of OQPSK. To see this, consider the QPSK signal specified by Equations (5.74) and (5.75). Let the line-code signal $g(t)$ be the half sine given by

$$g(t) = \sin\left(\frac{\pi t}{2T_b}\right), \quad 0 \le t \le 2T_b. \tag{5.119}$$

Then we have

$$\begin{aligned} x(t) = A\sum_{i=-\infty}^{\infty} a_{Ii} g(t - iT_s)\cos(2\pi f_c t) \\ -A\sum_{i=-\infty}^{\infty} a_{Qi} g(t - iT_s - T_b)\sin(2\pi f_c t), \end{aligned} \tag{5.120}$$

where $T_s = 2T_b$ and the T_b-second delay in the sine coefficient produces the offset for offset QPSK. Substituting Equation (5.119) into Equation (5.120) gives

$$\begin{aligned} x(t) &= A\sum_{i=-\infty}^{\infty} a_{Ii} \sin\left[\frac{\pi(t-2iT_b)}{2T_b}\right]\cos(2\pi f_c t) \\ &\quad - A\sum_{i=-\infty}^{\infty} a_{Qi} \sin\left[\frac{\pi(t-2iT_b-T_b)}{2T_b}\right]\sin(2\pi f_c t) \\ &= A\sum_{i=-\infty}^{\infty}\left\{\pm\sin\left(\frac{\pi t}{2T_b}-i\pi\right)\cos(2\pi f_c t)\mp\sin\left(\frac{\pi t}{2T_b}-i\pi-\frac{\pi}{2}\right)\sin(2\pi f_c t)\right\} \\ &= A\sum_{i=-\infty}^{\infty}\left\{\pm\sin\left(\frac{\pi t}{2T_b}-i\pi\right)\cos(2\pi f_c t)\pm\cos\left(\frac{\pi t}{2T_b}-i\pi\right)\sin(2\pi f_c t)\right\} \\ &= A\sum_{i=-\infty}^{\infty}\pm\sin\left(2\pi f_c t\pm\frac{\pi t}{2T_b}\mp i\pi\right) \\ &= \pm A\sum_{i=-\infty}^{\infty}\sin\left[2\pi\left(f_c\pm\frac{1}{4T_b}\right)t\mp i\pi\right]. \end{aligned} \tag{5.121}$$

This shows that the signal $x(t)$ is in fact a sequence of pulsed sinusoids at frequencies $f_c + \frac{1}{4T_b}$ or $f_c - \frac{1}{4T_b}$. A careful analysis shows that the phase is continuous regardless of the data pattern.[9]

Recognizing that MSK is equivalent to a form of QPSK is not just an interesting mathematical curiosity. The well-established techniques for coherent demodulation of QPSK can be applied to MSK, leading to a reduction in the E_b/N_0 needed to achieve a given probability of bit error.

Since Equation (5.121) gives an explicit formula for an MSK signal, it is possible to find an explicit formula for the power spectrum $S_x(f)$ by computing the Fourier transform of $x(t)$. We omit the details, but the result is

$$S_x(f) = \frac{4A^2T_b}{\pi^2}\left[\frac{\cos 2\pi(f+f_c)T_b}{1-16(f+f_c)^2T_b^2}\right]^2 + \frac{4A^2T_b}{\pi^2}\left[\frac{\cos 2\pi(f-f_c)T_b}{1-16(f-f_c)^2T_b^2}\right]^2. \tag{5.122}$$

9. For more detail see, for example, J. G. Proakis and M. Salehi, *Communication System Engineering*, 2nd ed. (Upper Saddle River, NJ: Prentice Hall, 2001), 708–11.

Note that even though the OQPSK signal of Equation (5.120) uses "half-sine" line-code pulses, the equivalent FSK signal of Equation (5.121) assumes a polar NRZ (rectangular) message. Owing to the rectangular message, we would expect the MSK signal to roll off with frequency as $1/f^4$. An examination of Equation (5.122) shows this to be the case. The "first null to first null" bandwidth of the MSK signal can be found from Equation (5.122) to be

$$B_{MSK} = \frac{3}{2T_b}. \tag{5.123}$$

Carson's rule overestimates this as $B_{MSK} = 2(\Delta f + B_m) = 2(\frac{1}{4T_b} + \frac{1}{T_b}) = \frac{5}{2T_b}$, which is the "second null to second null" bandwidth.

Minimum-shift keying is often used in connection with Gaussian filtering of the message signal. The result is called **Gaussian MSK** (GMSK). Gaussian MSK is the version of GFSK used in the GSM cellular telephone system.

Spread-Spectrum Signaling

Overview

Spread-spectrum signaling was developed during World War II as a method for providing clandestine communications that were resistant to jamming. The attributes that make this signaling method jamming resistant also make it resistant to unintentional interference and the degrading effects of multipath fading. Hence, the method has achieved some importance in the design of present-day wireless systems. We will see in Chapter 6 that spread-spectrum techniques are also an important way of sharing channel access among multiple users.

In order for a system to be called a spread-spectrum system, two important conditions must hold.

1. The transmission bandwidth must be much greater than the information bandwidth. Typically the bandwidth is expanded by a factor ranging from 100 to 1 million. The degree of bandwidth expansion is measured by a parameter called the **spreading factor** (SF) or processing gain.
2. The transmission bandwidth increase must be caused by modulation using a waveform not related to the information stream. The receiver knows the spreading waveform in advance.

Given a transmitted signal of fixed average power, spreading the bandwidth lowers the power spectrum over any particular frequency span. In extreme cases the power spectrum of the transmitted signal may be lower than the noise power spectrum, making the transmitted signal extremely hard to detect by an eavesdropper. Furthermore, a narrowband receiver will receive very little of the signal power, so that the spread-spectrum signal will cause little interference to conventional receivers sharing the same frequency band. Of course, these performance advantages come with a cost in the form of increased system complexity, especially in the receiver. In particular, synchronization of the spreading waveform becomes a necessary receiver function.

The advantages of spread-spectrum techniques, however, justify the increased complexity for many applications.

There are two widely used techniques for generating spread-spectrum signals, namely, **frequency hopping** and **direct-sequence spreading**. We will discuss both of these in turn and show how they provide resistance to interference from a narrowband source. There are also hybrids of the two methods in use that we will not discuss, as no new principles would be introduced.

Frequency-Hopping Spread Spectrum

A conceptually straightforward means of spreading the spectrum of a transmitted signal is to "hop" the carrier frequency rapidly in a seemingly random fashion. Of course, the hopping pattern cannot actually be random, as the receiver must know the pattern in order to demodulate the signal. During any hop the transmission is narrowband, but averaged over time, the transmitted power is spread over a wide bandwidth. A frequency-hopping spread-spectrum (FHSS) transmitter is shown in Figure 5.47(a) and the corresponding receiver in Figure 5.47(b).

The block marked "PN Sequence" generates a pseudonoise or pseudorandom bit sequence. A PN sequence is a repeating pattern of bits with a long period that appears to be a random sequence of 1s and 0s when viewed over the entire sequence. It can easily be generated by using a small

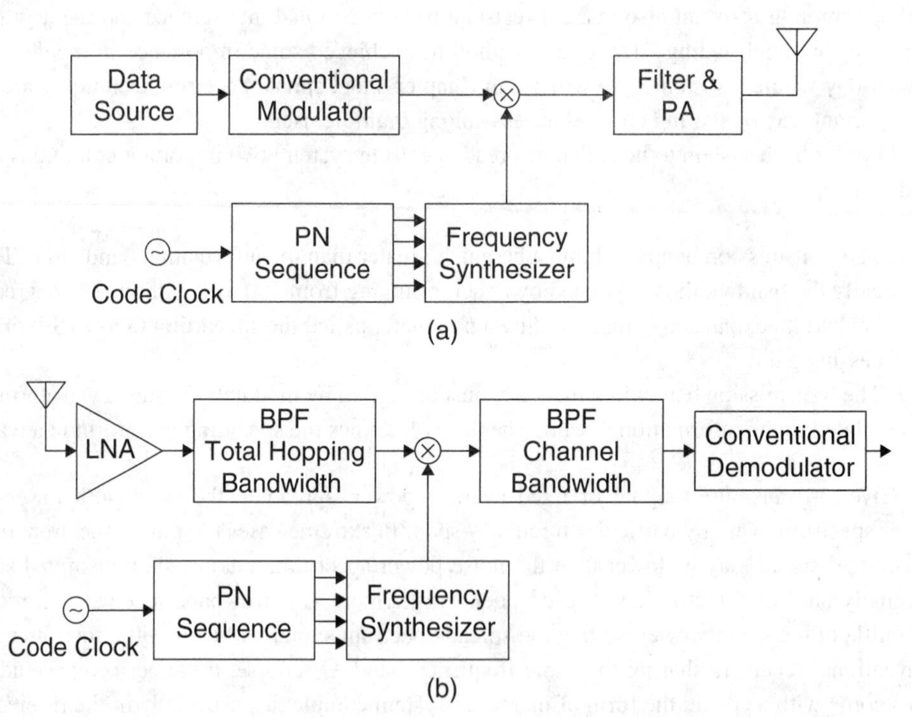

Figure 5.47 Frequency-Hopped Spread-Spectrum System: (a) Transmitter; (b) Receiver

finite-state machine consisting of a shift register with some feedback. Typically, if the shift register has a length of n bits, the PN sequence will have a period of $2^n - 1$ bits. The receiver has an identical PN sequence generator. The PN sequence controls a frequency synthesizer that generates a sinusoid at the carrier frequency. Thus, every time the PN sequence generator is clocked by the "code clock," the carrier frequency at the transmitter will "hop" to a new value. The receiver's synthesizer hops as well, so that the received signal is always properly demodulated. Clearly, the PN sequences at the transmitter and receiver must be synchronized. The synchronizing circuits are not shown in the figure. A well-designed PN sequence will cause the transmitter to hop among all of the available carrier frequencies, spending an approximately equal amount of time at each frequency.

Frequency hopping works well with any modulation method. It has been used with analog AM and FM, as well as with advanced digital modulation techniques. It is common practice to use FSK in frequency-hopping systems, as the hopping does not impair the constant envelope of the FSK signal.

A frequency-hopping system can be characterized by an instantaneous, or channel, bandwidth B, a total hopping bandwidth B_{ss}, and a hopping interval T_{hop}. The instantaneous, or channel, bandwidth is the bandwidth of the conventionally modulated signal that is generated before hopping. If the carrier hops among N hopping frequencies and the hopping frequencies are spaced B apart, then the total hopping bandwidth will be $B_{ss} = NB$. Of course, the total hopping bandwidth can be wider, if the hopping frequencies are spaced at wider frequency intervals. The code clock determines the hopping interval T_{hop}. Two cases can be identified. If the hopping interval T_{hop} is greater than the data pulse time T_s, the system is called a "slow hopper." Alternatively, if the hopping interval is less than the data pulse time, the system is said to be a "fast frequency hopper."

Frequency hopping is attractive in wireless applications because it provides resistance to narrowband interference. To see why this is the case, suppose that there are N hopping frequencies and a narrowband interference source is transmitting on or near one of these frequencies. Suppose that the spread-spectrum signal arrives at the receiver with power P_s and the interfering signal arrives at the receiver with power P_i. We can define the signal-to-interference ratio at the receiver input as

$$SIR_{in} = \frac{P_s}{P_i}. \tag{5.124}$$

Now the spread-spectrum signal hops, but the interference does not. Thus, the spread-spectrum signal shares a frequency with the interference only $1/N$ of the time. Interference will be present at the output of the "channel" bandpass filter only when spectra occupied by the spread signal and interference have significant overlap. Thus, at the output of the channel bandpass filter, the spread-spectrum signal power is still P_s, but the average interference power is P_i/N. The signal-to-interference ratio after processing is therefore

$$SIR_{out} = \frac{P_s}{P_i/N} = N \cdot SIR_{in}. \tag{5.125}$$

The parameter N in this context is called the **processing gain**. For the special case in which the N hopping frequencies correspond to contiguous channels spaced B apart, the processing gain is given by

$$N = \frac{B_{ss}}{B}. \tag{5.126}$$

In this case, the processing gain is equal to the bandwidth spreading factor.

Example

Determine the ability of a Bluetooth link to contend with interference from a microwave oven. Assume that the Bluetooth signal and the interfering signal arrive with equal power at the receiver input.

Solution

Bluetooth operates in the 2400 MHz band as an unlicensed user and therefore must contend with interference from a number of sources, including relatively strong signals from microwave ovens. To reduce the interference effects, the Bluetooth wireless standard (IEEE 802.15.1) uses frequency-hopping spread spectrum. There are 1600 hops per second among 79 channels spaced 1 MHz apart. The underlying modulation is GFSK with pulses at 1 Mbit/s, $BT_s = 0.5$ and $h = 0.32$.

Microwave ovens emit radiation at 2450 MHz and typically sweep a narrowband emission across about a 10 MHz range of frequency in a time interval on the order of 10 ms. This sweeping makes it impossible to simply avoid the microwave frequency.

Assume that the microwave oven's instantaneous bandwidth is less than 1 MHz, so that it interferes with only one Bluetooth channel at a time. If we model the Bluetooth signal as hopping at random in a pattern that is uniformly distributed among all 79 frequencies, then interference will be experienced only 1/79 of the time. We can write the processing gain as $N = 79$, or in decibels, $10\log 79 = 19.0$ dB.

The assumption of equal signal and interference power gives $SIR_{in} = 0$ dB without the spread-spectrum processing gain. After processing we have $SIR_{out} = 0 + 19 = 19$ dB.

The ability of a frequency-hopping spread-spectrum system to "hop away" from an interference source also makes the frequency hopper robust in the presence of frequency-selective fading, a common impairment in wireless systems discussed in Chapter 3. The analysis is similar to that given above for narrowband interference. Since each frequency is visited only $1/N$ of the time, a single frequency of low strength will have a reduced effect. For a fast hopper, the effect will be a slight reduction in the average signal-to-noise ratio. For a slow hopper, the series of bits transmitted during a single hop will occasionally be lost. These bits can be restored through the use of an error-correcting code, provided that bursts of errors do not occur too frequently. Error correction coding will be introduced in Chapter 7.

Direct-Sequence Spread Spectrum

Direct-sequence spread spectrum (DSSS) also uses a PN sequence to spread the bandwidth of the transmitted sequence. The mechanism, however, is entirely different from that used in frequency-hopping systems. A direct-sequence spread-spectrum transmitter is shown in Figure 5.48(a) and the corresponding receiver is shown in Figure 5.48(b). Direct-sequence spread spectrum is normally used with digital modulation methods. We will assume BPSK modulations in the following discussion since it is widely used in contemporary spread-spectrum systems.

To generate a direct-sequence signal, the incoming data is first encoded into a polar line code. Raised-cosine pulses are normally used to control spectral rolloff, but to simplify the discussion we will assume rectangular pulses are used. The polar line code has a bit interval of T_b seconds. The spreading code generator creates a PN sequence that appears to an outside observer to be a random sequence of 1s and 0s. Unlike the frequency hopper, there is no synthesizer and the direct-sequence spreading code generator produces the code as a serial stream encoded in polar NRZ format. The 1s and 0s of the PN sequence are known as **chips**, to distinguish them from information bits. The chip period T_{chip} is determined by the code clock. It is always much smaller than T_b and is usually an integral submultiple of T_b. The message line code $m(t)$ carrying the information is multiplied by the spreading sequence $p(t)$. The resulting waveform $m(t)\,p(t)$ is in polar NRZ format and also appears to an outside observer to

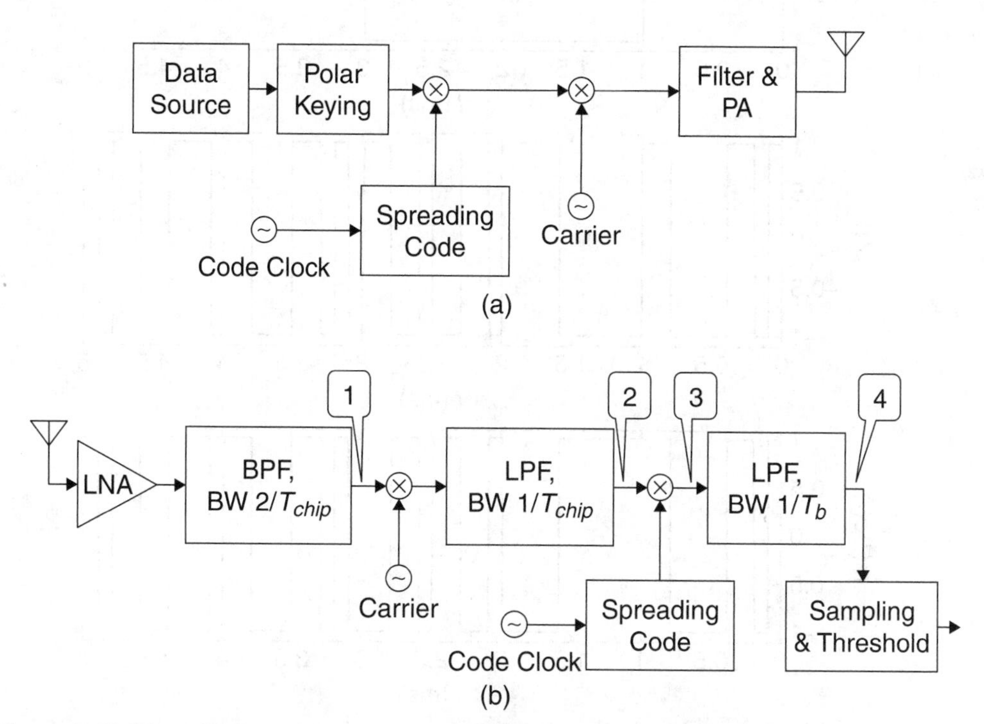

Figure 5.48 Direct-Sequence Spread-Spectrum System: (a) Transmitter; (b) Receiver

be a random sequence of 1s and 0s. The effect of multiplying the message by the spreading code is to spread the bandwidth from $1/T_b$ (the first-null bandwidth of $m(t)$) to $1/T_{chip}$ (the first-null bandwidth of $p(t)$).

Example

A sequence of 2000 random 1s and 0s was generated and represented in polar NRZ format to simulate $m(t)$. An additional 20,000 random 1s and 0s were generated to simulate $p(t)$. The simulated waveforms have $T_b = 1$ ms and $T_{chip} = 0.1$ ms. Figure 5.49 shows $m(t)$, $p(t)$, and $m(t)\,p(t)$ during a 5 ms interval.

Figure 5.50 shows the estimated power spectra of $m(t)$, $p(t)$, and the product $m(t)$ $p(t)$. The first-null bandwidth of $m(t)$ is clearly 1 kHz, and the first-null bandwidth of

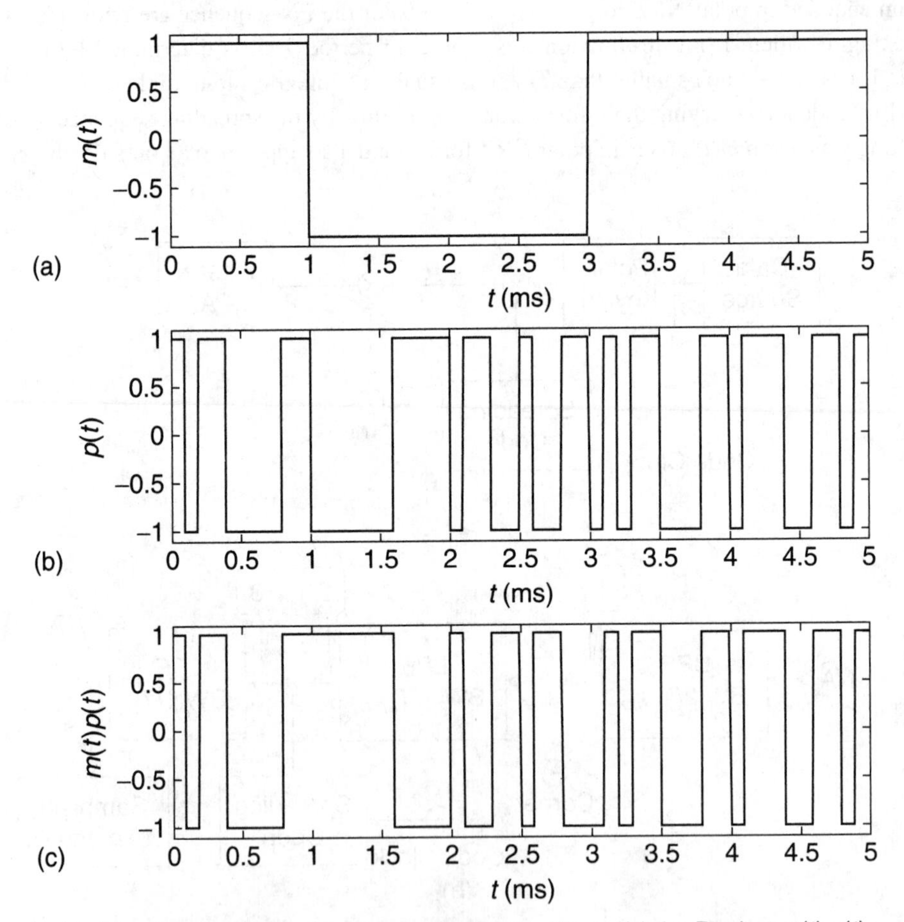

Figure 5.49 (a) A Message $m(t)$; (b) a Spreading Code $p(t)$; and(c) the Product $m(t)$ $p(t)$

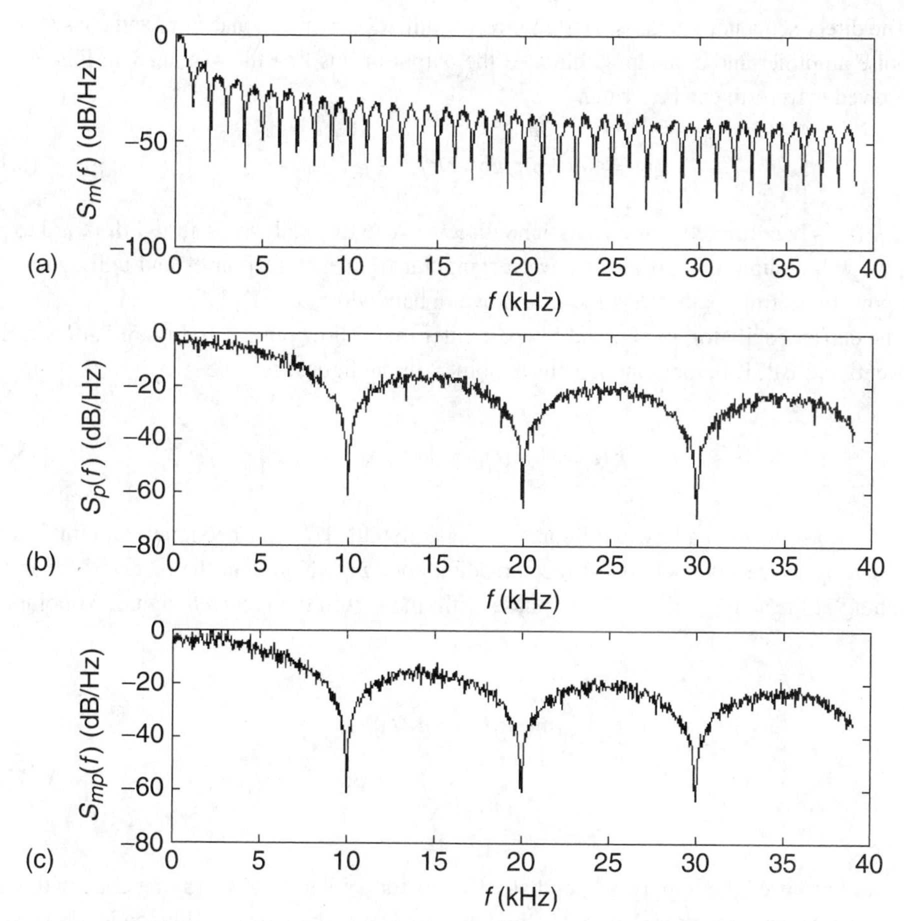

Figure 5.50 Estimated Power Spectra of (a) Message, (b) Spreading Code, and (c) Product

both $p(t)$ and $m(t)$ $p(t)$ is 10 kHz. It is interesting to note that the spreading code and the product waveform have identical power spectra. It is clear that multiplying the message by the spreading code results in a widened bandwidth for the product. (The slow rolloff in spectral level after the first null is a consequence of our assumption of rectangular pulses. In practice, raised-cosine pulses might be used to reduce the out-of-band spectral emissions.)

Once the baseband message has been "spread" by multiplication with the spreading code, the carrier multiplies the resulting polar NRZ waveform to produce a BPSK modulated signal. The bandwidth of the modulated signal is twice the bandwidth of the spreading code, $2/T_{chip}$, for rectangular pulses. The final transmitter stages, filtering and power amplification, are typical of most systems.

The direct-sequence spread-spectrum receiver utilizes a conventional front end consisting of a low-noise amplifier and a bandpass filter. At the output of this first filter (point 1 in Figure 5.48), the received waveform can be written

$$r_1(t) = Am(t)p(t)\cos(2\pi f_c t) + n_1(t), \tag{5.127}$$

where $n_1(t)$ is bandlimited white Gaussian noise. To keep the analysis as straightforward as possible, we will assume that all filters have rectangular frequency responses and unity gains. The noise power spectrum is then $S_{n_1}(f) = N_0/2$ over a bandwidth of $2/T_{chip}$.

The carrier oscillator, mixer, and lowpass filter that follow point 1 in Figure 5.48 constitute a conventional BPSK demodulator. Thus at point 2 in the figure we have

$$r_2(t) = \frac{A}{2}m(t)p(t) + n_2(t), \tag{5.128}$$

where the noise $n_2(t)$ is a baseband noise, with bandwidth $1/T_{chip}$. The signal at point 2 is next multiplied by the receiver's copy of the spreading code $p(t)$. Note that the receiver's code must be running at the same rate, and be in phase with, the $p(t)$ in the received signal. At point 3 we have

$$\begin{aligned} r_3(t) &= \frac{A}{2}m(t)p^2(t) + n_2(t)p(t) \\ &= \frac{A}{2}m(t) + n_3(t), \end{aligned} \tag{5.129}$$

where we have used the important fact that $p^2(t) = 1$ for a polar NRZ spreading code with values of ± 1. It is interesting to note that multiplying the noise $n_2(t)$ by the spreading code has no statistical effect. Thus $n_3(t)$ has the same spectrum and bandwidth as $n_2(t)$.

Example

White Gaussian noise was generated and lowpass filtered to a bandwidth of 10 kHz to match the bandwidth of the spreading code of Figure 5.49. Figure 5.51 shows the lowpass noise $n_2(t)$, the spreading code $p(t)$, and the product $n_3(t) = n_2(t)\,p(t)$. Although $n_2(t)$ and $n_3(t)$ are different waveforms, their statistical similarity is apparent.

The remaining lowpass filter, sampler, and threshold detector constitute a conventional receiver for a polar NRZ line code. For optimal performance, the final lowpass filter should be a filter matched to the pulses used in $m(t)$. Because the "despreading" step in Equation (5.129) has no effect on the noise, the performance of the system is exactly the same as that of a conventional BPSK system, $P_e = Q(\sqrt{2\frac{E_b}{N_0}})$.

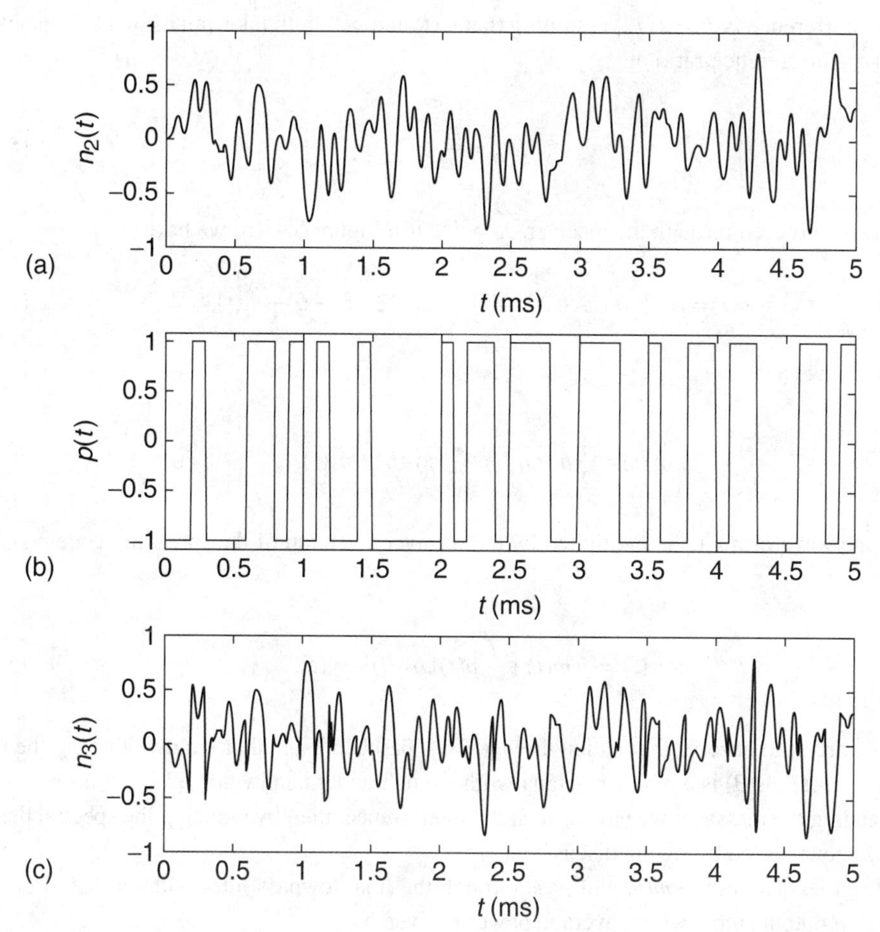

Figure 5.51 (a) Lowpass Noise; (b) Spreading Code; (c) Lowpass Noise Multiplied by Spreading Code

Direct-sequence spread spectrum provides suppression of narrowband interference, just as frequency-hopping spread spectrum does. It is instructive to see how this works. Suppose the waveform at the input to the receiver is given by

$$r(t) = Am(t)p(t)\cos(2\pi f_c t) + J\cos(2\pi f_c + \theta) + n(t), \tag{5.130}$$

where $J\cos(2\pi f_c t + \theta)$ is the interfering signal and $n(t)$ is the usual white Gaussian noise. The interference has been placed at the carrier frequency to produce a worst-case situation. The phase angle θ is included because it is highly unlikely that the interfering signal would be phase locked to the received carrier, unless the interferer were very malicious indeed. From Equation (5.130), the average power in the received signal is $P_s = A^2/2$, and the average power in the interference is $P_i = J^2/2$, assuming that $m(t)$ and $p(t)$ both take values of ±1. This gives a signal-to-interference ratio of

$$SIR_{in} = \frac{A^2}{J^2}. \tag{5.131}$$

Now let us proceed through the receiver. At point 1 in Figure 5.48(b) we have

$$r_1(t) = Am(t)p(t)\cos(2\pi f_c t) + J\cos(2\pi f_c t + \theta) + n_1(t). \tag{5.132}$$

At point 2 we have

$$r_2(t) = \frac{A}{2}m(t)p(t) + \frac{J}{2}\cos(\theta) + n_2(t). \tag{5.133}$$

Next, the waveform $r_2(t)$ is multiplied by the receiver's version of the spreading code $p(t)$. This gives

$$r_3(t) = \frac{A}{2}m(t) + \frac{J}{2}p(t)\cos(\theta) + n_3(t). \tag{5.134}$$

It is important to observe that although $\frac{A}{2}m(t)$ is a baseband signal of bandwidth $1/T_b$, the interference $\frac{J}{2}p(t)\cos(\theta)$ is a baseband signal with a much wider bandwidth $1/T_{chip}$. It is clear that in despreading the message, we have *spread* the interference, thereby reducing the spectral density. This is shown in the following discussion.

The message term $\frac{A}{2}m(t)$ will pass through the final lowpass filter without distortion, producing an output signal whose average power is given by

$$P_{so} = A^2/4. \tag{5.135}$$

To find the interference output power we must pass the interference term $\frac{J}{2}p(t)\cos(\theta)$ through the lowpass filter of bandwidth $1/T_b$. The first step is to find the power spectrum of the interference term.

A rectangular pulse of unit amplitude and duration T_{chip} has Fourier transform T_{chip} $\text{sinc}(T_{chip}f)$. The spreading sequence $p(t)$ is a train of such pulses, with amplitudes that are independent and equally likely to be 1 and −1. This means that the power spectrum $S_p(f)$ of $p(t)$ is given by

$$S_p(f) = T_{chip}\,\text{sinc}^2(T_{chip}f). \tag{5.136}$$

An example of such a power spectrum is shown in Figure 5.50(b). The spectrum has an amplitude value close to T_{chip} for frequencies $|f| \ll \frac{1}{T_{chip}}$ and a first null at $1/T_{chip}$. To find the power spectrum $S_{i3}(f)$ of the interference term, we multiply the power spectrum of Equation (5.136) by $J^2 \cos^2(\theta)/4$. This gives

$$
\begin{aligned}
S_{i3}(f) &= \frac{J^2 \cos^2(\theta) T_{chip}}{4} \operatorname{sinc}^2(T_{chip} f) \\
&= \frac{J^2 T_{chip}}{8} \operatorname{sinc}^2(T_{chip} f),
\end{aligned}
\tag{5.137}
$$

where we have averaged $\cos^2(\theta)$ over all possible values of θ to give 1/2. Now if $H(f)$ is the frequency response of the lowpass filter, then the power spectrum $S_{i4}(f)$ of the interference at the filter output is

$$
\begin{aligned}
S_{i4}(f) &= |H(f)|^2 S_{i3}(f) \\
&= |H(f)|^2 \frac{J^2 T_{chip}}{8} \operatorname{sinc}^2(T_{chip} f) \\
&= \begin{cases} \dfrac{J^2 T_{chip}}{8} \operatorname{sinc}^2(T_{chip} f), & |f| \le \dfrac{1}{T_b} \\ 0, & |f| > \dfrac{1}{T_b}, \end{cases}
\end{aligned}
\tag{5.138}
$$

where we have used the assumption that the filter has unity gain and a rectangular cutoff at a bandwidth of $1/T_b$. Since $\frac{1}{T_b} \ll \frac{1}{T_{chip}}$, very little of the spectrum of Equation (5.137) passes through the lowpass filter. To a good approximation we can write

$$
S_{i4}(f) = \begin{cases} \dfrac{J^2 T_{chip}}{8}, & |f| \le \dfrac{1}{T_b} \\ 0, & |f| > \dfrac{1}{T_b}. \end{cases}
\tag{5.139}
$$

The average power in the interference is the area under the power spectrum, that is,

$$
P_{io} = \int_{-\infty}^{\infty} S_{i4}(f)\,df = \int_{-1/T_b}^{1/T_b} \frac{J^2 T_{chip}}{8}\,df = \frac{J^2 T_{chip}}{4 T_b}.
\tag{5.140}
$$

We can now combine Equations (5.135) and (5.140) to find the signal-to-interference ratio at the output. We have

$$SIR_{out} = \frac{P_{so}}{P_{io}} = \frac{A^2/4}{J^2 T_{chip}/4T_b} = \frac{A^2}{J^2}\frac{T_b}{T_{chip}} \\ = SIR_{in}\frac{T_b}{T_{chip}}. \tag{5.141}$$

The term T_b/T_{chip} is the processing gain for the direct-sequence spread-spectrum system. Note that since the bandwidth of the spread signal is $2/T_{chip}$ and the bandwidth of conventional BPSK is $2/T_b$, the processing gain is equal to the bandwidth expansion, just as in a frequency-hopped system.

Example

In the cdmaOne forward link, voice data is provided at 19.2 kbits/s, a rate that includes "parity" bits added for error correction. The chip rate is 1.2288 Mchips/s. For this system we have $T_b = 1/19{,}200 = 52.1\ \mu s$ and $T_{chip} = 1/1.2288 \times 10^6 = 814$ ns. The processing gain is $T_b/T_{chip} = 52.1 \times 10^{-6}/814 \times 10^{-9} = 64$.

Direct-sequence and frequency-hopping spread-spectrum systems both provide protection against narrowband interference. Although we have not discussed this for the direct-sequence case, both methods provide robustness in the presence of frequency-selective fading. Naturally, there is a question of which method to use in a given application. It turns out that the advantage of choosing one over the other depends to a great extent on the technology that will be used to implement the transmitter and receiver, as well as the overall system constraints and service objectives. As an example of the dilemma systems engineers face in making a choice, the IEEE 802.11 standard for wireless local-area networks contained *both* frequency-hopping and direct-sequence options when it was first released. We will encounter both frequency-hopping and direct-sequence spreading again in Chapter 6, when we discuss code-division multiple access methods.

Conclusions

Our focus in this chapter has been on choosing a modulation method so that data can be transferred from a source to a destination over the wireless link within certain important constraints. A data stream at this level is an unstructured sequence of bits. A modulator and demodulator do not impose any structure on groups of bits, nor do they attach any meaning to the bits. The performance parameters that characterize the service to be provided to the source and destination are data rate and bit error rate.

We began the chapter with an investigation of baseband transmission, as this allowed us to focus on line-code options and bit error rate. Bit error rate is predicted by the probability of error, and we developed the matched filter and correlation receivers as structures that minimize the probability of error when bits are received in the presence of additive white Gaussian noise.

The central part of this chapter introduced a set of alternative modulation methods. Modulation serves the purpose of converting the information bit stream into a waveform that will propagate over a wireless channel. From the point of view of the wireless link, modulation methods can be characterized by channel bandwidth, spectral efficiency, out-of-band spectral rolloff, and implementation complexity. We examined several alternative modulation techniques in detail. Our purpose was not to provide an exhaustive catalog of all known alternatives, but to illustrate how various representative modulation methods perform. We restricted our attention to digital modulation schemes typical of those used in cellular telephone and wireless networking systems.

Modulation methods are conventionally divided into linear and nonlinear categories. Phase-shift keying and its variations represent the linear categories. We investigated binary and quadrature phase-shift keying. Quadrature phase-shift keying has twice the spectral efficiency of binary phase-shift keying. Both of these modulation methods require synchronization at the receiver, which adds to the receiver's complexity. Differential versions of both these modulation methods that do not require synchronization were also discussed. We showed that there is a small (less than 1 dB) performance penalty associated with these differential (noncoherent) methods. We demonstrated that both BPSK and QPSK methods can make effective use of pulse shaping to significantly limit out-of-band spectrum. To maintain the amplitude shape of the transmitted signal, however, the power amplifiers in the transmitter must be linear. We described how distortion or clipping of the transmitted signal produces "growth" in the side lobes of the transmitted power spectrum, raising the possibility of adjacent-channel interference. We introduced offset quadrature phase-shift keying as a technique that limits the most extreme amplitude variations in the transmitted waveform, thus reducing the sensitivity to out-of-band spectral growth.

There are other linear modulation schemes that we have not discussed. Phase-shift keying can be extended to more than four phase values. Eight-phase PSK, for example, allows transmission of three bits per pulse, increasing the spectral efficiency. Quadrature amplitude modulation (QAM) allows both the amplitude and phase of each pulse to carry information, increasing the spectral efficiency even more. The latest third-generation cellular systems employ both 8PSK and 16-QAM (four bits per symbol) modulation. Fourth-generation systems employ 64-QAM, which allows the transmission of eight bits in a single symbol. The cost of increased spectral efficiency is an increased sensitivity to noise, thereby requiring a higher value of E_b/N_0 to achieve a given probability of bit error.

In our discussions, nonlinear modulation techniques were represented by frequency-shift keying and its variations. We stressed the importance of maintaining a continuous phase to keep out-of-band spectral emissions low. All of the versions of frequency-shift keying maintain a constant transmitted signal envelope and are therefore insensitive to nonlinearities in the transmitter's power amplifiers. This is a major advantage to FSK in narrow-channel cellular telecommunications applications. In Gaussian FSK, the message line code is smoothed using a Gaussian filter prior to modulating the carrier. The resulting FSK signal is very smooth and has a very rapid out-of-band spectrum rolloff.

The last variation on frequency-shift keying that we investigated was minimum-shift keying. This modulation method can be viewed as a special case of offset quadrature phase-shift keying, with "half-sine" pulse shaping. This modulation method blurs the distinction between the linear and nonlinear techniques.

We showed that the choice of modulation method has a significant impact on receiver sensitivity. Every modulation method produces a relationship between probability of error and E_b/N_0, usually displayed as a "waterfall" curve. Given the required probability of error, the bit energy E_b relative to the noise power spectral density N_0 can be determined. At the frequencies used for wireless telecommunications and networking, the noise power spectral density N_0 is set by the receiver's noise figure. The receiver sensitivity is the necessary average power, determined by the bit energy multiplied by the data rate in bits per second. From a systems-engineering perspective, relating the receiver sensitivity to the probability of error is an important design tool. In Chapter 3 we investigated the relationships between various properties of the wireless link such as path loss, antenna gain, antenna height, and noise figure. A systems engineer uses these relationships to create a link budget whose purpose is to ensure that the received signal power exceeds the receiver sensitivity by an appropriate margin that accounts for fading, multipath interference, component variability, and so on. Given these discussions, we should be able to perform a system analysis that relates the various system parameters to the data rate and bit error rate performance measures.

The last section of this chapter provided an introduction to spread-spectrum techniques. Spread spectrum is a class of modulation methods in which the bandwidth of the transmitted signal is artificially increased through the use of a signal unrelated to the transmitted message. Spread spectrum significantly reduces the sensitivity of signals to interference and frequency-selective fading. The cost is additional complexity, particularly in the receiver. It is not intuitively clear that a scheme that increases bandwidth would be of interest in designing cellular telephone systems, which traditionally sought to narrow the channel bandwidth. In Chapter 6 we will show some of the distinct advantages of spread-spectrum systems, especially in regard to increased user capacity.

Problems

Problem 5.1

Draw the waveforms representing the bit pattern 1 0 1 1 0 0 1 0 in each of the following line-code formats:

A. Unipolar NRZ

B. Polar NRZ

C. Unipolar RZ (25% duty cycle)

D. Manchester code

E. 2B1Q (use -3 V as an initial value)

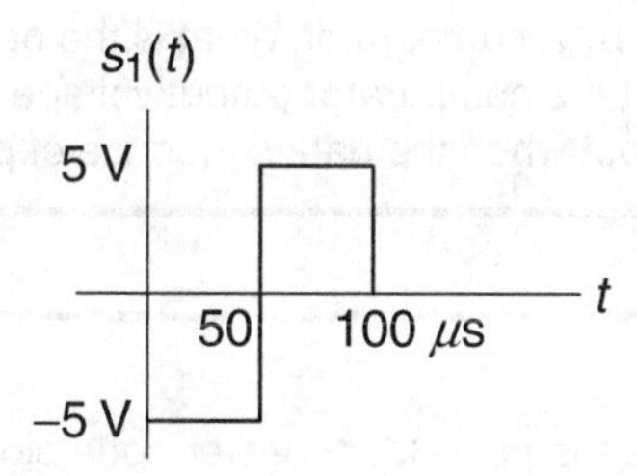

Figure P 5.1 Pulse for Polar Keyed Signal

Problem 5.2

Compare the Manchester code with the polar NRZ line code with respect to

A. Edges for synchronization
B. DC component (Is there a DC component? If so, is the DC component dependent on the data pattern?)
C. Bandwidth (For this problem define "bandwidth" as the frequency of the first null in the power spectrum.)

Problem 5.3

Data is transmitted using a unipolar NRZ line code. The noise at the output of the receiver filter is found to have an RMS value of 0.6 mV when no signal is received. When a 1 is transmitted, the signal at the receiver filter has a value of 3.5 mV at the sample time. Find the probability of error.

Problem 5.4

A polar line code uses the pulse shape shown in Figure P 5.1. Draw the impulse response of a filter matched to this pulse shape. Find and plot the output of the matched filter when a pulse representing a 1 is received in the absence of noise. Determine the filter output value at the optimum sample time.

Problem 5.5

Consider the pulse and matched filter of Problem 5.4.

A. Find the output $y(t)$ of the matched filter when the data sequence 1 1 0 1 0 1 is received. Plot your result, showing all significant voltages and times.
B. The filter is followed by a sampler. Indicate on your plot of filter output the times at which the waveform should be sampled.

C. The sampler is followed by a comparator. What is the optimum comparator threshold level? Assuming that the comparator output voltage levels are 3.5 V and 0.2 V, plot the comparator output when the data sequence of part A is received.

Problem 5.6

A communication system uses polar Manchester code signaling with the pulse $s_1(t)$ given by Figure P 5.1 and a matched filter receiver.

Suppose that the pulse $\pm s_1(t)$ is received along with additive Gaussian noise $n(t)$. The noise has power spectrum $S_n(f) = 110 \times 10^{-6}$ V²/Hz.

A. Find the average noise power at the output of the matched filter.
B. Find the probability of error.

Problem 5.7

One form of Barker code is the bit pattern 1 1 1 0 0 0 1 0 0 1 0 encoded in polar NRZ format. Find and plot the autocorrelation function of the Barker code. Compare your result with the autocorrelation function plotted in Figure 5.12.

Problem 5.8

Find the cross-correlation $R_{12}(\tau)$ between the waveforms $x_1(t)$ and $x_2(t)$ shown in Figure P 5.2. Plot your result.

Problem 5.9

Find the energy in a unit-amplitude rectangular pulse of duration 10 µs. Find the fraction of total energy contained in the first-null bandwidth. Find the fraction of total energy contained in the second-null bandwidth.

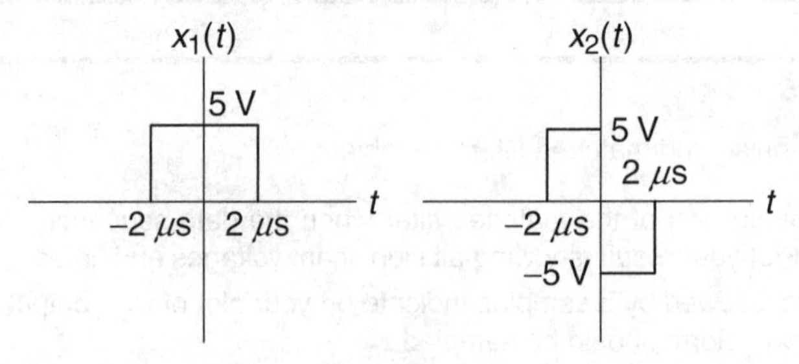

Figure P 5.2 Waveforms for Cross-Correlation

Problem 5.10

Accurately plot the pulse given by Equation (5.62) over the interval $-3T_s \le t \le 3T_s$ for $\alpha = 0.5$ and $\alpha = 1$. What happens to the pulse shape as α increases from 0 to 1?

Problem 5.11

A binary phase-shift keyed signal has a bit rate of 100 kbits/s and a carrier frequency of 2400 MHz. The message waveform uses a root-raised-cosine line code with $\alpha = 0.5$.

A. Find and plot the power spectrum of the BPSK signal. Plot the spectrum over a frequency range of $\pm$ 400 kHz about the carrier.
B. Find the absolute bandwidth and the Nyquist bandwidth of the BPSK signal.

Problem 5.12

A binary phase-shift keyed signal has a bit rate of 100 kbits/s and a carrier frequency of 2400 MHz. The message waveform uses a root-raised-cosine line code with $\alpha = 1$. The receiver is shown in the block diagram of Figure 5.24. The bandpass filter in the receiver front end has a rectangular frequency response with a bandwidth equal to the absolute bandwidth of the received signal.

Suppose a probability of error of $P_e = 10^{-5}$ is required.

A. Find the minimum value of E_b/N_0.
B. The antenna noise temperature is 290 K, and the LNA has a noise figure of 1.6 dB. Find the noise power spectrum $S_n(f) = N_0/2$ referred to the LNA input.
C. Find the received signal power needed to ensure the required probability of error (that is, find the receiver sensitivity).
D. Find the signal-to-noise ratio at the bandpass filter output.

Problem 5.13

The data sequence 1 0 1 0 1 1 0 0 is to be transmitted using differential binary phase-shift keying.

A. Assuming the preprocessor has an initial state of 0, find the bit sequence at the preprocessor output.
B. Draw the message line code if rectangular pulses (polar NRZ) are used.
C. Suppose the receiver makes an error on the fourth data bit. Find the bit sequence at the receiver output.

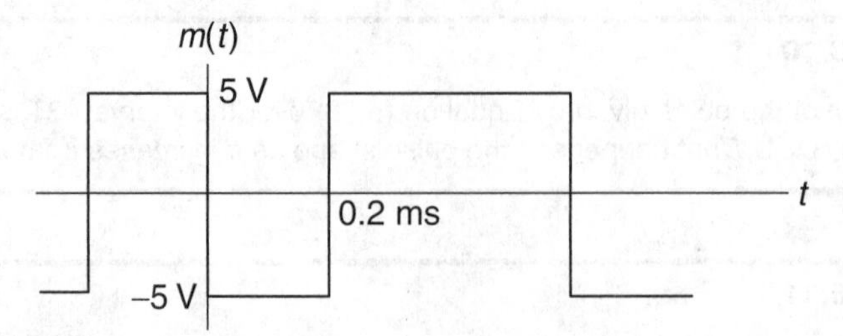

Figure P 5.3 Message Signal for FSK

Problem 5.14

Find the required E_b/N_0 for a probability of error of $P_e = 10^{-4}$ for BPSK and DPSK. How much more signal power (in decibels) is needed for DPSK?

Problem 5.15

A data stream sent using BPSK has a probability of error of 10^{-7}. Suppose the average received signal power, noise power spectrum, and transmission bandwidth are kept the same, but the modulation is changed to QPSK. (Note: The source bit rates are not the same.) Calculate the probability of bit error for the QPSK system.

Problem 5.16

The message signal $m(t)$ shown in Figure P 5.3 frequency modulates a carrier.

Draw the FSK signal $x(t)$ if the carrier amplitude is 10 V, the carrier frequency is 10 MHz, and the modulation index is $h = 4$. Indicate on your plot all significant frequencies, amplitudes, and times.

Problem 5.17

For a probability of error of $P_e = 10^{-5}$, calculate E_b/N_0 for coherent and noncoherent FSK. How much penalty (in decibels) is incurred if we choose the less complex noncoherent receiver?

Problem 5.18

A carrier frequency of 400 MHz is frequency modulated by a polar NRZ message of amplitude ± 3 V and rate 200 kbit/s. The frequency sensitivity of the modulator is $k_f = 50$ kHz/V.

A. Determine the approximate bandwidth of the FSK signal using Carson's rule.

B. Find the modulation index h.

C. Find the spectral efficiency.

D. Find the approximate bandwidth of the FSK signal assuming that the amplitude of the data signal has doubled.

E. Restore the amplitude to its original value. Find the approximate bandwidth assuming that the data rate has doubled.

Problem 5.19

The GFSK modulation specified in the Bluetooth wireless standard uses a modulation index $h = 0.32$. This is smaller than the "minimum" value of $h = 0.5$ specified for minimum-shift keying.

A. What would be the advantage of using such a small modulation index?

B. What would be the disadvantage of using a value of modulation index less than $h = 0.5$?

Problem 5.20

A frequency-hopping system uses GMSK modulation with $h = 0.5$, $BT_s = 0.5$, and a data rate of 1 Mbit/s. There are 250 hopping frequencies spaced $1.5B_{MSK}$ apart, where B_{MSK} is the bandwidth of the GMSK modulated waveform.

A. Use Carson's rule to estimate the bandwidth B_{MSK}.

B. Use Equation (5.123) to make another estimate of B_{MSK}. Compare with your answer to part A.

C. Find the instantaneous bandwidth and the total hopping bandwidth of the frequency-hopped signal.

Problem 5.21

Consider the frequency-hopping system of Problem 5.20. A receiver is shown in Figure 5.47(b). Assume that each of the bandpass filters has a rectangular frequency response. Suppose that the received signal power is $P_s\big|_{dB} = -110$ dBm. The antenna noise temperature is $T_0 = 290$ K and the LNA has a noise figure of $F = 2$ dB. Find the signal-to-noise ratio at the output of each of the bandpass filters.

Problem 5.22

A direct-sequence spread-spectrum system transmits at a bit rate of 1 Mbit/s. There is a source of narrowband interference transmitting on the carrier frequency for which the signal-to-interference ratio is $SIR_{in}\big|_{dB} = 3$ dB. Find the processing gain and chip rate if the signal-to-interference ratio is to be raised by 20 dB.

CHAPTER 6

Access Methods

Introduction

In this chapter we return to a theme that we explored at some length in Chapter 4, the provision of wireless service to the largest possible number of users. In that chapter we focused on geographic coverage and introduced the frequency reuse technique to show how coverage can be provided to a practically unlimited number of users spread over a practically unlimited geographic area. In the present chapter we take a somewhat narrower perspective. We explore techniques for accommodating a maximum number of users in a fixed geographic area such as a cellular cluster, given a limited spectral allocation.

As we observed in the earlier chapter, radio spectrum is a limited resource, with certain specific bands allocated to cellular telephone and other bands to other wireless services. In our discussions we assumed that the system bandwidth B_{sys} allocated to a specific cellular system was subdivided into N_{chan} channels, each of bandwidth B_{chan}. The bandwidth B_{chan} was selected to provide one subscriber with a link having enough bandwidth to support an adequate quality of service for the subscriber's intended application. Thus, in one cellular cluster, using all of the channels, N_{chan} subscribers can enjoy simultaneous access. This technique of dividing the system bandwidth into channels by frequency is known as **frequency-division multiple access**. As we shall see in this chapter, a systems engineer has alternatives in the way that multiple users can be provided simultaneous service. The two additional techniques that we shall explore are known as time-division multiple access and code-division multiple access.

As an alternative to the frequency-division technique, **time-division multiple access** (TDMA) allocates the entire bandwidth B_{sys} to a single user, but only for a limited time. The various users take turns transmitting, usually in a round-robin fashion. Because individual users can transmit only momentarily, although at regular intervals, the signals from each user are effectively sampled during transmission. Time-division multiple access is most easily used with digital signaling schemes, since digital signals are already pulse based and hence single transmissions are limited in time. It is necessary to ensure that the round-robin proceeds fast enough that the successive pulses from each user can be transmitted as they become available.

All multiple-access schemes depend critically on the receiver's ability to separate the signals from various users. The receiver in a frequency-division system uses a filter to pass the desired signal and reject signals from other subscribers. The receiver in a time-division system uses time gating for that purpose; the receiver listens to the communication channel only during the appropriate time slots. It turns out that frequency selection and time gating are special cases of a more general technique. If signals from distinct users are transmitted using signal sets that are orthogonal, in the sense defined in Chapter 5, then these signal sets can be separated at the receiver by correlation. The use of specially designed orthogonal signal sets to support multiple users is called **code-division multiple access** (CDMA). Some of the orthogonal signal sets that have proved successful are derived from the same pseudorandom signals introduced in the previous chapter in connection with spread-spectrum signaling. We will examine code-division techniques in some detail later in this chapter.

When introducing techniques for sharing spectral resources among multiple users, it is helpful to distinguish between **multiplexing** and **multiple-access** techniques. These two terms refer to similar concepts and denote primarily a difference in emphasis. When the set of users is fixed in advance, or remains stable for long periods of time, the schemes for providing service are usually referred to as *multiplexing*. When users come and go at random moments, the term *multiple-access* is used. When describing multiple-access systems there is generally a greater emphasis placed on the techniques used by subscribers to gain access to a communication channel or a time slot. As an example, the architecture of the wired telephone network is based on multiplexing. Channels or time slots are assigned when a call is set up, and the assignments remain in effect until the call is released. The duration of a call is long enough that channel or slot assignments appear stable. A parallel situation exists in wireless telephone systems, where voice channels are assigned for the duration of a call, or at least until a mobile subscriber moves into an adjacent cell. An example of the multiple-access situation can be found in both wired and wireless data networks such as Ethernets or WiFi networks. These are packet networks in which subscribers transmit packets of several dozen to several thousand bytes of data at irregular intervals. Because each user's transmissions are irregular and infrequent, there is no point in assigning users to time slots. Users contend for access to the channel every time they wish to transmit a packet; the "winner" gets to transmit right away and the "losers" transmit later. As we will describe later in this chapter, this same contention-based time-division multiple-access situation appears in wireless telephone systems in the context of enabling users to access a control channel when initiating or answering a call.

From the point of view of a systems engineer, the choice of a multiple-access or multiplexing scheme is, like all other system design issues, governed by a desire to provide high-quality service to a maximum number of subscribers at a cost that will provide an appropriate return on investment to the service provider. Now, at the highest level of abstraction, the multiplexing scheme does not matter. A communication channel with a given bandwidth operating at a given signal-to-noise ratio can carry a given amount of information. In principle, the same number of users can be served whether the channel is divided in frequency, in time, or by code. In practice, however, the costs of the various multiplexing alternatives are very technology dependent. Thus, given a particular state of technology, one or another of the multiplexing schemes may be most cost-effective. As an example, we have mentioned that time-division multiple access is most effectively used with digital signaling. The cost of a given multiplexing scheme depends on available filter technology, on the modulation method used by the signals to be multiplexed, and on the processing power available at the transmitting and receiving points. A brief historical summary describing multiple-access methods in cellular systems will illustrate the point.

The original AMPS cellular system, introduced in the United States in the early 1980s, used frequency-division multiple access. The available spectrum was divided into 30 kHz channels as described in Chapter 4. The AMPS system used analog frequency modulation, in which the transmitted signals were continuous and not sampled. Consequently, frequency-division multiple access was virtually the only cost-effective choice for serving multiple simultaneous users. The 30 kHz channel bandwidth was a legacy from earlier mobile telephone systems and was originally dictated by the technology used to implement the receiver filters.

By the early 1990s AMPS systems were reaching capacity in some major cities, and it was desirable to find a way to support an increased number of simultaneous users in the available system bandwidth. Speech compression techniques had evolved to allow almost-telephone-quality voice signals to be transmitted at bit rates below 10 kbits/s, as compared with the wired telephone standard rate of 64 kbits/s, and it became possible to consider digital modulation methods to replace the analog FM. The U.S. Digital Cellular (USDC) system, introduced in 1991, used $\pi/4$-DQPSK (differential quadrature phase-shift keying) modulation in the existing 30 kHz channels. Pulse streams from three subscribers were time-division multiplexed in each channel, allowing an immediate increase in capacity. By keeping the existing channels, the operating companies were able to minimize their investment in new radio frequency hardware, making the migration to second-generation systems economically attractive. From a systems-engineering perspective it is important to note that advances in speech compression, modulation methods, and microprocessor technology during the 1980s made the use of time-division multiplexing feasible for cellular applications. The GSM system, the first digital cellular system introduced in Europe in the early 1990s, used GMSK modulation to time-division multiplex signals from eight subscribers onto one 200 kHz channel.

Also in the early 1990s Qualcomm, Inc., introduced a CDMA alternative to the USDC system. This system used QPSK modulation with direct-sequence spreading to support multiple users in 1.25 MHz channels. Although the capacity of the CDMA system is larger than that of USDC, the unfamiliar code-division technique along with the larger channel bandwidth initially led to a slow adoption of the system. The cdmaOne system (also known as IS-95) was successfully deployed in the mid-1990s and clearly demonstrated a significant improvement in spectral efficiency. As a result, CDMA techniques became universally adopted as the fundamental approach for third-generation (3G) cellular systems. The two major systems that are being deployed worldwide are cdma2000 and Wideband-CDMA (W-CDMA) (also known as the Universal Mobile Telecommunications System, UMTS). From a systems perspective, it is the increase in signal-processing capability of the mobile unit that has been the enabling technology in these third-generation systems.

This chapter begins with a high-level description of how cellular subscribers get access to a channel when a call is placed or received. This description is intended to provide an overview, in very general terms, of how multiple access actually works. Next we proceed to specifics of frequency-division, time-division, and code-division schemes. As frequency-division and time-division systems are conceptually straightforward, we will present one or two examples of classical systems to illustrate the concepts. Both frequency-division and time-division systems have firm upper limits on the number of users that can be accommodated in a given system bandwidth. We will examine some practical considerations that may cause the maximum number of users to be smaller than the upper limit.

A significant part of this chapter is an introduction to code-division multiple-access techniques. We begin by describing a CDMA system using frequency hopping and then turn to direct-sequence methods of providing CDMA. In each case we estimate the number of users that the system can service. We show that there is not a firm upper limit to the number of users. Instead, the signal-to-interference ratio for all users degrades gradually as the number of users increases. The practical limit to the number of users is governed by how much the service quality is allowed to degrade. We consider two specific examples of direct-sequence CDMA. First we explore the possibility of using orthogonal waveforms to separate the channels. This provides essentially perfect separation, at the cost of demanding requirements for synchronization. Next we examine the use of pseudonoise spreading codes to separate the channels. In this case the channel separation is not perfect, and each additional signal adds to the interference experienced by all of the users. On the other hand, the synchronization requirements are relaxed. It turns out that both of these CDMA methods are used in the cdmaOne and cdma2000 cellular systems. Orthogonal waveforms are used in the forward links and pseudonoise spreading codes in the reverse links.

In the final section of the chapter we examine contention-based access methods. We introduce the Aloha and slotted Aloha access techniques. The latter is widely used for mediating access to the control channels in cellular telephone systems. We conclude with an introduction to carrier-sense multiple access, a technique widely used in wireless data networks. For each of the contention-based techniques we carry out a performance analysis, so that the relative merits of the alternatives can be quantified.

Channel Access in Cellular Systems

Regardless of whether channels are provided by frequency division, time division, or code division, a cellular user wishing to place or receive a telephone call must obtain access to a voice channel pair. In this section we describe how that access is accomplished in very general terms. In our description we omit many details that are particular to one specific system or another and focus on the general multiple-access procedure. It is important to note that establishing access to a cellular system voice channel is essentially a circuit-switching operation. That means that once a channel, time slot, or code is assigned to a subscriber, the subscriber has exclusive use of that

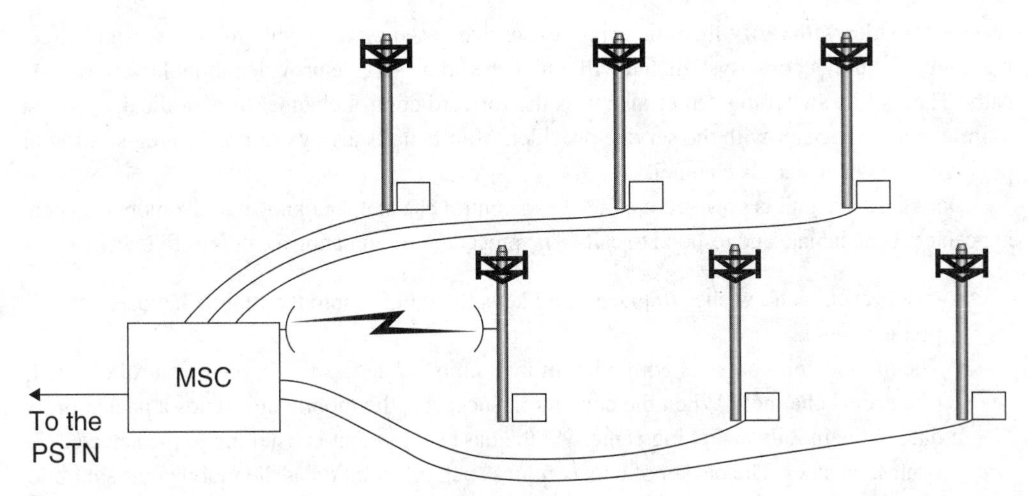

Figure 6.1 Base Stations Connected to a Mobile Switching Center

channel, time slot, or code, and the assignment remains in effect for the duration of the call. It is interesting to contrast the circuit-switching model with the packet-switching or contention-based access model that we will present later in this chapter.

A simplified diagram showing the configuration of a cellular system was presented in Figure 4.20 of Chapter 4, repeated here as Figure 6.1. This diagram shows a set of base stations connected to a mobile switching center (MSC). The mobile switching center manages all of the channel assignments in the cellular system and also manages connections with the public switched telephone network (PSTN). Within a cellular system there is a large set of voice channels (dozens per cell) and a small set of control channels (a few per cell). (To make the discussion more readable, the term *channel* refers here to a frequency assignment, a time slot, or a code, as appropriate for the particular system.) There may in fact be several kinds of control channels. Some of these provide beacons containing synchronization information and system-dependent information needed by the mobile units. These are often called "sync" channels. Others, called "paging" channels, are used to alert a mobile unit to the arrival of an incoming call. Yet others are the "access" channels, which are reverse control channels used to implement the multiple-access procedure that we are about to describe. In this discussion we will not attempt to distinguish the various types of forward control channels in any precise way but will refer to them in the aggregate as "control channels." Access channels are a universally adopted reverse control channel type.

When a mobile phone is turned on, it scans a preprogrammed set of forward control channels looking for the strongest signal. The mobile unit tunes to the strongest control signal and obtains information that it needs to operate on the local system. This information may include access codes, location of the paging channels, and identification of the service provider. After it obtains sufficient information about the system, the mobile unit will register with the system using an access channel. If the mobile unit is not in its home system—that is, it is roaming—the service provider will verify the unit's right to service based on a suitable mutual service agreement with the unit's home system and will notify the home service provider about how to reroute calls. The mobile switching center also uses the forward control channel to periodically request mobile units to register with the service provider. Mobile units always respond to registration or paging alerts via an access channel.

Once a mobile unit is powered up, tuned to a control channel, and known to the mobile switching center, it can initiate and respond to calls. The procedure for initiating a call is as follows.

- A subscriber who wishes to place a call keys the number into the mobile handset and presses "send."
- The mobile unit contends with other mobile units for access to a reverse control channel (the access channel). When the contention succeeds, the mobile unit sends a packet of data to the mobile switching center via the base station that is listening to the selected control channel. The packet of data contains such information as the mobile unit's mobile identification number (i.e., telephone number), its electronic serial number (functioning as a password), and the destination telephone number.
- The mobile switching center verifies that the mobile unit is eligible for service, identifies the base station best situated to handle the call, selects an available voice channel pair (one channel in each direction), and directs the mobile unit and base station to switch to the selected voice channel pair.
- Finally, the mobile switching center places the call on the public switched telephone network.

This procedure is depicted graphically in Figure 6.2.

The procedure for responding to an incoming call is similar.

- When a call arrives at the mobile switching center for a mobile unit registered in the system, the mobile switching center instructs the appropriate base station or base stations to page the mobile unit over a forward control channel.
- Here, too, the mobile unit must contend with other mobile units for use of the access channel to answer the page.
- Once the mobile unit answers, the mobile switching center identifies the base station that is best situated to handle the call, selects an available voice channel pair, and directs the mobile unit and base station to switch to the selected voice channel.
- Next, the mobile switching center sends a command (over the voice channel now) causing the mobile phone to "ring."
- When the subscriber answers, the ringing is stopped and the voice connection is completed.

This procedure is depicted graphically in Figure 6.3.

The common theme that applies both to placing and to receiving calls is this: Mobile units must contend with each other to use an access channel, thereby allowing communication with

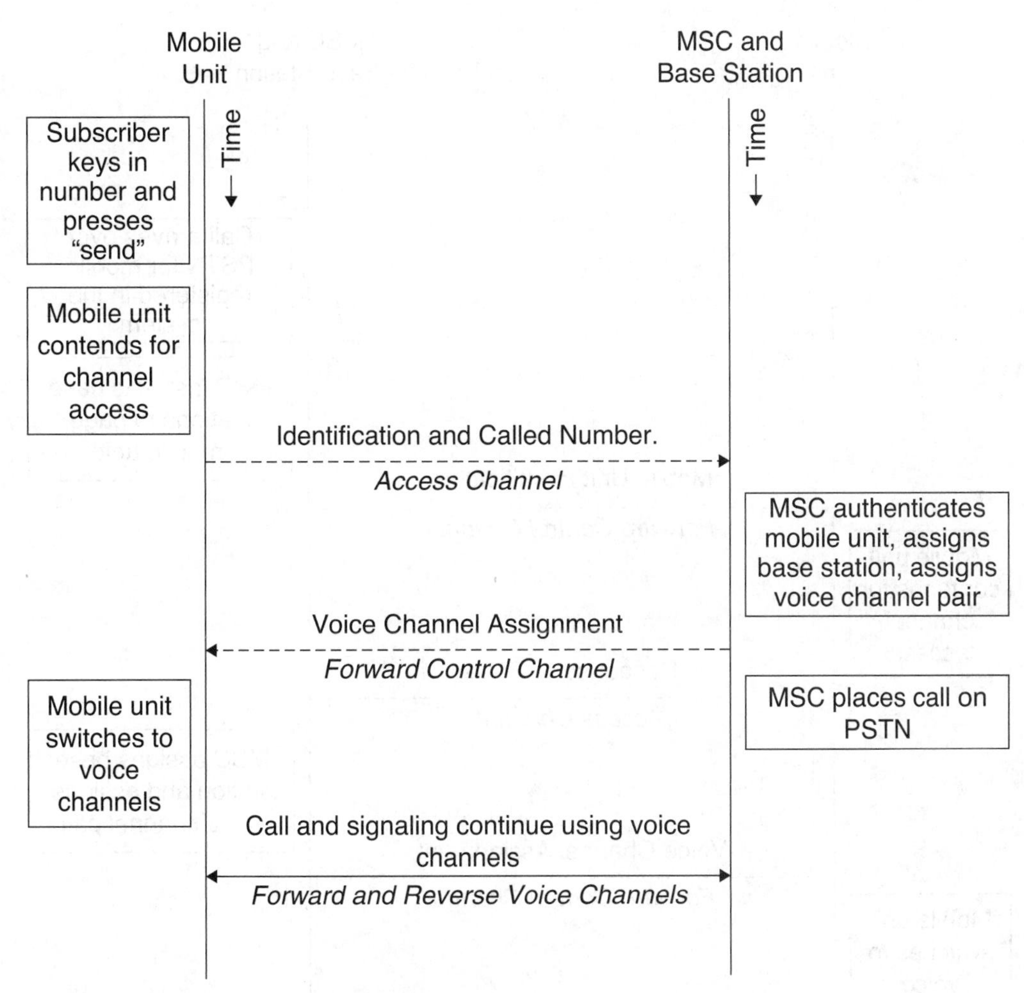

Figure 6.2 Procedure for Mobile Unit to Initiate a Call

the system. For most system implementations, access channels are used only when a voice channel has not been assigned to the mobile unit. Access channels are shared among multiple users, each of whose contribution to the overall control channel throughput is small. We will discuss contention-based access later in this chapter. Access to voice channels is not based on contention. Instead, voice channels are assigned to subscribers by the mobile switching center. Forward control channels are also controlled by the system, which can apportion messages to each user as necessary and to all users simultaneously when common messages are "broadcast" to all units. Therefore, even though all mobile units share the forward channels, there is no contention. When a mobile unit is operating on a traffic (voice) channel, all signaling messages with the system are carried by the traffic channel.

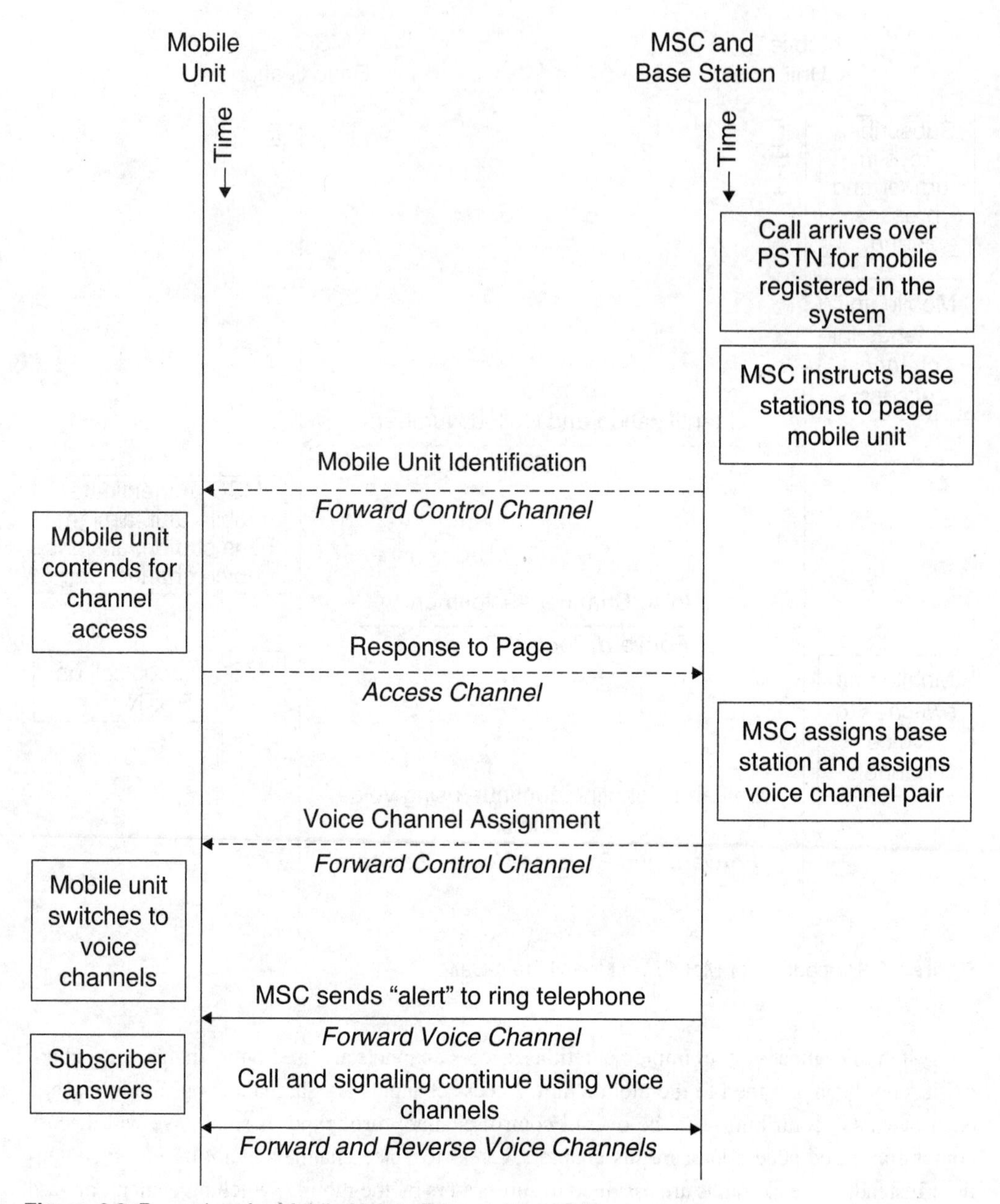

Figure 6.3 Procedure for Mobile Unit to Respond to a Call

Frequency-Division Multiple Access

Frequency-division multiple access is the oldest of the multiple-access methods and was virtually the only option for circuit-switched or long-term users in the days before digital modulation became practical. From a system design perspective, the central issue is maximizing the number of independent users that can be simultaneously supported in a given system bandwidth. In this section we will identify the factors that limit the number of users and give several examples of frequency-division systems to illustrate the ideas.

As we observed in the introduction to this chapter, a simple calculation suggests that a system bandwidth B_{sys} divided among users each of whom requires a bandwidth B_{chan} can support $N_{chan} = B_{sys}/B_{chan}$ simultaneous users. The challenge is determining an appropriate value to use for the parameter B_{chan}. At a minimum, the channel bandwidth B_{chan} must be wide enough to pass a modulated signal without unreasonable distortion. It is also the case, though, that the bandwidth B_{chan} determines how close in frequency adjacent signals can be placed. The channels must be separated by enough bandwidth to protect against adjacent-channel interference. Controlling this kind of interference can be the most important factor in determining a value for the channel spacing.

In Chapter 4 we provided an example in which we supposed that a base station received two signals from mobile transmitters using adjacent channels 30 kHz apart. We showed that under plausible assumptions about filter properties and propagation conditions a nearby interfering transmitter could provide a signal 50 dB stronger than the signal from a more distant cochannel transmitter. In the example this 50 dB difference was enough to overcome the stopband attenuation of the channel-selection filter. This example illustrates that to avoid such adjacent-channel interference it may be necessary to establish **guard bands** between channels. These guard bands constitute a kind of "overhead" and reduce the number of usable channels in the system bandwidth B_{sys}. As the example from Chapter 4 shows, the width needed for guard bands depends largely on two factors: the selectivity of the filters that can be used at the receivers to separate desired from undesired signals, and the range of relative amplitudes of the desired and interfering signals. The out-of-band spectral occupancy of the transmitted signal may also be a factor.

In the remainder of this section we provide several examples in which a bandwidth B_{sys} is subdivided to allow simultaneous transmission by a number of independent users. The first example is AM broadcast radio, the second is the frequency-division multiplexing scheme adopted by the North American telephone network prior to the introduction of digital transmission, and the third is the AMPS cellular telephone system. You will observe that these are all examples of legacy systems. Frequency-division multiplexing has been in use for a very long time, and the principles are the same whether one looks at historical or at modern communication systems. After we explore these classical frequency-division systems, we present an example that illustrates the consequences of poor spectral design of the transmitted signal. The section concludes with a discussion of frequency-division duplexing, a technique for providing two-way radio conversation.

The AM Broadcasting Band

A broadcast AM signal has a baseband (message) bandwidth of about 5 kHz. Since AM is a double-sideband modulation method, the transmitted signal bandwidth is 10 kHz. The AM broadcast band extends from 530 kHz to 1700 kHz. It is divided into channels, with carrier frequencies allowed every 10 kHz across the band. A naive calculation of B_{sys}/B_{chan} would suggest that 118 independent stations can be supported on this band in a given geographic location.

AM radios are mass produced and are generally optimized to minimize manufacturing costs. Consequently the filters that are typically used are not highly selective, certainly not selective enough to reject a strong station in an adjacent channel. As a result it is normal practice not to assign adjacent channels to stations operating in the same geographic area. As an illustration of actual practice, a quick look at the FCC Web site[1] shows that New York City has 16 active AM broadcast stations. These stations operate at carrier frequencies spread over the entire band, and in no case are stations assigned to carrier frequencies closer than 40 kHz apart.

Although there are no specific guard bands, it appears that assignment of stations on the AM band is made with $B_{chan} \geq 40\text{ kHz}$. Using this wider channel bandwidth, we can determine that a fully populated AM band can support at most about 30 stations.

There is another consideration that further reduces the number of stations that can operate together on the AM broadcast band. A "Class A" AM station is authorized to transmit at a power level of up to 50,000 W. This is adequate to ensure good reception by receivers of possibly limited quality under poor propagation conditions over a large metropolitan area. It turns out, however, that when propagation conditions are good, such as on cold, clear winter nights when atmospheric noise is at a minimum, these powerful AM signals can be received over distances of a thousand miles or more. To avoid cochannel interference between these stations, a number of AM channels are designated as "clear" channels. Clear channels can be assigned to more than one station, but stations assigned to the same clear channel are located at least half a continent apart. (If this were a cellular system, think of the cell size!) As an example, station WBZ transmits from Boston, Massachusetts, on 1030 kHz. The same frequency is used by KTWO in Casper, Wyoming, but the clear-channel frequency of 1030 kHz is not available for assignment at other locations on the East Coast.

This example clearly shows how considerations of adjacent-channel and cochannel interference reduce the number of stations that a frequency-division multiple-access system can support in a given system bandwidth. Stations could be spaced more closely together if filters having greater selectivity were used in the receivers. This is not a cost-effective option for AM broadcasting, but our next example will show a different application in which selective filters were effectively used.

Frequency-Division Multiplexing in the Telephone Network

Perhaps the classic example of efficient frequency-division multiplexing is the system used by the telephone companies for long-distance telephone transmission. This system became highly

1. www.fcc.gov/mb/audio/amclasses.html, accessed June 4, 2007.

developed during the 1950s and persisted into the 1970s, when it began to be gradually replaced by time-division multiplexing.

The baseband input to the telephone multiplexing hierarchy was (and still is) the voice signal. Voice signals are filtered as they enter the switching office, using a filter with a 4 kHz stopband bandwidth. The passband bandwidth is somewhat narrower, in the range of about 3200 to 3600 Hz. To form the multiplexed signal, 12 voice signals were modulated onto distinct carriers using single-sideband modulation. Given the strict 4 kHz message bandwidth, the bandwidth of a modulated signal is also 4 kHz. It was then possible to space the carrier frequencies 4 kHz apart so that the combined signal, known as a "group," occupied a total bandwidth of 48 kHz. As all 12 signals were transmitted from the same location and received at the same location, signal levels could be balanced, so that there was no problem with interference to a weak in-channel signal from a strong adjacent-channel signal. The filters used to generate the single-sideband modulated signals and also to separate signals at the receiver were specialized sharp-cutoff filters constructed from mechanical resonators. Hence adjacent 4 kHz channels could be used, without allocating guard bands or leaving empty channels between each pair of occupied channels.

Once 12 voice channels were combined into a group, the group could be treated as a "message" signal with a 48 kHz bandwidth. Five groups could then be combined into a "supergroup," also by using single-sideband modulation. As in forming a group, adjacent channels were spaced at the minimum spacing, in this case 48 kHz, with all channels used. The supergroup bandwidth was 240 kHz. Subsequently ten supergroups were combined into a "mastergroup," and six mastergroups were combined into a "jumbo group." A jumbo group contained 3600 voice signals. The jumbo group signal could be transmitted over coaxial cable or used as the baseband input into a microwave transmitter.

The AMPS Cellular Telephone System

The AMPS cellular telephone system was introduced in Chapter 4. The AMPS system uses 30 kHz channels, with 395 voice channels available to be assigned to each of "A side" and "B side" operating companies. Table 4.3 shows part of a channel plan assuming seven-cell clusters, with each cell divided into three 120° sectors. As was pointed out in Chapter 4, the channel plan ensures that channels assigned to the same cell are spaced at least 15 channels apart, and channels assigned to the same sector are spaced by at least 21 channels. Thus we see that although there are no guard bands between channels, and all channels are used, significant spacing is guaranteed between channels that will be used in the same geographic area. It is important to note that the same minimum channel spacing could not be guaranteed if a smaller cluster size were used, as more channels would then be assigned to each cell.

As we observed previously, the second-generation U.S. Digital Cellular system uses the same 30 kHz channels used in the AMPS system. The cdmaOne system, however, uses 1.25 MHz channels. It is possible for cdmaOne and AMPS systems to coexist in the same geographic area if 1.25 MHz of the cellular band is allocated to the cdmaOne system. In this case, though, a guard band of nine AMPS channels must be set aside on each side of the cdmaOne subband to avoid interference between the systems.

Effect of Transmitted Signal Design

The efficiency of a frequency-division multiple-access system can be adversely affected by poor spectral shaping of the transmitted signal. The following example illustrates what can happen.

Example

In a certain frequency-division multiple-access system the transmitters transmit data using binary phase-shift keying with a bit rate of 10 kbits/s. Rectangular pulses are used with no filtering at the transmitter. Channels are spaced every 20 kHz. Suppose the receivers use highly selective filters that we can take as ideal brick-wall (i.e., "rectangular" frequency response) filters, each having a bandwidth of 20 kHz centered on the appropriate carrier frequency.

Suppose a receiver receives two signals of equal amplitude. One of these is the desired signal, and the other has a carrier frequency n channels away from the carrier frequency of the desired signal.

How large must we make n so that the signal-to-interference ratio is at least 40 dB?

Solution

A binary phase-shift keyed signal can be written

$$g(t) = Am(t)\cos 2\pi f_0 t, \tag{6.1}$$

where A is the amplitude, $m(t)$ is the message consisting of a train of polar NRZ rectangular pulses, and f_0 is the carrier frequency of channel "zero." The power spectrum of this signal is given by

$$S_g(f) = \frac{A^2 \cdot 10^{-4}}{4}\operatorname{sinc}^2[10^{-4}(f+f_0)] + \frac{A^2 \cdot 10^{-4}}{4}\operatorname{sinc}^2[10^{-4}(f-f_0)]. \tag{6.2}$$

The power in this signal is the area under the power spectrum, that is,

$$P_g = \int_{-\infty}^{\infty} S_g(f)df = A^2/2. \tag{6.3}$$

The received signal power is the power in the passband of the receiver filter. The receiver filter passband occupies one channel and extends from $f_0 - 10^4$ to $f_0 + 10^4$, giving

$$\begin{aligned} P_s &= \int_{-f_0-10^4}^{-f_0+10^4} S_g(f)df + \int_{f_0-10^4}^{f_0+10^4} S_g(f)df \\ &= 2\int_{f_0-10^4}^{f_0+10^4} S_g(f)df \\ &= 0.451A^2, \end{aligned} \tag{6.4}$$

that is,

$$P_s = 0.903 P_g. \tag{6.5}$$

We see that the receiver filter passes more than 90% of the power available. If the interfering signal has a carrier frequency of $f_n = f_0 + 2 \times 10^4 n$, then the power received from the interfering signal is

$$P_i = 2\int_{f_0-10^4}^{f_0+10^4} \frac{A^2 \cdot 10^{-4}}{4} \operatorname{sinc}^2\left[10^{-4}(f - f_n)\right] df. \tag{6.6}$$

The signal-to-interference ratio is given by

$$SIR = 10\log\left(\frac{P_s}{P_i}\right). \tag{6.7}$$

Numerical evaluation of the integral in Equation (6.6) and substitution in Equation (6.7) shows that the signal-to-interference ratio reaches 40 dB when $n = 17$.

This example shows that even with perfect brick-wall receiver filters, geographically colocated transmitters must use a channel spacing of at least 17 when the transmitted signals have a spectrum that falls off as slowly with frequency as does the spectrum of a rectangular pulse. The efficiency of the multiple-access system can be improved significantly if the pulses are shaped, for example, with a raised-cosine spectrum.

Frequency-Division Duplexing

A radio cannot transmit and receive simultaneously on the same frequency, unless separate and widely spaced transmitting and receiving antennas are used. The transmitted power level is so much higher than the power levels that the receiver is designed to expect that any leakage of power from the transmitter directly into the receiver can desensitize the receiver or possibly even damage components in the receiver front end. Mobile radios such as cellular telephones use a single antenna for both transmitting and receiving, and even cellular base stations locate their transmitting and receiving antennas on a common tower.

The traditional method of separating transmission and reception in a mobile unit is to operate **half duplex**. This means that the user alternately transmits and listens, as on a walkie-talkie. Now half-duplex operation is effective for data transmission, but most subscribers find the mode awkward for conversation. Land telephones provide **full-duplex** service, in which the subscriber can talk and listen at the same time (at least the telephone can do this). To provide full-duplex service using a mobile radio, either it is necessary to interleave the transmitting and receiving functions in time in a way that is transparent to the user, or it is necessary to transmit and receive on different frequencies.

Frequency-division duplexing provides two separate bands of frequencies for each user. A forward channel in one of the bands carries information from the base station to the mobile unit, and a reverse channel in the other band carries information from the mobile unit to the base station. The two channels together are sometimes referred to as a single full-duplex channel. A bandpass filter called a "duplexer" is used in each mobile unit and in the base station to prevent energy from the transmitter from reaching the receiver input. The forward and reverse channels must be separated in frequency by enough bandwidth to allow the duplexer to attenuate the transmitted signals in the receive band, but not so far in frequency that a common antenna cannot be used for transmitting and receiving. In multichannel systems the separation in frequency between the forward channel and the reverse channel is often a fixed constant to simplify the design of the duplexer. Forward and reverse channels are assigned in pairs that preserve the fixed frequency spacing.

In the AMPS cellular telephone system frequency-division duplexing was used with a spacing of 45 MHz between corresponding forward and reverse channels. Frequency-division duplexing with the 45 MHz spacing was preserved in the migration to U.S. Digital Cellular, cdmaOne, and also in the third-generation cdma2000. Frequency-division duplexing is also used in the GSM system.

Time-Division Multiple Access

Suppose, as above, that we have a multiple-access system with an allocated bandwidth of B_{sys} that is to support communications among users whose signals each have bandwidth B_{chan}. According to the sampling theorem, the signal produced by a user can be sampled without loss of information, provided that the sampling rate is greater than $2B_{chan}$ samples per second. Also according to the sampling theorem, a smooth waveform of bandwidth B_{sys} or less can be passed through a sequence of arbitrary samples, as long as the samples are provided at a rate less than $2B_{sys}$ samples per second. Now, if this sequence of arbitrary samples is generated by taking samples in round-robin fashion from a group of N_{chan} users, then the number of users is bounded by

$$N_{chan} < \frac{2B_{sys}}{2B_{chan}} = \frac{B_{sys}}{B_{chan}}. \tag{6.8}$$

Thus, according to this simple argument, the maximum possible number of users that can be supported by a time-division multiple-access system is the same as the maximum number of users that can be supported by a frequency-division multiple-access system.

Where this argument runs into difficulty is that in digital transmission systems the samples of the transmitted signal are quantized, and each quantized sample is represented as a series of bits. If, in the round-robin, each user supplies a value that represents a bit rather than a value that represents a sample, the user may be supplying values at many times the $2B_{chan}$ rate. The number of such users that the system can support will be correspondingly reduced.

Example

We have seen that in wired telephone systems voice signals are filtered to a stopband bandwidth of 4 kHz when the signals enter the switching office. A 4 kHz signal can be sampled at 8000 samples per second. Now standard practice in the wired telephone industry is to represent each sample as an eight-bit number. Thus a single voice signal produces a 64,000 bit/s bit stream. If this bit stream were encoded using a polar-keyed line code, transmission would require 64,000 pulses/second. The pulses having minimum bandwidth are sinc-function shaped. Using sinc-function pulses, the bandwidth of the line code would be 32 kHz. Of course, sinc-function pulses are not necessarily practical, as they are very susceptible to errors caused by timing jitter at the receiver. Using raised-cosine-shaped pulses would produce a more robust system but would increase the bandwidth to a value between 32 and 64 kHz.

This example suggests that the use of time-division multiple access with digital transmission can reduce the number of users by a factor of between 8 and 16 compared with frequency-division multiple access. There are two ways by which the efficiency of the time-division scheme might be improved. First, if a multilevel line code were used, each user would be able to transmit fewer pulses per second. There is a trade-off, however, since for a given signal-to-noise ratio, a multilevel line code experiences a higher bit error rate than does a binary line code. Second, a more efficient way of digitizing the source data might be found. In cellular telephone systems, the introduction of time-division multiple access depended critically on the availability of speech processors that could sample and quantize "telephone-quality" voice signals at much lower bit rates than the 64,000 bits/s used in wired telephone. The use of sophisticated speech processing in cellular systems is a consequence both of improvements in the processing algorithms and also of improvements in digital signal processor technology that allow powerful processors to be included in small mobile handsets.

In a time-division multiple-access system signals from multiple users are sampled in rotation. The composite signal containing one contribution from each user is called a **frame**. A typical frame structure is shown in Figure 6.4.

Frames are repeated cyclically. Each user is assigned to an individual slot, and the user transmits at regular intervals when that slot is available. It is apparent in the figure that more information is contained in the frame than just the data from each user. The frame preamble is needed so that the receiver can identify the start of the frame. Correct identification of the start of the frame is needed so that the receiver can correctly distribute data to the various users. In the figure each slot is shown as containing a preamble as well. These preambles allow the receiver to locate the start of data within each slot. Also, in a multiple-access system the user assigned to a given slot may change when, for example, a call terminates. The preamble may contain addressing information that identifies the user transmitting in the slot. In a wireless system individual users must key their transmitters when the slot time to which they have been assigned begins. The "guard time" allows time for radios to switch from receive to transmit or transmit to receive and for the transmitting radio to power up. The "sync" interval allows the radio starting a transmission to transmit a periodic signal so that the receiver can synchronize to the carrier and bit clock.

All of the guard, sync, and preamble intervals constitute overhead in the time-division multiple-access system. The presence of this overhead reduces the time that can be allocated to user data and thus reduces the number of users that can be supported in a given system bandwidth. The

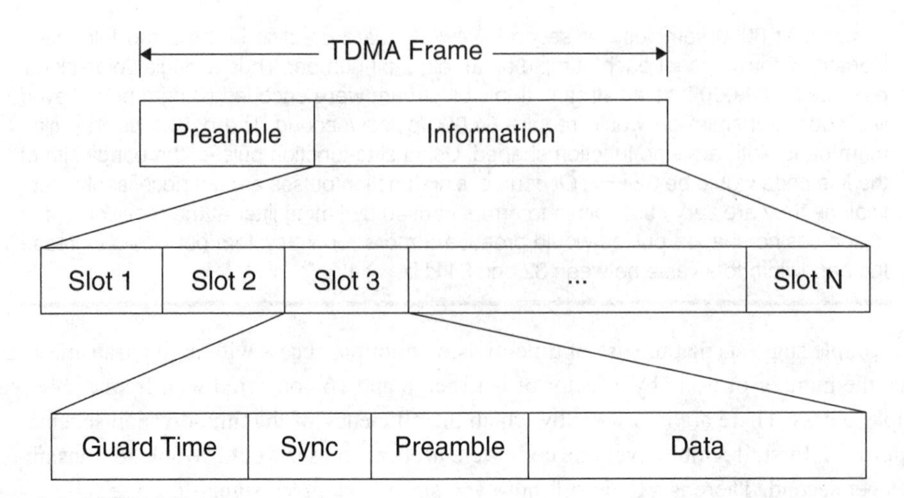

Figure 6.4 A TDMA Frame
(Adapted from Figure 9.4 of T. S. Rappaport, *Wireless Communications*, 2nd ed. (Upper Saddle River, NJ: Prentice Hall, 2002).

need for guard and sync intervals is a consequence of the application of time-division multiple access to a wireless system with multiple transmitters. These intervals would not be needed for a wire-based system with a single transmitter and receiver. The length of the preamble is determined to some extent by the signal-to-noise ratio and by the consequences of incorrect frame reception. In the DS-1 (digital signaling, level 1) time-division multiplexing system introduced for wired telephone systems in the 1960s, there were no guard or sync intervals at all, and the "preamble" was reduced to a single bit per frame!

In the remainder of this section we provide two examples of practical time-division multiple-access systems. We show the frame and slot structure for the U.S. Digital Cellular and GSM systems. These examples provide a look at some practical transmission rates and illustrate the number of users that can be supported by these systems in a given channel bandwidth. The examples also show clearly how much and what kind of overhead practical time-division systems require. Following the examples, we introduce the concept of time-division duplexing. This discussion complements the discussion of frequency-division duplexing in the previous section and shows, again by example, an alternative means of avoiding the necessity of transmitting and receiving on the same frequency at the same time.

The U.S. Digital Cellular (USDC) System

The U.S. Digital Cellular (USDC) system is a second-generation cellular telephone system that was introduced in the early 1990s to offer an increase in capacity over that achievable with AMPS. The USDC system was designed to coexist with AMPS systems to make upgrading to the new system as easy as possible for operating companies. The USDC system uses the same

30 kHz channels that were allocated for AMPS, but transmission is digital. The modulation method is $\pi/4$ - DQPSK at a pulse rate of 24,300 pulses/second. To restrict the signals to the allocated bandwidth, square-root raised-cosine pulse shaping is used, with a rolloff factor of 0.35. Quadrature modulation at 24,300 pulses/second gives a bit rate of 48.6 kbits/s. The transmitted signal is divided into frames of 40 ms duration, with each frame containing 1944 bits. Each frame is divided into six slots of 324 bits each. The USDC system can carry six "half-rate" or three "full-rate" voice signals per 30 kHz channel. For full-rate service, the first user occupies slots 1 and 4, the second user slots 2 and 5, and the third user slots 3 and 6.

For a cellular system to operate properly, a certain amount of "signaling" or supervisory information must be sent along with the voice signals. The 48.6 kbit/s bit stream in one 30 kHz channel actually carries four "logical" channels for each user. The first of these is the "digital traffic channel" (DTC) that carries the digitized voice signals or other user data. Next there is a "coded digital verification color code" (CDVCC), a "slow associated control channel" (SACCH), and finally a "fast associated control channel" (FACCH). The CDVCC is a 12-bit coded message sent in every time slot. The message is originated at the base station and echoed back by the mobile unit. The CDVCC helps to ensure that the mobile unit is receiving signals from the correct base station. If the correct CDVCC is not echoed back, the mobile unit transmitter is turned off and the time slot reassigned. The SACCH is also implemented as a 12-bit message sent in every time slot. This channel is used to communicate power level changes to the mobile unit and to allow the mobile unit to report the signal strength of nearby base stations. The FACCH, when it is needed, uses the field in each slot normally allocated to user data. The FACCH is used to transmit dual-tone multifrequency (DTMF) dialing information, call release instructions, and handoff requests.

Figure 6.5 shows the structure of one slot in the forward and in the reverse link. In the forward link, 28 bits of synchronization are followed by the SACCH bits. The 260 bits of data are divided into two 130-bit segments with the CDVCC between them. There are 12 bits "left over" at the end of the slot. In the reverse link, 6 bits each are allocated for guard and ramp-up of the mobile transmitter. The 260 bits of data are divided into three segments and interleaved with the synchronization bits, the SACCH bits, and the CDVCC as shown. Note that the FACCH is not shown. When needed, the FACCH uses the bits otherwise assigned to data.

It is interesting to note how much of each time slot is allocated to "overhead" in the form of supervisory channels, synchronization, guard, and ramp-up time. In fact there is even more overhead than first appears, since of the 260 bits allocated to data, 101 are used for error control.

One time slot of 324 bits actually has room for only 159 bits of digitized voice. Now at "full" rate, each user is given two time slots per frame. Twice 159 bits is 318 bits, and at 318 bits per 40 ms frame, the voice signal bit rate is 7.95 kbits/s.

From a systems perspective it is important to recognize where the increase in capacity that USDC offers over AMPS comes from. First, the $\pi/4$ - DQPSK used by USDC is a linear modulation method, and linear modulation would allow more than one voice signal to be transmitted in a 30 kHz channel, even without digitization. The digital modulation restores the noise immunity provided by FM. Second, quadrature modulation allows the transmission of two bits per pulse.

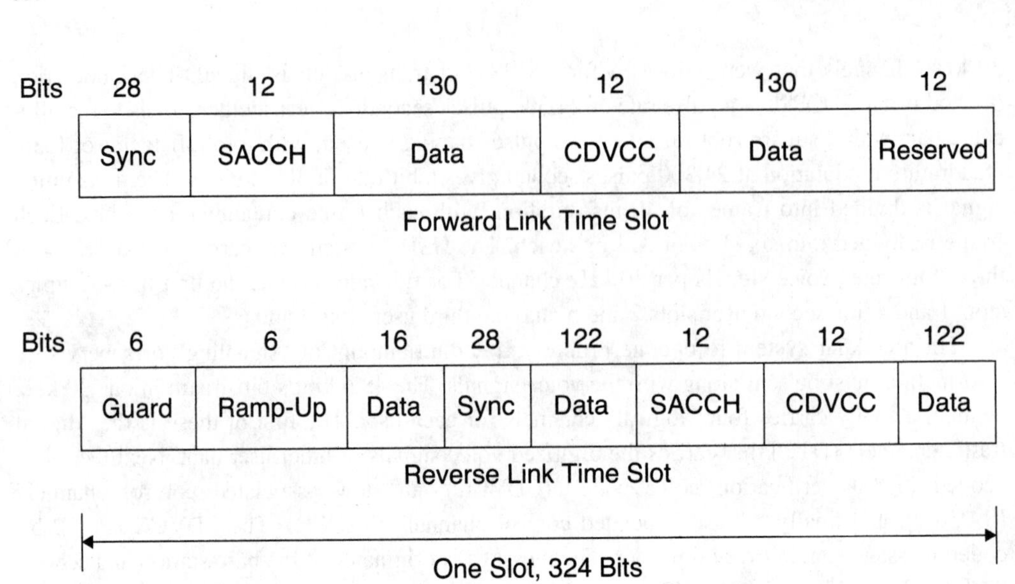

Figure 6.5 Structure of a USDC Time Slot

Third, and perhaps most significant, the speech processor encodes voice signals at 7.95 kbits/s, which is a significantly lower bit rate than the 64 kbits/s associated with speech transmission over wired telephone systems.

The GSM System

GSM is a second-generation cellular system introduced in Europe in 1991 but now used worldwide. GSM was designed to provide a common cellular system throughout Europe, to replace the incompatible national first-generation systems that then existed. GSM was originally assigned to the 900 MHz cellular band in Europe, but it is also used in the 1800 MHz PCS band in the United States and elsewhere. GSM transmits in 200 kHz channels using GMSK modulation at a bit rate of 270.833 kbits/s. Frames have a duration of 4.615 ms and carry 1250 bits.

GSM has a variety of frame structures, depending on whether the frames carry traffic or control signals. There are six formats for traffic channels; the format used depends on whether the channel carries full-rate speech, half-rate speech, or data at any of several bit rates. In this discussion we describe only the format used for full-rate speech, as this is adequate to illustrate the time-division multiple-access principle. There are a number of excellent references that can provide further information.[2]

2. See, for example, T. S. Rappaport, *Wireless Communications: Principles and Practice*, 2nd ed. (Upper Saddle River, NJ: Prentice Hall, 2002), 549–66; M. Schwartz, *Mobile Wireless Communications* (Cambridge: Cambridge University Press, 2005), 201–7.

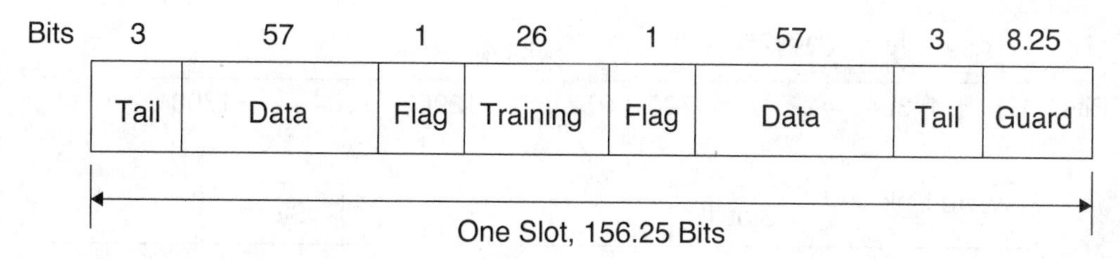

Figure 6.6 Structure of a GSM Slot, Format Used for Full-Rate Speech

A GSM frame is divided into eight slots, each 576.92 μs or 156.25 bit times in duration. With eight slots, up to eight subscribers can share one 200 kHz frequency assignment. The "normal" format for a single time slot of full-rate speech is shown in Figure 6.6. In the format shown, the three tail bits at the start and end of the slot and the guard interval of 8.25 bit-times duration allow for transmit-receive switching, transmitter ramp-up, and synchronization. The slot carries 114 bits of data divided into two 57-bit segments. The 2 flag bits are used to indicate that subscriber data is present in the data segments. As was the case in the USDC system, the data fields can be used as a fast associated control channel (FACCH) when needed. A feature that we have not encountered in our examination of the USDC system is the 26-bit "training" field in the center of the slot. This field carries a known bit pattern that the receiver uses to assess the frequency response of the communication channel. The 200 kHz channel bandwidth is wide enough that frequency-selective fading can cause significant intersymbol interference. Using the bits in the "training" field, the receiver can adjust an equalizing filter to reduce the frequency-selective effects.

Twenty-six GSM frames are grouped together into a "multiframe." In each multiframe, the thirteenth and twenty-sixth frames do not carry data. The thirteenth frame carries eight slots of slow associated control channel (SACCH) data, one control channel for each of the eight traffic channels carried in the remaining frames. The twenty-sixth frame is idle. (It is used for SACCH data in the "half-rate" speech format.) Given the slot, frame, and multiframe structure, we can calculate the overall rate at which user traffic is transmitted. One user is allocated one slot per frame. This slot carries 114 traffic bits in every frame time of 4.615 ms. Dividing gives a bit rate of 24.7 kbits/s. This bit rate must be reduced by a factor of 12/13, since only 24 of every 26 frames carry user traffic. The result is a user traffic rate of 22.8 kbits/s.

The subscriber traffic that fills the "data" fields of each slot is digitized speech. The speech coder processes speech in 20 ms blocks, producing 260 bits in each block. This gives a data rate of 13 kbits/s. Prior to transmission, each block of 260 bits has an additional 196 bits added for error correction. The resulting 456 bits per 20 ms gives the 22.8 kbit/s user traffic rate referred to in the preceding paragraph.

Time-Division Duplexing

The USDC system is usually described as a frequency-division duplexed system, since the forward and reverse links use frequencies in separate bands. Since USDC operates using the AMPS

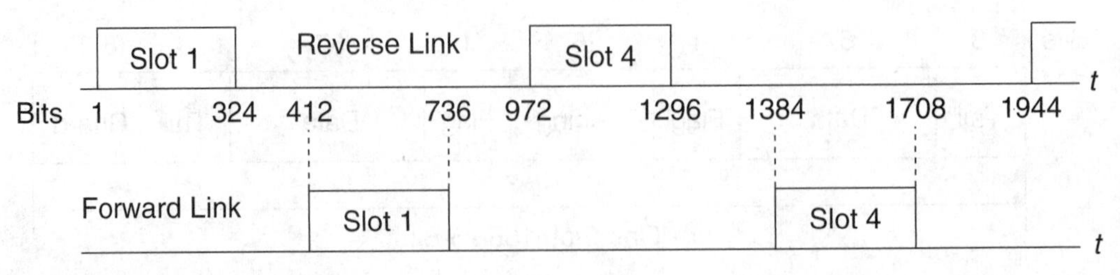

Figure 6.7 Timing of Reverse and Forward Links in USDC

channels, the forward and reverse transmissions are always assigned frequencies separated by 45 MHz. In the USDC system, however, there is a second duplexing method that takes advantage of the fact that each transmission is a burst of 324 bits followed by a period of silence during which other users transmit. In this system the forward and reverse channels are synchronized so that transmission of slot 1 in the forward direction begins 412 bit times later than transmission of slot 1 of the corresponding frame in the reverse direction. The timing is shown in Figure 6.7. It is apparent from the figure that a station never actually transmits and receives at the same time. It should also be clear that the same relative timing would apply for a station using slots 2 and 5, or 3 and 6. A radio that does not transmit and receive at the same time does not need a duplexer. The staggered time relation allows the cost of the radio to be reduced.

Generalizing the preceding discussion, we see that a radio that transmits and receives during nonoverlapping slots does not actually need to use separate frequencies for the forward and reverse links. A scheme in which duplexing is achieved by separate transmit and receive time intervals rather than separate frequencies is called **time-division duplexing**. Time-division duplexing differs from half-duplex operation. In the former case the transmission slots are short and spaced closely enough together that the user perceives communications in both directions as simultaneous. In the latter case the user is aware of separate time intervals allocated for talking and listening.

Time-division duplexing is used in a number of cordless phone systems. It tends to work best over short transmission distances, as significant propagation delay can interfere with the slot coordination. Even more damaging is the time-varying propagation delay that can result when one or both of the radios is moving rapidly.

Code-Division Multiple Access

Code-division multiple-access techniques are based on spread-spectrum modulation. Spread-spectrum modulation was introduced at the end of Chapter 5, where we showed two such techniques: frequency-hopping spread spectrum and direct-sequence spread spectrum. Spread spectrum is defined by the use of a signal, unrelated to the information content of the transmitted message, to spread the bandwidth of the transmitted signal. We showed in Chapter 5 that spread-spectrum modulation offers advantages in providing resistance to narrowband interference as well as providing robustness in the presence of frequency-selective fading.

Because a spread-spectrum signal is created to have an artificially wide bandwidth, it might seem that this kind of signaling is unsuited for a densely channeled system such as a cellular telephone system designed to support a maximum number of subscribers. In fact, spread-spectrum systems have traditionally been used in situations in which the information data rate is very low, the channel bandwidth is relatively unrestricted, or the spreading is so broad that narrowband users have been able to coexist with the spread-spectrum system without being aware of its presence. We show in this section that multiple spread-spectrum transmitters can share a common communication channel if their spreading codes are distinct and properly chosen. This ability of spread-spectrum signals to be distinguished by their spreading code leads to the code-division multiple-access method that can be used as an alternative to frequency-division or time-division multiple access.

Both frequency-hopping and direct-sequence spread-spectrum systems can be used for code-division multiple-access communications. Bluetooth, for example, is a frequency-hopped system. Multiple Bluetooth "piconets" can exist with overlapping coverage areas if distinct spreading codes are used. Cellular telephone applications of code-division multiple access use direct-sequence spread spectrum. Direct-sequence systems include the second-generation cdmaOne as well as the third-generation cdma2000 and Wideband CDMA. We will describe both frequency-hopping and direct-sequence systems in this section. We begin with a somewhat brief description of frequency-hopping CDMA. Frequency hopping is easier to understand than direct sequence, and it provides an effective framework for introducing some of the important properties and limitations of the technique. We will then move to the direct-sequence case. Our purpose in the discussion is twofold: On the one hand we describe how CDMA systems work, and on the other we wish to determine the number of subscribers that a system can support, given system parameters such as system bandwidth and available signal-to-noise ratio. Our discussions will focus on "generic" CDMA systems. Space does not permit getting into the implementation details of commercial second- and third-generation cellular systems. For those details the reader is referred to the standards documents or to textbooks written at a more advanced level.

Frequency-Hopping CDMA Systems

A frequency-hopping spread-spectrum system is shown in simplified block diagram form in Figure 6.8. Recall that the PN sequence generator controls the frequency synthesizer in such a way that the frequency of the transmitted carrier hops at regular intervals among a predetermined set of frequencies. The receiver is equipped with an identical PN sequence generator and frequency synthesizer. The receiver tracks the carrier of the transmitted signal so that the received signal can be continuously demodulated. The PN sequence is usually designed so that the transmitted carrier makes equal use of all of the available frequencies. To an outside observer the hopping pattern appears random, though of course it is not random, as the receiver knows the pattern.

Figure 6.8 is simplified in that it does not show the circuits needed to synchronize the PN generators in the transmitter and the receiver.

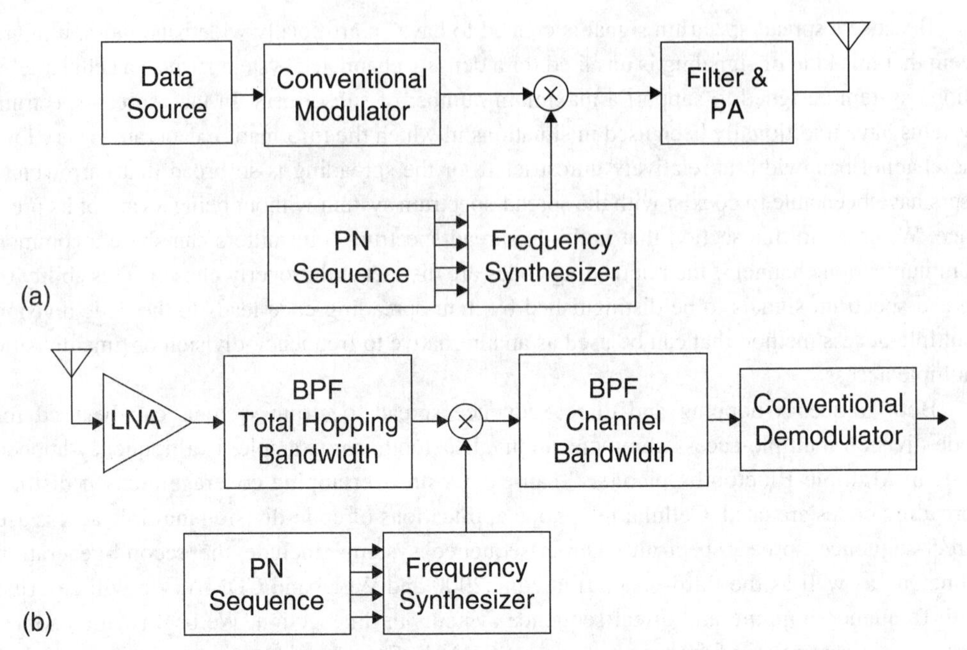

Figure 6.8 Frequency-Hopped Spread-Spectrum System: (a) Transmitter; (b) Receiver

Frequency hopping works well with any modulation method. It has been used with analog AM and FM, as well as with advanced digital modulation techniques. It is common practice to use FSK in frequency-hopping systems, as the constant envelope of the FSK signal is not impaired by the hopping.

Example

Suppose noncoherent FSK is used in a frequency-hopping spread-spectrum system. If the channel noise is white, and fading is not frequency selective, then the signal-to-noise ratio will be the same at each of the possible carrier frequencies in the hopping set. This means that the probability of error will not be affected by the fact that the carrier frequency changes at regular intervals. For noncoherent FSK we have

$$P_e = \frac{1}{2} e^{-\frac{E_b}{2N_0}}, \tag{6.9}$$

where E_b is the energy per bit, and $N_0/2$ is the noise power spectral density.

Now let us suppose that two frequency-hopping systems share the same set of carrier frequencies. If the two systems use different PN sequences to control their hopping, then each transmitter-receiver pair will perceive the other transmitter to be hopping at random. Every once in a while the two transmitters will hop onto the same carrier frequency. This is called a "collision" or a "hit," and when a collision occurs we can expect that the signals obtained by both receivers will be severely impaired. In fact, for modeling purposes, let us make the worst-case assumption that when a collision occurs, the probability of error becomes $P_e = 0.5$. If the probability of collision is p_{coll2}, then the average probability of error experienced by one of the receivers is

$$P_e = p_{coll2} \times \frac{1}{2} + (1 - p_{coll2}) \times \frac{1}{2} e^{-\frac{E_b}{2N_0}}, \quad (6.10)$$

where we have continued to assume for purposes of illustration that noncoherent FSK is the modulation method used.

To complete our calculation of the probability of error for this two-transmitter-receiver-pair system, we need to obtain a value for p_{coll2}. Suppose there are q carrier frequencies in the hopping set. For the purposes of example, let us assume that a slow-hopping system is used, with N_b bits of data transmitted between hops. We will also make the reasonable assumption that the two transmitter-receiver pairs are not synchronized but hop independently of each other.

For purposes of discussion, let the two transmitter-receiver pairs be designated as "pair A" and "pair B." We calculate the performance of pair A in the presence of pair B. Let us now pick a particular moment in time when pair A is occupying a carrier frequency designated as f_0. There are two ways that pair B can collide with pair A. One possibility is that at the start of the current bit time, pair B is also using carrier frequency f_0. The other possibility is that at the start of the current bit time pair B is using some other carrier frequency but hops to frequency f_0 during the current bit time. Now the probability that pair B is using frequency f_0 at the start of the current bit time is $1/q$. Consequently, the probability that pair B is not using frequency f_0 must be $1 - 1/q$.

Each transmitter-receiver pair hops once every N_b bit times. Since the two systems are not synchronized, the time at which pair B hops is uniformly distributed over the N_b bit-time interval. Therefore, the probability that pair B will hop during just one bit time is $1/N_b$. Thus we see that the probability that pair B is not using frequency f_0 at the start of the current bit time and hops during the current bit is $(1-\frac{1}{q})\frac{1}{N_b}$. Now when pair B hops, it may land on any of $q-1$ destinations, assuming that it cannot hop back onto its original frequency. We see that the probability that pair B is not using frequency f_0, but hops during the current bit time landing on f_0 (thereby causing a collision), is $(1-\frac{1}{q})\frac{1}{N_b}\frac{1}{q-1} = \frac{1}{qN_b}$. Finally, we put the pieces together. The probability p_{coll2} for a two-transmitter-receiver-pair system is

$$p_{coll2} = \frac{1}{q} + \frac{1}{qN_b} = \frac{1}{q}\left(1 + \frac{1}{N_b}\right). \quad (6.11)$$

Having reached this point, we can easily generalize to the case in which there are more than two transmitter-receiver pairs. Suppose that there are K such pairs. To find p_{coll} for this larger system we reason as follows. We begin as before, by singling out one transmitter-receiver pair and calling the frequency on which it is transmitting f_0. The probability that a second transmitter-receiver pair collides during the current bit interval is p_{coll2}, given by Equation (6.11). The probability that this second transmitter-receiver pair does *not* collide during the current bit interval is then $1 - p_{coll2}$. The probability that none of the other $K-1$ transmitter-receiver pairs collide

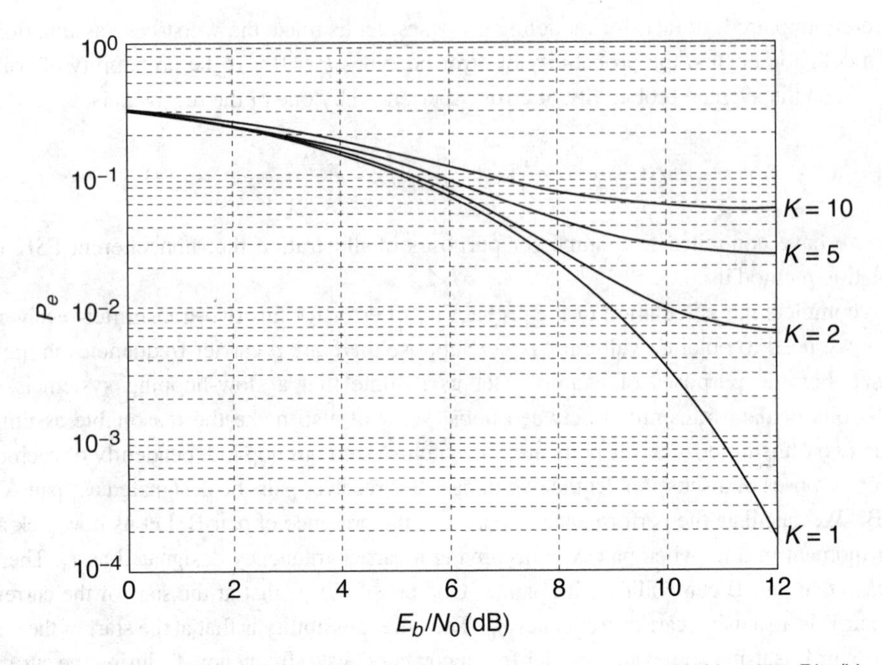

Figure 6.9 Probability of Error versus E_b/N_0 for Frequency Hopping with $N_b = 625$ Bits/Hop, $q = 80$ Frequencies, and $K = 1, 2, 5$, and 10 Users

during the current bit interval is $(1 - p_{coll2})^{(K-1)}$. Therefore, the probability of one or more collisions during the current bit interval is

$$p_{coll} = 1 - (1 - p_{coll2})^{(K-1)}. \tag{6.12}$$

The probability of error for this K transmitter-receiver pair system is given by Equation (6.10), with p_{coll} substituted for p_{coll2}, that is,

$$P_e = p_{coll} \times \frac{1}{2} + (1 - p_{coll}) \times \frac{1}{2} e^{-\frac{E_b}{2N_0}}. \text{[3]} \tag{6.13}$$

Example

Consider a slow-hopping system with $N_b = 625$ bits/hop and $q = 80$ hopping frequencies. Figure 6.9 shows probability of error plotted against E_b/N_0 from Equations (6.13), (6.12), and (6.11), assuming that there are $K = 1, 2, 5$, and 10 users.

3. E. A. Geraniotis and M. B. Pursley, "Error Probabilities for Slow-Frequency-Hopped Spread-Spectrum Multiple-Access Communications over Fading Channels," *IEEE Transactions on Communications* COM-30, no. 5 (May 1982): 996–1009.

Several observations can be made from this simple example. First, when there is more than one user, the probability of error does not continue to decrease as E_b/N_0 is increased. Instead, there is a so-called irreducible error rate that is reached when E_b/N_0 becomes large enough. For large E_b/N_0, virtually no bit errors are caused by the channel noise. Instead, nearly all errors are caused by collisions. In this case the probability of error can be reduced only by increasing the number of hopping frequencies or reducing the number of users.

A second observation is that for the numbers used in this example, the probability of error achieved is very high. In practice, CDMA systems used for voice transmission often use error-correcting codes to lower the probability of error. The errors caused by collisions tend to occur in bursts lasting from 1 to N_b bits. There are error-correcting codes that are very effective at correcting bursts of errors, provided the burst duration is not too long. To deal with long bursts, data is often "interleaved" before transmission. This means that blocks of data are shuffled, like a deck of cards, prior to transmission and unshuffled after reception. The interleaving (and de-interleaving) has no effect on the data, but bursts of errors become spread out by the de-interleaving, so that the error-correcting code can deal with the errors more effectively.

In our preceding discussions we assumed that the frequency-hopping system was a slow hopper. A similar analysis can be done for the fast-hopping case. When fast hopping is used there are several hops per bit time, which means that there are multiple opportunities per bit time for collisions. If the hopping is very fast, so that the hop time is much shorter than the bit time, a single collision may not cause enough disturbance to produce a bit error. Thus a fast-hopping system may in fact be more robust than a calculation based on a simple model might indicate.

For our final observation we return to the fundamental question: How many subscribers can a frequency-hopped multiple-access system support? It is apparent from Figure 6.9 that each user added to the system causes an increase in the probability of error for all users. Thus the number of users is not limited by a firm bound as is the case for frequency-division and time-division multiple-access systems. For a frequency-hopping system the number of users is limited by the quality of service that the system is designed to provide to each user. This "soft" limit in number of users provides an operational advantage, as there may be cases when it is desirable to add a user beyond the quality of service limit, if the user will be present for only a limited time. We will see in the next section that the soft limit on number of users applies to direct-sequence multiple-access systems as well.

Direct-Sequence CDMA Systems

A direct-sequence spread-spectrum system is shown in block diagram form in Figure 6.10. Direct-sequence spread spectrum was introduced in Chapter 5. We will review briefly how the system works and then proceed to the main issue before us: how multiple users can be supported on the same carrier frequency. We will show that multiple access can be very effective with this system, provided the spreading codes are chosen appropriately. Two cases will be investigated, both of practical importance in CDMA cellular telephone systems. First, we will investigate the case in which users are assigned orthogonal spreading codes. In this case the spreading codes

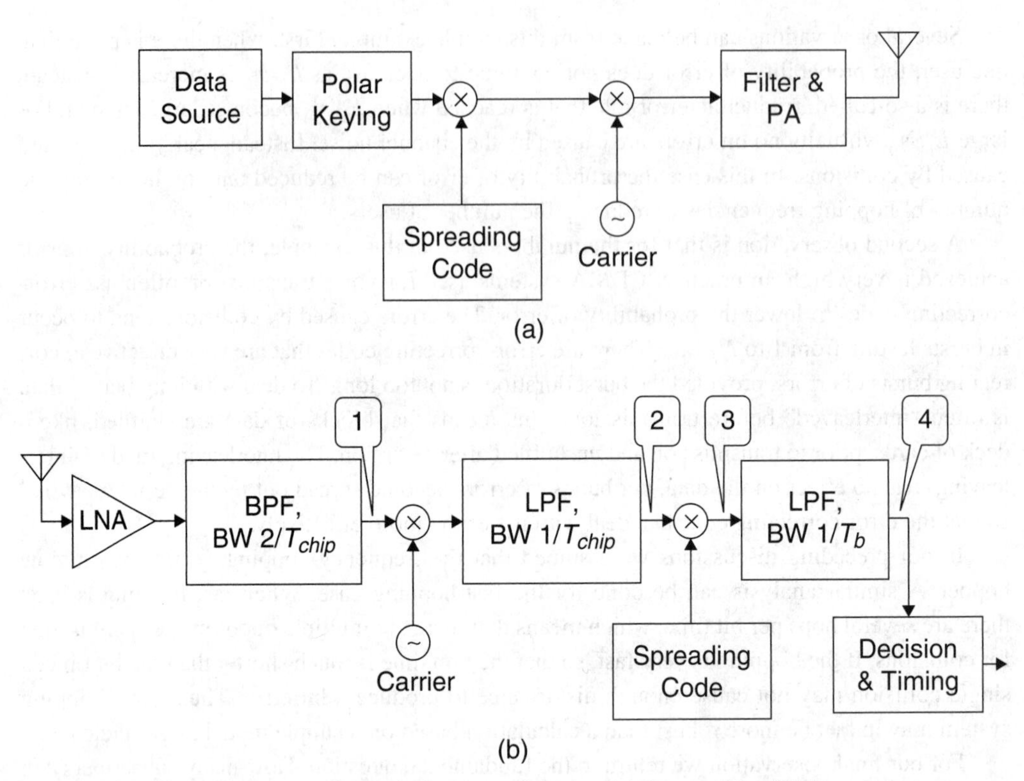

Figure 6.10 Direct-Sequence Spread-Spectrum System: (a) Transmitter; (b) Receiver

may not be very effective at spreading the spectrum, and distinct users must be synchronized. In return, however, multiple users can be perfectly separated and do not cause interference to each other. Second, we will explore the case in which the spreading codes are pseudonoise sequences. In this case the signal bandwidth will be effectively spread, leading to the benefits discussed in Chapter 5 relating to narrowband interference resistance and robustness in the presence of frequency-selective fading. For this case, distinct users do not have to be synchronized. We will show, however, that multiple users are not perfectly separated and that the overall interference level increases with each user added.

Referring to Figure 6.10, let us denote the signal at the receiver input by

$$r(t) = Am(t)p(t)\cos(2\pi f_c t) + n(t), \tag{6.14}$$

where $m(t)$ is the message, assumed to be a polar NRZ signal consisting of rectangular pulses of duration T_b; $p(t)$ is the spreading code, also assumed to be a polar NRZ signal of rectangular pulses of duration T_{chip}; f_c is the carrier frequency; and $n(t)$ is additive white Gaussian noise, with power spectral density $N_0/2$. Note that we are taking both the message and spreading code pulses as rectangular to simplify our discussion, even though in practice the pulses would be shaped to limit adjacent-channel interference. Further, we assume that both the message and the spreading code take values of ±1; we will use the parameter A to set the level of the received signal. For future reference the received signal power P_s is given by

$$\begin{aligned} P_s &= \langle A^2 m^2(t) p^2(t) \cos^2(2\pi f_c t) \rangle \\ &= \frac{A^2}{2}, \end{aligned} \tag{6.15}$$

where we have used the facts that $m^2(t) = p^2(t) = 1$ and $\langle \cos^2(2\pi f_c t) \rangle = \frac{1}{2}$. Finally, we recollect that in a direct-sequence spread-spectrum system, we always have $T_{chip} \ll T_b$.

The first stage of the receiver is conventional and consists of a low-noise amplifier and a bandpass filter. The filter bandwidth is approximately $2/T_{chip}$, which is wide enough to pass the spread signal without significant distortion. In our discussion we will assume that all filters have rectangular passbands and unity gain, unless we specifically state otherwise. At point 1 in the system we have

$$r_1(t) = Am(t)p(t)\cos(2\pi f_c t) + n_1(t). \tag{6.16}$$

Only the noise level is affected by the bandpass filter. The noise power spectrum is still $S_{n_1}(f) = N_0/2$, but the noise is now bandlimited to a bandwidth of $2/T_{chip}$.

The carrier oscillator, mixer, and first lowpass filter constitute a demodulator for the BPSK modulation. At point 2 in the system we have

$$r_2(t) = \frac{A}{2} m(t)p(t) + n_2(t). \tag{6.17}$$

The lowpass filter bandwidth is determined by the bandwidth of the spreading code and is approximately given by $1/T_{chip}$. The noise power spectrum at point 2 is given by $S_{n_2}(f) = N_0/4$. This noise has a baseband power spectrum with a bandwidth of $1/T_{chip}$.

The spreading code generator, the second mixer, and the second lowpass filter constitute the "despreading" circuit. Multiplying Equation (6.17) by the spreading code $p(t)$ produces

$$\begin{aligned} r_3(t) = r_2(t)p(t) &= \frac{A}{2} m(t)p^2(t) + n_2(t)p(t) \\ &= \frac{A}{2} m(t) + n_3(t), \end{aligned} \tag{6.18}$$

where we have again made use of the fact that $p^2(t) = 1$. We recall from Chapter 5 that multiplying the noise $n_2(t)$ by the spreading code makes no change in the noise from a statistical point of view. The signal $r_3(t)$ is thus an ordinary polar NRZ line-coded message in additive (bandlimited) white Gaussian noise.

Figure 6.10 shows the filter following point 3 as a lowpass filter of bandwidth $1/T_b$. We know, however, from the discussion in Chapter 5, that optimal receiver performance will be obtained if this filter is matched to the pulse shape used to represent the message bits. For rectangular pulses,

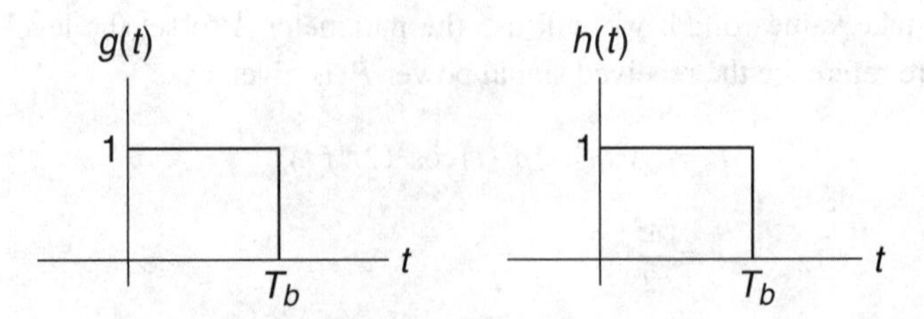

Figure 6.11 Data Pulse and Matched Filter Impulse Response

the matched filter will have a rectangular impulse response. The data pulse and the matched filter impulse response are shown in Figure 6.11.

We can write the filter output using convolution. We have at point 4

$$\begin{aligned} r_4(t) &= \int_{-\infty}^{\infty} h(t-\tau) r_3(\tau) d\tau \\ &= \int_{t-T_b}^{t} r_3(\tau) d\tau, \end{aligned} \tag{6.19}$$

where we have used the fact that the impulse response is a rectangular pulse of duration T_b. If the filter output is sampled at $t = T_b$, when the signal component at the filter output is maximum, we obtain the decision statistic

$$r_4(T_b) = \int_0^{T_b} r_3(\tau) d\tau. \tag{6.20}$$

The decision statistic will have a signal component $s_4(T_b) = \pm \frac{A}{2} T_b$ and a noise component $n_4(T_b)$. If the matched filter has frequency response $H(f)$, then the power spectrum $S_{n_4}(f)$ of the noise at the matched filter output is

$$S_{n_4}(f) = S_{n_3}(f) |H(f)|^2 = \frac{N_0}{4} |H(f)|^2 . \tag{6.21}$$

The variance $\sigma_{n_4}^2$ of the noise component $n_4(T_b)$ is

$$\begin{aligned} \sigma_{n_4}^2 &= \int_{-\infty}^{\infty} S_{n_4}(f) df \\ &= \frac{N_0}{4} \int_{-\infty}^{\infty} |H(f)|^2 df \\ &= \frac{N_0}{4} \int_0^{T_b} 1^2 dt \\ &= \frac{N_0}{4} T_b, \end{aligned} \tag{6.22}$$

where we used Parseval's theorem to replace the frequency response integral with an impulse response integral. Recall that the probability of error for the system depends on the signal-to-noise ratio

$$\frac{s_4^2(T_b)}{\sigma_{n_4}^2} = \frac{A^2T_b^2/4}{N_0T_b/4} = \frac{A^2T_b}{N_0}. \tag{6.23}$$

From Equation (6.15) we have the average power in the received signal as $P_s = A^2/2$. The energy per bit is then $E_b = P_sT_b = A^2T_b/2$. Substituting in Equation (6.23) gives the result

$$\frac{s_4^2(T_b)}{\sigma_{n_4}^2} = \frac{2E_b}{N_0}, \tag{6.24}$$

which leads to the familiar expression for probability of error,

$$P_e = Q\left(\sqrt{2E_b/N_0}\right). \tag{6.25}$$

Multiple Users

Let us now suppose that there is a second user sharing the same carrier frequency. The received signal is

$$r(t) = A_1m(t)p_1(t)\cos(2\pi f_ct) + A_2p_2(t-t_d)\cos(2\pi f_c(t-t_d)) + n(t), \tag{6.26}$$

where $p_1(t)$ and $p_2(t)$ are distinct spreading codes and t_d is an unknown delay, representing the fact that the two received signals may not be synchronized. Note that we have not explicitly included a message signal $m_2(t)$ in the second component of the received signal $r(t)$. To a receiver that does not know the spreading code $p_2(t)$, both $m_2(t)p_2(t)$ and $p_2(t)$ appear to be random sequences of T_{chip}-second rectangular pulses. As these sequences are indistinguishable, we can simplify the notation by omitting explicit reference to $m_2(t)$.

After filtering, the signal at point 1 becomes

$$r_1(t) = A_1m(t)p_1(t)\cos(2\pi f_ct) + A_2p_2(t-t_d)\cos(2\pi f_c(t-t_d)) + n_1(t). \tag{6.27}$$

At point 2 we have

$$r_2(t) = \frac{A_1}{2}m_1(t)p_1(t) + \frac{A_2}{2}p_2(t-t_d)\cos(2\pi f_ct_d) + n_2(t). \tag{6.28}$$

Note that the factor $\cos(2\pi f_ct_d)$ is not a function of time but is a constant that represents the effect of a receiver that is not synchronized to the second user's carrier signal. The receiver next multiplies by the spreading code $p_1(t)$, giving

$$r_3(t) = \frac{A_1}{2}m_1(t) + \frac{A_2}{2}p_1(t)p_2(t-t_d)\cos(2\pi f_ct_d) + n_3(t). \tag{6.29}$$

Passing $r_3(t)$ through the matched filter gives the decision statistic

$$r_4(T_b) = \pm\frac{A_1}{2}T_b + \frac{A_2}{2}\cos(2\pi f_c t_d)\int_0^{T_b} p_1(t)p_2(t-t_d)dt + n_4. \tag{6.30}$$

The polarity of the first term in Equation (6.30) depends on the value of the message bit. The second term in the equation represents the interference from the second user, and it is this term we wish to evaluate. Notice the similarity between the integral in this term and the cross-correlation function defined in Chapter 5. In the following discussion we will exploit this similarity to evaluate this term.

From this point the discussion can proceed in two directions, depending on the assumptions we make about the design of the spreading codes. As described previously, we will follow both paths. We first investigate the possibility of designing the spreading codes so that the cross-correlation in Equation (6.30) is identically zero.

Orthogonal Spreading Codes

In the forward link of a cellular telephone system, a single base station transmits to all of the mobile units in the cell. In this case the carriers and spreading codes of all of the transmitted signals can be synchronized. At any given mobile receiver, all of the received signals arrive with the same delay, particularly if we assume that flat, slow fading is the only multipath effect. To model this case in which received signals intended for distinct users are synchronized, we set the delay parameter t_d equal to zero in Equation (6.30).

The cross-correlation integral in Equation (6.30) extends from time zero to time T_b. In many spread-spectrum systems the spreading code is designed to appear as a random sequence of chips and hence has a period much longer than the bit time. It is also possible, however, to design short spreading codes. Let us assume that the entire spreading code repeats every T_b seconds, so that each data bit is multiplied by an identical spreading sequence. In this case the integral

$$\int_0^{T_b} p_1(t)p_2(t)dt \tag{6.31}$$

$$H_1 = [0] \qquad H_2 = \begin{bmatrix} 0 & 0 \\ 0 & 1 \end{bmatrix}$$

$$H_4 = \begin{bmatrix} 0 & 0 & 0 & 0 \\ 0 & 1 & 0 & 1 \\ 0 & 0 & 1 & 1 \\ 0 & 1 & 1 & 0 \end{bmatrix} \qquad H_{2N} = \begin{bmatrix} H_N & H_N \\ H_N & \overline{H_N} \end{bmatrix}$$

Figure 6.12 Structure of the Hadamard Matrices

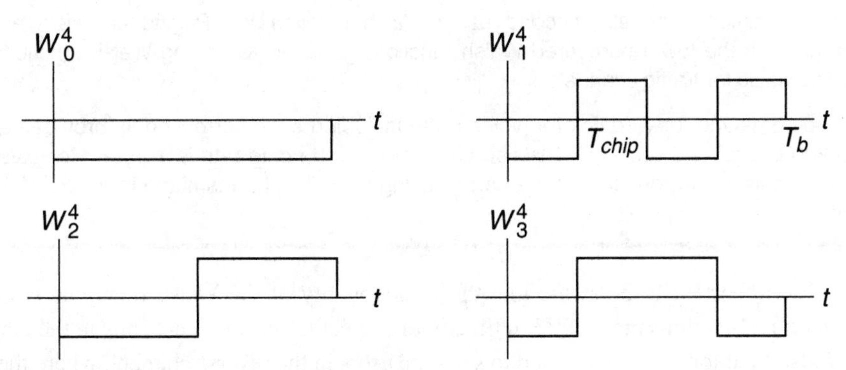

Figure 6.13 The Walsh Functions $W_0^4,\ldots,W_3^4$

is precisely the cross-correlation of $p_1(t)$ and $p_2(t)$ at zero lag. Two waveforms whose zero-lag cross-correlation is zero are said to be "orthogonal." To support noninterfering multiple users, we seek sets of waveforms that are mutually orthogonal.

One useful set of orthogonal waveforms is the set of so-called **Walsh functions**. These are obtained from the rows of the **Hadamard matrices**. The Hadamard matrices are defined recursively as shown in Figure 6.12. Hadamard matrix H_1 consists of a single Boolean zero. To construct Hadamard matrix H_{2N}, form a matrix of four elements. Three of the elements are Hadamard matrices H_N, and the element in the lower right-hand corner is the Boolean complement $\bar{H}_N$. Figure 6.12 shows the progression H_1, H_2, and H_4.

Once a Hadamard matrix of an appropriate size has been constructed, the Walsh functions are simply the rows of the Hadamard matrix coded as polar-keyed NRZ signals. If the Hadamard matrix H_N is taken as the Walsh function generator, then the Walsh functions can be designated $W_0^N, W_1^N, \ldots, W_{N-1}^N$. Figure 6.13 shows the Walsh functions generated by the matrix H_4.

The important attribute of the Walsh functions from our current perspective is that distinct Walsh functions are orthogonal, that is,

$$\int_0^{T_b} W_i^N(t) W_j^N(t)dt = 0, \quad \text{for } i \neq j. \tag{6.32}$$

Thus we can assign N users to the same carrier frequency and assign each user a separate Walsh function as a spreading code; according to Equation (6.30), each receiver will be able to receive its own assigned signal without interference from any of the other users.

Example

The cdmaOne cellular system is a second-generation system based on CDMA. In the forward link the Walsh functions W_i^{64} are used as spreading codes. Walsh function W_0^{64} designates the pilot channel, which provides carrier synchronization. There is also a synchronization channel, encoded using Walsh function W_{32}^{64}. Paging channels are assigned to the lower-numbered Walsh functions, and the remaining Walsh functions are available for traffic channels.

In the cdmaOne forward link the voice data, including error correction, is provided at 19,200 bits/s. As there are 64 Walsh chips per bit, the chip rate is 1.2288 Mchips/s. The signals are modulated onto a carrier using BPSK and transmitted in a 1.25 MHz channel.

It is very important to be aware that the orthogonality of the Walsh functions is not preserved if distinct functions are shifted in time with respect to one another. Thus in the cdmaOne system, Walsh functions cannot be used to separate users in the reverse channel, where the transmissions from different mobile units, located at various distances from the base station, are harder to synchronize. It is also important to note that not all of the Walsh functions are effective at spreading the bandwidth of the message signal. Walsh function W_0^N, for example, does no spreading at all. For this reason, Walsh functions used to separate channels are properly called a **cover**, rather than a spreading code. In the cdmaOne system, as well as in other systems in which the benefits of bandwidth spreading are desirable, additional spreading using a pseudonoise sequence is employed.

Pseudonoise Spreading Codes

In this section we return to Equation (6.30) and pursue the alternative line of inquiry: We investigate the effect of multiple users sharing a carrier frequency when PN sequences are used as the spreading codes. Recall that a PN sequence is a deterministic binary sequence of finite duration that has the appearance of a random train of bits. The PN sequences used as spreading codes cannot actually be random sequences, however, because a given transmitter-receiver pair must generate the same code in order to communicate. Often, to enhance the appearance of randomness, the PN sequences used have a duration that is very long compared to a bit time.

PN sequences are commonly used to separate users in multiple-access systems when the signals from distinct users cannot be synchronized. In CDMA-based cellular systems mobile stations transmitting on the reverse channel can be difficult to synchronize. Even when a pilot signal transmitted on a forward channel by the base station is used to provide a common clock for the mobile transmitters, the propagation distances from mobile stations to a common receiver at the base station are all different. The propagation differences translate to differences in arrival times for the received signals. We have modeled that unpredictable propagation delay as the parameter t_d in Equation (6.26).

If $p_1(t)$ and $p_2(t)$ represent distinct PN sequences in Equation (6.26), then the integral in Equation (6.30) will not be identically zero. Thus we see that the presence of a second user adds interference to the signal we are trying to demodulate. Our immediate task is to quantitatively estimate the amount of this interference. To make this estimate without getting lost in mathematical detail, we will use the simplest model for the interference that is consistent with obtaining a reasonably accurate result. We observe that to receiver number 1, the spreading code $p_2(t)$ is indistinguishable from a random sequence of bits. When a delayed version of $p_2(t)$ is multiplied by $p_1(t)$ and passed through a lowpass (matched) filter, the output is noiselike. We will therefore use a noise model to determine the average power in the interference at the output of the matched filter.

Let us begin with the desired signal. Equation (6.30) gives the desired sample at the matched filter output as $\pm\frac{A_1}{2}T_b$. Squaring this value gives

$$P_{s4} = \frac{A_1^2 T_b^2}{4}. \tag{6.33}$$

Now from Equation (6.15), the average power in the desired part of the signal at the receiver input is given by

$$P_s = \frac{A_1^2}{2}. \tag{6.34}$$

Combining Equations (6.33) and (6.34) gives

$$P_{s4} = \frac{P_s T_b^2}{2}. \tag{6.35}$$

Now let us model the interference. From Figure 6.10 and Equation (6.29) the interference signal at the input to the matched filter is given by

$$r_{i3}(t) = \frac{A_2}{2}\cos(2\pi f_c t_d) p_1(t) p_2(t - t_c). \tag{6.36}$$

The PN sequences $p_1(t)$ and $p_2(t)$ can be treated as random sequences of polar NRZ T_{chip}-second pulses. If we treat the product $p_1(t)p_2(t-t_d)$ the same way, then the power spectrum of this product is approximately

$$S_{p1p2}(f) \cong T_{chip}\,\mathrm{sinc}^2(T_{chip} f). \tag{6.37}$$

This gives the power spectrum of the interference term as

$$\begin{aligned} S_{i3}(f) &= \frac{A_2^2}{4}\left\langle \cos^2(2\pi f_c t_d)\right\rangle S_{p1p2}(f) \\ &= \frac{A_2^2}{4}\frac{1}{2} T_{chip}\,\mathrm{sinc}^2(T_{chip} f) \\ &= \frac{A_2^2}{8} T_{chip}\,\mathrm{sinc}^2(T_{chip} f), \end{aligned} \tag{6.38}$$

using Equation (6.37) and the fact that $\langle \cos^2(2\pi f_c t_d)\rangle = \frac{1}{2}$. Now the matched filter in the receiver is essentially a lowpass filter with a bandwidth of approximately $1/T_b$. Since $1/T_{chip} \gg 1/T_b$, the spectrum $S_{i3}(f)$ is very broad compared with the bandwidth of the matched filter. Over the matched filter's limited bandwidth we can take

$$S_{i3}(f) \approx \frac{A_2^2}{8} T_{chip}, \tag{6.39}$$

since $\text{sinc}^2(T_{chip} f) \cong 1$ for $f \ll 1/T_{chip}$. If the frequency response of the matched filter is $H(f)$, then the power spectrum of the interference at the matched filter output is

$$S_{i4}(f) = |H(f)|^2 \frac{A_2^2}{8} T_{chip}. \tag{6.40}$$

The average interference power at the matched filter output is obtained by integrating Equation (6.40) over all frequencies. We have

$$\begin{aligned} P_{i4} &= \int_{-\infty}^{\infty} S_{i4}(f) df \\ &= \frac{A_2^2}{8} T_{chip} \int_{-\infty}^{\infty} |H(f)|^2 \, df \\ &= \frac{A_2^2}{8} T_{chip} \int_{-\infty}^{\infty} h^2(t) dt \\ &= \frac{A_2^2}{8} T_{chip} T_b, \end{aligned} \tag{6.41}$$

where we have once again used Parseval's theorem to replace the frequency response integral with the impulse response integral. The impulse response of the matched filter was given in Figure 6.11. Finally, referring back to Equation (6.26), we see that the average power in the interfering signal at the receiver input is $P_i = \frac{A_2^2}{2}$. Substituting in Equation (6.41) gives

$$P_{i4} = \frac{P_i}{4} T_{chip} T_b. \tag{6.42}$$

Ultimately, the effect of the interfering signal on system performance can be characterized by the signal-to-noise-and-interference ratio at the matched filter output. If we combine Equations (6.35), (6.42), and (6.22) we obtain

$$\begin{aligned} SNIR &= \frac{P_{s4}}{P_{i4} + \sigma_{n_4}^2} = \frac{\frac{P_s T_b^2}{2}}{\frac{P_i}{4} T_{chip} T_b + \frac{N_0}{4} T_b} \\ &= 2 \frac{P_s}{P_i} \frac{T_b}{T_{chip}} \frac{1}{1 + \frac{N_0}{P_i T_{chip}}}. \end{aligned} \tag{6.43}$$

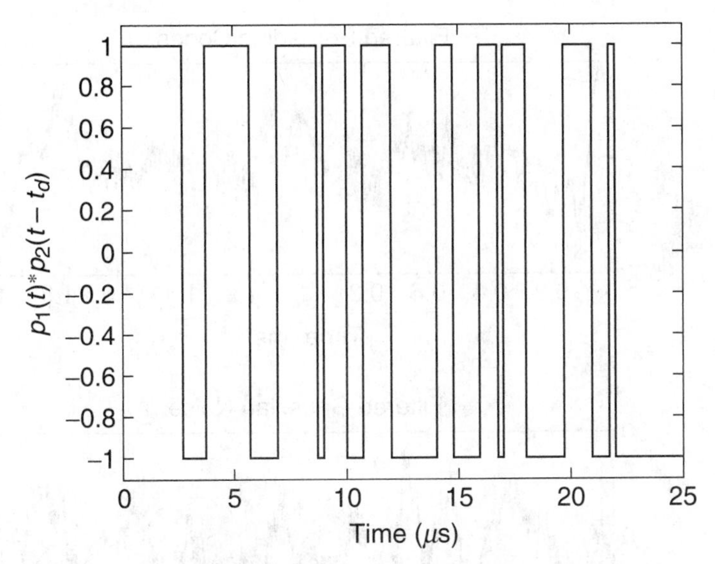

Figure 6.14 A Segment of the Product $p_1(t)\ p_2\ (t - t_d)$ of Two PN Sequences

Thus, the signal-to-interference ratio at the matched filter output is proportional to the signal-to-interference ratio at the receiver input multiplied by the spread-spectrum processing gain.

Example

Our derivation of signal-to-interference ratio depends on the assumption that when the spreading code product $p_1(t)p_2(t - t_d)$ is applied to a matched filter, the output can be modeled as noise. As an illustration to help make this assumption plausible, two long PN codes were generated, each having a chip duration of $T_{chip} = 1\ \mu s$. A segment of the product $p_1(t)p_2(t - t_d)$ is plotted in Figure 6.14, where the delay was arbitrarily taken as $t_d = T_{chip}/4$. The product of the two spreading codes was then passed through a filter having a rectangular impulse response of duration T_b, where T_b was taken as $64T_{chip}$. The filter output is plotted as Figure 6.15(a). For comparison, white Gaussian noise having the same power spectrum at low frequencies as the product $p_1(t)p_2(t - t_d)$ was passed through the same filter. The output is shown as Figure 6.15(b). A comparison of the two graphs suggests that the noise model for the filtered PN sequence product might be reasonable.

Several important conclusions can be drawn from the signal-to-interference ratio expression given by Equation (6.43). If $P_s = P_i$, so that the desired signal and the interfering signal are received at the same signal strength, then the spread-spectrum processing gain can reduce the interfering signal to insignificance. If, however, a base station is trying to receive a distant desired signal in the presence of interference from a nearby station, the ratio P_s/P_i can become very small. In this case the processing gain may not be sufficient to overcome the interference. This latter case is an example of the so-called near-far problem. In practical CDMA cellular systems the transmitted power of

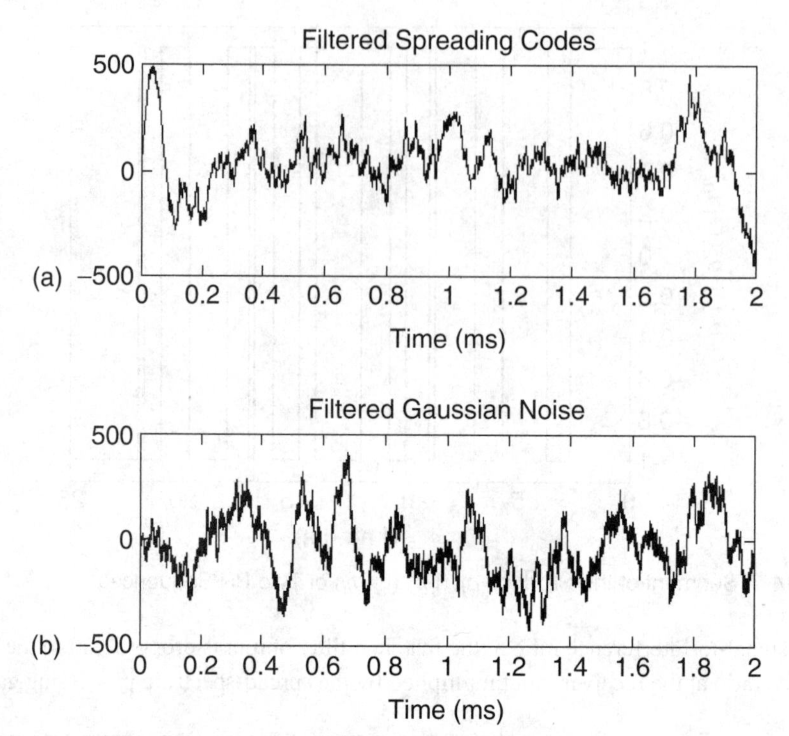

Figure 6.15 (a) Filtered Spreading Code Product Compared with (b) Filtered White Gaussian Noise

each mobile unit is carefully controlled so that all signals are received at the base station with equal power. For example, in the cdmaOne system, mobile unit transmitter power is adjusted in 1 dB steps every millisecond to maintain received power equality. For the case in which the desired and interfering signals are received at equal power levels, Equation (6.43) becomes

$$SNIR = 2\frac{T_b}{T_{chip}}\frac{1}{1+\frac{N_0}{P_sT_{chip}}}. \tag{6.44}$$

When there are more than two interfering signals, the powers of all of the interference components can be added. If we suppose that K stations share a carrier frequency, each of the stations uses its own spreading code, and all of the signals are received at equal power, then

$$SNIR = 2\frac{T_b}{T_{chip}}\frac{1}{(K-1)+\frac{N_0}{P_sT_{chip}}}. \tag{6.45}$$

We see that in a direct-sequence multiple-access system the number of users is ultimately limited by the signal-to-interference ratio, just as in a frequency-hopping multiple-access system.

When direct-sequence spread spectrum is used for code-division multiple access in a cellular telephone system, the maximum number of subscribers depends primarily on the allowable signal-to-interference level. Several steps can be taken to reduce the level of interference. First and most important is power control. As we have mentioned in the discussion leading to Equation (6.44), it is essential to ensure that all mobile signals are received by the base station at nearly equal power levels. Power control is not as critical an issue in the forward channel, as signals to all mobile units are sent at the same power level from a single transmitter. A second technique for reducing interference is sectoring. If a cell is divided into three 120° sectors by directional base station antennas, then the number of interfering signals is reduced by a factor of 3. Sectoring thus virtually allows the capacity of the cellular system to be tripled. A third technique for interference mitigation is voice activity detection. It turns out that a speaker in a telephone conversation is active only about 3/8 of the time. Thus, if transmission from each user is interrupted during quiet intervals, the interference power will be reduced by a factor of about 3/8. In practice, a decrease in interference power by a factor of about 2 is realized.

Example

Gilhousen et al.[4] analyzed the capacity of a direct-sequence CDMA cellular system. They assumed a single 1.25 MHz channel, a bit rate of 8 kbits/s from the speech encoder, three-sector cells, and a voice activity factor of 3/8. They characterized propagation by a path-loss exponent of 4 and log-normal fading with a standard deviation of 8 dB. They assumed that with two receiving antennas and an error-correcting code the reverse link requires an E_b/N_0 of 7 dB to produce a probability of error of at most 10^{-3} at least 99% of the time. On the forward link a pilot carrier can be transmitted on a control channel to synchronize the receivers, allowing E_b/N_0 to be reduced to 5 dB for the same performance. By modeling the interference from users within a cell, and also from users in nearby cells, they determined that the reverse link can support 108 users/cell. The forward link can support 114 users/cell.

The same study compared direct-sequence CDMA with the AMPS system that uses frequency-division multiple access. For AMPS, a 1.25 MHz band can support just under 42 30 kHz channels. Assuming seven-cell frequency reuse with three sectors per cell, there is a maximum of 6 users per cell. Extending the analysis to USDC, time-division multiple access allows 3 users to be combined in each channel. This gives a total of 18 users per cell. The CDMA system exceeds this capacity by a factor of 6.

Example

Suppose that the reverse link of a certain cellular system has the following attributes for a mobile unit at the cell boundary:

---

4. K. S. Gilhousen, I. M. Jacobs, R. Padovani, A. J. Viterbi, L. A. Weaver, and C. E. Wheatley III, "On the Capacity of a Cellular CDMA System," *IEEE Transactions on Vehicular Technology* 40, no. 2 (May 1991): 303–12.

- Transmitted power $P_t\big|_{dB} = 23$ dBm
- Transmitting antenna gain $G_t\big|_{dB} = 0$ dBi
- Receiving antenna gain $G_r\big|_{dB} = 9$ dBi
- Average path loss $L_{path}\big|_{dB} = 135$ dB
- Connector losses $L_c\big|_{dB} = 2$ dB
- Fade margin $M_{fade}\big|_{dB} = 8$ dB
- Receiver noise figure $F\big|_{dB} = 5$ dB
- Data rate $R = 9600$ bits/s
- Spreading code chip rate $R_{chip} = 1.2 \times 10^6$ chips/second
- Factor for interference from other cells $f = 0.6$
- Voice activity factor $\alpha = 0.45$
- Maximum probability of error $P_e = 10^{-3}$

Direct-sequence spread spectrum is used for code-division multiple access. Assume that the modulation is binary phase-shift keying using rectangular pulses. Find the number of subscribers that can be supported per sector.

Solution

For binary phase-shift keying we can relate the probability of error to signal-to-noise-and-interference ratio at the matched filter output by

$$P_e = Q\left(\sqrt{SNIR}\right). \tag{6.46}$$

Substituting $P_e = 10^{-3}$ gives $SNIR = 9.55$. This gives the minimum SNIR that the receiver requires.

Next let us calculate the received signal power. From the given data

$$\begin{aligned} P_s\big|_{dB} &= P_t\big|_{dB} + G_t\big|_{dB} + G_r\big|_{dB} - L_{path}\big|_{dB} - L_c\big|_{dB} - M_{fade}\big|_{dB} \\ &= 23 \text{ dBm} + 0 \text{ dB} + 9 \text{ dB} - 135 \text{ dB} - 2 \text{ dB} - 8 \text{ dB} \\ &= -113 \text{ dBm}, \end{aligned} \tag{6.47}$$

that is,

$$P_s = 5.01 \times 10^{-15} \text{ W}. \tag{6.48}$$

The thermal noise power spectrum referred to the receiver input is

$$\begin{aligned} \frac{N_0}{2} &= \frac{kT_0}{2}F \\ &= \frac{1.38 \times 10^{-23} \times 290}{2} 3.16 = \frac{12.7 \times 10^{-21}}{2} \text{ W/Hz}, \end{aligned} \tag{6.49}$$

where the noise figure of $F\big|_{dB} = 5$ dB has been converted out of decibels to give $F = 3.16$.

Finally, using $T_b = \frac{1}{9600}$ seconds and $T_{chip} = \frac{1}{1.2\times10^6}$ seconds we can substitute in Equation (6.45) to obtain

$$SNIR = 2\frac{T_b}{T_{chip}}\frac{1}{(K-1)(1+f)\alpha + \frac{N_0}{P_s T_{chip}}}$$
$$9.55 = 2\frac{1.2\times10^6}{9600}\frac{1}{(K-1)(1.6)(0.45) + \frac{12.7\times10^{-21}}{5.01\times10^{-15}}1.2\times10^6}. \qquad (6.50)$$

In Equation (6.50) we have multiplied the term representing the total interference power by the factor $(1+f)$ to account for interference from other cells, and by the factor α to account for voice activity of less than 100%. Solving gives

$$K = 33.1 \qquad (6.51)$$

users per sector.

We conclude by noting that a three-sector-per-cell organization will provide for 99 users per cell. We also note that direct-sequence CDMA can allow all frequency assignments to be reused in every cell. Recall from Chapter 4 that the number of users is larger when the cluster size is smaller. Using a cluster size of one gives an additional enhancement to the system capacity.

Contention-Based Multiple Access

In our discussions up to this point we have assumed that access to a channel is granted to a user for an extended duration, and that this access is exclusive. In other words, whether a "channel" is provided by frequency division, time division, or code division, a single user holds possession of that channel for the duration of a call. For direct-sequence CDMA systems we saw that it can be advantageous for a transmitter to cease transmission during pauses in voice activity, but the channel is not reassigned during these pauses. This strategy of long-term channel assignment is based on a "circuit-switching" principle that corresponds to provision of a "circuit" to support a telephone call in the wired telephone network. This strategy makes sense for transmission of voice, video, and other "streaming" signals, in which a user spends a significant fraction of the call duration transmitting data. There are other applications, however, in which data transmission is very different and an alternative switching strategy makes sense.

Consider a data application in which a user transmits short bursts of data at infrequent and irregular intervals. There are many applications for which this kind of bursty transmission is typical. Transmission of e-mail, browsing the World Wide Web, and terminal-based computer activity are applications of this sort. In cellular telephone applications, use of the control (particularly the paging and access) channels is a bursty transmission activity. A mobile unit, when answering or placing a call, sends to the base station a packet of data containing identification and authorization information and perhaps a destination telephone number. These packets are sent only during call setup and hence are sent infrequently. All of these application scenarios share a common feature: It is very wasteful of communication resources to assign a channel on an exclusive long-term basis to a user who sends infrequent, short bursts of data.

The common strategy for providing communication service to users who transmit and receive data in infrequent bursts is to require a number of such users to share a communication channel. User data is presented as "packets," and each packet includes destination and source addresses so that it can be delivered correctly. In many systems there is a maximum allowable packet size, to prevent a single user from hogging the channel for an extended period. This "packet-switching" model characterizes transmission over the Internet, wireless data networks such as those based on the IEEE 802.11 standards, and the control channels of cellular telephone systems.

A major issue for packet-switching systems is how to provide unsynchronized and infrequent users with access to a communication channel in an orderly way. In this section we describe a family of multiple-access methods derived from a simple and elegant procedure called the Aloha protocol first put into practice at the University of Hawaii in the early 1970s. These multiple-access methods are said to be "contention based" for reasons that will become apparent. They are very effective for combining users with bursty data but are most efficient when the aggregate of users only lightly loads the communication channel. We will describe three variations on the contention-based multiple-access theme. We first describe the original Aloha protocol. Then we describe a variation known as "slotted Aloha" that has found application in cellular telephone networks. Finally we describe the "carrier sense" variations that form the basis for wireless data networks. For each of these variations, we will carry out some analysis to determine the efficiency with which the communication channel is used.

The Aloha Multiple-Access Protocol

The Aloha Network was developed at the University of Hawaii to provide data access to the university's central computer from a number of remote sites. Two radio channels were provided: a downlink to the remote sites and an uplink to the central computer. The multiple-access protocol was needed on the uplink, where there were several stations supporting users with data to send at unpredictable times. The access protocol that was developed for this network is known as the "Aloha" or "pure Aloha" protocol.

An Aloha protocol is very simple. A station (user) with data to send just sends it. Every packet received on the uplink is acknowledged by transmission of a short acknowledgment packet on the downlink. When a station transmits a packet, that station starts a timer. The timer is set so that the acknowledgment should be received before the timer expires, at which point the timer is stopped. If the timer times out, the station assumes that its packet was not received and retransmits it.

Since stations transmitting on the uplink are uncoordinated, it will occasionally happen that two or more stations transmit at once, or nearly at once, so that their packets overlap at the receiver. This occurrence is called a "collision," and a collision normally results in incorrect reception of all of the transmitted packets. In this case there will be no acknowledgments, and all of the stations involved in the collision will time out and retransmit.

Now if two stations transmit at nearly the same time, causing a collision, and both stations time out and retransmit, some mechanism must be provided to ensure that a second collision does not occur. The Aloha protocol requires a station not receiving an acknowledgment to follow a "backoff" procedure. When a station's timer expires, the station generates a random number generated uniformly from a small range of possible values. The station waits ("backs off") a number of time units specified by the random number and then retransmits. A second collision might still occur, but then the procedure is repeated. Eventually the stations involved in a collision will choose random numbers that are sufficiently different that the retransmissions do not collide. Random numbers are used in the backoff procedure instead of fixed assignments in the interests of fairness. Every station has an equal chance of getting its retransmission in first.

We can estimate the efficiency of an Aloha network as follows. Suppose the network has a large number of users who offer packets for transmission at random times and without collusion. Suppose for simplicity that all packets have the same T-second duration. The "offered load" is defined as the number of packets offered to the link for transmission by all users during a T-second interval. The offered load includes packets carrying new data and also packets that are being retransmitted because of collisions and errors. The random backoff procedure ensures that retransmission times are not correlated with the original transmission times, so retransmitted packets are indistinguishable from new packets to the network. Let the offered load be designated by G. The units of offered load are "packets per packet time," or erlangs in telephone terminology.

In Figure 6.16 we show a packet, designated packet 1, transmitted at time t_0. Packet 1 is shown colliding with two other packets. Packet 2 was in progress when packet 1 began transmission, and packet 3 is beginning transmission while packet 1 is still in progress. Now to avoid collisions of the packet 1–packet 2 variety, no packet can be offered for transmission during the T-second interval immediately preceding time t_0. Similarly, to avoid collisions of the packet 1–packet 3 variety, no packet can be offered for transmission during the T-second interval that immediately follows t_0. Thus successful transmission requires that there be an interval of $2T$ seconds during which no packets are offered to the link for transmission.

To calculate the probability that $2T$ seconds will elapse without an offered packet, we need a model for packet arrivals. It turns out that we can use the same Poisson process model that we

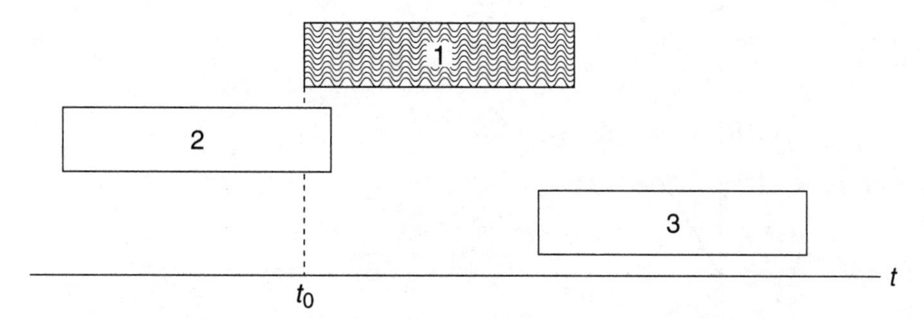

Figure 6.16 Packets 2 and 3 Colliding with Packet 1

introduced in Chapter 4 and described in detail in Appendix B to represent call arrivals at a telephone switch. Equation (B.1) gives us the probability P_k that k packets arrive during an interval of Δt seconds. Repeating the formula here,

$$P_k = \frac{e^{-\lambda \Delta t}(\lambda \Delta t)^k}{k!}, \ k = 0, 1, 2, \ldots, \tag{6.52}$$

where λ is the average packet arrival rate in packets per second. If we substitute $\lambda = G/T$ and $\Delta t = 2T$, we can find the probability $P[success]$ that there is no collision when packet 1 is transmitted. We have

$$P[success] = P_0 = \frac{e^{-\frac{G}{T}2T}(\frac{G}{T}2T)^0}{0!} = e^{-2G}. \tag{6.53}$$

Given the offered load and the probability of success, we can find the rate at which packets are successfully delivered to their destinations. This quantity, which we will designate S, is called the "throughput" by computer network engineers and the "carried load" by telephone engineers. The units of S are the same as those of offered load, packets per packet time, or erlangs. We have

$$\begin{aligned} S &= G \cdot P[success] \\ &= Ge^{-2G}. \end{aligned} \tag{6.54}$$

A plot of throughput versus offered load is shown in Figure 6.17. We see that the throughput is at a maximum when $G = 0.5$. At this offered load the maximum throughput is $S = 0.5e^{-2\times 0.5} = 0.184$. Thus an Aloha network is at most about 18% efficient. For $G \ll 0.5$throughput is low because the link is underutilized. For $G \gg 0.5$ the link is congested, and throughput is low because of an excess of collisions.

The Slotted Aloha Protocol

The Aloha protocol has the advantage of simplicity but the disadvantage of making very inefficient use of the communication channel. One enhancement, known as "slotted Aloha," assumes

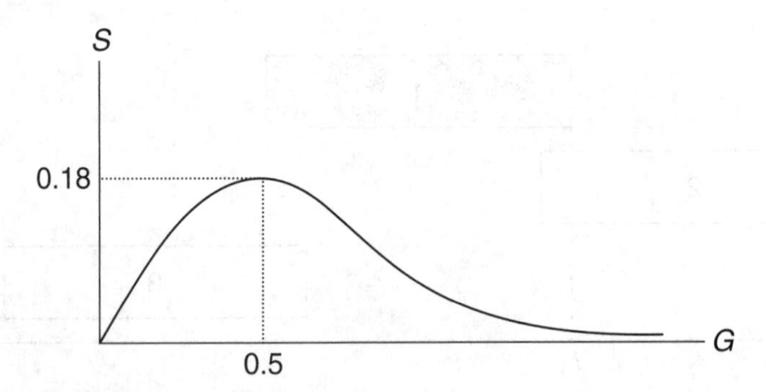

Figure 6.17 Throughput versus Offered Load for an Aloha Network

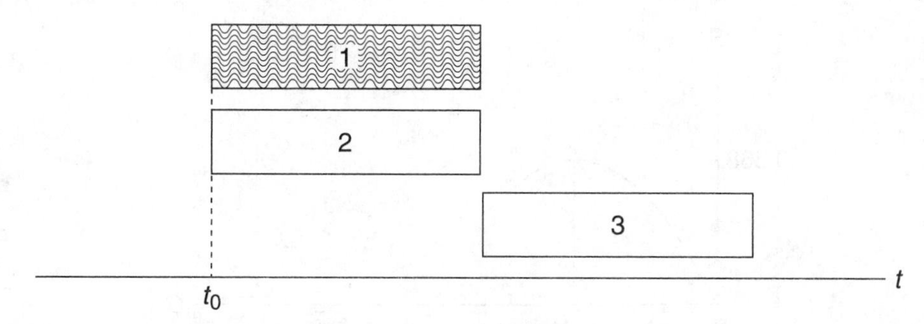

Figure 6.18 Packet 2 Colliding with Packet 1 in a Slotted Aloha System

that the time axis is divided into "slots" of length T. Transmission of a packet may commence only at the beginning of a slot. Acknowledgments and backoff are handled exactly as for a pure Aloha system.

To evaluate the efficiency of a slotted Aloha system, consider the scenario shown in Figure 6.18. In this figure the time t_0 represents the beginning of a slot. In this scenario, packet 2 "arrived" for transmission during the T-second interval preceding t_0. Transmission was delayed until the beginning of the slot, and packet 2 is shown colliding with packet 1. Packet 3, on the other hand, "arrived" during transmission of packet 1. Transmission was delayed until the start of the next slot, and packet 3 does not collide with packet 1. We see that successful transmission of packet 1 requires that there be no packets like packet 2. Packets like packet 3 do not impact the transmission of packet 1. In other words, we require that no packets be offered for transmission during the T-second interval preceding t_0.

As in the pure Aloha case, we use Equation (6.52) to evaluate the probability of a successful transmission. Substituting $\lambda = G/T$ and $\Delta t = T$ gives the probability of no collisions as

$$P[success] = P_0 = \frac{e^{-\frac{G}{T}T}\left(\frac{G}{T}T\right)^0}{0!} = e^{-G}. \tag{6.55}$$

In this case the throughput or carried load is

$$\begin{aligned} S &= G \cdot P[success] \\ &= Ge^{-G}. \end{aligned} \tag{6.56}$$

Throughput is plotted against offered load for slotted Aloha in Figure 6.19. The slotted Aloha system has the added complexity of requiring synchronization of all of the stations to a common slot clock, but in return, the maximum throughput is about 37%, double that of a pure Aloha system.

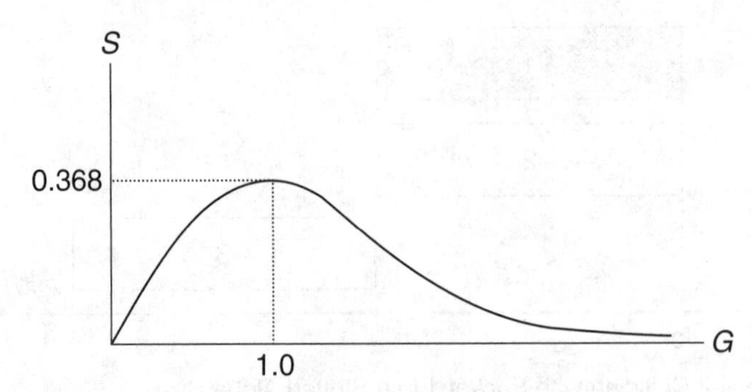

Figure 6.19 Throughput versus Offered Load for a Slotted Aloha Network

---

Example

Suppose in a slotted Aloha system that the offered load is $G = 1$ erlang. To an outside observer, the probability of seeing an empty slot is $P_0 = e^{-G} = e^{-1} = 0.368$. The probability of seeing a slot carrying only a single packet is $P_1 = \frac{e^{-G}G^1}{1!} = Ge^{-G} = S = 0.368$. Then the probability of seeing a slot containing a collision is $1 - 2 \times 0.368 = 0.264$. Thus at maximum throughput, about one-quarter of all of the slots carry collisions.

The slotted Aloha protocol is used for control channel access in GSM, USDC, and cdmaOne second-generation cellular systems. Variations on slotted Aloha can be found in third-generation cellular systems both for control channel access and for transmission of packet-based data.

Carrier-Sense Multiple Access

The likelihood of collision in an Aloha system can be reduced if stations with packets to send listen to the channel before transmitting. This operation is called "carrier sense." If the channel is in use, a station must defer before transmitting. A system that implements carrier sense and deferring is called a **carrier-sense multiple-access** (CSMA) system. Several disciplines govern how long to defer and what to do afterward. These disciplines differ in how aggressive a station is when trying to gain access to the communication channel.

- The most aggressive discipline is known as "1-persistent" CSMA. A station with a packet to send senses the channel. If the channel is idle, the packet is transmitted. If the channel is busy, the station defers and monitors the channel continuously until it becomes available. When the channel becomes available, the station immediately transmits its packet. The 1-persistent discipline can be applied both to pure and to slotted systems.
- The least aggressive discipline is known as "nonpersistent" CSMA. A station with a packet to send senses the channel. If the channel is idle, the packet is transmitted. If the channel is busy, the station uses the backoff procedure to delay a random length of time. After the backoff, the station senses the channel and repeats the procedure. This discipline can also be applied to either pure or slotted systems.

- The "p-persistent" CSMA system is applied only to slotted systems. A station with a packet to send senses the channel. If the channel is busy, the station waits until the next slot and repeats the procedure. If the channel is idle, the packet is transmitted with probability p. This means that with probability $(1-p)$ the station waits until the next slot and repeats the procedure.

The 1-persistent CSMA discipline offers the least delay in gaining access to the channel when the offered load is very light. When the load is heavy, there will nearly always be a collision when the channel becomes idle following a transmission. It is typical for a transmission to be followed by a "contention period" with multiple collisions. Eventually only one station will transmit. The successful transmission will be followed by another contention period. Nonpersistent and p-persistent CSMA are intended to spread out transmission attempts so that collisions are less likely. The p-persistent discipline can be more like 1-persistent or more like nonpersistent CSMA, depending on whether the value of p is closer to 1 or closer to 0. As an example of the use of these access disciplines, 1-persistent CSMA is the basis of the Ethernet wire-network protocol. A variation on nonpersistent CSMA is used in the IEEE 802.11 (WiFi) local-area network standard. Finally, p-persistent CSMA is used in the **packet reservation multiple access** (PRMA) system used for combining voice and data in cellular telephone systems.

To illustrate how CSMA improves efficiency over Aloha, we can estimate the efficiency of a CSMA system.[5] Nonpersistent CSMA is the simplest case and the only one we will consider. The random deferral in the nonpersistent case avoids the bunching up of offered packets at the end of a transmission. This allows us to represent the process of packets arriving for transmission as a Poisson process, without having to treat either deferrals or retransmissions in any special way. We will continue to assume that all packets have the same length T. An outside observer would observe that the communication channel experiences busy intervals followed by idle intervals. During a busy interval at most one packet can be transmitted successfully. We will estimate the average duration of a busy interval and the average duration of an idle interval. This will allow us to calculate the average packet transmission rate. We can also calculate the probability that a transmitted packet is successful. This will allow us to calculate the throughput. Let the maximum propagation delay of the wireless channel be t_p. A transmitted packet is vulnerable to collisions only for t_p seconds after the start of transmission. After this time all other stations will hear the transmitted carrier and defer.

Figure 6.20 shows a collision scenario. Referring to the figure, suppose a busy interval begins at time t_0. The transmitted packet, packet 1, lasts T seconds. If there is a collision, the colliding packet will add an additional Δt seconds to the busy interval. Now

$$\begin{aligned} P[\Delta t<\tau] &= P[\text{no packets arrive for transmission between } t_0+\tau \text{ and } t_0+t_p] \\ &= e^{-\frac{G}{T}(t_p-\tau)}, \text{ for } 0\le\tau\le t_p. \end{aligned} \tag{6.57}$$

5. See S. Haykin and M. Moher, *Modern Wireless Communications* (Upper Saddle River, NJ: Prentice Hall, 2005), 245–48.

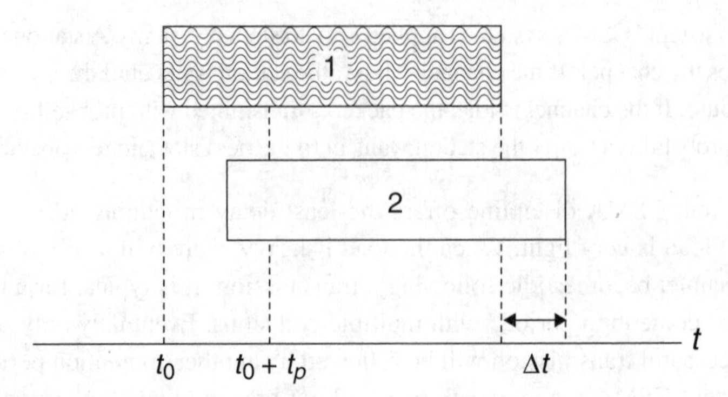

Figure 6.20 Packet 2 Colliding with Packet 1

Then the probability density function for Δt is

$$\begin{aligned} p_{\Delta t}(\tau) &= \frac{dP[\Delta t < \tau]}{d\tau} \\ &= \frac{G}{T} e^{-\frac{G}{T}(t_p - \tau)}. \end{aligned} \tag{6.58}$$

The average value of Δt is then

$$\begin{aligned} \overline{\Delta t} &= \int_0^{t_p} \tau \frac{G}{T} e^{-\frac{G}{T}(t_p - \tau)} d\tau \\ &= t_p - \frac{T}{G}\left(1 - e^{-\frac{G}{T}t_p}\right). \end{aligned} \tag{6.59}$$

Finally, once transmission of packet 1 and any colliding packets ends, the channel remains busy for t_p seconds while the various packets propagate through the link. The total average length of a busy interval is

$$\begin{aligned} \overline{T}_{busy} &= T + t_p - \frac{T}{G}\left(1 - e^{-\frac{G}{T}t_p}\right) + t_p \\ &= T + 2t_p - \frac{T}{G}\left(1 - e^{-\frac{G}{T}t_p}\right). \end{aligned} \tag{6.60}$$

We now estimate the average duration of an idle interval. Suppose a busy interval ends at time t_1. An idle interval begins and lasts until transmission of the next packet. Let the idle interval have duration T_{idle}. The probability that the idle interval lasts longer than τ is the probability that no packets arrive for transmission between t_1 and $t_1 + \tau$, that is,

$$P[T_{idle} \geq \tau] = e^{-\frac{G}{T}\tau}. \tag{6.61}$$

Then

$$P[T_{idle} < \tau] = 1 - e^{-\frac{G}{T}\tau}. \tag{6.62}$$

The probability density function for T_{idle} is given by

$$\begin{aligned} p_{T_{idle}}(\tau) &= \frac{dP[T_{idle} < \tau]}{d\tau} \\ &= \frac{G}{T} e^{-\frac{G}{T}\tau}. \end{aligned} \tag{6.63}$$

The average value of T_{idle} is given by

$$\begin{aligned} \bar{T}_{idle} &= \int_0^{\infty} \tau \frac{G}{T} e^{-\frac{G}{T}\tau} d\tau \\ &= \frac{T}{G}. \end{aligned} \tag{6.64}$$

Now refer back to Figure 6.20. In that figure only frames of type 1 can be successful. Frames of type 2 are in collision and cannot be successful. Every busy interval begins with transmission of a frame of type 1. The average transmission rate R_{avg} is

$$\begin{aligned} R_{avg} &= \frac{1}{\bar{T}_{busy} + \bar{T}_{idle}} \\ &= \frac{1}{T + 2t_p - \frac{T}{G}\left(1 - e^{-\frac{G}{T}t_p}\right) + \frac{T}{G}} \\ &= \frac{1}{T + 2t_p + \frac{T}{G} e^{-\frac{G}{T}t_p}} \text{ packets/second.} \end{aligned} \tag{6.65}$$

Multiplying by T seconds/packet gives the average transmission rate of type 1 frames in erlangs:

$$R_{0avg} = \frac{1}{1 + 2\frac{t_p}{T} + \frac{1}{G} e^{-\frac{G}{T}t_p}} \text{ erlangs.} \tag{6.66}$$

The probability that a type 1 frame is successful is the probability that no frame arrives for transmission during the t_p-second interval following time t_0, that is,

$$P[success] = e^{-\frac{G}{T}t_p}. \tag{6.67}$$

We can now calculate the throughput S as

$$
\begin{aligned}
S &= R_{0avg} P[success] \\
&= \frac{e^{-\frac{G}{T}t_p}}{1+2\frac{t_p}{T}+\frac{1}{G}e^{-\frac{G}{T}t_p}} \\
&= \frac{G}{1+G\left(1+2\frac{t_p}{T}\right)e^{G\frac{t_p}{T}}}
\end{aligned}
\tag{6.68}
$$

The throughput depends on the offered load G and on the ratio of propagation delay to packet length.

When the propagation delay t_p is a very small fraction of the packet duration T, the throughput can be very high. In fact, the only reason that S does not approach unity in this case is that the nonpersistent deferral strategy always produces some idle time. When propagation delay is large, however, the throughput can be very poor. For this reason wireless networks using a CSMA strategy are usually limited in geographic size. Figure 6.21 shows throughput plotted against offered load for several values of the ratio t_p/T. The Aloha curve from Figure 6.17 is included for comparison.

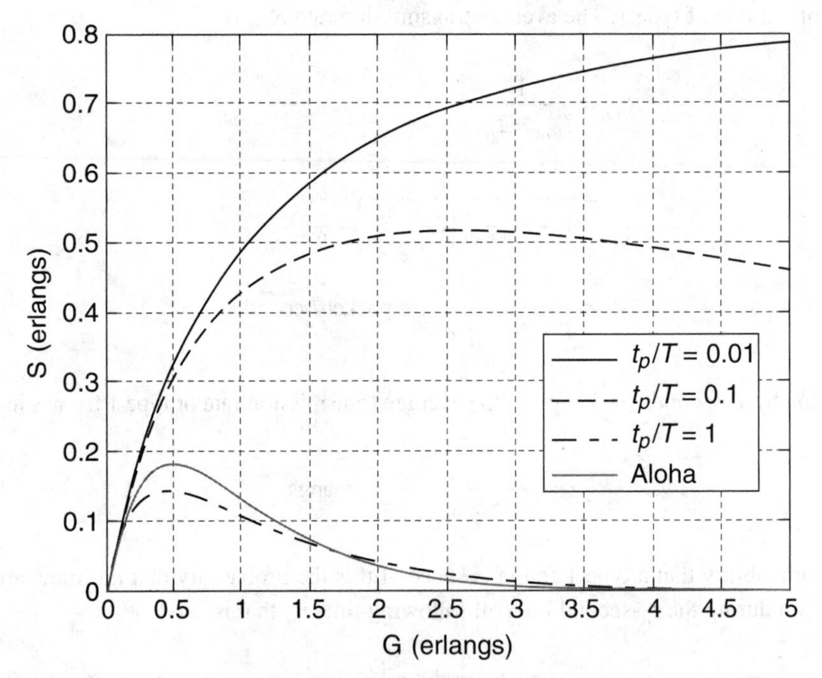

Figure 6.21 Throughput versus Offered Load for CSMA for Several Values of t_p/T

Example

Consider a wireless network that transmits 1500-byte packets at a rate of 10 Mbits/s. The offered load is $G = 1$ erlang. Find the throughput if the maximum transmission distance is 30 m. Repeat for a distance of 1 km. Finally, repeat for a "national network" of 5000 km.

Solution

For 1500-byte packets at 10 Mbits/s, the packet duration T is given by $T = 1500 \times 8/10\times10^6 = 1.20$ ms. For a propagation distance of 30 m, the propagation delay t_p is $t_p = \frac{30 \text{ m}}{3\times10^8 \text{ m/s}} = 100.0$ ns. Substituting in Equation (6.68) gives $S = 0.5$ erlangs.

For a propagation distance of 1 km we have $S = 0.498$ erlangs. For 5000 km, $S = 32.3\times10^{-9}$. This example makes it clear why we do not have national CSMA networks!

In a wireless link it is possible that all of the stations cannot hear each other. A station may sense the channel and not hear the carrier of another station that is transmitting. Unfortunately, the receiver may hear both stations, resulting in a collision. This is called the "hidden station" problem. The IEEE 802.11 wireless networking standard includes a number of specialized techniques for dealing with the hidden station problem.

When the data link is from a number of mobile units to a single base station, the base station can send a "busy" or "idle" signal on a control channel that can be monitored by all of the mobile units. When any mobile unit is transmitting, the base station signals "busy." Another mobile unit with a packet to send detects the "busy" signal rather than the carrier of the station that is actually transmitting. This access technique is sometimes known as **data-sense multiple access**. Data-sense multiple access was used in the **cellular digital packet data** standard introduced in 1993 to allow first- and second-generation cellular telephone systems to provide data services.

On wire networks such as Ethernet, efficiency can be further improved if stations stop transmitting as soon as a collision is detected. Detecting a collision requires the ability to receive on the transmitting frequency while transmitting, however. As this ability is not common on wireless links, the collision detection option is not available in wireless applications. In Ethernet systems the technique is known as **carrier-sense multiple access with collision detection**, or CSMA/CD.

Conclusions

Wireless systems such as cellular telephone systems and wireless data networks have the important common attribute that multiple users must share access to the communication medium. An important task for a systems engineer is to maximize the number of users to whom simultaneous service of acceptable quality can be provided. In this chapter we have examined a number of techniques for implementing shared access and compared their relative merits. Our focus for most of the chapter was on methods for providing channels that a subscriber would use for extended periods of time, such as for the duration of a telephone call. In the last section we considered ways of sharing a single channel among users transmitting sporadic, bursty data.

We began the chapter with a high-level overview of how a subscriber obtains access to a voice channel in a cellular system. We described the use of a reverse control channel (universally called an access channel) for this purpose. The access method, at least when described in general terms, is independent of the specific multiplexing scheme that is used to provide voice channels for multiple simultaneous users.

Next we explored three techniques for creating multiple user channels in an overall system bandwidth allocation. The first technique we examined was frequency-division multiple access. Frequency-division schemes date back almost to the beginnings of radio. The overall system bandwidth is divided into disjoint frequency subbands, with each such subband available for assignment to an individual user. The width of each subband, and hence the number of subbands that can be provided, is determined primarily by the bandwidth of the users' modulated signals. Although it seems that the system bandwidth divided by the user bandwidth should give the number of users that can be accommodated, several inefficiencies reduce the maximum number of users in practice. These inefficiencies are primarily a consequence of adjacent-channel interference. The amount of interference experienced is determined by the spectral shaping of the transmitted signals and by the selectivity of the filters used in the receivers.

To illustrate what kind of user density is achievable in practice, we presented several examples of frequency-division schemes. We examined the frequency allocations used in AM radio, an application in which minimizing receiver cost is a primary concern. Next we examined the frequency-division multiplexing scheme developed by AT&T for long-distance telephone communication. This scheme used single-sideband modulation and was characterized by a high level of efficiency. Finally, we reviewed the AMPS first-generation cellular telephone system, whose channel allocations were introduced in Chapter 4.

The second multiple-access method that we examined was time-division multiple access. In this method time is divided into slots, and a number of users take turns using the slots in rotation. Since the users are transmitting sampled data, the slot provided to a given user must become available with sufficient frequency. We saw that application of time-division multiple access to cellular telephone systems did not become practical until speech-processing technology became sufficiently advanced that speech could be digitized at rates below 10 kbits/s.

As is the case with frequency-division multiple access, time-division multiple access is subject to inefficiencies that limit the number of users. These inefficiencies manifest themselves as time slots that cannot be assigned to user data because they are needed for synchronization, transmitter "ramp-up," and other forms of overhead. As illustrations of practical systems we examined time-division multiple access in the voice channels of USDC and GSM cellular systems.

The third multiple-access method, code-division multiple access, offers a higher user density in today's technology than either frequency-division or time-division multiple access and is therefore the method of choice for third-generation cellular telephone systems. We examined frequency-hopping CDMA and direct-sequence CDMA. Direct-sequence CDMA was presented in more detail, as this method is used in both the second-generation cdmaOne and the third-generation cdma2000 and W-CDMA systems. We introduced the Walsh functions as orthogonal spreading codes, and we examined the use of pseudonoise waveforms as nearly orthogonal spreading codes. In the latter case we developed the relation between the number of users and the signal-to-noise-and-interference ratio. This relation allowed us to calculate the number of users that a given system can support.

In the last section of this chapter we examined a different kind of multiple-access problem. In this section we introduced the Aloha, slotted Aloha, and CSMA techniques that allow users generating infrequent bursty data to share a single communication channel. These multiple-access methods are known as "contention-based" methods, because the users compete directly for channel access. The slotted Aloha system has been widely used to provide access to the control channels in cellular telephone systems.

In describing the contention-based access methods we have subtly changed the focus of our discussion of wireless systems. The operation and efficiency of methods such as Aloha, slotted Aloha, and CSMA are not strongly linked to the design of the physical link and can work well on wireless links and on wired links of various kinds. A systems engineer would say that we have moved up a "layer" in our discussion. Providing power to overcome noise, shadowing, and fast fading; designing modulation methods to convey bits with an acceptable error rate and avoid adjacent-channel interference; accommodating many users over a wide geographic area; and providing multiple channels in a given spectral allocation are all aspects of system design at the "physical" layer. The contention-based multiple-access methods, as well as the use of control channels to arrange assignment of traffic channels, are aspects of system design at the "medium-access control" layer. We will encounter layering considerations in more detail in the next chapter, when we examine the kinds of user information that are communicated over wireless systems, how quality of service is specified, and how user information can be packaged for transmission.

Problems

Problem 6.1

When a call is initiated on a wired telephone system, access to the switch is obtained as soon as the handset goes "off hook." The switch receives the dial tones one at a time and proceeds to connect the call when it has all of the necessary digits. In a cellular telephone system, the destination telephone number is keyed into the handset, and access to the switch is not attempted until the "send" key is pressed. Speculate intelligently on why the basic procedure for accessing the switch was changed for cellular telephone systems.

Problem 6.2

What is the spectrum allocation for the FM broadcast band? What is the channel bandwidth of a single FM station? How close together are FM channels actually assigned in a given location? How many FM broadcast stations are active in your location? You may need to refer to the FCC Web site (www.fcc.gov) for the information you need to answer these questions.

Problem 6.3

Suppose 400 MHz of spectrum is allocated to a wireless service in which users transmit in 200 kHz channels.

A. How many users can be serviced if a guard band of 50 kHz must be provided between each pair of channels?

B. How many users can be serviced if there are no guard bands, but only every second channel can be used?

Problem 6.4

The FCC spectrum allocation for mobile telephone service consists of the bands 824–849 MHz for reverse channels and 869–894 MHz for forward channels. Suppose a system consists of one 1.25 MHz CDMA channel and as many 30 kHz AMPS channels as possible. Remember that a guard band of nine AMPS channels must be set aside on each side of the CDMA channel. How many AMPS channels can the spectrum allocation support?

Problem 6.5

In a certain frequency-division multiple-access system the transmitters transmit data using QPSK modulation with a data rate of 400 kbits/s. Rectangular pulses are used with no filtering at the transmitter. Channels are spaced every 400 kHz. Suppose the receivers use highly selective filters that we can take as ideal brick-wall filters, each having a bandwidth of 400 kHz centered on the appropriate carrier frequency.

Suppose a receiver receives two signals. One of these is the desired signal, and the other has a carrier frequency n channels away from the carrier frequency of the desired signal. The interfering signal is 15 dB stronger than the desired signal.

How large must we make n so that the signal-to-interference ratio is at least 20 dB?

Problem 6.6

Suppose a voice signal having an absolute bandwidth of 4 kHz is sampled at 8000 samples/second, and each sample is quantized to eight bits. Suppose the resulting bit stream is encoded using a 2B1Q line code.

A. If the line code uses rectangular pulses, find the first-null bandwidth.

B. If the line-code pulse shape is designed for minimum bandwidth, what will be the shape of the pulses? What is the minimum bandwidth?

C. Find the bandwidth if the line code uses raised-cosine pulses with rolloff parameter $\alpha = 0.25$.

Problem 6.7

In a USDC time slot, what percentage of the time is available for transmission of user data, exclusive of error control bits? What percentage of the slot time is "overhead" (i.e., everything else)?

Problem 6.8

In a GSM time slot carrying full-rate speech, what percentage of the time is available for transmission of user data, exclusive of error control bits? What percentage of the slot is "overhead" (i.e., everything else)?

Problem 6.9

A slow-hopping CDMA system uses noncoherent FSK modulation. There are 1000 hopping frequencies and 20 bits of data are transmitted per hop time. Plot probability of error versus E_b/N_0 for one, five, and ten users.

Problem 6.10

A frequency-hopped spread-spectrum system uses 50 kHz channels over a contiguous 20 MHz spectrum. Fast frequency hopping is used, where two hops occur for each data bit. Assume for simplicity that all users hop at the same time. If binary noncoherent FSK is the modulation method used, determine

A. The probability of error for a single user operating at $E_b/N_0 = 20$ dB
B. The probability of error for a user operating at $E_b/N_0 = 20$ dB with 20 other frequency-hopping users sharing the channels
C. The probability of error for a user at $E_b/N_0 = 20$ dB with 200 other frequency-hopping users sharing the channels

Problem 6.11

A. Find the Hadamard matrix H_8.
B. Find and plot the Walsh functions $W_0^8(t), \ldots W_7^8(t)$, assuming a polar NRZ format.

Problem 6.12

Find and plot the cross-correlation function $R_{37}(\tau)$ of Walsh functions $W_3^8(t)$ and $W_7^8(t)$. Assume the Walsh functions are represented in polar NRZ format.

Problem 6.13

Figure P 6.1 shows a four-stage shift register connected with exclusive-or gates to form a state machine that can generate a PN sequence. The three branches marked with an *X* may or may not be present, so the figure actually represents eight possible circuits. Imagine that the shift register is preloaded with any nonzero pattern of bits. As the register is clocked, the state machine will cycle through a sequence of states, periodically returning to its original state.

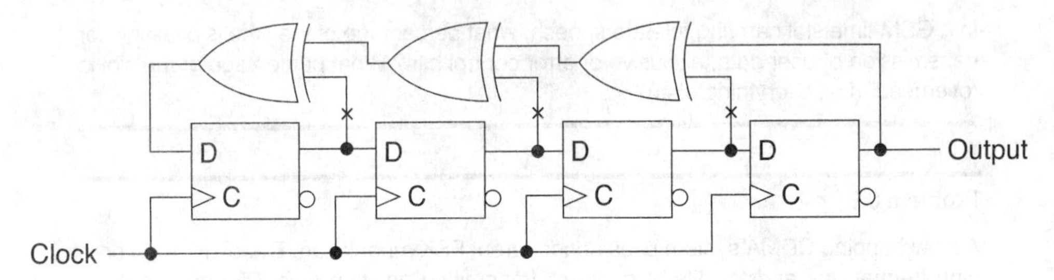

Figure P 6.1 A Four-Bit PN Sequence Generator

A. Prove that regardless of which of the *X* wires is present, the state machine will eventually return to its original state.
B. Find a set of *X* connections that will cause the state machine to cycle through all of the nonzero states (the order does not matter) before returning to the original state. Find the corresponding output sequence. This output is a PN sequence. (Note: There are two possible answers.)
C. What output is generated if the shift register is initially in the all-zero state?

Problem 6.14

The sequences p_1 and p_2 given below are PN sequences.

$$p_1 = 100010011010111$$

$$p_2 = 100011110101100$$

Represent each sequence in polar NRZ format, then find and plot the autocorrelation function $R_{11}(\tau)$ and the cross-correlation function $R_{12}(\tau)$. How nearly orthogonal are these two sequences?

Problem 6.15

A cellular system uses direct-sequence spread spectrum for CDMA. All of the users transmit on nearly the same carrier frequency, and each user is assigned a unique spreading code. Suppose that the spread-spectrum system has a processing gain of 511. Suppose further that there are 63 users transmitting simultaneously in the cell.

A. Find the signal-to-interference ratio at the output of the base station receiver. Ignore thermal noise and assume that signals from all the users are received at the same power level.

B. Now suppose that there are only 2 users. Suppose that the 2 users transmit at the same power level and the path-loss exponent is $\nu = 4$. One of the users is 3 km from the base station and the other is 0.5 km from the base station. The base station is attempting to demodulate the signal from the more distant user. Find the signal-to-interference ratio. Ignore thermal noise.

Problem 6.16

A certain contention-based multiple-access system uses a pure Aloha access protocol. Suppose the offered load is $G = 0.4$ erlang. An outside observer monitors the communication channel for T seconds, where T is the packet duration.

A. Find the probability that the observer sees no transmissions occur.

B. Find the probability that the observer sees a collision.

Problem 6.17

A certain contention-based multiple-access system uses a slotted Aloha access protocol. Users transmit at 64,000 bits/s and packets are 1000 bytes long. Find the maximum possible throughput in packets per second.

Problem 6.18

A contention-based wireless system uses a nonpersistent CSMA access protocol. Packets of 1000 bytes are transmitted at a 100 Mbit/s rate. The offered load is $G = 0.3$ erlang, and the propagation distance is 30 m. Find the throughput S in erlangs.

CHAPTER 7

Information Sources

Introduction

The ultimate purpose of any communication system is to facilitate information transfer between users. In most modern communication systems the information is digitally encoded, and so the ability to transfer bits between users is a prerequisite to the transfer of information. Up to this point we have focused on the system considerations necessary to establish this transfer of bits between users. In Chapters 2 and 3 we explored the factors a systems engineer would take into account to ensure that enough carrier power reaches a receiver to produce an adequate signal-to-noise ratio. The signal-to-noise ratio must be maintained even in the presence of shadowing and multipath reception. In Chapter 5 we examined techniques for modulating the carrier so that a stream of data bits could be communicated between users at a predictable bit error rate. Chapters 4 and 6 were concerned with the design of a communication service that can support a maximum number of users over a virtually unlimited geographical area.

The bit streams that the communication links described in Chapter 5 convey between users are simply undifferentiated sequences of bits. The link does not impose any structure on the bit stream other than framing needed for multiple access. The link is not aware of any information content that the bit stream may convey, and it functions from the users' point of view as a somewhat imperfect "bit pipe" between source and destination. As we show in this chapter, however, digitized information sources do impose structure on the bit streams they produce, and this structure must be recognized in the system design if the system is to provide the users with an acceptable quality of service.

Figure 7.1 shows a high-level view of a communication system, modeled as a data link connecting a source to a destination. There are many kinds of information sources, and in this chapter we examine a small set of examples, chosen to illustrate the variety of challenges that must be met in order to satisfactorily communicate information between users. Ultimately we want to explore how the users' expectations of service quality can be expressed in engineering terms and how these high-level service requirements impose design requirements on the physical data link.

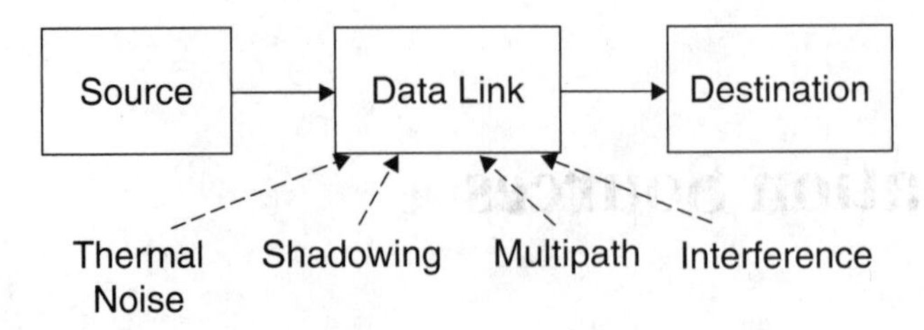

Figure 7.1 A Communication System Model

We show that performance requirements can be met in part by processing the signals generated by the information source and in part by appropriate design of the physical link.

The users of modern communication systems communicate information in the form of speech, music, images, video, and digital data. Even cellular telephone systems, which in the first generation were designed for communication of speech, are now used to communicate information in all of these forms. It must be recognized that even this list does not include all possible information sources. Our goal in this chapter is to illustrate some of the possibilities and not to provide a complete catalog of sources. It is interesting to note that four of the sources we have listed produce analog signals at the transducer output. Only the fifth, data, is inherently digital. As we discussed in Chapter 5, modern communication systems transmit information digitally, and so efficient digitization of the analog information signals is an important design issue. In this chapter we focus on digitization of speech as a detailed example. Space does not permit a complete treatment of the system design factors involved in digitization of all of the aforementioned sources. The reader is referred to specialized textbooks for more information.

When describing "satisfactory" communication of information from a source to a user, we must take into account both the characteristics of the source and the expectations of the user. For example, speech and music both produce audio signals at a microphone output, but a listener would probably expect a music signal to be reproduced with a significantly wider bandwidth than a speech signal would be expected to have. Further, the user's expectations might vary with time and experience. We would not be surprised to find that a listener who was delighted with a "high-fidelity" monaural music reproduction in 1955 might today be barely satisfied with "CD-quality" stereo. As a practical matter, users may have to be satisfied with a quality of information delivery that can be provided at an acceptable cost, although improvements in technology will often allow a source processor or data link to be redesigned to provide a more satisfactory level of user experience. Given the characteristics of the various sources, a systems engineer must translate the expectations of the user into detailed design requirements for the data link under the constraints of system cost and complexity.

If the characterizations of a source and of user experience are to be of help in designing a data link, they must be expressed in engineering terms. User satisfaction is often expressed in subjective terms, or in terms that are difficult to relate to system parameters. For instance, a telephone customer might feel that it is important to recognize Grandma's voice. A music listener might want the music to sound "like it does in a concert hall." Somehow these expectations must be translated into engineering terms; the engineering terms will ultimately determine requirements to be met by the data link. As examples of engineering terms, analog sources are often described in terms of bandwidth, signal-to-noise ratio, and dynamic range. User requirements may be described in terms of total harmonic distortion and maximum delay. Digital sources can be characterized by peak and average bit rate. "Bursty" sources may have a maximum burst duration. User requirements may include error rate, delay, and delay jitter.

The design requirements of the data link are often expressed as minimum performance guarantees. The term **quality of service** (QoS) is used to describe sets of parameter values that guarantee acceptable performance for particular sources and users. Typical QoS parameters at the link level might include average bit rate, maximum bit rate, bit error rate, delay, and delay jitter. It is the systems engineer's job to identify constraints on the physical link that may be expressed in terms of bandwidth and received power level, and then to design the link so that it can provide the necessary quality of service guarantees.

Ultimately the wireless network is an economic venture. Its capacity is meant to be used efficiently to generate revenue by providing services to users. This means that there will always be pressures to increase the number of users, increase the quality of information provided, and increase the range of services provided. Pressures to increase the range of services led to the introduction in the 1950s of data transmission over a public switched telephone network that had been designed and optimized for voice transmission. Similar pressures led to the introduction of data transmission over the cellular telephone network, which had also originally been designed for voice transmission. Systems engineers often find themselves generating new uses for legacy communication systems.

When providing an information link between a source and a user in an economical fashion, it is not always possible to convey to the user all of the information present in the source. A variety of methods of "source processing," "source coding," or "data compression" have been developed with the goal of easing the requirements on the data link without seriously compromising the user's experience. For analog sources bandwidth reduction is a common processing step. Speech signals that allow speaker recognition require much less bandwidth than music signals that give a concert-hall experience. Digital sources can be processed in a variety of ways to reduce the rate at which bits must be transmitted. For the systems engineer there is a trade-off between efforts to make the source processing more efficient and efforts to increase the quality of service of the data link.

As we have noted, most of the wireless data links being designed today are digital links. Digital links are generally less expensive than analog links to implement and maintain, they offer superior performance in the presence of noise and other channel impairments, and they facilitate a variety of services such as error control and power control. To transmit data from an analog source over a digital link, the source output must be sampled and digitized. Techniques for digitizing analog sources are intimately related to source processing. Also, the characteristics of the bit stream from digitized sources are different from the characteristics of the bit stream from data sources, and thus digitized analog sources require a different quality of service than do data sources.

In this chapter we begin with an overview of the five example sources listed previously, a brief survey of specific source attributes, and user requirements. We include properties of analog, digitized, and digital sources. Next we introduce the quality of service concept and identify, at least in qualitative terms, the requirements that communication links must meet to properly convey information from each of the different types of sources. We particularly emphasize the distinction between "streaming" and "bursty" sources and discuss the different quality of service demands these types of sources impose. The major part of this chapter is about efficient digitization of speech. We discuss two general categories of speech "coding": waveform coding and source coding. As an example of the waveform coding technique we examine pulse code modulation (PCM). The PCM digitization technique is in fact not limited to speech signals, and so this section has application to other types of analog sources as well. Our example of source coding is linear predictive coding (LPC). This technique applies to speech only, but it has been extremely successful in providing high-quality, low-bit-rate digitization. Linear predictive coding is the technique used in all of the second- and third-generation cellular systems. Following the discussion of speech coding, the chapter concludes with a presentation of convolutional coding as an effective error control technique. Convolutional coding is incorporated into all of the digital cellular telephone systems. We introduce the coding method and the widely used Viterbi decoding algorithm.

Information Sources and Their Characterization

In this section we give a brief overview of the significant characteristics of a number of familiar information sources. To emphasize the interplay between properties of the source and expectations of the user, the discussion of each source is divided into "source characteristics" and "user requirements." In some cases the distinction between source characteristics and user expectations is made arbitrarily. The intent in making the distinction is to facilitate the discussion, and not to provide definitive rulings in this matter. The important point is to recognize that to a communication system engineer, source attributes must be interpreted in terms of user needs. Ultimately the system design must provide user satisfaction, within the constraints of available resources.

Following the discussion of source properties, we provide a brief look at the QoS requirements that transmission systems must meet to support various information sources. Our purpose is to show how source requirements flow down to influence network requirements and eventually to affect design requirements on the data links. We conclude the section with a discussion of factors a systems engineer must consider when data from a streaming source such as speech is to be transmitted over a packet network.

Speech

Source Characteristics

The original purpose of the telephone network was to transfer speech. Analog speech signals are produced at the output of a handset microphone. Wired "telephone-quality" speech is carried over what is euphemistically called a "4 kHz channel." In fact, the incoming speech signal is filtered when it enters the switching office, using a filter whose stopband edge is 4 kHz. The filter passband edge is slightly lower, about 3.5 kHz (frequencies below 300 Hz are also excluded).

The ratio of the envelope level of the loudest speech signal to that of the quietest speech signal is called the **dynamic range**. The human ear is capable of distinguishing a dynamic range of about 120 dB, but the dynamic range of speech is smaller and may be from 30 to 60 dB, depending on how it is measured.

Speech signals are often considered "streaming" signals and are treated as if they are generated continuously, without interruption (at least for the duration of a call). We noted in our discussion of CDMA in Chapter 6, however, that a telephone speaker is active somewhat less than 50% of the time. Some telephone systems—early transatlantic cable systems and CDMA cellular systems, for example—have been able to take advantage of the speech "activity factor" to optimize performance.

User Requirements

It is generally expected that telephone-quality speech will be intelligible to the listener and that the listener will be able to recognize the talker's voice. Since telephone networks support two-way conversations, the delay between generating the speech and delivering it must not be so long as to be annoying. It is also expected that the delivered speech will be relatively free of irritations such as noise, cross talk (other conversations in the background), choppiness, echo, and "dead" intervals between words. A system that conveys speech signals must be able to accommodate both male and female voices and a suitable dynamic range. The standard 4 kHz bandwidth has been determined through experimentation and long experience to be adequate for speaker recognition. For applications where bandwidth is at a premium and speaker recognition can be sacrificed, narrower bandwidths can be used to transmit speech.

Signal-to-noise ratio (SNR) is another parameter that characterizes the quality of a speech signal. There are studies showing that word recognition is not adversely affected by noise for signal-to-noise ratios above 10 dB, although noise is clearly audible at a 10 dB signal-to-noise ratio. The signal-to-noise ratio of a good-quality telephone connection can approach 40 dB.

There is another factor concerning the transmission of speech signals which is important to the user: In telephone systems the one-way delay must be kept below 250–300 ms. Delays longer than this make a free two-way conversation difficult, and speakers begin talking over each other. The 250–300 ms figure represents total audio-to-audio delay and includes propagation delay and delay in switching, as well as signal-processing delay at both ends of the connection. Telephone systems are often designed using a "delay budget," in which a part of the maximum delay is apportioned to each element of the total transmission system.

Delay jitter is the variation of delay with time. This is a particularly significant impairment in packet-based digital transmission systems, in which the delay may vary from packet to packet. To provide a continuous voice signal at the destination, delay jitter must be removed by buffering. Thus delay jitter translates to an increase in overall delay. The allowable delay jitter is an extremely important system design constraint for streaming applications.

Music

Source Characteristics

Music, like speech, is represented by an analog signal generated at the output of one or more microphones. As such, it is characterized by a bandwidth and a dynamic range. For music, as for speech, the questions of "how much" bandwidth and "how much" dynamic range can be answered only by appeal to the user. Unlike a speech signal, however, a music signal may consist of several channels, two for stereo and more for theatrical effects. Music is a "streaming" signal, generally considered to require continuous and uninterrupted service for a period of time.

User Requirements

Music of "entertainment quality" has proved to place more stringent requirements on communication service than does speech. Communication of speech is often acceptable at a utilitarian level, whereas most people prefer high-quality music reception. This preference, coupled with the fact that music as a source is more complex than speech, creates different design criteria for a systems engineer.

Although it is generally accepted that music has a wider bandwidth than speech, what is usually meant is that a satisfactory "user experience" requires that the music be presented at a wider bandwidth than the bandwidth used to present speech. Music contains frequency components over the entire range of human hearing, a much broader range than is required for voice recognition. The human auditory system can respond to signals in the range from about 20 Hz to about 20 kHz. AM broadcast radios utilize a message bandwidth of about 5 kHz. FM broadcasting was developed with music transmission in mind, and the message bandwidth was increased to 15 kHz. For compact disc recording, signals are sampled at 44,100 samples per second. This allows a signal to be accurately reproduced at a theoretical maximum bandwidth of up to 22,050 Hz. In practice, a 20,000 Hz bandwidth is achievable.

Dynamic range is an important factor in music reproduction, and the dynamic range of a music signal can be larger than that of speech. As noted above, the human auditory system can detect a dynamic range of about 120 dB. In contrast, good vinyl records and analog magnetic tapes can provide a dynamic range of about 45 dB. In compact disc recording, the left and right channels are each digitized using 16 bits per sample. This degree of quantization is capable of producing a dynamic range of over 90 dB.

Users prefer that music be transmitted without noticeable distortion and without discernible noise. Like speech transmission, music transmission also requires avoiding cross talk, choppiness, and dead intervals. Music transmission is usually not conversational (i.e., not two-way), and so a much greater amount of absolute delay can be tolerated than is acceptable for a speech signal. For musicians coordinating a rehearsal over an audio link, however, one-way delays may have to be kept below 50 ms.[1] As music is a streaming application, the amount of delay jitter must be carefully controlled using buffering at the receiver. Without such buffering there may be discernible stops and starts in the audio presented to the listener.

Images

Source Characteristics

Images, because they are two-dimensional and may be in color, can include vast amounts of data. A single five-megapixel image from a digital camera, with three primary colors each represented as an eight-bit number, represents roughly 15 MB of data. Often such an image is stored as a file at its source, and, as such, the communication of an image is no different from the communication of any large data file.

To turn a two-dimensional image into a one-dimensional waveform, it is common practice to scan the image. Typically the scan proceeds in rows from left to right, working from the top to the bottom of the image in a pattern called a **raster**. The number of scan lines per inch is an important design parameter, as is the speed of the scan. As an example, the raster scan used in NTSC[2] television scans 525 lines every 33.3 ms and produces a waveform with a bandwidth of about 4.4 MHz. Scanning an image usually requires the addition of synchronization information indicating where scan lines begin and where the image begins and ends. This overhead increases the information content of the resulting signal.

User Requirements

In some applications the full content of the image must be communicated, but in other contexts only certain features of the image are important to the user (e.g., edges only, black and white only). An image that will be enlarged and printed requires more fine detail than an image that will be e-mailed and viewed on the small screen of a cellular telephone handset. Thus for some applications there is considerable opportunity for source coding to reduce the amount of data that must be sent.

In many applications a certain amount of delay can be tolerated in the reception of an image. In applications such as downloading a Web page, however, transmission delays that stretch into minutes can be frustrating to the user. Since an image is not usually viewed until it has been received in its entirety, delay jitter in transmission is not normally an issue.

1. M. Anderson, "Virtual Jamming," *IEEE Spectrum* 44, no. 7 (July 2007): 53–55.
2. The National Television System Committee color transmission standard was adopted for the United States in 1954.

Video

Source Characteristics

A video signal consists of a sequence of images sent at a rate that varies from application to application. Motion pictures, for example, generate 24 images per second. All of the attributes of still images continue to apply, but there is an additional requirement that the video images arrive at intervals that are nearly equally spaced and at the proper rate. Overhead may be added to the video signal to synchronize the images. There may also be an associated audio signal, which must also be synchronized with the images.

The NTSC television standard transmits luminance (brightness of the black-and-white signal) and chrominance (color saturation and hue) as separate signals. Both luminance and chrominance are able to occupy the same frequency band because the luminance and chrominance spectra are cleverly interleaved. A television signal is a sequence of images, with a new image transmitted every 1/30 second. To avoid annoying flicker, each image is scanned twice. One scan covers the even-numbered lines of the raster; the other scan covers the odd-numbered lines in a process called "interleaved" or "interlaced" scanning.[3] The two scanned "fields" are sent sequentially and are combined by the receiver to reproduce the image.

For accurate presentation to the viewer, successive images of a video sequence must arrive at the same rate at which they were transmitted. Thus video transmission is a streaming application with stringent requirements on delay jitter. Excessive delay jitter is manifest as a "jerkiness" in the motion of the images presented to the user. If the video is not interactive, long transmission delays may be tolerated.

User Requirements

We have mentioned that image transmission times can be reduced if source coding or compression is used prior to transmission. The same observation applies to video transmission times. Even in as unsophisticated a system as NTSC analog television, channel bandwidth is reduced by transmitting the chrominance signal in a narrower bandwidth than the luminance signal. Because effective image compression can require a great deal of processing time, it may be necessary to compress a video sequence in its entirety before transmission. If compression is to be done in real time during transmission, the processing time may be limited by the time interval between frames. In most video signals there is very little change between one image and the next, and compression schemes take advantage of this property by transmitting only the updates.

Videoconferencing is a use of video transmission that is increasing in importance. In this application the transmission is two-way (or multiway) and interactive, so there are significant limits on transmission delay. The limitation on delay ripples back through other transmission parameters. There is only a limited time for processing, so only modest amounts of image compression can be achieved. There is a limit on buffering at the receiver,

3. "Progressive" scanning, in contrast, scans all of the lines of the image in sequential order.

so the limitation on delay imposes a limit on delay jitter. Videoconferencing is a very challenging application.

Data

Source Characteristics

The term *data* generally refers to information that originates in digital form. A variety of applications generate data—for example, paging calls, short-message service ("texting"), instant messaging, e-mail, Web browsing, remote login, file transfer, telemetry. Most of the data these applications generate, except for file transfer and possibly telemetry data, is bursty in nature. When a user checks e-mail, for instance, there is a burst of traffic while the mail is downloaded, followed by a long pause as the messages are read. There may be subsequent bursts of data as messages are answered. The burstiness of many data sources is a significant attribute and is to be contrasted with the "streaming" nature of the information produced by several of the sources examined previously. We saw in Chapter 6 that whether the information to be transmitted comes from a bursty or a streaming source can have an effect on network design. Bursty sources often contend for access to shared communication channels, whereas streaming sources tend to be assigned exclusive use of channels for the duration of their need.

User Requirements

For many types of data the users have very high expectations of accuracy. We expect to be able to download computer code that works when we install it. We expect dollar amounts to be transferred correctly in electronic business transactions. We expect e-mail to be delivered to the correct address.

The significant parameters for data transmission from the user's point of view are throughput and error rate. Throughput is the number of bits per second received. (This parameter is often erroneously called "bandwidth" in the computer industry.) Telemetry (health or status information from equipment) may require a throughput of hundreds of bits per second. In contrast, the Internet backbone, or main transport system, carries hundreds of gigabits per second. Most home users subscribe to Internet access links that are asymmetric, with the uplink having average rates in the few hundreds of bits per second and the downlink rates being a few hundred kilobits per second.

Bit error rate (BER) is a measure of transmission quality at the physical link level. This is normally a design and test parameter evaluated by considering how many bits are received in error divided by the total number of bits transmitted. To identify some extremes, we have mentioned that some wireless links (e.g., Bluetooth) have error rates that can be as high as 10^{-3}. To make these links usable in data applications, error control procedures must be applied outside the physical link. Error control will be discussed in some detail later in this chapter. At the other extreme, some optical fiber links are specified at error rates as low as 10^{-15}. Many links operate quite well with bit error rates in the 10^{-5} to 10^{-8} range. A bit error rate of 10^{-12} is effectively error-free for many applications.

Most of the indicated data applications are not sensitive to delay. An exception is remote computer login, where excessive transmission delay can cause unwanted user behavior like pressing the "enter" key multiple times. Insensitivity to delay allows designers to use error control techniques such as error detection and retransmission that can be very effective at eliminating transmission errors.

Transmission of long messages or long files can cause difficulties in two ways. First, long messages can tie up available transmission resources, delaying access by other users. Second, the probability of message error increases with message length. If a large message must be retransmitted because of an error, the delay experienced by a user can become unacceptably long. As a consequence, many networks divide long messages into shorter "packets" for transmission. Typical packet sizes vary from a few dozen bytes to a few thousand bytes.

Quality of Service (QoS)

In a different context, the term *quality of service* also refers to guarantees offered by a transmission network to ensure satisfactory conveyance of various kinds of information. The term is usually associated with digital transmission networks, particularly networks that are designed to carry information from a wide variety of sources. It is possible to identify several transmission parameters that influence the QoS. In some networks the values of these parameters may be negotiated at the time a network connection is set up. Establishing a particular set of QoS parameter values is intended to guarantee that the network connection will be suitable for a given source and a given user application. Network usage fees may also depend on the quality of service guarantees. It is important to note that QoS parameters are associated with the transmission system and not with the source; however, sources may be prepared for transmission in a variety of ways that may influence the quality of transmission service needed. For example, when speech and images are processed to reduce the number of bits that must be transmitted, the sensitivity to bit and/or packet errors increases. Source processing for speech signals will be discussed in greater detail in the remaining sections of this chapter.

Some of the common transmission system QoS parameters are

- Average data transmission rate (throughput)
- Peak transmission rate
- Frame loss rate
- Delay
- Delay jitter
- Priority

An examination of the typical information sources described previously leads to the observation that QoS can be offered as "packages" of parameter settings. The following are examples of the way a package of QoS parameters might configure a link for a particular information source.

Speech

Speech requires transmission at a guaranteed average data rate. For full-duplex communications there are strict limitations on the maximum delay as well as on the maximum delay jitter. Maintaining a constant delivery rate, however, may be even more important than minimizing delay. Voice signals are relatively tolerant of errors. These properties lead to a fixed-bit-rate, low-delay, moderate-error-rate service.

Music

Music is also a fixed-data-rate service with strict limitations on the maximum delay jitter. The data rate needed may be higher than that needed for speech. For noninteractive applications, the requirements on delay may be considerably relaxed compared with speech. Music signals may also be relatively tolerant of errors. Music transmission requires a fixed-bit-rate, moderate-error-rate service.

Images

Transmission of a digitized image is essentially the same as transmission of a file, with the proviso that image files may be particularly large. QoS requirements for file transmission are discussed under "Data."

Video

Some compressed video signals vary in bit rate during transmission, owing to the way the compression algorithm functions. For these signals the transmission network may be able to allocate resources to other users during times of low bit rate, but the service must provide for peaks in transmission rate when they are needed. Variable-bit-rate video may require strict control of delay jitter. These considerations lead to a variable-bit-rate, low-delay, moderate-error-rate service package of QoS parameters.

Data

Data sources usually require a high peak transmission rate and a very low error rate. Many data sources are very bursty, and so the high transmission rate may be needed for only a small fraction of the time. Delay and delay jitter are often not significant for data sources. The "data" quality of a service package would include high peak transmission rate and effective error control.

Some data sources can take advantage of an "available-bit-rate" service in which data is stored and transmitted when the transmission facility is lightly loaded. A lower throughput rate will be accepted when the facility must be shared with other users. Network operators sometimes offer available-bit-rate service at a significantly lower cost than fixed-bit-rate service, since the customer will not get to determine when, or whether, high throughput rates may be available.

The error parameter in this list of quality of service parameters was specified as a frame loss rate rather than as a bit error rate. The processing that is used to prepare source data for transmission often results in a natural organization of bits into "frames" or "packets."[4] In our discussion of speech processing later in this chapter we will show how the frame organization comes about for the speech application. Also, we have already noted that large files are often divided into frames or packets to improve error control. When the source data is organized into frames, the frame loss rate becomes a natural measure of performance. In addition, frame loss rate may be relatively easy to measure when the receiver is organized to process data one frame at a time. Now, frames may contain bits for addressing and synchronization, and checksums or "parity bits" for error control, in addition to source data. It turns out that an error in, say, a synchronization bit will have a different effect on performance than an error in a data bit. Consequently, it can be difficult to relate the bit error rate of the communication link to the frame error rate in a simple analytic way. Often, to determine the maximum bit error rate that will guarantee a desired frame loss rate, a systems engineer must resort to simulation of the communication system.

Smooth versus Chunky

The source attribute that has the greatest effect on design of the transmission system is the streaming versus bursty nature of the source data. In our discussion of access methods in Chapter 6 we observed that a streaming source is often assigned a dedicated channel, whereas it is a more efficient use of resources to allow bursty sources to contend for access to a common channel.

The wired telephone network is the classic example of a network that was designed for transmission of data from a streaming source. A channel ("circuit") is assigned when a call is placed, and the resources associated with that circuit are dedicated to that call until they are released.

The Internet is the classic example of a network that was designed for transmission of bursty data. Data bits are divided into packets (called "datagrams") that are transmitted independently over the data links. Packets from different users are interleaved for transmission, so each packet must carry its own destination and source addresses.

It is possible to carry bursty data over a circuit-switched network, but it may be inefficient to do so. A dedicated circuit must be set up for each packet and then released once the packet is transmitted. The frequent setup and release operations may involve a significant amount of overhead for the transmission network. Likewise, it is possible to carry streaming data over a packet-switched network, but it may be inefficient to do so. The destination and source addresses that must be placed in each packet represent a significant overhead when packets in a continuous stream are intended for the same destination. Because packets are independently stored and routed, they may arrive after very variable propagation delays and may arrive out of order. Packets that arrive out of order must be either buffered until they can be sorted into the correct order for delivery, which increases delay, or discarded, which increases the frame loss rate.

4. The terms *frame* and *packet* have identical meanings. We use the two terms interchangeably.

It has been a long-standing dream of network designers to create a single network that can handle streaming as well as bursty applications. The Integrated Services Digital Network (ISDN) is one such network that was created for the wired telephone system. Cellular telephone networks are also evolving to support bursty applications in addition to the traditional speech application.

The current trend in network design is to use packet switching to accommodate both bursty data applications and streaming speech applications. To design a packet-switched network that works well in a streaming application, a systems engineer must ensure that appropriate quality of service guarantees are in place. Specifically, streaming applications require that delay and delay jitter be controlled. One way that this can be done is to give packets from streaming applications priority at the switches, so that those packets will be forwarded promptly. An alternative is to "overdesign" the transmission and switching capabilities of the network so that *all* packets are forwarded promptly. A second guarantee is to ensure that error control techniques are used that do not increase packet delay. Error control for streaming applications will be discussed in detail later in this chapter.

Digitization of Speech Signals

The preparation of signals from information sources for transmission over digital communication links is a vast topic that could serve as the subject of its own textbook. In this section we illustrate the important ideas by restricting our attention to the particular case of speech signals. In preparing analog signals for digital transmission there are two major issues. The first is converting the analog signals that appear at the terminals of an input transducer to binary form. The second major issue is minimizing the bit rate of the converted signals without significantly reducing the quality of the reconstructed signal at the receiver. With these goals in mind, we can identify two distinct approaches to the challenge of digitizing speech. In the first approach, sometimes referred to as **pulse code modulation** (PCM), the speech waveform is sampled and digitized with the intent of reconstructing the actual analog waveform as accurately as possible at the receiver. In principle, accurately reproducing the waveform will reproduce the information content. When the PCM approach is used, any subsequent bit rate reduction, or "compression," is achieved by processing the digital signals. The PCM approach is very general and can be used to digitize music, images, and other analog signals in addition to speech. In the second approach the focus is on the information content rather than on the waveform. This approach is sometimes called **source coding**, or the **vocoder** (voice encoder) approach. At the transmitter a model of the human vocal tract (the vocoder) is set up, with a variety of adjustable parameters. The parameters of this model are then varied so that the model reproduces the incoming speech signal. The time-varying parameters of the model are transmitted over the data link. At the receiver the parameters control an identical model, which reproduces, within limits, the original speech signals. The vocoder approach is specific to speech and has been very successful in creating good speech reproduction at comparatively low bit rates.

This section begins with a detailed examination of PCM analog-to-digital conversion. We examine the basic techniques of sampling and quantization. Next we discuss compressing and expanding ("companding") the speech waveform. This is a method of nonuniform quantization that increases the dynamic range of the conversion system. We also introduce "differential" methods that reduce the bit rate needed to reproduce the waveform. Following the discussion of PCM and its differential extensions, we introduce the vocoder approach to digitizing speech. Our approach here is less detailed, as the subject is broad and there are many related techniques. We examine the "linear predictive coding" approach to give a sense of how the method works.

Pulse Code Modulation

Converting an analog signal to digital form is a three-step process. The steps of sampling, quantization, and coding are shown in Figure 7.2. We will discuss these three steps as if physically distinct devices perform them, although in practice the steps are often implemented in a single device.

The analog waveform $x(t)$ is first sampled at a sampling frequency f_s to produce a discrete-time version $x[k] = x(kT_s)$, where $T_s = 1/f_s$ is the sampling interval. The samples $x[k]$ take values over a continuous range and cannot be represented by binary numbers having a finite word length. As we shall describe later in a brief review, sampling does not destroy information.

In the second step of the conversion process each of the samples $x[k]$ is replaced with an approximation $\hat{x}[k]$. The "quantized" samples $\hat{x}[k]$ can take values only from a finite set of numbers. The larger this set of possible values is, the more closely $\hat{x}[k]$ can approximate $x[k]$; however, more bits will be needed to identify the sample value. Quantization is inherently destructive of information, and $x[k]$ can never be exactly reconstructed from the quantized sequence $\hat{x}[k]$.

In the final conversion step each possible value of the quantized sample $\hat{x}[k]$ is represented by a binary number. An n-bit binary number can take only $N = 2^n$ distinct values, and this is why the quantization step is required. This third conversion step is often called "coding," and it is important not to confuse this use of the term with the error correction "coding" that we will discuss later in this chapter, or with encryption that may be applied to the digital signal to ensure privacy.

Sampling

Before discussing source processing in detail, a review of the Nyquist sampling theorem is in order. The primary objective of this review is to recall the basic requirement on sampling rate

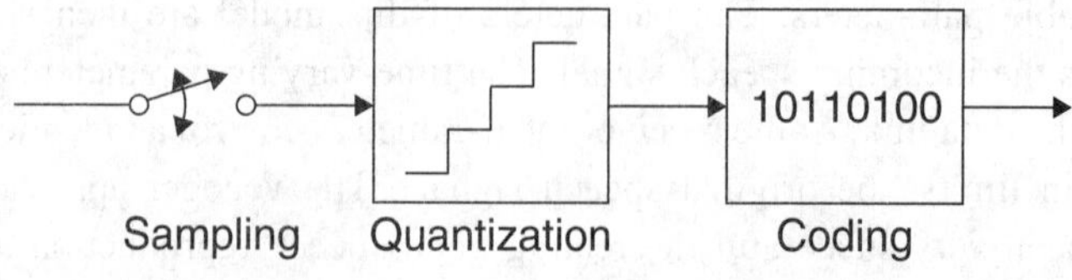

Figure 7.2 The Three Steps of Nyquist Analog-to-Digital Conversion

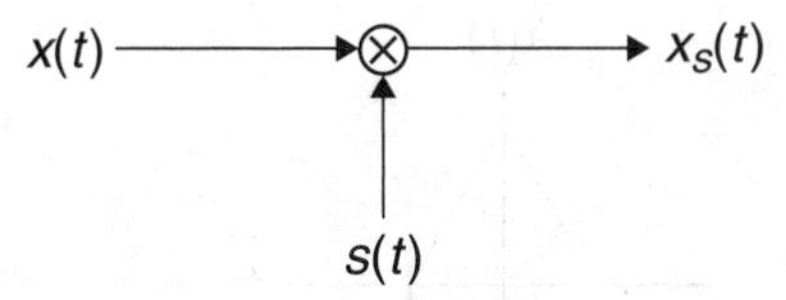

Figure 7.3 Model of a Pulse Sampler

and the impact of the sampling process on the waveform's digital representation. A model of a pulse sampler is shown in Figure 7.3. The sampling waveform $s(t)$ is a train of narrow pulses whose shape is determined by the physical characteristics of the sampling device. In the simplest model, $s(t)$ is a train of ideal impulses. We have

$$s(t) = \sum_{k=-\infty}^{\infty} \delta(t - kT_s). \tag{7.1}$$

The output of the sampler can be obtained by multiplying the input by the sampling waveform. This gives

$$x_s(t) = x(t)s(t) = x(t)\sum_{k=-\infty}^{\infty} \delta(t - kT_s) = \sum_{k=-\infty}^{\infty} x(kT_s)\delta(t - kT_s), \tag{7.2}$$

where $x(kT_s)$ is the desired sequence of samples. Insight comes from viewing the sampling process in the frequency domain. If we Fourier transform Equation (7.2), multiplication will be replaced by convolution. If $X(f)$ is the Fourier transform of $x(t)$, we have

$$\begin{aligned} X_s(f) &= X(f) * \left[f_s \sum_{k=-\infty}^{\infty} \delta(f - kf_s) \right] \\ &= f_s \sum_{k=-\infty}^{\infty} X(f - kf_s). \end{aligned} \tag{7.3}$$

We see that the spectrum $X_s(f)$ of the sampled signal is, up to a scale factor, equal to the spectrum $X(f)$ of the input signal repeated at harmonics of the sampling frequency f_s. The sampling process is depicted in the frequency domain in Figure 7.4. Figure 7.4(a) shows the spectrum of the input signal. The input signal is shown having an absolute bandwidth B. Figure 7.4(b) shows "undersampling," where the sampling frequency f_s is low and the harmonics of $X(f)$ overlap. Figure 7.4(c) corresponds to a higher value of f_s for which there is no overlap of harmonics of $X(f)$.

If the sampled signal $x_s(t)$ is passed through a lowpass filter, a continuous-time output will be obtained. For the case of Figure 7.4(c), a lowpass filter with a passband bandwidth greater

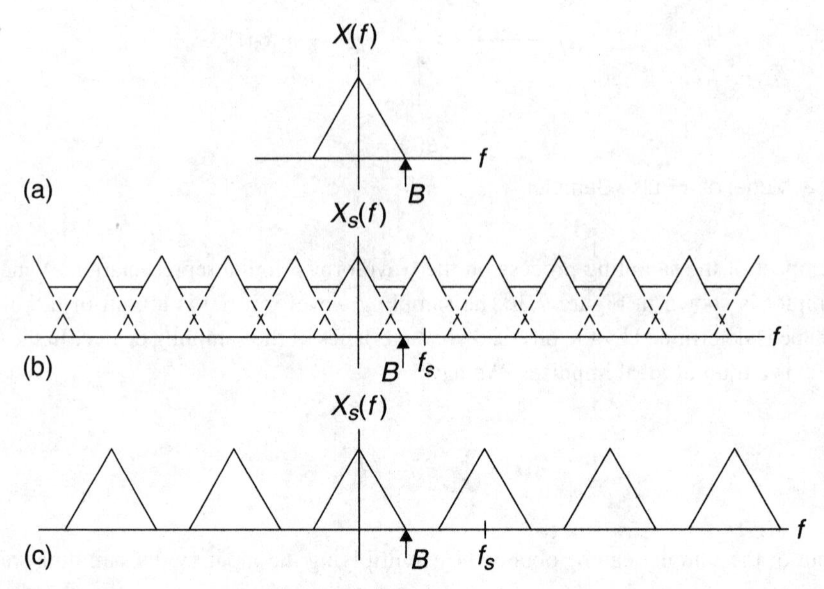

Figure 7.4 Sampling as Seen in the Frequency Domain: (a) Continuous-Time Signal; (b) Undersampled Signal; (c) Appropriately Sampled Signal

than B and a stopband bandwidth less than $f_s - B$ will produce an output whose spectrum exactly matches that of Figure 7.4(a). Thus the original signal $x(t)$ can be recovered without distortion from its samples. For the case of Figure 7.4(b) no lowpass filter can recover the original signal. The output of any such filter will be distorted by components of the harmonic of $X(f)$ centered at f_s. This distortion is called **aliasing**. We see that a necessary and sufficient condition for a signal $x(t)$ of bandwidth B to be recoverable from its samples is that the sampling frequency be large enough. Specifically, the sampling frequency must exceed the Nyquist rate $2B$, that is,

$$f_s > 2B. \tag{7.4}$$

As a practical matter, signals encountered in real applications are seldom absolutely bandlimited in as clear a manner as the signal in Figure 7.4(a). Also, input signals are often accompanied by noise, which may have a wide bandwidth. As a consequence, the bandwidth B is set by a lowpass filter called an "antialiasing" filter that precedes a well-designed sampler. This filter will introduce some distortion, but the sampling and reconstruction process will not cause additional distortion.

It is apparent in Figure7.4(c) that setting the sampling frequency too close to the Nyquist rate $2B$ will require a reconstruction filter with a very sharp cutoff. Increasing the sampling rate eases the specifications on the reconstruction filter. As we shall see, however, increasing the sampling rate will increase the number of bits per second in the digitized data stream that represents the input waveform. Consequently there is a trade-off between the data rate of the digitized signal and the complexity of the reconstruction filter.

Figure 7.4 shows spectra of a signal sampled using an ideal impulse sampler. If the sampling waveform $s(t)$ is a pulse train instead of an impulse train, the harmonic copies of $X(f)$ in the spectrum $X_s(f)$ will not all be the same size. The amplitude of the harmonic copies follows a pattern that depends on the shape of the sampling pulses. Generally, harmonic copies decrease in size at higher multiples of the sampling frequency. The fact that the harmonics of $X(f)$ are of different sizes does not affect our ability to reconstruct $x(t)$ from the sampled waveform $x_s(t)$ by lowpass filtering.

Example

As we mentioned in Chapter 6, in the wired telephone system speech signals are lowpass filtered when they enter the switching office. The lowpass filters have a stopband bandwidth of 4 kHz. This lowpass filter can function as an antialiasing filter. The 4 kHz stopband bandwidth allows a speech signal to be sampled at 8000 samples per second without causing any additional distortion. The 8000 samples/second rate is a standard throughout the wired telephone system.

Quantization

The second analog-to-digital conversion step, converting a sequence of analog samples $x[k] = x(kT_s)$ to a sequence of quantized values $\hat{x}[k]$, is always destructive of information to some extent. Ultimately each sample will be represented by a binary word, and a binary word of fixed length can take only a finite set of values. For example, an 8-bit binary number can take $2^8 = 256$ distinct values, whereas a 12-bit binary number can take $2^{12} = 4096$ distinct values. Analog samples, on the other hand, take on a continuum of values. Thus when converting the analog samples into binary it is necessary to approximate each sample by the nearest value that can be represented by a binary code. This approximation makes the reverse operation impossible to carry out exactly. It is never possible to infer from the binary code the exact value of the original sample.

The block in Figure 7.2 labeled "Quantization" can be modeled as an ideal quantizer with input $x[k]$ and output $\hat{x}[k]$. The quantizer is memoryless, so we can write

$$\hat{x}[k] = Q\{x[k]\}, \quad k = \ldots, -1, 0, 1, 2, \ldots. \tag{7.5}$$

To define the operation of the quantizer, refer to Figure 7.5. The figure shows an amplitude range divided into L **quantization intervals** $I_n, n = 1, \ldots, L$. Each quantization interval is bounded by a pair of **decision levels** $x_1, x_2, \ldots, x_{L+1}$. Within each quantization interval is a **quantization level** $\hat{x}_n, n = 1, \ldots, L$. The operation of the quantizer is simple: If the sample $x[k]$

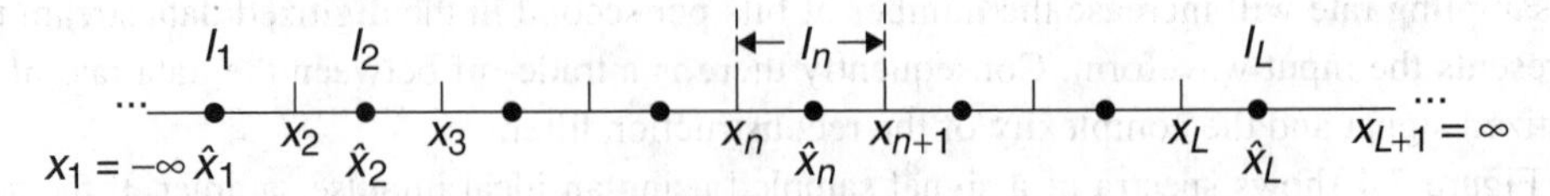

Figure 7.5 A Continuous Amplitude Range Divided into Quantization Intervals

takes a value in quantization interval I_n, then

$$\hat{x}[k] = Q\{x[k]\} = \hat{x}_n. \tag{7.6}$$

In many signal-processing systems the quantization is uniform. This means that all of the quantization intervals are of equal size. We can write

$$x_{n+1} - x_n = \Delta, \quad n = 2, \ldots, L-1. \tag{7.7}$$

It is also customary to place the quantization levels in the centers of the quantization intervals. As a result, the quantization levels are also equally spaced, that is,

$$\hat{x}_{n+1} - \hat{x}_n = \Delta, \quad n = 1, \ldots, L-1. \tag{7.8}$$

(Note that the intervals I_1 and L_L at the ends of the range are very large, and the corresponding quantization levels $\hat{x}_1$ and $\hat{x}_L$ are not at the centers of these intervals.)

When the range of values that the sample $x[k]$ can take includes both positive and negative values, there are two schools of thought about where to put the zero. Figure 7.6 shows the two options applied to a four-level uniform quantizer. In this figure we show the quantizer input along the horizontal axis and the output along the vertical axis. Remember that in the coding step each quantizer output value will be represented by a binary word. Figure 7.6(a) shows a

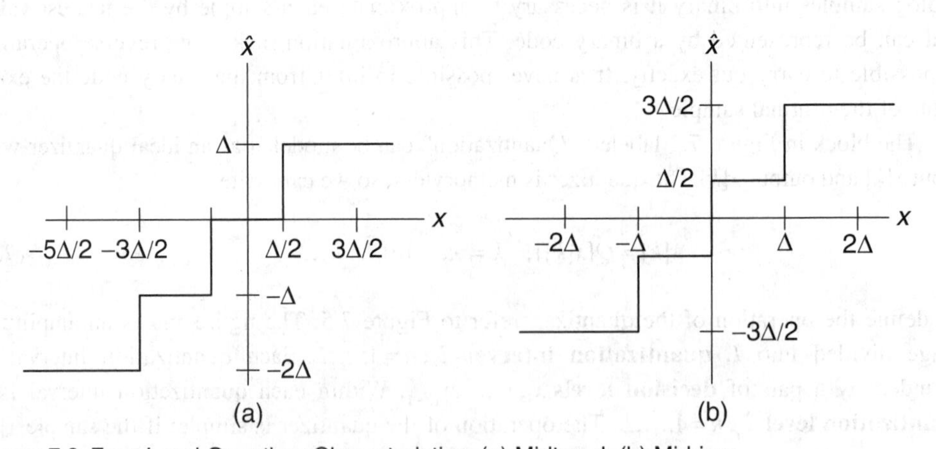

Figure 7.6 Four-Level Quantizer Characteristics: (a) Midtread; (b) Midriser

"midtread" quantizer. The quantization levels are $\hat{x}_1 = -2\Delta, \hat{x}_2 = -\Delta$, $\hat{x}_3 = 0$ and $\hat{x}_4 = \Delta$. For this case zero is included as a quantization level. Figure 7.6(b) shows a "midriser" quantizer. The quantization levels are $\hat{x}_1 = -3\Delta/2$, $\hat{x}_2 = -\Delta/2$, $\hat{x}_3 = \Delta/2$, and $\hat{x}_4 = 3\Delta/2$. The midtread quantizer is sometimes preferred over the midriser quantizer in practice, because with a midtread quantizer very small changes in input around zero (possibly caused by noise) will not produce changes in the quantizer output.

Although all possible values of x are theoretically acceptable as input values, practical quantizers have a finite input range. If we set the sizes of I_1 and I_L each to be Δ, we can identify a maximum input x_{max} and minimum input x_{min} that the quantizer can handle without overload. The **full-scale range** of the quantizer is taken to be $x_{max} - x_{min} \approx 2x_{max}$. Input values that exceed the full-scale range are said to be "out of range." For the midtread quantizer of Figure 7.6(a), the full-scale range is $\frac{3\Delta}{2} - \left(-\frac{5\Delta}{2}\right) = 4\Delta$. The full-scale range is $2\Delta - (-2\Delta) = 4\Delta$ for the midriser quantizer of Figure 7.6(b).

Coding

There are many ways of assigning binary words to quantization levels. The most important consideration is that the transmitter and receiver both use the same code. For instrumentation applications it is common to use binary numbers, so that larger signal values are given larger binary representations. Two's complement, one's complement, and sign magnitude are common numbering systems used for this purpose. In communications applications it is often useful to assign binary words so that words assigned to adjacent quantization levels differ in only a single bit. This way an input sample whose value lies on a boundary between two quantization intervals causes uncertainty in only one bit. A binary code of this form is called a Gray code. Table 7.1 lists the sign-magnitude, one's complement, two's complement, and Gray code binary words that could be used to represent quantization levels in an eight-level quantizer.

Table 7.1 Alternative Codes for Representing Quantization Levels

	Sign Magnitude	One's Complement	Two's Complement	Gray Code
3	011	011	011	100
2	010	010	010	101
1	001	001	001	111
0	000	000	000	110
–0	100	111		
–1	101	110	111	010
–2	110	101	110	011
–3	111	100	101	001
–4			100	000

PCM Performance

In this section we develop a measure of the error introduced by the quantization process and relate this measure to properties of the input signal and the quantizer. Given an input sample $x[k]$, we can define the quantization error $e_q[k]$ by

$$e_q[k] = \hat{x}[k] - x[k], \tag{7.9}$$

where $\hat{x}[k]$ is given by Equation (7.6). An examination of Figure 7.6 shows the very important property that if the input sample $x[k]$ is within the full-scale range of the quantizer, then the quantization error is bounded by

$$-\frac{\Delta}{2} < e_q[k] \le \frac{\Delta}{2}. \tag{7.10}$$

To determine the effect of quantization error on the communication system performance, we need a mathematical model for $e_q[k]$. We observe that by rearranging Equation (7.9) we obtain

$$\hat{x}[k] = x[k] + e_q[k]. \tag{7.11}$$

Written this way, the quantization error appears as if it were an additive noise term. Once we recognize quantization error as a kind of noise, we can measure the amount of quantization error generated by a given quantizer by a signal-to-quantization-noise ratio (SQNR). This ratio is usually expressed in decibels and is defined as

$$SQNR\Big|_{dB} = 10\log\frac{\overline{x^2[k]}}{\overline{e_q^2[k]}}, \tag{7.12}$$

where $\overline{x^2[k]}$ and $\overline{e_q^2[k]}$ are the mean-square values of $x[k]$ and $e_q[k]$, respectively.

The mean-square values needed to determine the SQNR are usually obtained by statistical analysis. In general this analysis is difficult, as the statistical properties of the quantization noise depend on the properties of the input signal. Under certain conditions, however, usually satisfied in practice, the statistical analysis can be greatly simplified. In the following we will assume that the quantization step size Δ is small compared to the full-scale range of the quantizer and that the input signal commonly traverses several quantization intervals between successive samples. Under these assumptions we assert the following.

- The quantization error $e_q[k]$ is *uniformly distributed* over its range $-\frac{\Delta}{2} < e_q[k] \le \frac{\Delta}{2}$. This means that in a large collection of error samples, each value in the range will occur about equally often. Mathematically, $e_q[k]$ can be modeled as a random variable with a probability density function $f_q(e)$ given by

$$f_q(e) = \begin{cases} \dfrac{1}{\Delta}, & -\dfrac{\Delta}{2} < e \le \dfrac{\Delta}{2} \\ 0, & \text{otherwise.} \end{cases} \tag{7.13}$$

- The error sequence $e_q[k]$ is a *white noise* sequence. This means that the random variables $e_q[k]$ and $e_q[i]$ are uncorrelated when $k \neq i$.
- The error sequence is not statistically correlated with the input signal sequence $x[k]$.

Given the statistical model for quantization error, we can calculate the mean-square value $\overline{e_q^2[k]}$ directly. We have

$$\begin{aligned}\overline{e_q^2[k]} &= \int_{-\frac{\Delta}{2}}^{\frac{\Delta}{2}} e^2 f_q(e)de \\ &= \int_{-\frac{\Delta}{2}}^{\frac{\Delta}{2}} e^2 \frac{1}{\Delta} de \\ &= \frac{\Delta^2}{12}.\end{aligned} \tag{7.14}$$

Let us suppose that the coding step represents quantizer samples as binary words of n bits. This means that there are $N = 2^n$ quantization levels. If the full-scale range of the quantizer is $2x_{\max}$, then the size of the quantization interval is $\Delta = \frac{2x_{\max}}{N} = \frac{2x_{\max}}{2^n}$. The mean-square quantization noise can then be written as

$$\overline{e_q^2[k]} = \frac{\left(\frac{2x_{\max}}{2^n}\right)^2}{12} = \frac{x_{\max}^2}{2^{2n} \times 3}. \tag{7.15}$$

We can now write the signal-to-quantization-noise ratio as

$$\begin{aligned}SQNR\Big|_{dB} &= 10 \log \frac{\overline{x^2[k]}}{\frac{x_{\max}^2}{2^{2n} \times 3}} \\ &= 10 \log \frac{2^{2n} \times 3 \times \overline{x^2[k]}}{x_{\max}^2} \\ &= 20n \log 2 + 10 \log 3 + 10 \log \frac{\overline{x^2[k]}}{x_{\max}^2} \\ &= 6.02n + 4.77 + 20 \log \frac{\sqrt{\overline{x^2[k]}}}{x_{\max}} \text{ dB}.\end{aligned} \tag{7.16}$$

Equation (7.16) gives the classical result that increasing the binary word length by one bit causes a 6 dB increase in SQNR.

Example

Suppose the sampled analog signal input to an eight-bit quantizer is a sinusoid of peak value x_{max}. Then $n = 8$ and $\sqrt{x^2[k]} = \frac{x_{max}}{\sqrt{2}}$. Equation (7.16) gives

$$SQNR\Big|_{dB} = 6.02 \times 8 + 4.77 + 20\log\frac{1}{\sqrt{2}} = 49.9 \text{ dB}. \quad (7.17)$$

Equation (7.16) shows that the SQNR is a function of the ratio $\frac{\sqrt{x^2[k]}}{x_{max}} = \frac{x_{rms}}{x_{max}}$, where x_{rms} is the RMS value of the input to the quantizer. To avoid excessive distortion, the peak value of the input signal should not exceed the maximum input x_{max} that the quantizer can handle. For a typical speech signal, the RMS-to-peak ratio is much smaller than it is for a sinusoid, and the ratio $\frac{x_{rms}}{x_{max}}$ will be smaller still. Consequently, the preceding example may be overly optimistic about the performance of an eight-bit quantizer.

Example

Figure 7.7 shows a few seconds of speech for which the RMS-to-peak ratio is measured as 0.095. For this waveform eight-bit quantization gives an SQNR of about 32 dB when we set x_{max} equal to the speech signal peak.

Nonlinear Quantization

A telephone system is designed to carry voice signals from loud speakers and from quiet speakers without requiring adjustment of gain settings. For systems carrying digitized voice the difference in amplitude between speakers, which can be as high as 40 dB, provides a special challenge. If the quantizer is adjusted so that the maximum "in-range" input signal x_{max} is equal to the peak voice signal for a loud speaker, then the signal-to-quantization-noise ratio for a quiet speaker will be unacceptably low. On the other hand, if x_{max} is set to give an acceptable SQNR for a quiet speaker, then the signal peaks will be clipped for a loud speaker, producing excessive distortion.

The dynamic range of the quantizer can be increased by increasing n, the number of bits used to encode each sample. Increasing the number of bits per sample, however, increases the rate in bits per second needed to transmit the digitized speech signal. An alternative way of increasing the dynamic range of the quantizer is to use nonuniform quantization. A common implementation of a nonuniform quantizer is shown in Figure 7.8. A memoryless nonlinearity called a **compressor** is placed before the quantizer. Of course at the receiver a corresponding **expander** must be used to reconstruct the speech. Figure 7.9 shows an example of a compression curve and a matching expansion curve. Since telephone communication is bidirectional, a speech

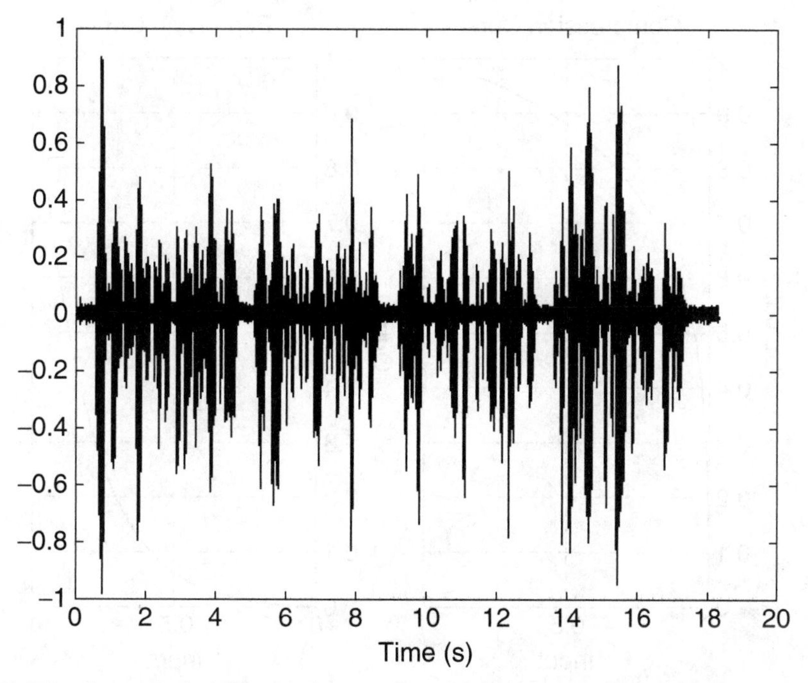

Figure 7.7 "The Cremation of Sam McGee" by Robert W. Service (First Verse)

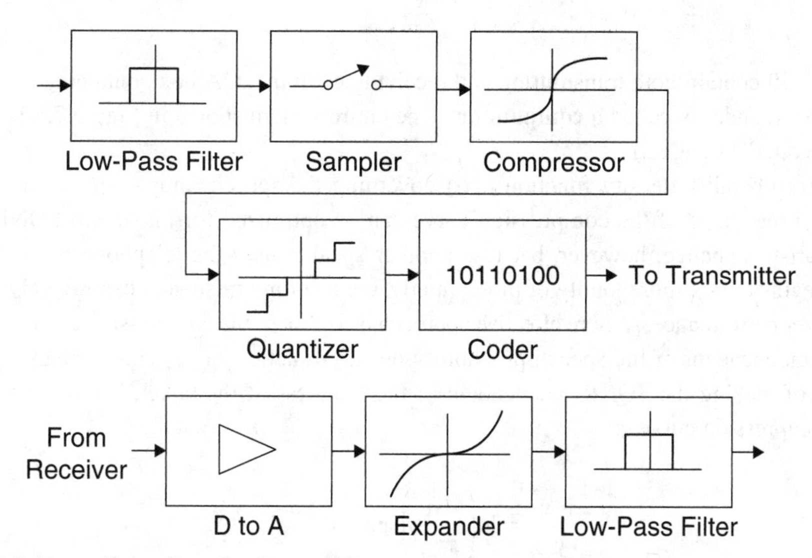

Figure 7.8 Nonuniform Quantization and Reconstruction

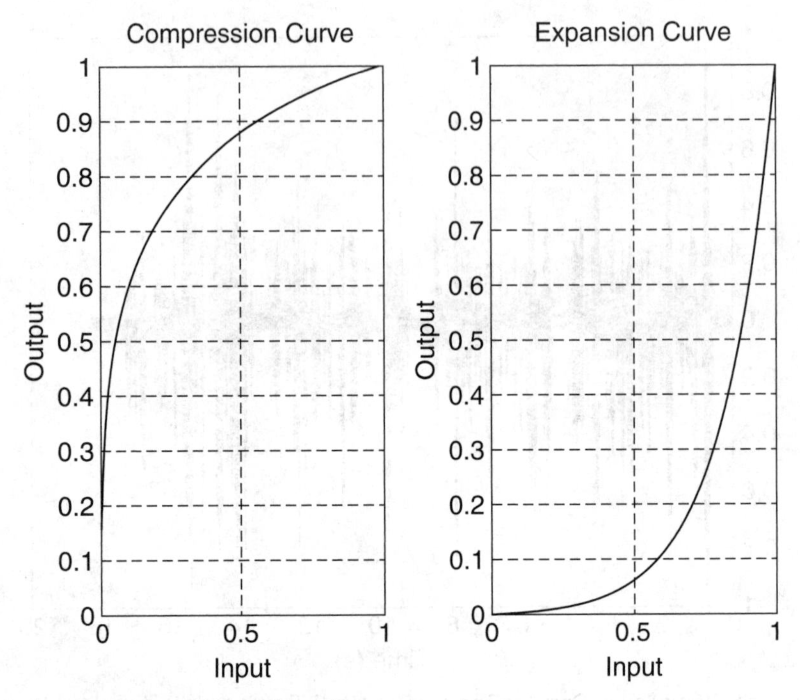

Figure 7.9 Compression and Expansion Curves

processor will contain both transmitting and receiving equipment. A box containing a compressor and an expander is called a **compandor**. (The entire system shown in Figure 7.8 is called a "coder-decoder" or **codec**.)

If the probability density function $f_x(\alpha)$ governing the speech sample $x[k]$ is known precisely, then the shape of the compression curve can be optimized for maximum SQNR. If the speech statistics change, however, because another speaker uses the telephone or the original speaker begins to talk more loudly or more quietly, the resulting mismatch can severely degrade the quantizer performance. A more effective approach is to choose the compression curve so that the SQNR is independent of the speech probability density function $f_x(\alpha)$. This strategy will have the effect of making the SQNR independent of the loudness of the talker. It turns out that the desired compression curve is

$$\frac{y}{y_{\max}} = \ln\left(\frac{|x|}{x_{\max}}\right)\operatorname{sgn}(x). \tag{7.18}$$

The logarithmic compression curve of Equation (7.18) has a problem for very small values of x, since $\ln(x) \xrightarrow[x\to 0]{} -\infty$. The curve is also discontinuous at the origin. A slightly modified logarithmic curve that is linear at the origin has been adopted as the North American standard. This is

$$\frac{y}{y_{\max}} = \frac{\ln\left(1+\mu\frac{|x|}{x_{\max}}\right)}{\ln(1+\mu)}\operatorname{sgn}(x). \tag{7.19}$$

This curve is called **μ-law compression**. The standard value of the parameter μ for eight-bit quantization is $\mu = 255$. The curve plotted in Figure 7.9 is a μ-law compression curve. In Europe a slightly different modified-logarithmic curve has been adopted as a standard. This is

$$\frac{y}{y_{\max}} = \begin{cases} \dfrac{A\left(\dfrac{|x|}{x_{\max}}\right)}{1+\ln(A)}\operatorname{sgn}(x), & 0 < \dfrac{|x|}{x_{\max}} \le \dfrac{1}{A} \\[2ex] \dfrac{1+\ln\left(A\dfrac{|x|}{x_{\max}}\right)}{1+\ln(A)}\operatorname{sgn}(x), & \dfrac{1}{A} < \dfrac{|x|}{x_{\max}} \le 1. \end{cases} \tag{7.20}$$

This curve is called **A-law compression**. A standard value of A is $A = 87.6$. With the standard parameter values, the A-law and μ-law curves are nearly indistinguishable for practical purposes.

Example

Recall from the previous example that for the speech sample of Figure 7.7, eight-bit quantization gives an SQNR of about 32 dB when we set $x_{\max}$ equal to the speech signal peak. Now if the signal of Figure 7.7 is reduced in amplitude by a factor of 20 (26 dB), then the signal-to-quantization-noise ratio also drops by 26 dB, from 32 dB to 6 dB. Figure 7.10 shows the speech sample of Figure 7.7 after μ-law compression.

Assuming eight-bit quantization, the signal-to-quantization-noise ratio of the compressed signal is about 45 dB. If the signal of Figure 7.10 is reduced by a factor of 20 and then compressed, however, the SQNR drops from 45 dB to 33 dB, a difference of only 12 dB. Experience with μ-law compression of speech signals over a wide range (more than 40 dB) of power levels shows that the SQNR tends to remain in the vicinity of 38 dB.

Differential PCM

In the PCM systems shown in Figure 7.2 and Figure 7.8, each sample of input signal is quantized independently, without any reference to the values of any previous input samples. It turns out that when a bandlimited signal is sampled at or above the Nyquist rate, successive samples are correlated. The dependence between successive sample values is particularly strong when

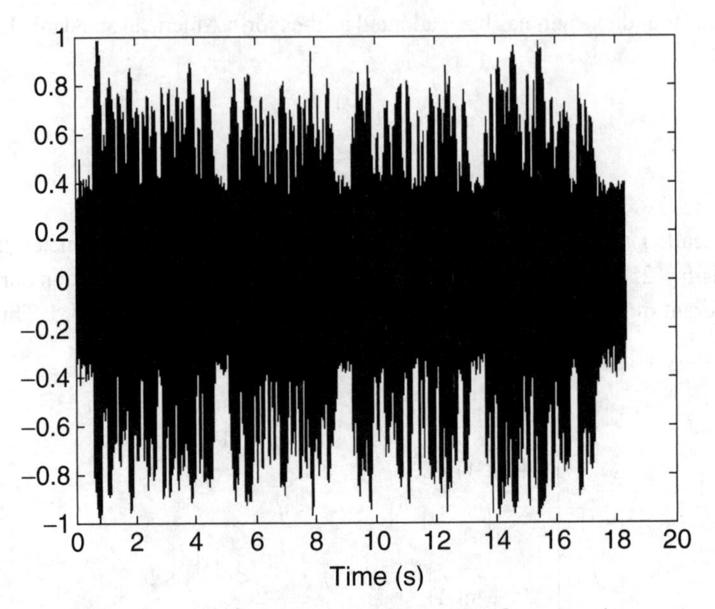

Figure 7.10 The Speech Sample from Figure 7.7 after μ-Law Compression

the sampling rate is much higher than the Nyquist rate. Figure 7.11 shows an oversampled bandlimited signal, which illustrates the point that successive samples tend to have similar values. A PCM system can take advantage of this signal behavior by quantizing the difference between successive sample values instead of quantizing the sample values. The smaller range taken by

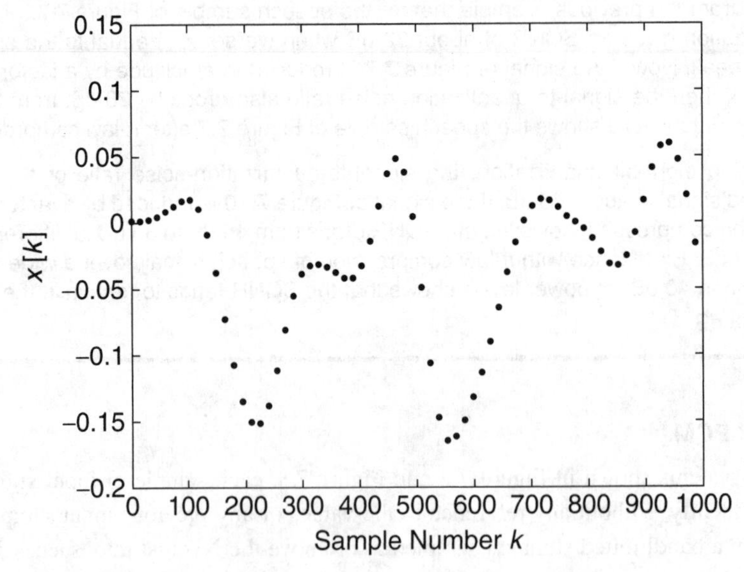

Figure 7.11 An Oversampled Bandlimited Signal

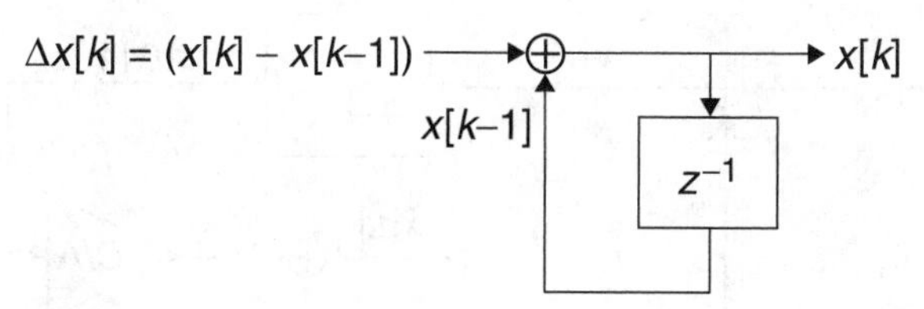

Figure 7.12 An Accumulator

the difference signal allows a given SQNR to be achieved with fewer quantization levels, which in turn implies that fewer bits are required to represent each sample. Alternatively, a higher SQNR can be achieved for a given number of bits per sample.

In this section we examine two PCM schemes that take advantage of the correlation between input samples. The first, and simpler, scheme is delta modulation. The second scheme is a generalization of delta modulation known as differential PCM (DPCM).

Delta Modulation

We first note that if the decoder were given the sequence of differences $\Delta x[0] = (x[0] - x[-1])$, $\Delta x[1] = (x[1] - x[0]), \Delta x[2] = (x[2] - x[1]), \ldots$, with initial condition $x[-1] = 0$, then it could reconstruct the input sequence by using an accumulator, as shown in Figure 7.12. In the figure the block labeled "z^{-1}" represents a one-sample delay. At each sample time k, the accumulator takes in the difference $\Delta x[k] = (x[k] - x[k-1])$ and adds the previous output $x[k-1]$ to produce the new output $x[k]$.

Now the decoder does not actually receive the sequence of differences $\Delta x[k] = (x[k] - x[k-1])$ but instead receives their quantized representation, which we will call $\widehat{\Delta x}[k]$. Therefore the best that the decoder can do is to accumulate the quantized representations, as shown in Figure 7.13

The encoder can maintain a duplicate copy of the decoder's accumulator and can thereby keep track of what the decoder is producing. The encoder's goal is to ensure that the decoder's reconstructed estimate $\hat{x}[k]$ is as close as possible to the input sample $x[k]$. Consequently, the signal that the encoder sends should actually be the difference signal $y[k] = x[k] - \hat{x}[k-1]$ rather than $\Delta x[k] = x[k] - x[k-1]$. Of course, the encoder sends a quantized version $\hat{y}[k]$ of this difference signal. The resulting delta modulation encoder and decoder are shown in Figure 7.14. For classical delta modulation, the quantizer is a two-level (one-bit) quantizer/coder that produces an output of 1 when its input is positive and an output of 0 when its input is negative. The blocks labeled "D/A" turn a 1 into a signal level of $+\Delta$ and a 0 into a signal level of $-\Delta$.

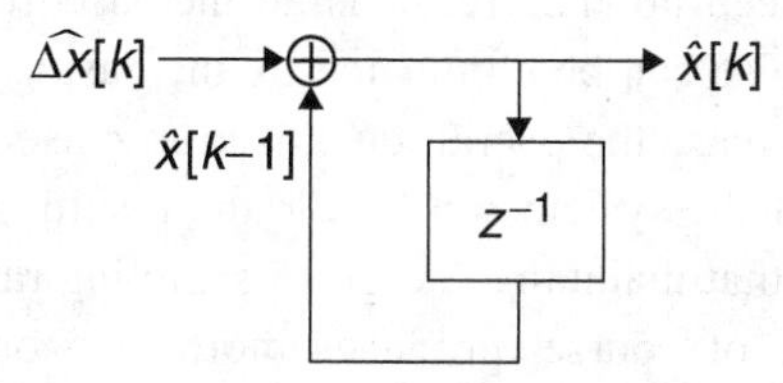

Figure 7.13 Accumulating Quantized Differences

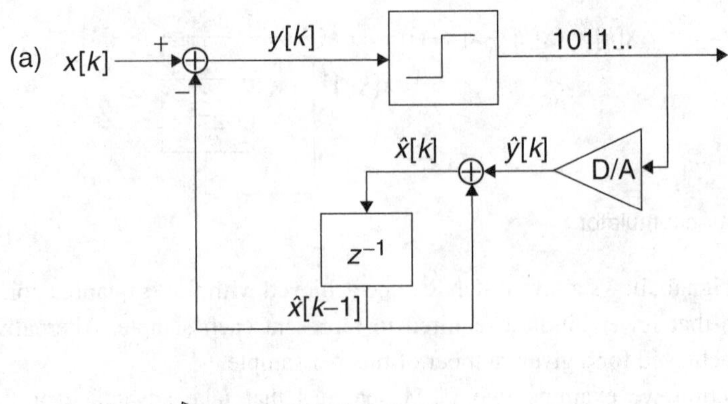

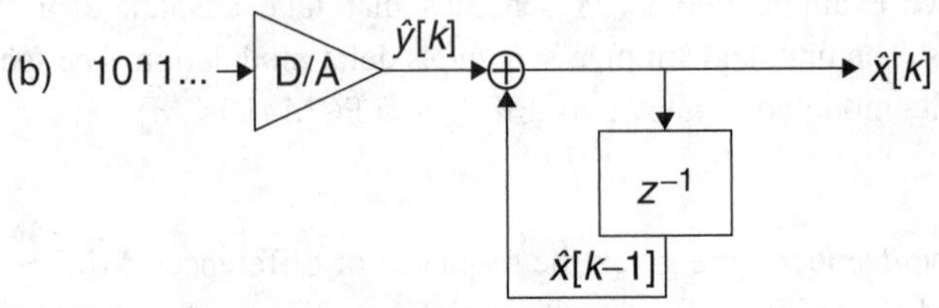

Figure 7.14 Delta Modulator: (a) Encoder; (b) Decoder

The quantization error produced by the delta modulator is given by

$$\begin{aligned} e_q[k] &= \hat{x}[k] - x[k] \\ &= (\hat{y}[k] + \hat{x}[k-1]) - x[k] \\ &= \hat{y}[k] - (x[k] - \hat{x}[k-1]) \\ &= \hat{y}[k] - y[k], \end{aligned} \tag{7.21}$$

which is the error produced by the quantizer itself. Since at the quantizer output $\hat{y}[k] = \pm\Delta$, the quantization error will depend on the values taken by the differences $y[k] = x[k] - \hat{x}[k-1]$. As long as the differences in the sequence $y[k]$ remain near zero, the quantization error will be dominated by $\pm\Delta$.

Figure 7.15 shows an input signal $x(t)$ and a delta modulation decoder output $\hat{x}[k]$. As long as the input signal changes gradually with time, the decoder output will "dither" around $x(t)$ in steps of $\pm\Delta$. The figure shows a segment in which $x(t)$ rises rapidly, and $\hat{x}[k]$, increasing in steps of Δ, is unable to keep up. The result is an increased error, called **slope overload noise**. The slope overload noise can be eliminated by increasing the size of the quantization parameter Δ, but this will increase the overall quantization noise level. A better solution is to increase the sampling rate, if the system can handle the resulting increase in transmitted bit rate. An alternative solution that maintains the given sampling rate is to increase the number of quantization levels. This, of course, produces more than one bit per sample and also increases the transmitted bit rate.

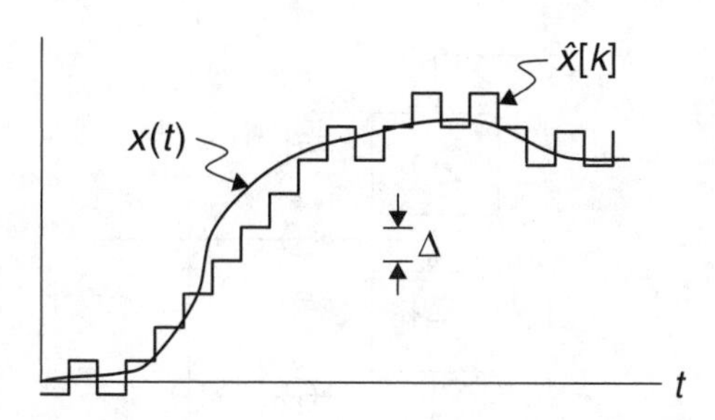

Figure 7.15 Continuous Input Signal and Reconstructed Delta Modulator Output

Differential PCM

The performance of the delta modulator shown in Figure 7.14 depends on the differences $y[k] = x[k] - \hat{x}[k-1]$ being small most of the time. We can think of $\hat{x}[k-1]$ as representing a *prediction* of the input sample $x[k]$. The more accurate the prediction, the smaller the quantization noise. Now the prediction can be made more accurate if we use several past samples of $\hat{x}[k]$ to predict $x[k]$, rather than just the one sample $\hat{x}[k-1]$; that is, let $x[k]$ be predicted by $\sum_{i=1}^{p} a_i\hat{x}[k-i]$, where p is a design parameter and the coefficients $a_i, i = 1,\ldots,p$ are chosen to minimize the prediction error.[5] One possible implementation of the predictor is shown in Figure 7.16. A block diagram of a general DPCM system is shown in Figure 7.17. Figure 7.17 is the same as Figure 7.14, with the z^{-1} block replaced by the predictor.

Differential PCM systems can be made adaptive by allowing the quantizer to adjust the step size Δ based on the dynamic range of the input. This allows the SQNR to be maintained at a constant level for both loud and quiet speakers. Adaptive DPCM systems that encode speech at 32 kbits/s are used in several cordless telephone systems.[6]

Vocoders

The vocoder concept begins with a model of the human vocal system. The vocal tract extends from the vocal cords to the mouth, a distance of about 17 cm in the average man, a few centimeters less in the average woman. The vocal tract acts as a lowpass filter with several broad resonances that speech physiologists call "formants." The speaker can adjust the formant frequencies to some extent by altering the shape of the vocal tract. In speech the shape of the vocal tract changes slowly. A lowpass filter model of the vocal tract needs to be updated every 20 ms or so. To produce sounds, the vocal tract has to be excited by an input. Voiced sounds are produced when periodic vibrations of the vocal cords excite the vocal tract. The pitch period is established

5. See J. G. Proakis and M. Salehi, *Communication System Engineering,* 2nd ed. (Upper Saddle River, NJ: Prentice Hall, 2002), 307–9, for details on how to choose the coefficients.
6. T. S. Rappaport, *Wireless Communications: Principles and Practice,* 2nd ed. (Upper Saddle River, NJ: Prentice Hall, 2002), 438.

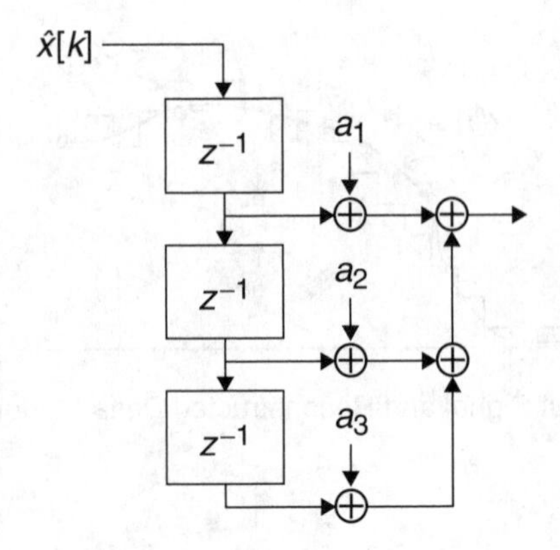

Figure 7.16 A Linear Predictor with $p = 3$

by the vocal cords and is typically between 2 and 20 ms. Unvoiced sounds are produced when air is forced at high speed through a constriction in the vocal tract.

A model of the speech-producing mechanism is shown in Figure 7.18. The model contains two blocks, one representing the excitation and one representing the filtering action of the vocal tract. Both blocks are adjustable. The excitation can produce tones of various frequencies or broadband noise, and the vocal tract model can be adjusted to track changes in the formant frequencies.

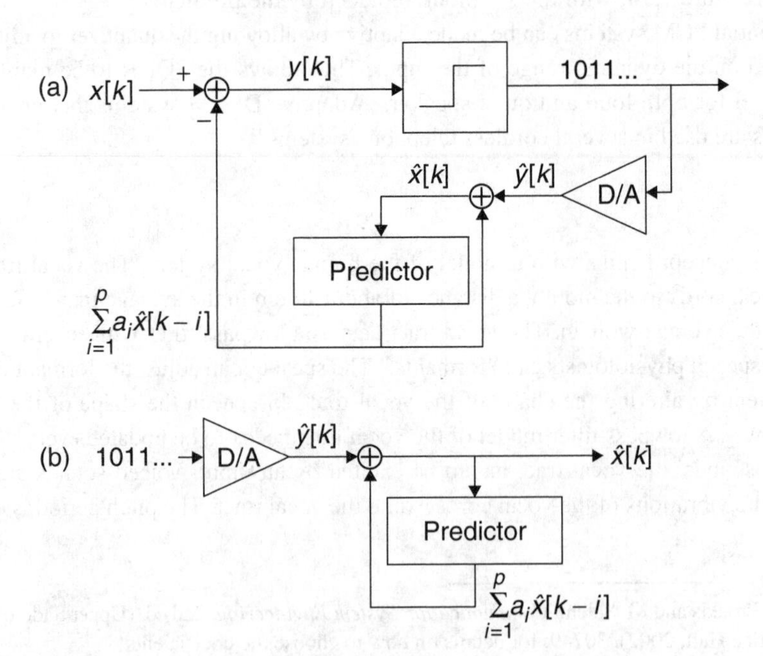

Figure 7.17 Differential PCM: (a) Encoder; (b) Decoder

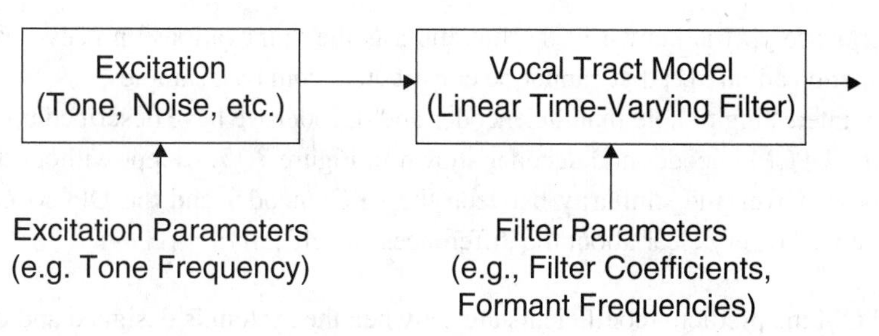

Figure 7.18 A Model of the Vocal System

In operation, the speech signal at the input to the encoder is sampled, and the samples are grouped into blocks, with each block spanning a period, typically 10 to 30 ms, during which the vocal system's parameters are not expected to change significantly. The encoder estimates the type of excitation and, if relevant, the pitch of the excitation tone. The encoder also estimates the parameters of the vocal tract filter. The excitation and filter parameters (not the speech samples!) are then transmitted over the communication link to the decoder. The decoder inserts the received parameters into a model like that shown in the figure to produce the recovered sound. The procedure is repeated for every block of input samples.

Linear Predictive Coding

The most popular form of vocoder at the present time is the **linear predictive coder** (LPC). The encoding side of a linear predictive coder takes advantage of the fact that successive samples of a speech waveform are highly correlated. Given the samples $x[k-1], x[k-2], \ldots, x[k-p]$ of the input speech signal, the linear combination $\sum_{i=1}^{p} a_i x[k-i]$ can be a very close prediction of the next sample $x[k]$, provided the coefficients $a_1, \ldots, a_p$ are chosen correctly. The encoder examines a block of speech samples and chooses the coefficients to minimize the mean-square prediction error $y[k]$ over that block, where

$$x[k] - \sum_{i=1}^{p} a_i x[k-i] = y[k]. \tag{7.22}$$

Note that the predictor is the same as that shown in Figure 7.16. The idea is that there is no point in transmitting the prediction $\sum_{i=1}^{p} a_i x[k-i]$ to the decoder; the decoder can make its own prediction of the reconstructed speech sample $\hat{x}[k]$ based on its prior reconstructed samples $\hat{x}[k-1], \hat{x}[k-2], \ldots, \hat{x}[k-p]$. The encoder need only provide the decoder with the error value $y[k]$, so that the decoder can correct its estimate. The decoder updates its estimate using

$$\hat{x}[k] = \sum_{i=1}^{p} a_i \hat{x}[k-i] + y[k]. \tag{7.23}$$

The error sequence $y[k]$ is known in the literature as the "innovations" process, as it contains "new" information about $x[k]$ that cannot be constructed from past samples.

It is very interesting to note that the encoder and decoder we have described up to this point are exactly the DPCM encoder and decoder shown in Figure 7.17, except without the quantizer and D/A blocks. Given the similarity between the LPC encoder and the DPCM encoder, it is important that we be very clear about the differences, specifically:

- In DPCM the predictor coefficients are set when the system is designed and do not change. In LPC a new set of coefficients is computed by the encoder for every block of input samples. The values of these coefficients must be transmitted to the receiver.
- In DPCM the innovations are quantized, and the quantized values are sent to the receiver. In LPC the innovations values are not sent to the receiver at all. The receiver is provided with only some general attributes of the sequence $y[k]$, such as an overall amplitude coefficient, whether $y[k]$ is "noiselike" or "tonelike," and, if tonelike, an estimate of the pitch.

Equation (7.23) is a difference equation that represents a discrete-time filter having input $y[k]$ and output $\hat{x}[k]$. This filter is the vocal tract model of Figure 7.18. Since the decoder is not provided with the actual innovations process $y[k]$, the excitation block is tasked with generating an approximation to that process.

Generating an effective excitation for the decoder's vocal tract model turns out to be a significant challenge. Historically there has been considerable research to find the most effective signal. Early LPC vocoders chose one of two excitation signals, a noiselike signal for unvoiced speech and a pulse train at the vocal cord pitch frequency for voiced signals. It turned out, however, that harmonics of the pulse train caused the reconstructed speech to have a "buzzy" quality. A multipulse-excited LPC encoder uses as many as eight pulses per period to excite the vocal track filter. Heights and positions of the pulses can be individually adjusted. A code-excited LPC vocoder has a codebook of potential excitation signals. The encoder tries each of the excitations in turn, looking for the one that produces the closest fit to the actual speech signal. The decoder has an identical codebook, so only the index number identifying the selected excitation signal needs to be transmitted to the decoder. It is easy to see that if the codebook is large, the encoder may use up a great deal of time in finding the optimum excitation for each segment of speech. A smaller codebook can be searched more quickly but will produce reconstructed speech of lesser quality.

The output of the encoder of an LPC vocoder is a frame of data. A new frame is generated for every block of input samples. A typical frame contains a binary representation of each of the predictor coefficients and a binary representation of each parameter of the excitation signal. To get a sense of the size of a frame, we might allow six bits for each of ten predictor coefficients. Voiced versus unvoiced excitation requires one bit. The pitch frequency requires six bits, and the overall gain coefficient five bits. If these parameters are updated every 15 to 30 ms, the resulting transmission rate is 2400 to 4800 bits/s.

Example[7]

The USDC system uses vector-sum excited linear predictive (VSELP) coding, a variation of code-excited LPC that uses an excitation signal formed as a combination of vectors from two codebooks. Recall that in Chapter 6 we described the USDC "full-rate" transmission format as allowing each user two slots in every 40 ms TDMA frame. Each slot carries 260 bits of data, so the vocoder can generate 260 bits every 20 ms.

The 260 data bits that make up one vocoder frame contain 159 bits of vocoder data and 101 additional "parity" bits for error correction.

The 159 bits of vocoder data in each frame are accounted for as follows:

- The exciter coefficients used to combine the codebook outputs are estimated four times per frame and each time are encoded together into an 8-bit parameter, for a total of 32 bits.
- Excitation code words are chosen four times per frame. Each of the two code words requires a 7-bit address. Together this accounts for 56 bits per frame.
- Pitch frequency is updated four times per frame. Each update generates 7 bits for a total of 28 bits.
- The vocal tract filter has ten coefficients. These are updated once per frame and together require 38 bits.
- An additional 5 bits encode the overall energy per frame.

For error control purposes the 159 data bits are divided into two groups. The "class 1" group contains 77 bits, and these are protected by error-correcting codes. The "class 2" group, containing 82 bits, is unprotected. As an example, all of the pitch control bits are designated class 1.

- Of the class 1 bits, 12 of the overall energy and filter coefficient bits are considered particularly sensitive. These have 7 parity bits added.
- All 77 class 1 bits plus the 7 parity bits plus 5 "tail" bits are encoded using a rate-1/2 convolutional code (described below). This expands the class 1 component to $(77 + 7 + 5) \times 2 = 178$ bits.
- Adding in the unprotected class 2 bits gives the vocoder frame of $178 + 82 = 260$ bits.

The USDC example is typical of vocoder applications and leads to two important observations. First, we now see how it is that digitized speech data is naturally organized into frames, even though speech is a streaming, rather than a bursty, application. Second, we see that the bits that make up a vocoder frame can differ in their sensitivity to transmission errors. The fact that certain bits are protected by error-correcting codes, while other bits are not, also complicates our ability to relate the bit error rate of the data link to the quality of service perceived by the user.

7. See M. Schwartz, *Mobile Wireless Communications* (Cambridge: Cambridge University Press, 2005), 251, for more detail.

Some bit errors are corrected by the codes and "don't count," whereas others remain and are detectable by the users.

Vocoder design involves many trade-offs, and ultimately user perception determines whether a design is a success. Vocoders are extensively tested on human subjects during the development stage. Since vocoders are specialized for encoding human speech, they are particularly sensitive to background noise, especially such "nonspeech" types of noise as background music. Another particular challenge comes from the fact that telephone speech signals are often encoded multiple times. This can arise when a speech signal originates at a cellular handset and is encoded using a vocoder for transmission over the wireless link. At the mobile switching center the signal will be decoded back to analog form and then reencoded using PCM for transmission over the wired telephone network. If the call terminates in another wireless connection, there may be an additional encoding, possibly using a different vocoder. Finally, we note that because vocoders do not reproduce the input waveform, they do not perform well with data or fax modems. In modern wireless systems, data and fax signals are identified at the handset and transmitted using data-specific protocols.

Coding for Error Correction

Wireless links in cellular telephone systems operate at bit error rates that can be as high as 10^{-3} or even 10^{-2}. These are very high error rates compared to those normally experienced on wired telephone and cable-based data networks. It has long been customary in data networks to protect the integrity of transmitted data with error correction mechanisms. This allows users of the Internet, for example, to be sure that received data is virtually error-free, even though the communication link may be noisy. The mechanism that is favored for error control on data networks is called **automatic repeat request** (ARQ). To implement this mechanism, transmitted bits are grouped into packets or frames, and each frame is transmitted along with a checksum or "parity check." The checksums allow detection of all likely errors and the vast majority of all possible errors. When a frame arrives with errors, the frame is discarded and eventually retransmitted. Thus virtually all frames eventually arrive without errors, but there may be unpredictable delays in the times at which frames arrive owing to the fact that some frames are transmitted more than once. In contrast to the data commonly transmitted over the Internet, the digitized voice signals that make up the primary information transmitted over cellular links are very sensitive to delay and to delay jitter. Consequently, ARQ techniques have not been effective for use with voice signals, and it has been preferable to allow errors to remain uncorrected rather than to introduce the unpredictable delays associated with ARQ error correction.

It turns out that there is an alternative mechanism for error control that does not require multiple transmissions and does not introduce unpredictable delays. This is **forward error correction**. Frames are protected with parity checks as in ARQ, but unlike in ARQ the parity checks allow the most common errors to be corrected without retransmission. Forward error correction is not as effective as ARQ in eliminating errors, but it is effective enough for use with voice signals that are somewhat tolerant of a residual error rate.

Example

As a simple illustration of how forward error correction works, consider the "triple redundancy" error-correcting code. Each bit is replicated three times, and a frame consists of blocks of three identical bits. At the receiver each frame is examined, and a majority vote determines the polarity of the bit provided to the user. If there are no transmission errors, there will be three identical bits of the correct polarity received. If there is a single transmission error, there will still be two bits of the correct polarity received out of three, and correct data will be provided to the user. Unfortunately two or three errors in a frame will result in incorrect data provided to the user. Thus the code is able to correct all single errors, but no double or triple errors.

Notice that if triple redundancy were used for error detection only, both single and double errors within a frame could be detected. This illustrates the general principle that codes can detect more errors than they can correct.

Triple redundancy requires that bits be transmitted on the link at a rate three times the rate at which information bits arrive. We say that the code has "rate" $R = 1/3$. The increased data rate is the price of the error protection.

The example illustrates a number of general properties of forward-error-correcting codes. Among these are the addition of "redundant" bits to a frame to allow errors to be corrected, the fact that not all errors are correctable, and most important, the trade of decreased error rate for increased transmission rate. Codes more sophisticated than triple redundancy will allow more than one error per three bits to be corrected in return for doubling or tripling the transmission rate.

Example

The cdmaOne cellular system uses a voice encoder that produces output data at a (maximum) rate of 9600 bits/s. In the forward link a rate $R = 1/2$ error-correcting code is used, so that data is provided to the link at 19.2 kbits/s. In the reverse link additional error protection is required owing to the lower transmitter power in the mobile units. A rate $R = 1/3$ code is used in this case, so that data is provided to the link at 28.8 kbits/s.

Convolutional Codes

A type of forward-error-correcting code that has proved highly effective for use in cellular systems is known as a **convolutional code**. In this section we describe how the encoding is done. In subsequent sections we discuss decoding and introduce an efficient optimal decoder known as the Viterbi decoder. We conclude with a brief discussion of the performance of convolutional codes.

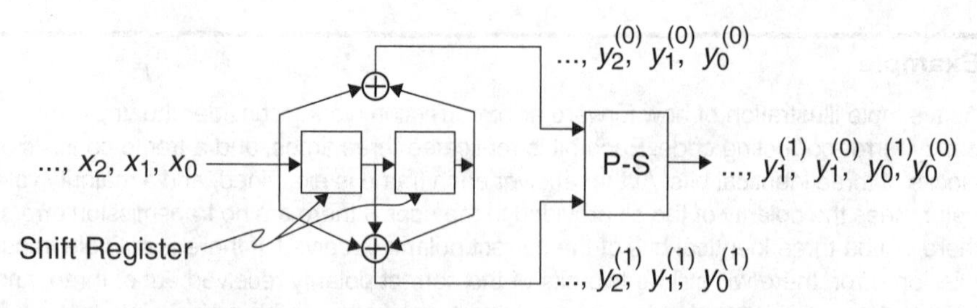

Figure 7.19 A Constraint-Length $K = 3$, Rate $R = 1/2$ Convolutional Encoder

Encoding

A convolutional encoder is a multiple-output, finite-impulse-response, discrete-time filter. An example is shown in Figure 7.19. In the figure the square boxes represent stages in a shift register, the ⊕ symbols represent modulo-2 addition ($0 + 0 = 0$, $0 + 1 = 1 + 0 = 1$, $1 + 1 = 0$), and the box labeled "P-S" is a parallel-to-serial converter. To illustrate the operation of the encoder, imagine that the shift register is in state 00, and that $x_0 = 1$ appears at the input. The upper branch will produce output bit $y_0^{(0)} = 1 + 0 = 1$, and the lower branch will produce $y_0^{(1)} = 1 + 0 + 0 = 1$. The output bits will then be 1,1, and the state of the shift register will advance to 10. In this example every input bit produces two output bits, and hence the rate of the encoder is $R = 1/2$. Additional adders and output streams can be included to produce a rate $R = 1/n$ code, with $n > 2$. This particular encoder is said to have memory $m = 2$, because the shift register has two stages. The **constraint length** is the maximum number of bits in any output stream that can be affected by any input bit. The constraint length is one greater than the memory, and in this example we have $K = 3$.

Convolutional encoders differ from one another in the rate, in the constraint length, and in how the shift register is wired to the output stream. We can specify the "wiring" by identifying the impulse response of each of the discrete-time filters that make up the encoder. In Figure 7.19 the upper branch has impulse response $g^{(0)} = 101$, and the lower branch has impulse response $g^{(1)} = 111$. Following tradition, these impulse responses are called "generators" of the code. They are sometimes written as octal numbers—for example, $g^{(0)} = 5$, $g^{(1)} = 7$—to save space.

Example

The forward links in the cdmaOne and cdma2000 systems use a $K = 9$, $R = 1/2$ convolutional encoder with $g^{(0)} = 753$ and $g^{(1)} = 561$. The encoder is shown in Figure 7.20.

The ability of a convolutional code to correct errors improves with decreasing rate and increasing constraint length. As the rate decreases, more bits must be transmitted for each bit of data. Increasing the constraint length does not affect the rate at which bits are transmitted, but, as we shall see, the complexity of the optimal decoder increases *exponentially* with increasing constraint length. The speed and size of the decoding processor thus determine the performance improvement that can be realized through use of a convolutional code.

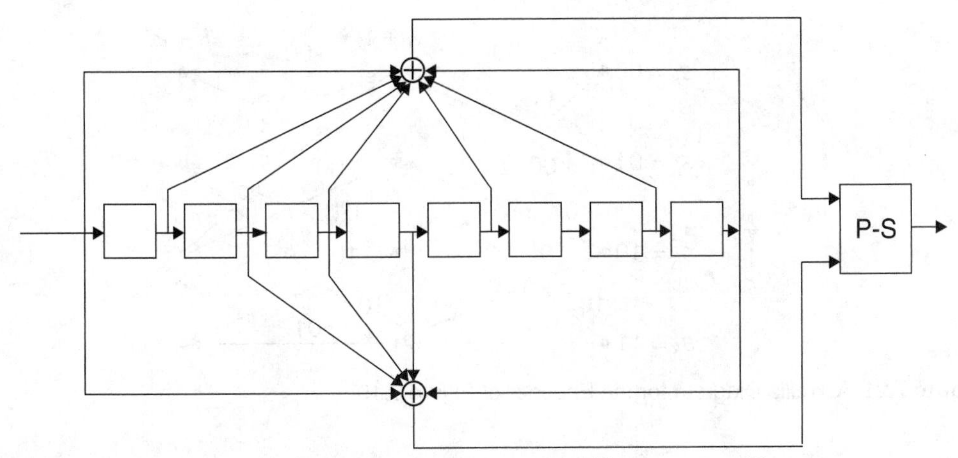

Figure 7.20 A $K=9$, $R=1/2$ Convolutional Encoder with $g^{(0)}=753$ and $g^{(1)}=561$

The Trellis Diagram

A convolutional encoder is a finite state machine, and because of this a state diagram is a particularly useful way of visualizing its operation. One way of drawing such a state diagram is to list the states vertically, showing time evolution in the horizontal direction. A diagram of this sort is called a **trellis diagram**. An example of a trellis diagram for the encoder of Figure 7.19 is shown in Figure 7.21. In this figure the states are shown as black dots arranged vertically. Time, indicated as $\ldots, k, k+1, k+2, \ldots$, increases to the right. The arrows show possible state transitions. Two possible paths leave each state. The upper path is the transition that will be taken when the input bit $x_k = 0$, and the lower path is the transition taken when the input bit $x_k = 1$. The numbers along each arrow are the output bits y_k^0, y_k^1 that result from the corresponding initial state and input.

Given the trellis diagram, an initial state, and an input sequence, we can trace a path through the trellis. This allows us to find both the sequence of states and the output sequence. For example, using Figure 7.21 with initial state $s_0 = 00$, the input sequence

$$x = 0,1,1,0,1,\ldots \tag{7.24}$$

produces the state sequence

$$s = 00,00,10,11,01,10,\ldots \tag{7.25}$$

and the output sequence

$$y = 00,11,10,10,00,\ldots \tag{7.26}$$

This path through the trellis is shown in Figure 7.22.

It is important to note that the process of tracing a path through the trellis can be reversed; that is, given a sufficiently long code sequence, we can match it up against the trellis and find the

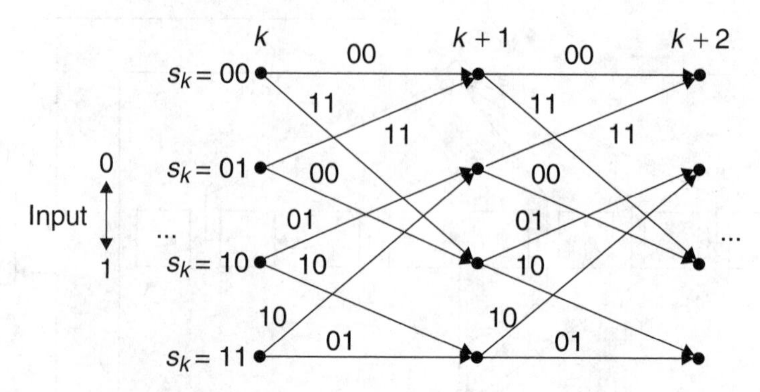

Figure 7.21 A Trellis Diagram for the Encoder of Figure 7.19

path corresponding to that sequence. For example, the code sequence

$$y = 11,10,10,00,01,\ldots \tag{7.27}$$

matches only the state sequence

$$s = 00,10,11,01,10,01,\ldots, \tag{7.28}$$

and this in turn matches the input sequence

$$x = 1,1,0,1,0,\ldots. \tag{7.29}$$

Note that we do not even need to know the initial state. The code sequence given by Equation (7.27) matches only one sequence of states, that given by Equation (7.28). Thus we see how decoding might be done, at least in the absence of bit errors.

Decoding

When a binary sequence is received over a noisy communication channel, occasional bits will be received incorrectly. This means that the received sequence will not correspond exactly with the transmitted sequence. Usually, the received sequence will not match *any* sequence that could

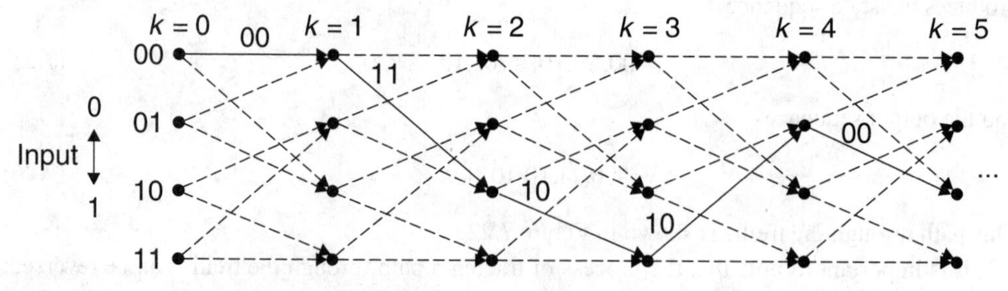

Figure 7.22 A Path through the Trellis

possibly have been generated by following a path through the code trellis. When this happens, the receiver's job is to determine the trellis path *most likely* to have produced the received sequence. When bit errors occur independently and with probability less than 0.5, a pattern with fewer errors is more likely to occur than a pattern with many errors.

Example

Suppose the probability of a bit error is $P = 10^{-3}$. The probability of no errors occurring in an eight-bit word is

$$P_0 = (1-10^{-3})^8 = 0.992.$$

The probability of a single error in any of the eight positions is

$$P_1 = \binom{8}{1} 10^{-3}(1-10^{-3})^7 = 0.00794.$$

The probability of a double error is

$$P_2 = \binom{8}{2}(10^{-3})^2(1-10^{-3})^6 = 2.78 \times 10^{-5}.$$

The probability of a triple error is

$$P_3 = \binom{8}{3}(10^{-3})^3(1-10^{-3})^5 = 5.57 \times 10^{-8}.$$

We see that the most likely event is that there are no errors. A single error is much more likely than a double error, and a double error is much more likely than a triple error.

The receiver finds the most likely transmitted sequence by locating the path through the trellis whose output sequence differs from the actual received bit sequence in the smallest number of places. Note that if an unfortunate pattern of errors occurs that causes the received sequence to exactly match a path through the trellis other than the one that was transmitted, the receiver will be unable to detect that errors have occurred. For a well-designed code, distinct paths through the trellis produce sequences that differ in many bits, so that the probability of an undetectable error occurring is small.

To help us quantify differences between bit sequences, let us define the **Hamming distance** between two bit sequences as the number of places in which the sequences differ. For example, the sequences $y_1 = 0,1,0,1,0,0,1,1$ and $y_2 = 0,0,0,1,0,1,1,1$ are separated by a Hamming distance of $d(y_1,y_2) = 2$.

Now given a received bit sequence r, the receiver finds the most likely transmitted sequence as follows. First, the receiver traces every possible path through the code trellis. For each path there will be a corresponding output sequence y. The receiver computes the Hamming distance $d(r, y)$ for each of these output sequences and chooses the trellis path for which the Hamming distance is smallest. Once a trellis path is found, the corresponding input sequence is easily deduced. This decoding procedure follows what is called a **maximum likelihood rule**. It is an optimum procedure that will produce the minimum probability of error for the case in which the input bits are independent and equally likely to be 1 or 0. In the following discussion, the Hamming distance $d(r, y)$ will be called the **path metric**.

Example

Suppose the encoder of 7.19 is used, and the sequence

$$r = 0,1,1,1,1,0,1,0$$

is received. Table 7.2 shows the comparison of r with each path through the trellis. Possible input sequences, corresponding encoder output sequences, and Hamming distances (path metrics) from the received sequence are shown. To save space, only paths starting at state $s_0 = 00$ are shown, but in practice the table would show paths beginning from each of the four states.

Table 7.2 Comparison of Trellis Paths (from State 00) with Received Sequence

Input Sequences *x*	Output Sequences *y*	Distance *d*(*r*,*y*)
0000	00000000	5
0001	00000011	5
0010	00001101	6
0011	00001110	5
0100	00110111	4
0101	00110100	4
0110	**00111010**	**1**
0111	00111001	3
1000	11011100	4
1001	11011111	4
1010	11010001	5
1011	11010010	3
1100	11101011	3
1101	11101000	3
1110	11100110	4
1111	11100101	6
	...	

In the table the row for input sequence $x = 0,1,1,0$ has been emphasized. Since the path metric $d(r,y)$ is minimum for y corresponding to input sequence $x = 0,1,1,0$, the decoder will choose this data sequence as its maximum likelihood output. Note that for this input sequence, r and y differ in only one bit. It is possible that the decoder is wrong and some other input sequence was actually applied to the encoder. If, for example, the actual input sequence were $x = 0,1,1,1$, then r and y would differ in three bits. This occurrence is certainly possible, but a three-bit error is much less likely than a one-bit error, and so the receiver, following the maximum likelihood rule, produces the output corresponding to the more likely error pattern.

The example illustrates the maximum likelihood decoding principle, but it also suggests the challenge that implementing such a decoder implies. For a 4-bit input sequence, there are 16 paths through the trellis starting from each state. For a 5-bit input sequence there are 32 paths starting from each state. The number of sequences that must be checked appears to double for each additional input bit! For a cellular telephone system that might take in 200 bits at a time, there would be on the order of $2^{200} = 1.6 \times 10^{60}$ paths to compare with the received sequence. It turns out, fortunately, that some clever shortcuts eliminate most of the paths. The most widely used decoding technique, which takes advantage of these shortcuts, was proposed by A. J. Viterbi in 1967[8] and is called the **Viterbi algorithm**. We present this procedure in the next section.

The Viterbi Algorithm

In using the Viterbi algorithm the decoder advances along the trellis one step for each set of n received bits. We describe the process the decoder uses to advance one step; the decoder applies this process over and over. It is convenient to imagine that the encoder begins in state $s_0 = 00$ and returns to state 00 at the end of transmission. To force the encoder into the zero state, the input data at the transmitter must end with a "tail" of $K-1$ zeros.

Refer, for example, to the trellis of Figure 7.21. Starting at the zero state, the transmitter may generate either of two branches, $y_0 = y_0^{(0)}, y_0^{(1)} = 00$ or $y_0 = y_0^{(0)}, y_0^{(1)} = 11$, ending on states $s_1 = 00$ or $s_1 = 10$, respectively. Given the received sequence $r_0 = r_0^{(0)}, r_0^{(1)}$, the receiver can compute the **branch metric**

$$M(r_0|y_0) = d(r_0^{(0)} r_0^{(1)}, y_0^{(0)} y_0^{(1)}) \tag{7.30}$$

for each branch. At an additional step into the trellis, there are now four possible paths,

$$y_1 = y_0^{(0)}, y_0^{(1)}, y_1^{(0)}, y_1^{(1)} = \begin{cases} 0000 \\ 1101 \\ 0011 \\ 1110 \end{cases}, \text{ ending on states } s_2 = \begin{cases} 00 \\ 01 \\ 10 \\ 11 \end{cases}, \text{ respectively.}$$

---

8. A. J. Viterbi, "Error Bounds for Convolutional Codes and an Asymptotically Optimum Decoding Algorithm," *IEEE Transactions on Information Theory* IT-13 (April 1967): 260–69.

The four branch metrics for the second step are

$$M(r_1|y_1) = d\left(r_1^{(0)}\, r_1^{(1)}, y_1^{(0)}\, y_1^{(1)}\right). \tag{7.31}$$

Adding the branch metrics for the first and second steps will give a **partial path metric** for each of the four paths beginning at state $s_0 = 00$ and ending on state s_2, that is,

$$M^2(r|y) = M(r_0|y_0) + M(r_1|y_1). \tag{7.32}$$

Example

Figure 7.23 shows the first two steps of the trellis of Figure 7.21. Suppose the received sequence is

$$r = r_0^{(0)}, r_0^{(1)}, r_1^{(0)}, r_1^{(1)}, \ldots = 0,1,1,1,\ldots \tag{7.33}$$

Then for the first step the branch metrics are $M(r_0|y_0) = d(01,00) = 1$ for state transition $s_0 = 00 \rightarrow s_1 = 00$ and $M(r_0|y_0) = d(01,11) = 1$ for state transition $s_0 = 00 \rightarrow s_1 = 10$. For the second step the brancn metrics are $M(r_1|y_1) = d(11,00) = 2$ for state transition $s_1 = 00 \rightarrow s_2 = 00$, $M(r_1|y_1) = d(11,11) = 0$ for state transition $s_1 = 00 \rightarrow s_2 = 10$, $M(r_1|y_1) = d(01,01) = 1$ for state transition $s_1 = 10 \rightarrow s_2 = 01$, and $M(r_1|y_1) = d(11,10) = 1$ for state transition $s_1 = 10 \rightarrow s_2 = 11$.

We can now compute the partial path metrics for each of the four paths shown. We have $M^2(r|y) = M(r_0|y_0) + M(r_1|y_1) = 1 + 2 = 3$ for path $s_0 = 00 \rightarrow s_1 = 00 \rightarrow$, $s_2 = 00$, and $M^2(r|y) = 1 + 0 = 1$, $1 + 1 = 2$, and $1 + 1 = 2$ for the remaining three paths.

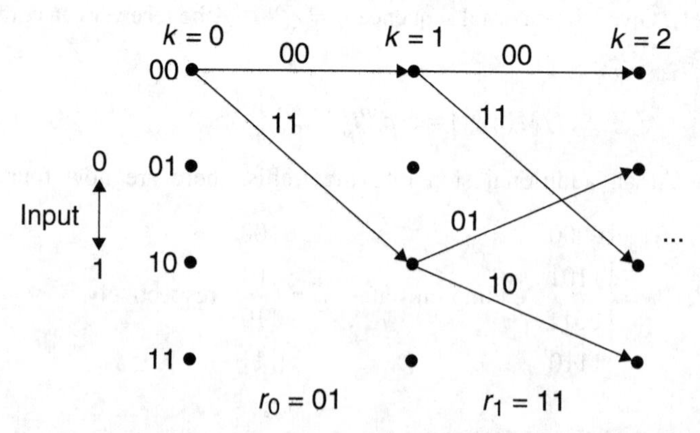

Figure 7.23 The First Two Stages of a Code Trellis and a Given Received Sequence

Note that as long as we know the initial state, we can represent a path through the trellis either by specifying a sequence of states or by specifying a sequence of inputs. Thus the path $s_0 = 00 \rightarrow s_1 = 10 \rightarrow s_2 = 01$ can also be written $x = 1,0$.

The procedure illustrated in the example can now be used incrementally as we advance through the trellis. At each step we can append a bit to the input sequence to specify a path and add an additional branch metric to update the path metric. Ultimately, when transmission is complete and the paths have all converged to state 00, the path with the smallest path metric is the maximum likelihood choice.

Example

Continuing the example of Figure 7.23, suppose the next two received bits are

$$r_2 = r_2^{(0)}, r_2^{(1)} = 1,1. \tag{7.34}$$

For the path $x = 1,0$ we already have $M^2(r|y) = 2$. Then for the path $x = 1,0,0$ we can calculate $M^3(r|y) = M^2(r|y) + M^2(r_2|y_2) = 2 + 0 = 2$, and for the path $x = 1,0,1$ we can calculate $M^3(r|y) = M^2(r|y) + M(r_2|y_2) = 2 + 2 = 4$.

Notice that once we get past the first two stages in the example of Figure 7.23, there will be two paths converging on every state. For each pair of paths that converge on a state, we can eliminate the path with the larger metric from further consideration. The path with the smaller metric is called the **survivor**. At each stage in our progress through the trellis, we will save only the survivor for each state. Thus, if there are four states, then the number of paths under consideration will never be more than four. This is the critical step in the Viterbi algorithm. As long as we keep only the survivors at each stage, the number of paths does not grow as we progress through the trellis. The decoder complexity will remain tractable no matter how long the surviving paths become.

Example

Figure 7.24 shows one more step in the example of Figure 7.23. Each of the states at $k = 2$ is marked with the value of the partial path metric $M^2(r|y)$ of the path leading to that state. The numbers adjacent to the branches that join $k = 2$ with $k = 3$ are the branch metrics $M(r_2|y_2)$ induced by the received bits given by Equation (7.34).

We see that two paths converge on state $s_3 = 00$. One of the paths is $x = 0,0,0$, with partial path metric $M^3(r|y) = 3 + 2 = 5$, and the other path is $x = 1,0,0$, with partial path metric $M^3(r|y) = 2 + 0 = 2$. Comparing the metrics, we choose $x = 1,0,0$ as the survivor and remove the other choice from the trellis. If we choose a survivor for every state, the result is shown in Figure 7.25.

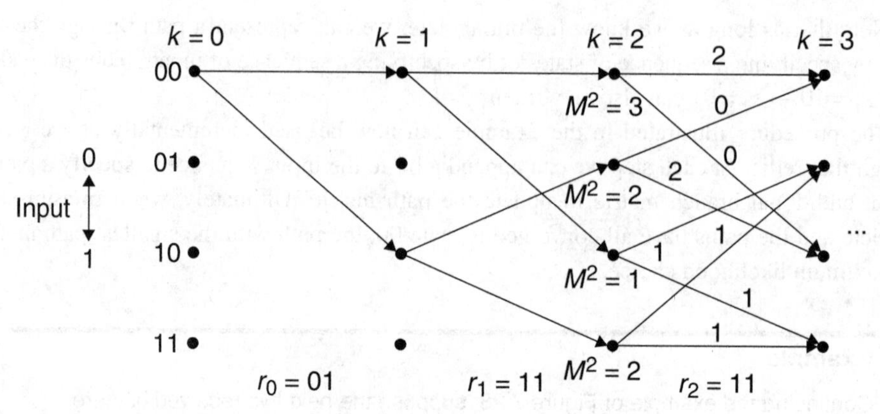

Figure 7.24 Two Paths Converge on Every State

We can now summarize the Viterbi algorithm as follows:

1. For each node in the trellis we maintain a surviving path to that node and a partial path metric for the surviving path.
2. Initially all of the paths are empty and all of the path metrics are zero. Initially $k = 0$.
3. Suppose the decoder is at stage k in the trellis. When the received bits $r_k = r_k^{(0)}, \ldots r_k^{(n)}$ arrive, the decoder computes a branch metric $M(r_k|y_k)$ for each branch joining stage k with stage $k + 1$.
4. The decoder updates the partial path metrics according to $M^{k+1}(r|y) = M^k(r|y) + M(r_k|y_k)$ for each path leading to a state at stage $k + 1$.
5. There will be two paths entering each state at stage $k + 1$. The decoder chooses the path with the smaller metric as the survivor and discards the other path. Ties can be broken by choosing either path arbitrarily.
6. The decoder sets the stage to $k + 1$ and repeats steps 3, 4, and 5. The decoder continues advancing through the trellis until transmission ends. The survivor at state 00 is the maximum likelihood path.

At each stage k, starting from $k = 0$, the decoder must calculate two branch metrics, update two path metrics, and perform one comparison for each state in the trellis. Thus the total storage the decoder needs and the number of computations it must carry out at each stage is proportional to the number of states. Since the number of states is $2^{(K-1)}$, where K is the constraint length, we see that the complexity of the decoder increases exponentially with the constraint length. In the preceding examples we have four states, but the $K = 9$ convolutional coders used in cdmaOne and cdma2000 systems have $2^8 = 256$ states. Fortunately, modest values of constraint length are sufficient to obtain satisfactory levels of error control.

Typically as the decoder advances through the trellis, the partial path metrics for incorrect paths will grow at about one for every two bits received, and the partial path metric for the correct

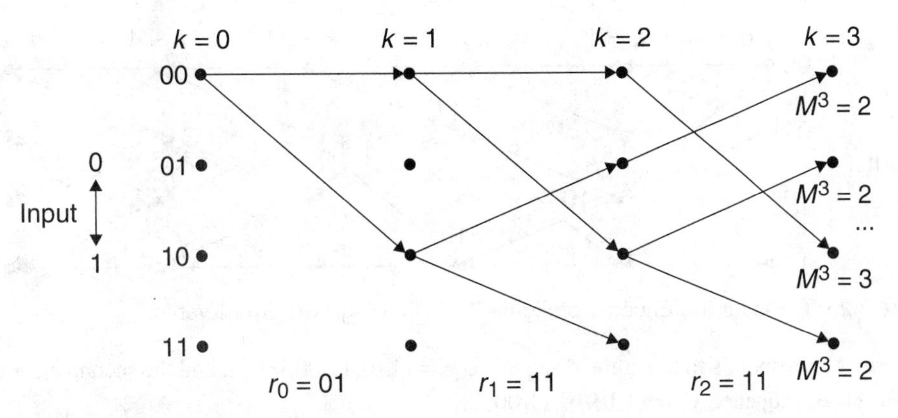

Figure 7.25 The Survivors at Stage $k = 3$

path will grow at about the bit error rate (e.g., one for every 1000 bits received for an error rate of 10^{-3}). Consequently, no incorrect path will last as a survivor for very long. If we were to continue the example of Figure 7.25 for a number of additional steps, it would become apparent that all of the paths tend to merge together at about four or five constraint lengths back from the present value of k. The fact that the surviving paths tend to merge has useful consequences. One of these is that the encoder does not have to begin in state $s_0 = 00$, and the decoder does not have to know the value of the initial state. Once the decoder has advanced four or five constraint lengths into the trellis, all of the surviving paths will, with very high probability, branch from the same initial state. A second useful consequence is that once the paths have merged together, the decoder can inform the user about the single merged path and purge that path from its memory. In other words, the decoder does not have to wait for the entire information sequence to be transmitted and received before it can provide a decision to the user. The decoder can provide its estimate of the information bits to the user on a continuing basis, lagging about four or five constraint lengths behind transmission.

Performance

The decoder makes an error when the surviving path through the code trellis is different from the path that was taken by the transmitter. It turns out that because the convolutional encoder is a linear system, the probability of error is independent of which path the transmitter actually follows. To simplify the discussion, then, we are free to select a simple path for the transmitter when we calculate the probability of error. We will assume for discussion that the input data to the transmitter is the all-zero sequence, so that the transmitter remains in state $s_k = 00$ for $k = 0,1,2\ldots$. This means that the transmitted code sequence y is also the all-zero sequence. The received sequence r will of course differ from the all-zero sequence owing to occasional bit errors caused by noise and interference.

Figure 7.26 shows the trellis for the encoder of Figure 7.19. Three paths are indicated in bold: the "all-zero" path that we assume was actually transmitted, the path $s = 00 \rightarrow 10 \rightarrow 01 \rightarrow 00$ representing an "error event," and the path $s = 00 \rightarrow 10 \rightarrow 11 \rightarrow 01 \rightarrow 00$ representing a different error event. Along the correct path the transmitted sequence is $y = 00, 00, 00, 00, 00, \ldots$. The first

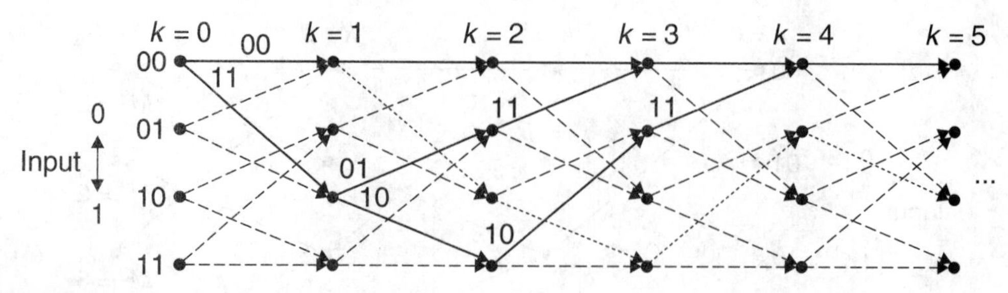

Figure 7.26 Trellis for the Encoder of Figure 7.19 Showing Two Error Events

error event corresponds to transmitted sequence $y' = 11,01,11,00,00,\ldots$, and the second error event to transmitted sequence $y'' = 11,10,10,11,00,\ldots$.

For an error to occur, the received sequence r must be closer in Hamming distance to an incorrect path than to the correct path when the two paths merge. The shorter Hamming distance will cause the incorrect path to be chosen as survivor. In Figure 7.26, the erroneous path y' will be chosen over the correct path y when the two paths merge at $k = 3$ if $d(r,y) > d(r,y')$. For example, if $r = 11,01,00,00,00$ we have $M^3(r|y) = d(r,y) = 3$, but $M^3(r|y') = d(r,y') = 2$. Note that in this example the error event requires at least three bit errors in the received sequence r. For the erroneous path y'' to be chosen instead of the correct path y, the received sequence r must be closer in Hamming distance to y'' than to y when the two paths merge at $k = 4$. This can happen if there are four or more bit errors in the received sequence.

An examination of the code trellis shows that there are many possible error events, so an exact calculation of the probability of error is difficult. We have seen, however, that patterns of few errors are much more probable than patterns of many errors. Thus the probability of error will be dominated by the error event caused by the smallest number of errors. In Figure 7.26 the most probable error event is the path y', and the probability of error can be approximated by the probability that this error event occurs.

In general, the path with the fewest number of ones that deviates from the all-zero path is the most probable error path. The Hamming distance between this error path and the all-zero path is called the **minimum free distance** d_{free} of the convolutional code. The larger the value of d_{free}, the lower the probability of error the code will have. The minimum free distance can be made larger for codes with lower rate and longer constraint length. It can be shown[9] that the bit error probability is approximately given by

$$P_e \approx B_{d_{free}} 2^{d_{free}} p^{d_{free}/2}, \tag{7.35}$$

where $B_{d_{free}}$ is the number of information bit errors along paths of distance d_{free} from the all-zero path, and p is the error rate of the transmitted bits.

9. S. Lin and D. J. Costello Jr., *Error Control Coding,* 2nd ed. (Upper Saddle River, NJ: Prentice Hall, 2004), 531.

Example

For the encoder of Figure 7.19, the error path with minimum distance from the all-zero path is $y' = 11,01,11,00,00,\ldots$, as shown in Figure 7.26. Then $d_{free} = 5$. The information sequence that produces this error path is $x = 1,0,0,\ldots$, giving $B_{d_{free}} = 1$. If $p = 10^{-3}$, then Equation (7.35) gives $P_e \approx 1.01 \times 10^{-6}$.

Typically, for a given constraint length and rate the wiring of the encoder that produces the largest d_{free} must be determined by trial and error. Good convolutional codes have been found by computer search and are tabulated in the literature.[10] The encoder of Figure 7.20 is one such tabulated encoder, and the code has $d_{free} = 12$.

Example

It is reported[11] that to achieve a probability of error of $P_e = 10^{-3}$ using binary phase-shift keying over a Rayleigh fading channel requires a signal-to-interference ratio of 24 dB. If a dual-diversity RAKE receiver[12] and convolutional coding are used, however, the required signal-to-interference ratio drops to 7 dB.

Conclusions

In this chapter we changed focus. Rather than concentrate on the physical data link that has been the center of our attention in previous chapters, we looked instead at the information sources that generate the data to be transmitted over that link. From an application perspective, the physical link is a somewhat imperfect bit pipe that connects an information source with a destination. Our interest in this chapter has been in identifying the requirements that the bit pipe has to meet to convey the source information in a manner that is acceptable to the user at the destination.

We began with an examination of five kinds of sources: speech, music, images, video, and digital data. In each case requirements are imposed on the communication link both by the nature of the source and by the expectations of the user. We carried out a brief survey of source requirements and user expectations for each of the source types. A key issue in stating both source requirements and user expectations is finding engineering terms to describe what is often a subjective user experience. Users might want a videoconference to make them feel "as if they were there," but systems engineers must work with pixel density, bandwidth, dynamic range, and frame rate.

10. For example, Lin and Costello, *Error Control Coding,* 539, 540.
11. Schwartz, *Mobile Wireless Communications,* 150.
12. RAKE receivers are described briefly in Chapter 8. See also R. Price and P. E. Green, "A Communication Technique for Multipath Channels," *Proceedings of the Institute of Radio Engineers* 46 (March 1958): 555–70.

Next we turned our attention to the communication link. To ensure that information from specific sources can be conveyed satisfactorily to users, links are designed to meet specific quality of service (QoS) requirements. QoS is often specified as a "package" of parameter values that configure a link for a specific source or group of sources. In some systems selection of QoS parameters is made by a negotiation when the link is set up. The price of the connection may depend on the QoS provided. Typical QoS parameters include average data transmission rate, peak transmission rate, frame loss rate, delay, delay jitter, and priority. QoS parameter selection can be used to distinguish communication links set up for streaming sources such as speech and video from links set up for bursty sources such as data. Streaming sources often require a high average bit rate, low delay, and low delay jitter. Bursty sources often require a high peak bit rate and a very low frame loss rate.

The physical data link that has been our concern in Chapters 2 through 6 is designed to carry a stream of undifferentiated bits. Its "quality" is described in terms of bit rate, delay, and bit error rate. The parameters that specify QoS to the users, however, often reflect the fact that source data contains some structure. Many information sources produce data that is naturally grouped into "frames" or "packets." Bits in certain positions in a frame may be particularly sensitive when source information is reconstructed at the receiver. QoS is often specified in terms of frame loss rate and frame delay jitter, rather than bit error rate and bit delay jitter. It is often difficult to relate the frame parameters to the bit parameters in a simple analytic way, and the flow-down of quality of service requirements to physical link requirements often involves simulation.

To show what is involved in preparing the signals from an analog source for transmission over a digital link, and to give a specific example showing how a frame structure arises naturally, even for data from a streaming source, we made a detailed examination of the digitization of speech signals. We first introduced the classical method of digitizing, or "coding," speech signals, pulse code modulation (PCM). Pulse code modulation can be used to digitize music, video, and telemetry signals in addition to speech. Consequently the discussion of PCM has wide applicability.

The PCM approach to speech coding is an example of "waveform coding." The goal is to reproduce the source waveform at the receiver, under the assumption that if the waveform is accurately reproduced, the sound will be accurately reproduced. Several steps are involved in this method of digitizing an analog waveform: lowpass filtering, sampling, quantization, and coding. Sampling, if the sampling rate is above the Nyquist rate, is a reversible process. The continuous-time signal can be reconstructed from the sequence of samples at the decoder. Our focus in this chapter was on the quantization step, since this step is inherently destructive of information and introduces an error that is usually characterized as "quantization noise."

When signals are digitized for instrumentation or signal-processing purposes, uniform quantization is normally used. For transmission of speech over wired telephone links, a nonlinear quantization method known as "companding" is normally used. The quantizer uses a logarithmic scale so that small signal values are more finely quantized than large signal values. The overall effect is that the signal-to-quantization-noise ratio remains approximately constant over a large signal dynamic range. This way a telephone system can accommodate both loud and quiet speakers without the need for "volume" or "mic gain" knobs on the handset. Standard wired telephone PCM represents a speech signal as a 64 kbit/s bit stream.

Because of the high correlation between successive sample values in many bandlimited signals (including speech), the signal-to-quantization-noise ratio can be enhanced if sample differences, rather than sample values, are quantized. We examined two schemes for quantizing differences, delta modulation and differential PCM. These are both practical schemes, and DPCM at 32 kbits/s is specified for use in the wireless link in several cordless telephone standards.

Next we turned our attention to an alternative method of speech digitization, one of a class of source-coding methods known as "vocoding" (voice encoding). A vocoder uses a model of the human vocal system to attempt to reproduce sounds at the decoder output without attempting to reproduce the specific waveform. Frames containing parameters of the vocal system model are sent over the communication channel. Updates to the model are typically sent every 10 to 30 ms. Vocoders have proved to be able to send remarkably high-quality speech at bit rates lower than can be achieved by DPCM. Bit rates as low as 2400 bits/s have been achieved (although at reduced quality), but bit rates in the 8000 to 16,000 bit/s range are more typical. We examined one technique, known as "linear predictive coding" (LPC), for estimating the parameters of the vocal system model. Linear predictive coding is used for digitizing speech in all of the second- and third-generation cellular telephone systems.

The final section of the chapter introduced a different kind of coding, coding for error control. Wireless links often have bit error rates that are much higher than those found in wired links, and some kind of error control mechanism is necessary to ensure an adequate quality of service. For communication of signals from streaming sources, forward error correction techniques are preferred, as they do not enhance delay jitter as automatic repeat request techniques do.

Our example of an error control coding technique was the convolutional code. Convolutional codes are easy to generate and easy to decode, and they provide a significant improvement in the bit error rate for a given signal-to-noise ratio. The cost of error-control coding, however, is a doubling or tripling of the bit rate. We examined the convolutional encoder and presented the trellis diagram used to represent its operation. We concluded with a detailed look at the Viterbi decoder. This is an optimum decoding algorithm that is easily implemented and widely used. There are other error control coding techniques, such as turbo codes, that are also used in wireless systems and are deserving of further study.

In Chapter 8 we bring our exploration of wireless systems to a close. After summarizing the principal lessons presented in Chapters 1 through 7, we conclude with a discussion of the evolution of cellular telephone systems. Cellular telephone systems began as relatively simple voice-oriented systems implemented in a largely analog technology. In a little over two decades the combination of improved technology and an almost unimagined consumer demand for new services has pushed cellular systems into their third generation. We briefly discuss the technological features that characterize these successive generations and end with a look at the fourth generation of cellular systems, a generation whose general technological outlines are just emerging.

Problems

Problem 7.1

An NTSC luminance signal has a spectrum that extends from DC to about 4.4 MHz. At what rate must it be sampled if the signal is to be perfectly reconstructed from its samples?

Problem 7.2

An NTSC color television signal consists of a luminance signal having a bandwidth of 4.4 MHz, an "in-phase" chrominance signal having a bandwidth of 1.6 MHz, and a "quadrature" chrominance signal having a bandwidth of 0.6 MHz. Suppose these signals are each sampled at the Nyquist rate and then quantized and coded at eight bits per sample. If the coded bit streams are multiplexed into a single bit stream, find the aggregate bit rate.

Problem 7.3

A signal $x(t)$ whose spectrum is shown in Figure 7.4(a) is sampled by the circuit shown in Figure 7.3. Draw the spectrum $X_s(f)$ of the sampled signal if the sampling waveform $s(t)$ is a train of rectangular pulses of amplitude 1 and 25% duty cycle rather than a train of impulses. Assume the sampling frequency f_s satisfies $f_s > 2B$.

Problem 7.4

An eight-bit midtread uniform quantizer has a full-scale range of 5 V.

A. Find the smallest positive input voltage that can be distinguished from zero.

B. Find the voltage of the smallest positive nonzero quantizer output.

Problem 7.5

Find the dynamic range of an eight-bit midriser uniform quantizer; that is, find the ratio of the peak value of the largest signal the quantizer can handle without overload to the peak value of the smallest signal that can be distinguished from zero. Express your answer in decibels.

Problem 7.6

The input to a 16-bit midtread uniform quantizer is the triangle wave shown in Figure P 7.1. If the quantizer has $x_{max} = 5$ V, find the SQNR. Repeat if $x_{max} = 10$ V.

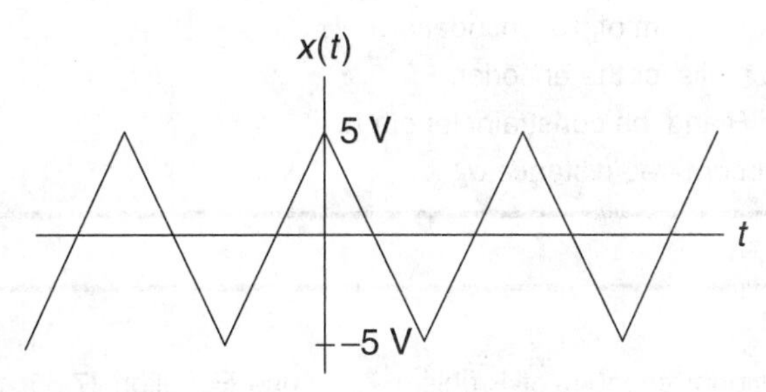

Figure P 7.1 A Triangle Wave

Problem 7.7

A sinusoid having a frequency of 1 kHz and a peak value of 5 V is digitized using a delta modulator with $\Delta = 0.1$ V. At what rate must the sinusoid be sampled if there is to be no slope overload? Compare with the Nyquist sampling rate.

Problem 7.8

A sinusoid having a frequency of 1 kHz and a peak value of 5 V is digitized using a delta modulator with $\Delta = 0.1$ V. The sampling rate is high enough that there is no slope overload. Calculate the SQNR. Express your answer in decibels.

Problem 7.9

Show that the linear predictive encoder defined by Equation (7.22) is exactly the DPCM encoder shown in Figure 7.17, but without the quantizer and D/A blocks.

Problem 7.10

Find the data rate in bits per second of the USDC VSELP vocoder:

A. Including the parity bits
B. Excluding the parity bits

Problem 7.11

A convolutional encoder has generators $g^{(0)} = 110$, $g^{(1)} = 101$, and $g^{(2)} = 111$.

A. Draw a block diagram of the encoder.
B. Draw a code trellis for the encoder.
C. Find the rate R and the constraint length K.
D. Find the minimum free distance d_{free}.

Problem 7.12

For the convolutional encoder of Problem 7.11, use Equation (7.35) to estimate the probability of bit error. Assume $p = 10^{-3}$.

Problem 7.13

A system uses the convolutional coder shown in Figure 7.19. Suppose the received sequence is

$$y = 10,01,11,11,10,01,10,11,00,00.$$

A Viterbi decoder is used to determine the maximum likelihood transmitted sequence. Draw a sequence of trellis diagrams showing the survivors to each state for $k = 0,\ldots,10$. Indicate the partial path metric for the survivor at each state in every diagram. When a tie occurs in choosing a survivor, choose the upper path. If it is known that the decoder ends in state $s_{10} = 00$, determine the decoder's estimate of the transmitted sequence x.

Problem 7.14

Most arbitrarily chosen convolutional encoders have good performance, but some are catastrophically bad. Consider the encoder with generators $g^{(0)} = 110$, $g^{(1)} = 101$.

A. Draw the trellis for this encoder.
B. Find the transmitted sequence if the input information sequence is $x = 1,1,1,1,\ldots$.
C. Suppose ten information bits following the sequence of part B are provided to the encoder. The encoded sequence is transmitted, and the received sequence contains errors in the first, second, and fourth bit. Find the maximum likelihood estimate of the transmitted sequence.
D. Explain why the behavior of this code is called "catastrophic."

Problem 7.15

A rate $R = 2/3$ convolutional coder takes in two bits at a time and produces three output bits for every pair of input bits. Such an encoder is shown in Figure P 7.2.

Find the output sequence y corresponding to the input sequence $x = 01,11,10,00$. Assume that the encoder starts in the zero state.

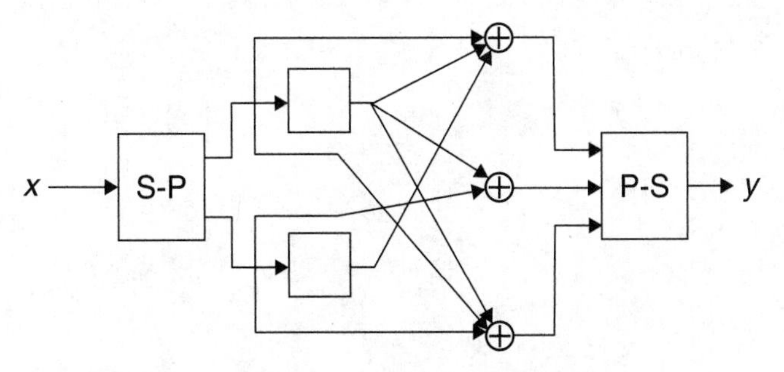

Figure P 7.2 A Rate $R = 2/3$ Convolutional Encoder

CHAPTER 8

Putting It All Together

Introduction

In Chapter 1 we introduced the discipline of systems engineering as we described some of the major aspects of wireless, and in particular, personal communication systems. Our approach has been to introduce topics as steps in solving a general systems-engineering problem:

To design a wireless telecommunications system that will

- Support the communication of information of various types, including speech, text, data, images, and video, in urban, suburban, and rural environments and with quality approximating that of wired communications
- Be capable of expanding in geographic coverage
- Allow for virtually limitless growth in the number of users
- Support endpoints that are not geographically fixed and, in fact, may be moving at vehicular speeds

Throughout our presentations we have endeavored to present underlying principles and how they relate to potential solutions within the problem framework. We have also presented some of the many trade-offs involved among possible approaches to solutions. Although our treatment is by no means comprehensive, we developed many of the major concepts that have had a profound effect in enabling the unprecedented growth that wireless communication has enjoyed over the past thirty years.

Our discussions have focused on the lowest layer of the air link, or what is commonly called the **physical layer**. The physical layer is directly concerned with the reliable communication of bits or packets as determined by the desired QoS. As such, the physical layer is not concerned with the content of the information but merely performs a service for higher layers of functionality. We have also seen that the **medium access control** (MAC) layer provides the protocols that allow users to gain access to a system and that organize users so that they can appropriately share the radio resources allocated by a higher layer. In many systems the MAC layer also determines the appropriate transmission parameters that the physical layer must use to meet the QoS requirements. Another higher layer, often called the **connection layer**, is responsible for managing and allocating resources, setting up and terminating connections, and managing/coordinating mobility. A still higher layer provides the functions necessary to establish and maintain the communication paths with land-based facilities or entities. Although the actual number of layers and their names vary significantly among systems of different types, the functions the layers perform are quite similar. Layers are organized in a hierarchy called a **protocol stack**, with each layer providing a specific set of services to the layer above it. The set of all layers that provides physical and logical connectivity between mobile units and the land-based facilities is usually called the **air interface**. Both in concept and in practice, the air interface hides all of the aspects of the radio environment from the upper layers. If a system is properly designed, existing public or private networks do not have to be modified to include wireless connectivity. Other higher-layer functions provide routing of traffic and control information (signaling) among the various elements of a service provider's network and even between different service providers. These functions include routing to the public switched telephone network (PSTN) or the packet-switched data network (PSDN, the Internet). As one might imagine, administrative functions are also necessary to manage system resources, create and maintain billing information, authenticate users and authorize their use of different services, collect performance and maintenance information, and so on. It should be evident that a complete system design must also encompass the design of a service provider's "core network" and the interconnection with the public networks as well as the design of the air interface. Systems engineering is required at each of these levels. It is in light of all this that we claim that the complete and detailed design of even a relatively simple cellular system is a daunting task. Nevertheless, we have provided a basis for understanding the principles underlying the design of current and future wireless systems.

Most of our presentations in this text have focused on wireless systems that are optimized for voice services. This approach was an expedient rather than a necessity. As cellular systems have evolved, so has our way of characterizing the nature of the data the systems must convey. From a systems-engineering point of view, the specific medium that a system conveys is relevant only insofar as it determines the set of QoS parameters that must be met. System designs must support the QoS parameters of any or all of the services they are to provide. Therefore, as the demand for new services evolves, so must the systems that are designed to support these services. As we look toward the future, system designs will be aimed at providing high-rate packet data services that are capable of supporting a number of multimedia services simultaneously. These systems must not only adapt instantaneously to the needs of the user but also achieve an order of magnitude improvement in spectral efficiency over that available today. This will require dramatic changes in system architecture from the lowest physical layer up to and including the interconnections with the public telephone network and the Internet. Although these evolving systems are dramatically more complex than those of the recent past, the same principles and concepts used in the design of legacy systems are still the bases for their design. In this chapter we describe some of the factors that have motivated cellular system evolution and how system architectures have changed to meet evolving user needs. We will conclude with a brief discussion of orthogonal frequency-division multiple access (OFDMA), which currently appears to be the access method of choice for all fourth-generation (4G) systems. Our discussion will not stress the analytic aspects of that evolution, but instead we will provide a historically based rationale for the architectural choices being made. Our hope is that this discussion will stimulate a sincere interest in wireless communications leading to further study at the graduate level.

Looking Backward

The First Generation

The first land-mobile telephone systems were introduced in 1946. Although the ability to conduct a telephone conversation from an automobile quickly captured the imagination of the public, limitations in technology as well as the scarcity of frequency spectrum made mobile telephone service a rare and rather expensive luxury. Even with technological advances and a quadrupling of spectrum efficiency over the next three decades, service could be provided to only a tiny fraction of the potential market. For example, in the mid-1970s the Bell System's New York mobile telephone system serviced about 550 customers on 12 channels. Over 3500 potential customers were on a waiting list for service, in a metropolitan area with a population of about 10 million people. With over 45 users per channel, availability (blocking probability) was unacceptably poor, especially when compared to the performance of the wire-based public telephone network.

The introduction of the first cellular systems in North America, Europe, and Japan in the early 1980s demonstrated both the viability of cellular technology and the explosive demand for wireless service. Indeed, the number of subscribers grew about tenfold over the first ten years of deployment. Today there are about 2.5 billion subscribers to wireless services worldwide; that is about 40% of the entire population of the Earth. This is an astounding market growth by any measure, but even more so given that the first systems were introduced fewer than 30 years ago.

The unpredictably rapid growth following the initial deployment of these first-generation (1G) systems quickly demonstrated the need for additional spectrum, greater spectral efficiency, and facilities for supporting efficient and seamless roaming among service areas. The need for seamless roaming was particularly acute in Europe. During the years 1981 through 1985, eight countries introduced their own versions of the European Total Access Communications System (ETACS), a European version of AMPS. Significant diversity and incompatibility among the various deployments, however, eliminated the possibility of roaming across national boundaries. Differences in both the air interfaces and spectrum channelization rendered it impossible for a mobile unit to obtain service outside its home service area. These differences were not surprising, given the history of incompatibility that existed for decades among the wired services of different European countries. With the burgeoning momentum for the formation of the European Economic Union and the contrasting fragmentation of telecommunications services as backdrop, the key goals of GSM, the first digital cellular system, were formulated in a spirit of cooperation and unification that began in 1981.

The Second Generation

The GSM standard is an outgrowth of multinational cooperative efforts among postal and telecommunications authorities (the Conference of European Postal and Telecommunications Administrations, CEPT) across Europe. It was developed by a commissioned workgroup of CEPT, the Groupe Spéciale Mobile (GSM) committee. This comprehensive standard encompasses more than 5000 pages and, at that time, was unique in that it specifies both the air interface and the interconnection with wired infrastructure elements. The first deployments of GSM occurred in Europe in the early 1990s, and the GSM standard was placed under the control of the European Technical Standards Institute (ETSI). Spectrum allocations and standards were initially defined for the 900 MHz band and later for the 1.8 GHz band. GSM is classified as a hybrid TDMA/FDMA system with frequency-division duplexing (FDD) as in the earlier analog systems. Technological advancements in speech compression, digital signal processing, and microprocessors were among the most important enabling factors in the development of the GSM system. The increased spectrum efficiency and seamless roaming capabilities of the GSM system provided attractive alternatives to the shortcomings of the first analog systems. Following its introduction, the GSM system grew to worldwide dominance through the 1990s and well into the first decade of the new millennium. The astonishing success of GSM on a global level is in no small measure due to its open and flexible standards framework, which ensures seamless and consistent coverage and services for subscribers while allowing service providers to create new services and choose among competitive subscriber units and infrastructure vendors.

In the United States, issues of incompatibility were not a concern given the imposition of FCC rules and standards. Roaming was not automated, however, and cooperation was required among service providers in the form of reciprocal arrangements to ensure that mobile units could receive service outside their home areas. Procedures often required subscribers to obtain roaming numbers in advance through a mobile operator. In 1991 the Telecommunications Industry Association (TIA) Interim Standard (IS) 41 Rev. A was issued to provide a consistent and uniform method for subscribers to enjoy seamless roaming without direct participation. The standard was a cooperative effort among service providers to define a wide-area network capable of providing all needed interservice-area roaming functions. These functions included authentication and handoff procedures among systems. This network was modeled after the GSM roaming architecture.

The immediate success of AMPS in the United States raised concerns that system capacity would reach limitations in major service areas by the early 1990s. Although cell splitting provides additional capacity, there are practical limitations imposed by cost, complexity, and performance and by the ability to locate and obtain cell sites. Within a few years after the first cellular frequency allocation of 40 MHz, the FCC granted an additional 10 MHz of spectrum for cellular services with the provision that no new spectrum allocations would be forthcoming soon. Given these constraints, the development of new and more efficient technology remained the only viable alternative for increasing system capacity on a global scale. In 1987 the FCC encouraged the development of more efficient technology by permitting alternative cellular technologies within the existing cellular band. In the late 1980s, industry forums considered a number of approaches, settling on a hybrid TDMA/FDMA system like GSM but maintaining the 30 KHz AMPS channelization. The system was designed to provide a smooth transition from analog AMPS to an all-digital (TDMA/FDD) system using the same frequency spectrum and frequency reuse factor as AMPS. In this arrangement, each AMPS channel supports six time slots that may be allocated according to the medium type. Full-rate speech encoding utilizes two slots per frame, resulting in a capacity improvement by a factor of 3. The standard provided for dual-mode handsets to allow an incremental evolution from analog to digital operation. The Digital AMPS (D-AMPS) system standards were initially issued as TIA/IS-54, the first North American TDMA (NA-TDMA) digital cellular standard. The standard was later expanded to encompass operation in both the original AMPS band and in the 1.9 GHz PCS band under ANSI/TIA/EIA-136, the United States Digital Cellular (USDC) standard.

The earliest proposals for using spread-spectrum techniques for cellular communications date back to the late 1970s. These proposals were of immediate interest to researchers, but it wasn't until approximately 1990 that Qualcomm, Inc., a U.S.-based company, demonstrated the world's first CDMA/FDD personal communication system operating in the AMPS band. The Qualcomm system was adopted by the TIA under IS-95 in 1993 and first introduced into service in 1995. Like USDC, the Qualcomm system was designed to operate in a compatible fashion within the AMPS cellular spectrum. The strong appeal of the CDMA approach was founded on the claim that capacity improvements of 20:1 could be obtained under ideal conditions. More realistic assessments revealed that improvements are actually closer to 14–16:1. In view of the staggering growth rates and the significant capacity advantages offered by CDMA, several service providers announced plans to adopt CMDA as their evolution path. IS-95A and other subsequent revisions provided for operation in the PCS band (standardized under American National Standards Institute [ANSI] J-Standard 008) and added a radio configuration capable of a 14.4 kbits/s maximum data rate. IS-95A, J-Standard 008, and other operability standards were subsequently consolidated into the IS-95B (or ANSI/TIA/EIA-95), and the family of related standards is designated as cdmaOne.

One of the most significant advantages of a CDMA approach is that it allows a reuse factor of unity; that is, all channels can be used in every cell. Recall from Chapter 4 that the cellular concept was founded on managing interference by geographic separation of cochannels and adjacent channels. Thus only a fraction of the channels could be deployed at any given cell site. As we saw in Chapter 6, all users share the same frequency in CDMA systems, a fact that significantly increases the radio resources available in each sector. Users are assigned codes that permit their signals to be separated. These codes are equivalent to channels in a very real sense.

We saw in Chapter 6 that Walsh codes are perfectly orthogonal and, in theory, can provide perfect separation (channelization) among users if the coded signals can be synchronized. In reality, performance is limited by system instabilities affecting synchronization and by multipath reception. Pseudonoise (PN) sequences, on the other hand, are nearly orthogonal, and although they do not provide perfect elimination of interfering signals, they can also be used to provide channelization. The key advantage of PN sequences is that they do not require synchronization among users. The use of PN sequences is limited by the amount of residual interference that a system can tolerate. The cdmaOne system employs long codes and short codes. The short codes are PN sequences that are 2^{15}–1 chips in length, repeating every 26.67 ms. Long codes are PN sequences that are 2^{42}–1 chips in length, derived from a 42-stage generator. At a 1.2288 Mchip/s chip rate, the long code sequence repeats approximately every 42 days. Although there is only one long code sequence, each user has a unique 42-bit mask based on its electronic serial number and a secret bit sequence that is used to generate a user-unique long code. This code is often called a **scrambling code**.

The cdmaOne system employs 64-bit Walsh codes to provide channel separation in the forward link (downlink). An 8.6 kbit/s full-rate speech coder provides a 9.6 kbit/s input to the channel, after 12 bits are appended to provide a parity check sequence for error detection (known as a **cyclic redundancy check** or CRC) and eight "tail" bits are appended to initialize the forward error correction (FEC) encoder. Forward error correction is implemented using a rate-1/2 convolutional encoder. The rate at the output of the FEC encoder is 2×9.6 kbits/s$=19.2$ kbits/s. The effectiveness of forward error correction can be reduced by burst errors. A **burst error** is a group of several adjacent bit errors. Burst errors are characteristic of most radio environments. Interleaving is a technique used to reduce the effects of burst errors by rearranging the ordering of bits in a sequence in a predetermined way. This rearrangement attempts to separate bits that are adjacent in the original sequence by some minimum number of bit periods prior to transmission. If the separation is sufficient, the likelihood of adjacent bits being corrupted by a burst error can be significantly reduced upon restoration of the original bit order at the receiver. The interleaver does not change the bit rate. The user-unique long code is modulo-2 added to the bits emerging from the interleaver to scramble the user data, thereby providing information security and privacy. User channels are then created by spreading the 19.2 kbit/s signal using a 64-bit Walsh code assigned to the user. Thus the 9.6 kbit/s speech data stream is spread by a factor of 128, yielding a 1.2288 Mchip/s chip rate. Prior to up-conversion, the encoded stream is then modulo-2 added with a short code associated with the serving sector. This short code is used to distinguish among signals in the different sectors. Because of the very sharp autocorrelation characteristic of PN sequences, the same sequence can be used in every sector by offsetting the starting point of the sequence differently for each cell sector and site. PN offsets are in multiples of 64 chips, and there are 512 different offsets that can be defined within a given system. The deployment of PN offsets requires system planning.

Because of complexities involved in synchronizing mobile unit transmissions, the user-unique long code is used to provide separation among users in the reverse link (uplink) instead of Walsh codes. As in the forward link, the output of a full-rate speech encoder is appended with parity and check bits to form a 9.6 kbit/s information bit stream. The information bits are then encoded by a rate-1/3 convolutional encoder and interleaved. The rate at the input and output of the interleaver is 3×9.6 kbits/s = 28.8 kbits/s. Every group of six adjacent bits from the output of the interleaver is mapped to a 64-bit Walsh code to provide a 64-ary orthogonal modulator. Here, the Walsh codes are used to improve the performance of the PSK modulation; they are not used to provide channelization. The rate at the output of the orthogonal modulator is $\frac{64}{6} \times 28.8$ kbits/s = 307.2 kbits/s. The user-unique long code is used to spread and scramble the reverse transmission, which results in a 1.2288 Mchip/s chip rate. This sequence also provides separation among users in the reverse direction. For both the forward and reverse directions, the spreading factor for a full-rate speech encoder is 128, yielding a theoretical processing gain of 128. Note that cdmaOne employs a variable bit rate (VBR) encoder, which allows the system to adjust the data rate to a half, a quarter, and an eighth of the full data rate (i.e., 4.8 kbits/s, 2.4 kbits/s, and 1.2 kbits/s) depending on voice or data activity. The VBR encoder allows voice activity detection, a scheme that exploits the intermittent characteristics of human speech by reducing the output data rate when a pause is detected. For lower rates the data blocks are repeated to maintain the 19.2 kbit/s bit stream. For example, an eighth-rate block is sent eight times; however, because the data blocks are repeated eight times, they will be received with eight times the apparent bit energy at the receiver. This allows the lower rate to be transmitted at a lower power level, thereby reducing interference and increasing capacity.

Pilot signals consisting of a $2^{15}-1$ bit PN sequence with a unique offset are transmitted from each base station. A pilot signal provides a reference for synchronization, a means for coherent demodulation, and a means for the mobile unit to discriminate among signals from different base stations. The pilot PN sequence covers all uplink and downlink transmissions in a given sector. Thus mobile units can distinguish among different sectors based on their offsets. On the uplink, the pilot PN sequence provides a means for the sector to synchronize with the mobile unit. In the downlink, power control is used to ensure that the power transmitted to each mobile unit is the minimum that meets the QoS requirements. In the uplink, power control is used to ensure that all signals arriving at the base station are approximately equal to mitigate the effects of the near-far problem described in Chapter 6. The base station measures the received signal strength from each mobile unit relative to the total interference and sends a high-low indication to each mobile unit on the forward link. Mobile units adjust their transmitted power in discrete steps for each bit. Power control bits are sent at an 800 Hz rate.

An interesting and important aspect of the cdmaOne system is related to the relatively high RF symbol (or chip) rate and correspondingly wide channel bandwidth. As noted previously, cdmaOne is based on 1.2288 Mchips/s with a 1.25 MHz channel bandwidth. In Chapter 3 we related the channel bandwidth to the standard deviation of the delay spread (see Equations 3.116 and 3.117). Using this relationship and a 1.25 MHz bandwidth, signals received over multiple paths should be resolvable if their separation is greater than about 160 ns. This corresponds to a path-length differential of about 157 feet. For typical outdoor environments, the cdmaOne channel bandwidth is sufficiently wide to permit multiple received signals to be resolved. We also showed that in such situations the multiple received signals are replicas of the transmitted pulse corrupted by statistically independent noise waveforms. Thus a gain can be achieved if these received signals are appropriately combined. The cdmaOne system was the first cellular system to exploit this phenomenon by incorporating RAKE[1] receivers. In the simplest view, a **RAKE receiver** is a bank of correlation receivers, each adjusted to receive one of a number of the multipath replicas of the transmitted signal. The outputs of the receivers are processed and combined using an appropriate algorithm to provide a significant improvement in system performance. Each correlation branch of a RAKE receiver is often referred to as a **finger**. The performance of a RAKE receiver depends on the number of fingers, the combining technique employed, and the accuracy of estimating the delay and path loss of the channel components. Recall that in cdmaOne a sequence of information bits is spread using a PN sequence, which possesses very sharp correlation properties. Thus, an interfering signal arriving late by as little as one chip time will be seen by a correlation receiver as noise. This property allows the receiver to separate different multipath components and provides a means for locating the various multipath components in time by simply shifting the sequence offset chip by chip. The cdmaOne system uses three fingers.

Recall that all cells and sectors in a cdmaOne system use the same short PN sequence, each sector having a different offset to provide a locally unique way of distinguishing among the sectors. As a consequence, it is possible for a given sector to "listen" to mobile units in adjacent sectors if the signal strength is sufficient. Since all transmissions are at the same frequency, a sector base station can receive the signals from a mobile unit in an adjacent sector by using the PN offset of that sector. By appropriately processing or combining the signals from each sector, a **diversity gain** is obtained. Processing may be as simple as picking the best signal or, with more sophistication, coherently combining the signals using RAKE receivers. Likewise, a mobile unit can receive multiple simultaneous forward link transmissions from adjacent sectors using the same techniques. Thus a mobile unit can receive and benefit from diversity gain in the forward link. This diversity scheme is termed **macro diversity** in reference to the large separation between the sector antennas involved. Perhaps the most important benefit of using multiple sectors is that it simplifies and significantly improves handoff performance among the sectors. Since the system is receiving mobile unit transmissions from multiple sectors at any given time, it can simply choose the sector associated with the best signal. In FDMA and TDMA systems, handoff requires breaking the connection, switching frequencies, and reestablishing the connection. This **hard handoff** technique requires considerable processing and coordination to avoid noticeable quality degradation. The **soft handoff** developed in cdmaOne is a simpler process that can be implemented by managing a list of the strongest sectors from the mobile unit's perspective. As a mobile unit moves, it continuously monitors the pilot signal strength of base station signals from the surrounding sectors. Depending on received strength, sectors are added or removed from the list of sectors usually referred to as the **active neighbor list** or **active set**. Members of the active set are managed and negotiated with the network over a signaling channel so that both ends of the radio link have a consistent view. All members of the active set implement macro diversity and are often referred to as being in soft

1. R. Price and P. E. Green, "A Communication Technique for Multipath Channels," *Proceedings of the Institute of Radio Engineers* 46 (March 1958): 555–70.

handoff. Macro diversity and soft handoff are fundamental techniques used by nearly all cdmaOne and current 3G systems.

GSM, USDC, and cdmaOne are all deemed second-generation (2G) systems. They were designed and developed to meet the demand for voice services, which were growing at rates as high as 50% per year. Although the communications industry heralded the coming integration of voice and data services, the concepts and potential applications were as yet unclear even in the wired world. The penetration of PCs and the Internet into our daily lives was only beginning, and the world could not have predicted just how pervasive an influence they would become. As a consequence, 2G systems were optimized for voice applications, and service providers scrambled to capitalize on the exponentially growing demand for voice services. Although all 2G systems did anticipate the need for communicating data, they were limited to data rates comparable to the rates used by the voice encoders (at most a few tens of kilobits per second). As noted earlier, 1G voice services were provided through an MSC that essentially housed a telephone network central office switch that connected to the PSTN. All 2G systems were based on similar circuit-switched access. The need for providing packet-switched interfaces to public and/or private data networks was certainly not apparent, if considered at all, given that the demand for wireless data services was as yet untested. As history demonstrated, the popularity of the Internet stimulated a worldwide demand for a wide range of communications capabilities, including simultaneous voice and data, imaging, streaming audio and video, conferencing, Web browsing, and so on. As this demand emerged for wired systems, the attractiveness of an "office on the move" for business applications and the ability to connect from anywhere without wires began to fuel predictions of enormous demand for wideband, packet-switched data communications. The capability to support this demand became the basis for the definition of third-generation (3G) systems.

Toward a Third Generation

The International Telecommunication Union (ITU), a standards organization of the United Nations, began the study and formulation of a global standard for mobile wireless communications in 1986. In the early to mid-1990s, these efforts culminated in the formulation of a plan called the Future Public Land Mobile Telecommunications System (FPLMTS)—renamed the International Mobile Telecommunications 2000 (IMT-2000) system in 1995—and the designation of spectrum required to support it on a global basis. The plan envisions a 3G system that integrates cellular, cordless, paging, and **low Earth orbit** (LEO) satellite systems into a single, globally compatible system capable of supporting these services with quality at least equal to that available from fixed networks. In all respects the plan envisions 3G systems as wireless extensions to fixed land-based networks, operating in any environment and with seamless worldwide roaming.

Among the many requirements and goals of IMT-2000 are

- Complete global coverage using terrestrial and satellite systems
- Common worldwide spectrum allocation to support global roaming

- Seamless evolution from the predominant installed base of 2G systems
- Support for high-quality multimedia services
- A common terminal for global use
- Support for both circuit-switched and packet-switched services
- Support for high-rate data services up to 2 Mbits/s (depending on speed, environment, and cell size)

Although the ITU's 3G vision was admirable, geopolitical, geo-economic, and intellectual property issues have rendered agreement on a single universal set of standards a dream yet to be realized. Instead, the worldwide diversity of wireless infrastructures drove the proposal of a number of approaches aligned along the predominant 2G architectures and the national and commercial interests that supported them. Although many system approaches were proposed, the world community focused on two different but compatible standards, both of which were based on CDMA architectures. Supporters of GSM (and later USDC and Japan's Pacific Digital Cellular system, or PDC) converged on a system called Universal Mobile Telecommunications Service (UMTS) and a particular member of the framework called Wideband CDMA (W-CDMA). Supporters of IS-95 (North American CDMA) or cdmaOne converged on cdma2000.

UMTS is an outgrowth of worldwide efforts to merge several diverse CDMA proposals. It originated with a proposal submitted by the European Telecommunications Standards Institute to the ITU as an evolution of GSM to 3G and was known at that time as UMTS Terrestrial Radio Access (UTRA). The 3G Partnership Project (3GPP), created in 1998 and working under the auspices of ETSI, is an international cooperative standards effort made up of regulators, developers, and service providers whose objective was, and still is, to develop UMTS standards that can be supported by GSM and North American TDMA access networks.

The evolution of cdmaOne converged on a standard developed under the aegis of the TIA in participation with the multinational technical membership of the 3GPP2, created in 1999. The evolutionary path for cdmaOne to 3G is a seamless and architecturally consistent sequence of incremental upgrades from 2G to the first generation of 3G and beyond. W-CDMA and cdma2000 were both aimed at providing significantly higher data rates through a packet-switched core network.

The efficacy of CDMA in providing significant improvements in capacity and spectrum efficiency was clearly demonstrated by cdmaOne. In addition, CDMA technology was shown to be more readily adaptable for packet-switching applications. In particular, CDMA allows a much simpler approach to allocating varying bandwidth among users, a property that is especially important for multimedia applications. As a result, CDMA became the architecture of choice for 3G evolutions. Although cdma2000 was designed to be backward compatible with cdmaOne, the architectural differences between 2G TDMA systems and 3G W-CDMA make upgrading more difficult and costly for service providers. Several alternatives were developed to provide packet-switched capabilities for TDMA systems that avoid the wholesale upgrade to W-CDMA, thereby allowing service providers to introduce and grow market share for packet-switched services incrementally. Systems that have implemented these enhancements are classified as 2.5G. The predominant evolutionary steps for GSM, USDC, and PDC are HSCSD, GPRS, and EDGE, which are very briefly described in the following sections. These systems are still in operation today, especially in venues where high-speed packet data services are not yet in great demand.

Generation 2.5

High-Speed Circuit-Switched Data (HSCSD)

All first- and second-generation cellular systems were optimized for the communication of voice. In particular, the system designs and cell layouts are designed to provide a specific minimum carrier-to-noise-and-interference ratio (CNIR) at a cell boundary for the 9.6 kbit/s data signal representing the output of a speech encoder. Recall that the probability of a bit error, and therefore the BER, is a strong function of E_b/N_0, that is, the ratio of bit energy to the noise power spectral density. Typically, interference is modeled as white noise, allowing E_b/N_0 to be directly related to CNIR. (Note that when we speak of "signal" in "signal-to-noise ratio," we are most often concerned with the measurement of the power in the carrier. We will not concern ourselves with the distinctions between the signal power and the power in the carrier for the ensuing discussions. In those discussions we will mention techniques requiring a mobile unit or base station to assess channel quality. This always entails the determination of the carrier-to-interference ratio, or CIR.) All 2G systems employed FEC coding as part of their digital signal-processing arrangements. By virtue of its error-correcting properties, FEC coding reduces the BER. Equivalently, one can obtain the same BER by eliminating the coder and increasing the transmit power. So in a very real sense FEC coding provides a signal gain relative to the noise and interference. For example, GSM employs a rate-1/2 convolutional encoder with a constraint length of 5. This is equivalent to a 4.6 dB signal enhancement. This coding gain is included in the link budget when developing a layout. Clearly, as a mobile unit moves toward its serving base station, the path loss decreases, or equivalently the CNIR increases, and the signal gain attributable to FEC coding becomes less important. Consequently, some or all of the coding bits can be used to transmit data bits, thereby allowing an increased data rate. This concept has profound consequences when viewed in more general terms. When signal conditions are poor (low CNIR), more protection is needed to meet QoS objectives. Protection may take one of several forms, but in general protection requires the use of parity bits for coding, or repetition of data, or changes in the modulation rate. Regardless of the form of protection, if the system bandwidth is fixed, the user data rate must decrease. Alternatively, when conditions are good, protection can be removed to allow the transmission of more user bits per unit time. This principle underlies the design of all 3G and 4G systems as well the 2.5G systems.

High-Speed Circuit-Switched Data (HSCSD) is an enhancement that provides higher-rate circuit-switched data capabilities with relatively simple software upgrades to the MSC, base stations, and handsets. HSCSD is specifically associated with GSM, although the techniques it uses can be employed in other TDMA systems. The HSCSD enhancement for GSM modifies the maximum physical layer payload data rate from 9.6 kbits/s to 14.4 kbits/s by modifying the effective FEC coding rate. This coding rate reduction is accomplished using a technique called **puncturing**. Puncturing entails the judicious removal of parity bits, in a manner that reduces, but does not eliminate, the protection. The data rate is increased accordingly. The rate is adjusted dynamically depending on the channel conditions. When channel conditions are poor, the maximum data rate is 9.6 kbits/s per slot, and 14.4 kbits/s per slot when channel conditions are good. HSCSD also allows the aggregation of up to four slots to produce a maximum data rate of 57.6 kbits/s, a rate comparable to switched wire-line circuit data rates.

Although the HSCSD enhancement to GSM provides higher-rate data services than GSM alone, it does not include interfaces to the packet-switched network as part of its architecture. As a result, packet-switched services can be provided only in the context of typical dial-up access. Data calls are charged as voice calls, and the use of multiple slots often incurs the cost of multiple voice connections. The allocation of resources between voice and data also presents a problem. Most often voice calls are given priority since they still represent the broadest base of user demand. Although circuit switching is spectrally inefficient for bursty data, it is excellent for low-latency or real-time applications.

General Packet Radio Service (GPRS)

The **General Packet Radio Service** (GPRS) was initially conceived and designed for GSM systems. Although the standard was later revised to support USDC and other TDMA systems under the umbrella of the 3GPP, we will limit our discussion to the evolution from GSM. GPRS was designed to provide full wireless packet-switched access for multimedia applications. In that regard, GPRS adds two new core network elements that perform functions for packet access that are analogous to functions that the MCS and associated components perform for circuit-switched access. As noted, bursty data may have long periods of inactivity between periods of intensive data transfer. Although HSCD offered higher-speed data access, users were charged at a voice service rate. GPRS allows several users to share a group of slots, so that users are charged based upon the amount of data they transfer (for example, on a per-megabyte basis). In addition, GPRS made provision for differing user QoS needs. Service classes were defined around four QoS parameters: priority, reliability, latency, and throughput. A user could subscribe to various QoS levels as defined by these parameter values stored in the user's QoS profile. Various classes of mobile devices are defined based upon their ability to use GSM and GPRS simultaneously.

GPRS for GSM preserves the frame and slot structure as well as the RF and modulation formats of the GSM system. Slots are usually pre-allocated to provide either voice or data, allowing service providers to engineer resources to meet varying user demand. As described in Chapter 6, GSM has a 120 ms multiframe structure consisting of 26 frames. Two frames in each multiframe are reserved for signaling purposes. Each frame consists of eight slots carrying 114 payload bits per slot. Therefore the maximum data rate per slot is 114 bits/slot $\times$ 24 frames/multiframe $\div$ 120 ms/multiframe = 22.8 kbits/s/slot. The multiframe structure consists of 52 frames (that is, two GSM multiframes). A radio block consists of four slots

Table 8.1 GPRS Radio Block Parameters

Code Scheme	Payload Size (bits)	Parity + Overhead (bits)	Encoder Input (bits)	Encoder Output (bits)	Punctured Bits	Effective Coding Rate	Data Rate (kbits/s)
CS-1	181	47	228	456	0	0.5	9.05
CS-2	268	26	294	588	132	0.64	13.4
CS-3	312	26	338	676	220	0.74	15.6
CS-4	428	28	456	456	0	1	21.4

arranged in contiguous frames, for example, slot 0 in frames 0, 1, 2, 3. Thus there are 12 blocks in a multiframe (four frames are used for signaling). Therefore a maximum of 456 payload bits can be transferred in a single radio block that takes an average (over 52 frames) of 20 ms for a raw data rate of 22.8 kbits/s per radio block. The data rate of 456 kbits/s does not include any parity coding for error detection or FEC coding. GPRS defines four coding schemes (CS) that map into four payload data rates per slot as shown in Table 8.1. Note that all coding schemes use a parity check sequence. The coding schemes provide varying amounts of code protection, which can be associated with different channel conditions. This is an extension of the concept that we saw for HSCSD. When conditions are good, coding scheme CS-1 is used, which has no FEC protection at all. When conditions are very poor, the full rate-1/2 coding is employed to enhance the signal relative to the noise and interference. CS-2 and CS-3 are punctured to achieve effective coding rates of 0.64 and 0.74 respectively. Note that the effective coding rate is the ratio of the number of bits into the coder to the number bits out of the coder. Thus the system can optimize the channel for the specific channel conditions. Note that multiple slots can be assigned depending on the amount of data to be transferred and the QoS. Theoretically, up to eight slots (radio blocks) can be assigned to a single user, thereby providing a maximum payload transmission data rate of 171.2 kbits/s.

Resources are dynamically allocated based on the packet size and required QoS. The reverse access channel (RACH) is used to request a radio block allocation. Since latency is directly affected by the performance of the reverse access channel, the availability and capacity of this channel must be engineered for the data traffic characteristics. A radio block allocation lasts for the duration of the packet transmission. Users may request changes or additional resources based on changing needs. Resource allocations may be assigned asymmetrically between uplink and downlink depending on the application and the associated user need. The reader should be aware that there is always some signaling overhead associated with a packet-switched communication scheme. Headers and trailers may be appended to the user data for various purposes, including routing and packet sequencing, as well as for controlling and maintaining the radio link. The data rates quoted here do not consider the percentage of the channel devoted to such signaling and overhead. Therefore, the actual user data rate will be some percentage less than the quoted rates. Furthermore, the average throughput that a user experiences decreases as the number of users increases. Realistic estimates of average throughput usually require simulation and characterization of the source statistics.

Enhanced Data for Global Evolution (EDGE)

The **Enhanced Data for Global Evolution** (EDGE) system provides significant improvements in throughput and flexibility for data services of GSM, IS-136, and PDC. (Note that prior to its applicability to other TDMA systems, the *G* in *EDGE* originally stood for "GSM.") The EDGE standard defines two enhancements: one for GPRS often called EGPRS, and one for HSCSD called ECSD. We will discuss the GPRS enhancement only.

EDGE introduces an 8-ary PSK (or 8PSK) modulation scheme and a rate-1/3 convolutional code that provide five additional rate schemes. Thus EDGE supports nine rate schemes, called **modulation and coding schemes** (MCS), which provide varying levels of protection based on channel conditions. These formats can be chosen autonomously and on a radio-block-by-block basis, thereby allowing the link to adapt to channel conditions in smaller increments than in GPRS. Recall that higher-order modulation schemes may produce a higher throughput but are more susceptible to noise and interference. Thus, when conditions are good, a speed improvement can be realized by changing to an 8PSK modulation scheme, which transmits one modulation symbol for each 3-bit sequence. All MCSs include a parity check sequence to detect transmission errors. A negative acknowledgment (NACK) is returned to the transmitting end whenever a packet is received in error. A technique called **incremental redundancy** (IR) is used to reduce the probability of an erroneous retransmission by successively increasing the redundancy on each retransmission until the packet is received without error. A measurement of channel quality determines the initial MCS value such that the first transmission of a packet contains a minimum amount of coding. If the first transmission is received successfully, the highest bit rate is achieved for the channel conditions. If the packet has an error, a new MCS is chosen with an incrementally higher code rate, thereby providing more protection. For each MCS the punctured bits are distinctly different. Therefore, as the number of retransmissions increases, the aggregate number of protection bits increases until the full complement of code bits is obtained. Thus the combination of coding rate, modulation scheme, and incremental redundancy provides great flexibility and opportunity for matching the transmission scheme to the channel conditions. Clearly, if the MCS can be chosen at the maximum rate with the minimum number of retransmissions, throughput will be optimum for the channel conditions.

The characteristics of the nine MCSs are provided in Table 8.2. Note that the original four rate schemes are slightly modified. Also note that rate-1/3 convolutional encoding is used exclusively with 8PSK modulation. As in the case of GPRS, up to eight slots (radio blocks) can be aggregated to support a maximum theoretical rate of $8 \times 59.2 = 473.6$ kbits/s. System providers typically limit the maximum number of radio blocks to four, yielding 286.8 kbits/s. As in the case of GPRS, the rates quoted for EDGE are dependent on system load and the air interface/packet-switching overhead.

Table 8.2 EDGE Radio Block Parameters

Modulation and Coding Scheme	Payload Size (bits)	Modulation Type	Effective Coding Rate	Data Rate (kbits/s)
MCS-1	176	GMSK	0.53	8.8
MCS-2	224	GMSK	0.66	11.2
MCS-3	296	GMSK	0.8	14.8
MCS-4	352	GMSK	1.0	17.6
MCS-5	448	8-PSK	0.37	22.4
MCS-6	592	8-PSK	0.49	29.6
MCS-7	896	8-PSK	0.76	44.8
MCS-8	1088	8-PSK	0.92	54.4
MCS-9	1184	8-PSK	1.0	59.2

Contemporary Systems and 3G Evolution

As we saw earlier, the evolution toward 3G is aligned with the worldwide dominant 2G architectures: one based on TDMA systems and GSM in particular, and one based on CDMA and IS-95 (cdmaOne) in particular. TDMA evolution was aimed at W-CDMA (UMTS) whereas cdmaOne is aimed at cdma2000. Although cdma2000 is designed to be backward compatible with IS-95, W-CDMA represents a much more difficult upgrade from GSM or other TDMA systems. In the previous section we briefly described some of the incremental changes made in 2G systems to introduce packet-switched data. The associated systems were popularly called 2.5G systems; however, there is no official definition or criterion for characterizing a system as 2.5G. Because IS-95B provides enhancements that increase the theoretical maximum data rate, it is sometimes classified as a 2.5G system. We will discuss some of the attributes of IS-95B as a preface to our discussion of cdma2000.

In this section we discuss the two predominant 3G systems, W-CDMA and cdma2000, and the evolution of their currently deployed configurations. In that regard, we will briefly discuss W-CDMA and two of its evolutionary enhancements: High-Speed Downlink Packet Access (HSDPA) and High-Speed Uplink Packet Access (HSUPA). For cdma2000 we will discuss some of the alternatives initially considered for evolution and the chosen path commencing with cdma2000 $1 \times$ EV-DO. We begin our discussion with W-CDMA.

Wideband CDMA (W-CDMA)

Among the many requirements imposed by IMT-2000 are that 3G wireless systems provide significantly higher data rates (2 Mbits/s), simultaneous information flows with differing QoS constraints, and efficient access to the public packet-switched data network (PPSDN). The GPRS core network architecture provided an excellent prototype for defining a packet-switched gateway to the PPSDN, but the adoption of a CDMA air interface represented a substantial difference from prior systems in the evolution path. Although cdmaOne provided an archetype for the air interface, systems engineers on each evolutionary path were confronted with the task of significantly improving the state of the art to provide the capabilities needed to meet the IMT-2000 requirements. Because of the dramatic changes in system architecture required to move from TDMA to CDMA in the first place, however, UMTS evolution was not as severely constrained by issues of backward compatibility as was the evolution from cdmaOne.

In order to provide some rationale for the architectural choices that systems engineers made for the W-CDMA system, it is helpful to view some of the alternatives from the perspective of cdmaOne.

Recall that in cdmaOne a 9.6 kbit/s information stream is first encoded by a rate-1/2 convolutional code for error control and then spread by a 64-bit Walsh code to provide orthogonal separation among users in the forward link. This combination of steps represents a spreading factor of 128, resulting in a 1.2288 Mchip/s transmission rate. As we mentioned earlier, cdmaOne was optimized for voice applications, and this particular configuration (usually called a radio configuration, or RC) is defined to provide adequate QoS at a cell boundary. For the given cdmaOne bandwidth of 1.25 MHz per channel, the data rate can be increased by any one or a combination of methods, provided the CNIR is sufficient to meet the error rate criteria. For example, increasing the FEC code rate by puncturing the code allows some of the coding bits to be replaced by information bits. Alternatively, the orthogonal spreading factor can be reduced, provided that shorter codes can be found that maintain orthogonality. Using a higher-order modulation method such as 8PSK or 16-QAM increases the data rate linearly with the order of the modulation. Each of these approaches requires a sufficient CNIR to ensure that a given BER objective is met. Reductions in coding rate or spreading factor, or an increase in the modulation order, require higher CNIR for a constant BER. Thus higher data rates are achievable by adapting the modulation, coding, and spreading rate to the channel conditions. As channel conditions improve, the data rate can be increased. The transmit power can also be increased to a limited extent to help adjust the CNIR, depending on the QoS constraints and the system interference budget. A linear increase in data rate can also be achieved by assigning multiple Walsh codes to a single user. Although this approach does not have any direct impact on CNIR, it does decrease the number of users that can be served. Ultimately, the maximum channel data rate is limited by the symbol rate and the achievable CNIR.

The preceding discussion assumed a bandwidth of 1.25 MHz, that is, the bandwidth of cdmaOne. To increase the data rate further, one could simply add more channels (carriers) to increase the overall bandwidth so that multiple data channels can be transmitted in parallel. (This multicarrier approach was proposed as one of the options for cdmaOne evolution.) Alternatively, one could increase the chip rate so that higher data rates can be afforded the benefits of a higher spreading gain. For example, the spreading gain for a 9.6 kbit/s information stream in the cdmaOne system is 128, yielding 1.2288 Mchips/s. If the chip rate were doubled, then a 19.2 kbit/s information stream could enjoy the same spreading gain. An increase in chip rate, however, requires an increase in channel bandwidth (all other processing such as FEC coding, modulation, etc., being the same).

The W-CDMA architecture supports both circuit-switched and packet-switched data and therefore requires the core network elements of both data types. For example, an MSC provides access to the PSTN for circuit-switched data and a packet data gateway provides access to the public PPSDN. Circuit-switched access is primarily used for voice services and low-speed data. For W-CDMA, systems engineers chose a spreading rate of 3.84 Mchips/s with a corresponding bandwidth of 5 MHz for each transmission direction. The uplink and downlink channels are separated by 190 MHz.

Like cdmaOne, W-CDMA uses orthogonal spreading and scrambling codes. Whereas cdmaOne uses 64-bit Walsh codes, W-CDMA uses **orthogonal variable spreading factor** (OVSF) codes to provide separation among users in the downlink. OVSF codes are essentially variable-length Walsh codes. In fact, every OVSF code is a member of a Walsh code set of the same order. User data is transmitted within 10 ms blocks containing 2560 chips. Signaling and control are time multiplexed within a given block. When conditions are poor, a high spreading gain is applied using a higher-order OVSF code to spread a relatively small number of bits. When conditions are good, a lower spreading gain is applied using a low-order OVSF code to spread a correspondingly larger group of bits. As we saw in GPRS and EDGE, the spreading gain is exchanged for protection as the conditions improve. Short scrambling codes are used to distinguish among sectors and cells as in cdmaOne; however, W-CDMA employs **Gold codes**[2] instead of PN sequences. Gold codes have correlation properties that are similar to those of PN sequences. Long scrambling codes based on Gold sequences are used to provide separation among users in the uplink. They too provide spreading. OVSF codes provide variable spreading within the user's channel, and several orthogonal codes can be transmitted in parallel to provide greater throughput or to support multiple information flows. It is important to emphasize that the OVSF codes provide separation only among multiple transmissions of a single user. The long Gold codes provide separation among users.

As defined in the standards, the system supports spreading rates in powers of 2 from 4 to 512 in the downlink and from 4 to 256 in the uplink. Because they use multiple codes, both the downlink and uplink are theoretically capable of achieving 2 Mbits/s depending on the speed of the mobile unit and the environment type. The system provides both convolutional and **turbo coders** (TC). Turbo coders are used for the higher data rates. The W-CDMA system relies on the processing gain obtainable with RAKE receivers. The higher chip rate (narrower symbol width) improves the RAKE receiver performance and allows an increase in the number of multipath components that can be resolved. QPSK is used for the forward link and BPSK is used in the reverse link. In the downlink, the raw single code channel data rates range from 15 kbits/s to

2. A Gold code is produced by adding two different PN codes together bit by bit, modulo-2. Families of "nearly orthogonal" Gold codes can be constructed that contain a larger number of codes than are available when ordinary PN codes are used.

1.92 Mbits/s in multiples of 15 kbits/s. The data rate DR is related to the spreading factor SF (or rate) according to $DR = 2 \times 3.84 \times 10^6 / SF$ kbits/s. Data rates in the reverse direction are half these values owing to the lower-order (BPSK) modulation. The actual user-attainable data rates are substantially lower depending on the FEC coding rate, error detection and tail bits, and control overhead. As stated in our earlier discussions on GPRS and EDGE, the data rate selected for a given transmission is based on the measured channel quality (related to CNIR). The system was designed to provide a maximum data rate of 384 kbits/s in outdoor (macrocell) environments at speeds of up to about 120 km/hr. Because the higher data rates are obtained at the expense of spreading gain, they are applicable only in microcell or picocell environments.

Although the W-CDMA architecture includes a packet-switched core network, the radio interface is not efficient in its treatment of high-rate bursty data. Channel resources are determined and allocated upon setup and, for the most part, remain fixed for the duration of the allocation. In W-CDMA, as it is a true CDMA system, power control is used to maintain the QoS as the channel conditions vary so that the channel resources do not change from packet to packet. As we have seen in Chapters 6 and 7, however, many packet-data-based services are bursty in nature, requiring channel resources for short periods of time. In that regard, the system is spectrally inefficient for many packet data applications. Furthermore, since CDMA systems in general attempt to minimize interference, transmit power is usually well below the maximum available in either direction. This suggests an underutilization of an important system resource.

High-Speed Packet Access (HSPA)

The standards developed for W-CDMA are comprehensive and rather complex. In most respects the system design meets or exceeds the IMT-2000 requirements, although the conditions under which the highest data rates apply are not normally encountered in practice. In any event, the systems as currently deployed do not support anywhere near the full functionality as defined in the original standard. The deployment of W-CDMA was based on a number of phased releases, each release providing additional and enhanced capabilities. In a very real sense, the system releases represent an evolution toward full capabilities. Initial W-CDMA systems are often referred to as UMTS Release 99. For a number of reasons, including performance limitations, spectral inefficiency, competition from other technologies, evolving user and service provider needs, and market experiences, the evolution shifted in favor of an architecture and two protocol enhancements termed **High-Speed Downlink Packet Access** (HSDPA) and **High-Speed Uplink Packet Access** (HSUPA). The combination of these enhancements is simply termed **High-Speed Packet Access** (HSPA). The standards for HSDPA were finalized in 2006 and those for HSUPA were finalized in 2007. Here, too, HSPA is being deployed in phases with well over 100 HSDPA systems deployed worldwide by mid-2007 and HSUPA deployments beginning in the fourth quarter of that year.

HSDPA, often referred to as UMTS Release 5, was designed to provide significant enhancements to the forward link only as compared to Release 99. These enhancements include significantly increased peak data rate and average user throughput, reduced latency for real-time applications, greater spectral efficiency, and better support for multimedia services. Release 5 does not provide improvements to the reverse link. A key attribute of Release 5 is full backward compatibility with Release 99. It supports both the circuit-switched facilities of Release 99 and the packet-switched capabilities of HSDPA simultaneously on a single downlink 5 MHz channel.

The performance of HSDPA is founded on a rather extensive modification to the packet data architecture. The major protocol enhancements that enable the performance of HSDPA include

- The addition of 16-QAM modulation
- Use of a highly efficient rate-1/3 turbo encoder for all packet data transmission
- Rapid feedback on channel conditions from the mobile unit to the base station
- Time-division multiplexing (TDM) of user transmissions using maximum available channel resources
- Dynamic maximization and adjustment of transmit power
- Use of multiple code channels
- Use of code-division multiplexing (CDM) for multiuser transmission during a TDM interval
- Incremental redundancy in the form of Hybrid ARQ (H-ARQ)

The radio link operation of HSDPA is much more closely aligned with that of a true packet data system. Users share the channel one at a time in TDM fashion for short periods of time. Each packet is transmitted using only the resources necessary to maintain the required QoS for the application. Resources may vary on a packet-to-packet and user-to-user basis, depending on channel conditions. In this way HSDPA optimizes system resources to meet the demand and QoS characteristics of the offered traffic and therefore is much more efficient in its use of the spectrum. Furthermore, data rate control and the allocation of resources are performed at the base station layer two (L2), the layer just above the physical layer. This reduces latency by improving responsiveness to changing conditions.

As already noted, W-CDMA Release 99 and cdmaOne were founded on the minimization of interference. These systems were optimized for the communication of voice, which was, and still is, the primary market demand. Both systems employ a power control philosophy, a technique that seeks to constrain the transmit power to the level required for adequate QoS. For voice applications, the QoS parameters of key concern are BER and latency. As discussed earlier, BER is related to CIR, so in particular the systems seek to maintain the CIR at an adequate level for the QoS. For access, latency is usually not an issue since the end-to-end facilities are not shared and are continuously available; however, voice over IP (VoIP) applications share facilities and are affected by traffic congestion, queuing delays, and other factors. HSDPA is based on a maximum available power, variable-rate philosophy. This approach seeks to maximize the data rate based on rapid estimates of the channel conditions and the available power. The approach and supporting techniques are a common basis for other systems to be discussed. A brief discussion follows.

In HSDPA, forward link data is transmitted using TDM. Each TDM time interval (TTI) or frame has a fixed duration of 2 ms and is dynamically assigned to one or more users. Unlike in Release 99 or the circuit-switched operation of Release 5, the spreading factor is fixed at 16. So each transmission in the forward link enjoys a spreading gain of 16. There are 16 spreading codes, 15 of which are assignable to users on a frame-by-frame basis. One spreading code is reserved for radio link control. One or more users may receive forward link transmissions in a single frame. Users are separated by their assigned spreading code so that multiple simultaneous transmissions can be supported by code-division multiplexing. HSDPA allows up to 15 spreading codes to be assigned to a single user or fewer codes to each of several users. In contrast to Release 99 or other CDMA systems, HSDPA does not support soft handoff in the forward direction. All transmissions to a mobile unit are from a single sector.

We have seen in earlier discussions that the achievable data rate is constrained by the CIR and by the system resources and capabilities. The achievable BER (packet error rate [PER] or block error rate [BLER]) is related to the radio channel quality, which depends on the CIR. We saw that when channel conditions are sufficiently good, a higher-order modulation can be used, thereby increasing the number of bits that can be transmitted per modulation symbol. Furthermore, when conditions are good, information bits can replace some or all of the FEC coding bits by puncturing. The effective coding rate of a punctured code is the ratio of the number of encoder input bits to the total number of coded output bits. The fact that the system supports multicode operation provides yet another method for increasing and/or controlling data rate. Clearly, if more power is available, the CIR can be increased to improve the channel conditions. The fact that CDMA systems seek to minimize transmit power most often results in a substantial margin of unused power. This unused or residual power is the difference between the maximum power available from the base station sector transmitter and the actually transmitted power. The amount of residual power is a function of the traffic within the sector and the geographic distribution of mobile units within the sector. In HSDPA the residual power is dynamically allocated on a frame-by-frame basis. The ability to adjust all of these parameters provides the basis for rate control.

The standard defines 12 categories of mobile unit (termed UEs or **user elements** in UMTS terminology), thereby allowing the manufacture and marketing of a diversity of handsets with a wide range of capabilities, and therefore a wide range of complexity and cost. The differences include, but are not limited to, the maximum number of simultaneous spreading codes the mobile unit can support, the maximum data block size it can handle in a 2 ms frame, and the optional support for 16-QAM. Depending on UE category, a series of transmission formats has been defined as a function of the channel quality. For example, for UE categories 1–6, 30 formats are defined that specify the data block size, the number of simultaneous code channels, the modulation type, and a reference power adjustment. The formats are identified by a channel quality indicator (CQI) value and are known by both the mobile unit and the base station. The CQI value is based on the measured pilot signal strength and local interference level, adjusted by the reference value in the CQI table and a user-specific value sent over the control link.

The procedures governing the operation of the forward link are rather simple in concept, consisting of four high-level steps described as follows:

1. Every 2 ms each mobile unit that has an active packet data connection measures and reports its channel quality in the form of a CQI value.
2. A software entity in the base station (termed Node B in UMTS terminology) called a **scheduler** considers a wide variety of factors, including the CQI values reported by all the mobile units, and determines which user(s) will be served during the next TTI.
3. Based on the reported CQI and buffer length (data queue), the base station determines the data rate and modulation type for each user selected for the next TTI and transmits the data using the selected formats.
4. Upon receiving the transmitted information, a mobile unit validates the data using the CRC parity bits associated with the data block. Each mobile unit sends an ACK (acknowledgment) or NACK accordingly.

The scheduler is a complex algorithm that is not specified by the standard. As such it is proprietary and varies depending on the equipment manufacturer. In addition to the reported CQI values, the scheduler may consider the length of the users' data queues, QoS parameters associated with the service type, UE category, subscriber profile, fairness treatment, time-varying traffic profiles, and so forth. Clearly, the manner in which users are served has a dramatic impact on user perception and service provider metrics.

During the establishment of a packet data session, users are assigned a temporary identifier. This identifier is transmitted along with a number of parameters, including the number and identity of the spreading codes, the number of CRC bits, H-ARQ parameters, modulation type and block size, and others, in the forward link control channel. (Recall that the control channel is defined by a specific 16-bit spreading code.) The control information for each data TTI is transmitted 1.3 ms ahead of the associated data TTI.

H-ARQ is implemented by transmitting the data over a sequence of three TTIs. The first contains the actual data and the second two each contain half the turbo coder parity bits. The CQI format chosen by the mobile unit is such that it should achieve a 90% probability of error-free reception. If the packet is received without error, the mobile unit sends an ACK and the transmission of the data block is complete. If the packet contains an error, the mobile unit sends a NACK and the base station sends the next block containing half the parity bits (that is, the first redundancy block). Upon reception, the mobile unit attempts to construct an error-free block by processing the data block and the first redundancy block. If successful, the mobile unit sends an ACK. If not, it sends a NACK, whereupon the base station sends the next redundancy block. The mobile unit then attempts to construct an error-free data block using the data block and both redundancy blocks. If upon receipt of the second redundancy block an error-free block cannot be constructed, the physical layer takes no further action. Higher layers of the protocol stack may initiate retransmission, depending on the nature of the data stream and its QoS.

The design of HSDPA was aimed at achieving data rates in excess of 10 Mbits/s. The maximum data rate achievable by a given mobile unit (UE) is determined by its category. The values range from 0.9 Mbits/s for category 11 through 14 Mbits/s for category 10. (Note that categories 11 and 12 have the lowest-level capabilities, capabilities increasing in the range 1 through 6.) The maximum theoretical rate of 14 Mbits/s assumes that all 15 codes are assigned to a single user with a category 10 UE and no other traffic is carried within the sector. This further assumes that no H-ARQ retransmissions are required. More practical limits on the peak rates are 10 Mbits/s and under, depending on the UE category. The average throughput experienced by a user is substantially lower given traffic conditions and the location of the mobile unit relative to the base station.

High-Speed Uplink Packet Access (HSUPA) is the common name for UMTS Release 6. Like HSDPA, Release 6 enhancements include significantly increased peak data rate and average user throughput, reduced latency for real-time applications, greater spectral efficiency, and better support for multimedia services. The concepts for achieving these performance enhancements are fairly simple, but the implementation details are rather complex. In keeping with our theme, we discuss the attributes of HSUPA only at the architectural level. Like HSDPA, Release 6 is backward compatible with Release 99, and it supports both the circuit-switched facilities of Release 99 and the packet-switched capabilities of HSUPA simultaneously on a single uplink 5 MHz channel.

As in HSDPA, the radio link operation of HSUPA is much more closely aligned with that of a true packet data system. Several users share the channel for short periods of time. Each packet is transmitted using only the resources necessary to maintain the required QoS for the application. Resources are granted by the system and may vary on a packet-to-packet and user-to-user basis, depending on channel conditions and within the constraints of the allocation. The mobile units determine the actual power level to be used in transmission. Resources are granted by the system based on requests from the UEs. Uplink transmissions can be as short as 2 ms in duration. Like HSDPA, HSUPA optimizes system resources to meet the demand and QoS characteristics of the offered traffic and therefore is much more efficient in its use of the spectrum.

The reverse link of HSUPA is interference limited as in most CDMA systems. Interference is predominantly a result of using scrambling codes to provide separation. Because scrambling codes are not perfectly orthogonal, each mobile unit introduces some degree of interference. Interference within a sector arises from mobile units both within and outside the sector. Therefore the ability of the system to meet QoS objectives for any given transmission is limited by the total interference within the sector. At a base station the interference in each sector is measured relative to the thermal noise level of the receiver. A maximum interference threshold, often termed the **rise-over-thermal threshold** (RoTT), is established for each sector. Each sector uses this threshold as a reference for controlling the reverse link power of the mobile units. All active mobile units transmit pilot signals. Pilot signals are used as a reference for measuring the power received from a mobile unit at the base station in its sector. Pilot signals are used as well for purposes of coherent demodulation.

Unlike the forward link defined for HSDPA, HSUPA uses macro diversity or soft handoff in the reverse link. As such, several sectors may be listening to an MU's reverse link transmissions at the same time. These sectors belong to what is known as a mobile unit's active set. Each mobile unit and the system cooperatively manage the members of an active set using control channel messages. The power level of a mobile unit is managed using low-layer signaling from each sector in its active set. For the purposes of power control, a base station sector's active set comprises all of the mobile units for which it is a member of that MU's active set. Each sector attempts to equalize the power received from all mobile units in its active set while minimizing the overall interference level within its sector. Each sector sends small up/down adjustments to each mobile unit in its active set as often as 1500 times per second. If any active set member sends a power down, the mobile unit powers down. Otherwise the MU powers up. This ensures that an MU will not create excessive interference to one of its active set members. For circuit-switched services, data signals are sent at a predetermined power level relative to the pilot signal. The level is determined based on the QoS parameters of the application. For packet-switched services, a mobile unit has the ability to adjust the power in the data signal relative to the pilot as the conditions warrant.

In HSUPA, data rate control is also performed at the base station in layer 2. The link operates on the basis of variable power and variable rate control. Like HSDPA, the system manages the available power to maximize the throughput without excessive interference. The process is similar to that described for HSDPA.

- Mobile units within a sector send scheduling requests to the system over a reverse link control channel. Each mobile unit's request indicates the amount of data in its buffer and the power headroom it possesses relative to pilot, that is, the amount of power it can transmit relative to its current pilot. Mobile units also send QoS information for their highest-priority data flow and the amount of data in the buffer that is associated with that flow.
- A scheduler at the L2 layer determines the amount of power to be allocated to each mobile unit. As noted in our discussion of the downlink scheduler for Release 5, the uplink scheduler for Release 6 is not specified in the standard. Among the many possible inputs, the scheduler considers the current channel conditions in the sector, requests from each mobile unit, and associated QoS information that might be available.
- The serving sector sends a "grant" message to the mobile unit over a forward control channel indicating the allocated power. Since other sectors may be adversely affected by an increase in the mobile unit's power level, however, they may also provide input to the determination of a power grant to the mobile unit. Other active sectors may send "up", "down," or "hold" relative grants to ensure that a mobile unit does not cause excessive interference to its neighbors.
- Based on the allocations, the mobile unit transmits its data in 2 ms (or 10 ms) bursts using an appropriately adjusted format and within the power constraints of the system. Upon receipt of

the burst, the network performs a parity calculation to check for errors. If there are no errors, the mobile unit sends an ACK; otherwise it sends a NACK to request additional information in an H-ARQ fashion.

The transmission format is determined based on the power allocation relative to pilot, the maximum block size, and TTI format (2 ms or 10 ms) to be used. A rate-1/3 turbo coder is used for all uplink transmissions. OVSF codes are used to adjust the spreading depending on the power allocation. Since scrambling codes are used to provide separation among users, the OVSF codes used within a mobile unit's uplink code channel are not visible to other mobile units (apart from the residual interference associated with the scrambling codes). The spreading codes used for uplink transmission are predetermined and depend on the configuration parameters and the allocated power. The spreading factors range from 2 to 256. When the power allocation is small relative to pilot, higher spreading factors are used, resulting in fewer payload bits being sent during a TTI. Under such circumstances the transmitted data rate is low. When the power allocation is large, lower spreading factors are used so that more payload bits can be sent during a TTI. Because OVSF codes are orthogonal regardless of order, multiple codes can be sent simultaneously. For example, a maximum rate of 5.76 Mbits/s can be achieved by transmitting two groups of information bits simultaneously, one with a spreading factor of 2 and one with a spreading factor of 4.

According to the evolutionary road map as it stands at the end of 2007, Release 6 is the last major release of UMTS under the classification of 3G systems as defined by the 3GPP. We now shift our attention to 3G systems evolving from cdmaOne as defined by the 3GPP2, namely, cdma2000 and its descendants.

cdma2000 Radio Transmission Technology (RTT)

The cdma2000 radio transmission technology (RTT) standards and framework were developed under the 3GPP2 and classified as a 3G system compliant with the intent of the IMT-2000 requirements. As originally conceived, the overall architecture supported a basic system and an enhanced system. The basic system utilized a single cdmaOne channel and the enhanced version supported three cdmaOne channels. These were classified as $1 \times$ RTT and $3 \times$ RTT respectively. Whereas $1 \times$ RTT provided significant enhancements to the spectral efficiency and maximum rate of IS-95B, $3 \times$ RTT was needed to approach the 3G requirement of 2 Mbits/s. $3 \times$ RTT presented significant challenges in terms of cost and complexity, however, especially for handsets. It was eventually abandoned in favor of other approaches that meet or exceed the 2 Mbit/s IMT-2000 requirements. Consequently, we will not discuss $3 \times$ RTT further.

Because the IS-95 system was designed primarily for voice service, it operates at a maximum data rate of 9.6 kbits/s; however, it is also capable of operating at the submultiple rates of 4.8 kbits/s, 2.4 kbits/s, and 1.2 kbits/s. In 1995 a revision of IS-95A provided support for higher-rate vocoders by adding a 14.4 kbit/s base rate and its submultiple rates of 7.2 kbits/s, 3.6 kbits/s, and 1.8 kbits/s. The 9.6 kbit/s rate and its submultiples constitute Rate Set 1 (RS 1) and the 14.4 kbit/s rate and its submultiples constitute RS 2. The fact that the system transmitted data in blocks or frames and that the rate could be changed from frame to frame made it an excellent basis from which packet data operation could evolve.

In order to accommodate a 14.4 kbit/s data rate, the radio configuration had to be modified, since the transmitted chip rate is fixed at 1.2288 Mchips/s. Recall that for the forward link in IS-95, the 9.6 kbit/s input data stream is encoded using a rate-1/2 convolutional encoder to produce a 19.2 kbit/s coded bit rate prior to Walsh coding. To support the 14.4 kbit/s rate the output of the convolutional encoder is punctured (2 out of every 6 bits are deleted) to maintain the 19.2 kbit/s coded bit rate. For the reverse link, RS 1 uses a rate-1/3 encoder, whereas RS 2 uses a rate-1/2 encoder prior to the Walsh code modulator.

IS-95B was designed to be fully backward compatible with its predecessors, IS-95, IS-95A, and subsequent revisions. As such, it is capable of operation in both the 800 MHz AMPS band and the 1900 MHz PCS band and includes support for both the 9.6 kbit/s and 14.4 kbit/s base rates. As the first evolutionary step toward high-rate packet data, IS-95B uses a multicode approach to substantially increase the data rates of the forward and reverse links.

Forward and reverse dedicated traffic channels are either **fundamental channels** (FCH) or **supplemental code channels** (SCCH). Fundamental channels are used to convey voice, low-rate data, or signaling and operate at any of the rates associated with the base rates. SCCHs are always associated with an FCH and operate at the maximum rate of the FCH, that is, the base rate. Up to seven SCCHs can be allocated to a given user so that the maximum theoretical rate is 8×14.4 kbits/s = 115.2 kbits/s for RS 2. Although IS-95B provides a significant increase in the maximum theoretical data rate, it is inefficient in its use of the spectrum. IS-95 and its successors reserve certain code channels for signaling and overhead. For example, every sector transmits a pilot channel, a sync channel, and at least one paging channel in the forward link. For high-density areas, up to seven paging channels may be configured. Furthermore, the system employs soft handoff so that there may be an average of only about 20 out of a possible 64 Walsh codes available to carry traffic. Furthermore, resources are allocated for fixed periods of time, and the data rate of an SCCH cannot vary from frame to frame. These characteristics make the approach especially inefficient for bursty data applications. Given that for most service providers voice services represent the major source of business revenue by a substantial margin, the allocation of many channels to a single data user is considered extravagant if not wasteful of system resources. IS-95B did not enjoy widespread deployment; however, several of the concepts were employed in cdma2000 RTT.

cdma2000 1 × RTT

The cdma2000 system is fully backward compatible with cdmaOne systems. In addition to a circuit-switched core network it also incorporates new network elements to provide packet-switched access to public or private packet-switched data networks. As in UMTS Release 99, however, radio resources do not support high-rate bursty data efficiently. cdma2000 uses both variable spreading and multiple codes to achieve higher data rates. User traffic may be conveyed on a dedicated traffic channel comprising a fundamental channel and supplementary channels. Fundamental channels are used to convey voice, low-rate data, and/or signaling, whereas high-rate data is always conveyed by supplementary channels.

A number of radio configurations have been defined for the cdma2000 system. (Several are specific to 3 × RTT and will not be discussed here.) For 1 × RTT, five radio configurations (RCs) were defined for the forward link and four were defined for the reverse link. RC 1 is based on RS 1 and RC 2 is based on RS 2, providing full backward compatibility with legacy cdmaOne systems. To support IS-95B functionality, RC 1 and RC 2 augment an FCH with up to seven SCCHs, preserving the maximum theoretical rate of 115.2 kbits/s for RS 2. RCs 3, 4, and 5 represent 1 × RTT 3G technology and are capable of achieving up to 614.4 kbits/s. We will limit our discussion to RC 3–RC 5.

Like IS-95B, dedicated traffic channels comprise an FCH and up to two supplementary channels (SCHs). (Note the distinction between SCH and SCCH: SCCHs are associated with the IS-95B legacy system whereas SCHs are associated with the cdma2000 enhancement.) SCHs, however, are capable of significantly higher rates than their associated FCH and SCCH counterparts. SCHs are always associated with an FCH or a dedicated control channel (DCCH) if no FCH is needed.

Table 8.3 summarizes the rates and characteristics of the forward link radio configurations. (RC 1 and RC 2 are included for completeness.) Observe that RCs 1, 3, and 4 are based on RS 1 whereas RCs 2 and 5 are based on RS 2. Fundamental channels may operate on RCs 3, 4, and 5. The maximum rate of an FCH is limited to the base rate of the rate set associated with an RC, that is, either 9.6 kbits/s or 14.4 kbits/s for RS 1 and 2, respectively. Fundamental channels may also operate at the rates below the base rates as defined in the table. Typically, FCH transmissions are conveyed in 20 ms frames; however, FCHs may also transmit in 5 ms frames. Each RC provides support for a 5 ms frame at the fixed rate of 9.6 kbits/s. Short frames are used to transmit short messages and signaling messages requiring rapid access. As in cdmaOne, the rate for FCHs may be varied on a frame-to-frame basis.

FCHs typically carry voice data but may also carry low-rate data and signaling. RC 3 and RC 4 support 8 kbit/s vocoders whereas RC 5 supports 13 kbit/s vocoders. RC 3 and RC 4 differ in the coding rate and Walsh code length. The more robust encoding scheme of RC 3 allows a lower transmission power, thereby reducing system interference. RC 4 has twice the number of

Table 8.3 Summary of Forward Link Radio Configurations

RC	Transmission Rate (kbits/s)	FEC Coding Rate (*r*)	Walsh Code Length	Modulation
1	9.6, 4.8, 2.4, 1.2	1/2	64	BPSK
2	14.4, 7.2, 3.6, 1.8	1/2	64	BPSK
3	153.6, 76.8, 38.4, 19.2, 9.6, 4.8, 2.7, 1.5	1/4	Rate ≤ 9.6 kbits/s, 64 Rate > 9.6 kbits/s, 4–32	QPSK
4	307.2, 153.6, 76.8, 38.4, 19.2, 9.6, 4.8, 2.7, 1.5	1/2	Rate ≤ 9.6 kbits/s, 128 Rate > 9.6 kbits/s, 128–8	QPSK
5	230.4, 115.2, 57.6, 28.8, 14.4, 7.2, 3.6, 1.8	1/4	Rate ≤ 14.4 kbits/s, 64 Rate > 14.4 kbits/s, 4–32	QPSK

Walsh codes but requires more transmit power. RC 3 can be allocated when a sector is interference limited, and RC 4 can be allocated when a sector is code limited.

SCHs are used exclusively for traffic. Related signaling information must be conveyed on an FCH or DCCH. The selected transmission parameters may be set according to the desired QoS of the associated application. When conditions are good, higher data rates can be obtained by choosing an RC with a lower coding rate and spreading factor. When conditions are poor, the rate is determined by the required coding and the spreading rate needed to support the error rate. More efficient turbo coders may be used for rates above 19.2 kbits/s for RCs 2 and 3 and above 28.8 kbits/s for RC 4. SCHs may operate on 20, 40, or 80 ms frames depending on the data rate and application. Larger frames permit a greater interleaving separation and improve the performance in the presence of burst errors. The maximum theoretical data rate is 614.4 kbits/s and can be achieved when two SCHs are allocated, both of which are operating at 307.2 kbits/s using RC 4.

SCHs are primarily used to convey high-rate data, which is typically bursty in character. The 1 × system was designed to handle bursty transmissions in a much more efficient manner than prior cdmaOne systems. SCHs may be allocated, redefined, and deallocated many times during a packet data session without releasing the connection or the associated FCH or DCCH. The system allocates resources based on a number of conditions, including the length of the incoming data queue, QoS parameters such as latency or frame error rate (FER), and channel conditions. The system allocates resources over an FCH or DCCH as necessary. The system computes a target data rate given the queue length and latency required. For applications that are insensitive to latency, default values may be established for the sector. Based on the target rate, the system determines the power necessary for the transmission and the power available from the power amplifier in the sector. Recall that higher data rates require higher transmit power because they trade bits required for protection or redundancy (such as FEC code bits, Walsh code spreading bits, etc.) for payload bits. The system grants the SCH resource if the required power is available given the channel conditions. If not, the data rate may be adjusted appropriately. The allocation message contains a start time, a time duration for which the allocation is valid, and a radio configuration. SCHs may be assigned to more than one user. In such circumstances the system schedules transmissions to each user based on an appropriate algorithm that attempts to treat users fairly. The 1 × system employs power control in both the forward and reverse directions. Information is sent uplink to the system over the reverse power control subchannel (R-PCSCH) to control the forward link transmit power and to help determine an appropriate allocation of resources.

Table 8.4 summarizes the rates and characteristics of the reverse link radio configurations. (Here, too, RC 1 and RC 2 are included for completeness.) Although the parameters are not shown in the table, the reverse link also supports 40 and 80 ms frames in addition to the 20 ms frames. RC 3 and RC 4 also support 5 ms frames at a fixed 9.6 kbit/s rate.

Observe that there are only four RCs for the reverse link as compared to five for the forward link. Reverse link RCs 1 and 2 provide backward compatibility with legacy cdmaOne systems.

Table 8.4 Summary of Reverse Link Radio Configurations (for 20 ms Frames Only)

RC	Transmission Rate (kbits/s)	FEC Coding Rate (r)	PN Sequence (chips/bit)	Forward Link RC Mapping
1	9.6, 4.8, 2.4, 1.2	1/3	128–1024	1
2	14.4, 7.2, 3.6, 1.8	1/2	85.33–682.67	2
3	307 153.6, 76.8, 38.4, 19.2, 9.6, 4.8 2.7, 1.5	1/2 1/4	2 4–256 455.11–819.2	3, 4
4	230.4, 115.2, 57.6, 28.8, 14.4, 7.2, 3.6, 1.8	1/4	5.33–682.67	5

Note that the mapping between forward and reverse link RCs is based on common rate sets. As in the case of the forward link, turbo coding may be used for rates of 19.2 kbits/s for RC 3 and 28.8 kbits/s for RC4. In the forward direction Walsh codes are used to provide separation, and variable-length Walsh codes provide sufficient spreading to achieve a 1.2288 Mchip/s output rate. In the reverse direction, a PN sequence is used to provide separation and spreading; however, the PN sequence is applied at the chip rate so that the number of chips per bit provides a more appropriate measure of spreading as shown in the table. All reverse link RCs use BPSK modulation.

Operation in the reverse link is directly analogous to the forward link. Up to two R-SCHs may be allocated to an R-FCH or R-DCCH. Several channels may be transmitted simultaneously within each frame by a given mobile unit. Each channel is assigned a unique and predefined (by standards) Walsh code to provide separation among the different channels. For example, an FCH and two associated SCHs are each covered by distinct orthogonal Walsh covers. All three channels are transmitted simultaneously by a given mobile unit. In the reverse direction, Walsh codes provide separation among channels transmitted by a given user, whereas PN sequences provide separation among mobile units. Apart from these differences, the reverse link transmissions operate in a manner analogous to that of the forward link.

As in its W-CDMA counterpart, the conditions under which the cdma2000 air interface can support relatively high data rates are often unrealistic. Because it is a true CDMA system, its performance depends on managing and minimizing interference. Consequently, adding data capabilities, especially high-rate data capabilities, increases interference and limits capacity and/or performance for voice services. Limiting voice services has negative financial implications for most service providers. In fact, capabilities beyond 153.6 Mbits/s have not been deployed. To effectively support high-rate bursty data, a system should be able to track and accurately determine channel conditions, quickly adapt the radio configuration parameters to match the channel conditions, and utilize the full system resources to support the desired QoS. cdma2000 as well as W-CDMA are both limited in these regards. Channel conditions are inferred from power control mechanisms so that actual channel conditions are not accurately determined by the receiver. Radio configuration adaptation does not include adjusting the coding or modulation parameters, so the system's ability to adapt is limited to variable spreading. Last, the system does not utilize the full power available within a sector so that the resources of the radio channel are underutilized.

Initially, the path for cdma2000 evolution to full IMT-2000 compatibility was based on $3 \times$ RTT technology. As we stated earlier, however, $3 \times$ RTT presented significant challenges in cost and complexity, especially for handsets. Furthermore, $3 \times$ RTT did not significantly enhance spectral efficiency. For these reasons the $3 \times$ RTT evolution was abandoned in favor of two alternative architectures, namely, cdma2000 $1 \times$ EV-DV and cdma2000 $1 \times$ EV-DO. cdma2000 $1 \times$ EV is an evolution framework for high-speed data that was originally conceived and developed by Qualcomm, Inc. As an internal project, this framework was simply termed HDR for High Data Rate, a phrase that was later adopted within various standards forums. The framework envisioned $1 \times$ channels that could be employed as an overlay to existing cdma2000 networks. The system was designed to support high-rate data only, however, and was subsequently called $1 \times$ EV for data only (DO). An alternative architecture was proposed that is similar in many respects to the HSDPA/HSUPA. This architecture was termed $1 \times$ EV for data and voice (DV).

The $1 \times$ EV-DV (or simply DV) standards were intended to be a fully backward-compatible evolution of cdma2000 (IS-2000). Downlink enhancements were developed under cdma2000 Rev. C, and uplink enhancements were developed under Rev. D. Because the underlying principles and techniques for enhancing system performance are so similar to HSDPA and HSUPA, which were discussed earlier, they will not be discussed here. $1 \times$ EV-DO (or simply DO), on the other hand, was not developed to be backward compatible with cdma2000. Instead, it is intended as an overlay to existing systems, sharing the RF architecture and perhaps the same geographic locations as the elements of deployed cdma2000 systems but using different channel center frequencies. The technical merits of one approach versus the other have been endlessly debated, but the decision to evolve along the DO path was made for other than technical reasons.

From about mid-2004 to early 2005 there were a number of market forces that had a significant influence on the choice of an evolution path. Among the most significant were competition from alternative technologies, namely UMTS; the growing interest in wireless high-speed data as demonstrated by the penetration of WiFi (Wireless Fidelity, referring to wireless data services based on IEEE 802.11 standards); and government mandates requiring the portability of wireless telephone numbers among service providers. The standards and equipment to support HSDPA and HSUPA were not yet available, so W-CDMA service providers were not yet in a position to offer efficient high-speed data services. The market interest in high-speed data was evident from the success of WiFi "hot spots" in public venues such as coffee shops, airports, and other places. Last, government mandates requiring that wireless telephone numbers be transferable removed a major obstacle for users who might wish to change service providers. At the end of 2004, a major U.S.-based cdma2000 service provider, having completed successful trials of DO, announced plans to deploy high-rate (up to 2.45 Mbits/s) data services nationwide using DO. This gave this provider a substantial advantage over its competitors. About six months later a second major service provider announced plans to deploy DO. These decisions, which were clearly aimed at capturing market share, essentially eliminated the market for DV infrastructure, and consequently the standard was abandoned. The success of DO is most assuredly attributable to the fact that the DO requirements had been finalized and that infrastructure was available. This is one of many examples in the history of technology evolution where the technical merits of a system are far less important than the market forces surrounding its deployment. Very often the most important element in the success of a technology venture is having a product that meets a market need at the right time and for the right price. Time to market, a clear identification of market need, an understanding of market forces and trends, and price must always weigh heavily in the development of a system concept. Sound systems-engineering practice requires a clear understanding of these and other market attributes to ensure that the "right" system is developed within the constraints of the intended market. Clearly, "right" may not always refer to the technical best when viewed in the context of a winning market strategy.

cdma2000 1 × EV-DO

Cdma2000 1 × EV-DO, Revision 0, was designed to convey non-real-time high-rate packet data wirelessly. In particular it was aimed at providing immediate support for simple "best effort" services like Web browsing, file transfer, and simple e-mail. By "best effort" we mean that the system did not provide any means for adjusting QoS parameters such as latency, error rate, and other criteria. Although IMT-2000 pointed to the eventual need for wireless real-time multimedia services as well as high-rate data, the most immediate market need was to extend the office environment by providing access to information sources and e-mail for "people on the move." Applications of this type are most often asymmetric, with relatively small amounts of data conveyed in the uplink and large amounts of data conveyed in the downlink. This is characteristic of Web browsing and file transfer and, often, of e-mail. DO Rev. 0 was aimed at meeting the demand for these types of services and in that regard emphasized the forward link. The forward link is capable of 2.45 Mbits/s, whereas the reverse link is capable of only a modest 156.3 kbits/s, a 16:1 asymmetry between forward and reverse links. DO was not designed to be backward compatible with cdma2000, and perhaps this was one of its major disadvantages in comparison to DV. Nevertheless, DO was designed to be an overlay to existing cdma2000 1 × RTT (hereafter referred to simply as 1 ×) systems, sharing the same RF channelization, RF architecture, and cell design. DO requires its own channels, however, so that voice and high-rate data can be conveyed only by using separate carriers. Interoperability can be achieved only by using hybrid handsets that are capable of operating in either system.

The DO forward link is a significant departure from 1 × insofar as it is based on TDMA and adaptive rate control. The reverse link is still CDMA based and very similar to the 1 × reverse link.

All transmissions in the forward direction are sent to one handset at a time using the full power allocated to the serving sector. The data rate is varied, however, based on the channel conditions reported by the handset (also known as an AT for "access terminal" in DO parlance). In the forward direction, an AT "listens" to the strongest sector in its active set. Only the strongest sector transmits to an AT in any given transmission interval. Soft handoff is not employed in the forward link; however, an active set and soft handoff are used in the reverse direction. The AT designates the strongest sector by coding its report of channel conditions using a short Walsh code assigned to the sector by the access network (AN). (The AN is the serving base station and its associated base station controller.) When an AT is granted a DO connection, it is given an identifier that is unique within the serving sector called a MAC Index or MAC-ID. Rev. 0 supports 64 (6 bits) MAC-IDs, 59 of which are for user data and the remainder for control information or other purposes. Distinct MAC-IDs are also allocated for all the members of the active set. Time slots are not allocated to a user in typical round-robin TDMA fashion. Instead, transmissions are addressed to a given AT using the MAC-ID encoded in a preamble that is sent at the beginning of a forward link packet. All transmissions are conveyed in 1.67 (5/3) ms slots, and there are 16 slots within a 26.67 ms frame; however, the concept of a frame is not particularly relevant for our brief discussion of the forward link. At the chip rate of 1.2288 Mchips/s, each slot contains 2048 chips, 1600 of which can be used to transmit control or user traffic data, and 448 of which are used to transmit pilot and signaling.

We stated that DO is based on adaptive rate control, a concept that has its roots in GPRS, EDGE, and systems we have already described. The concept involves the adjustment of the radio configuration (or transmission parameters) to maintain a 1% PER based on measurements by the AT. DO permits the adjustment of FEC coding rate, modulation, and number of repetitions to allow the link to be adapted to the channel conditions, while power level and chip rate remain fixed. (Note that DO also operates at 1.2288 Mchips/s.) When conditions are poor, coding rates and repetitions are increased, while the modulation order is reduced to yield a relatively low data rate. When conditions are good, less protection, fewer repetitions, and higher-order modulation are used to provide a higher data rate. In the forward link, DO employs rate-1/3 or rate-1/5 turbo coders and QPSK, 8PSK, or 16-QAM modulation schemes. Repetitions vary based on the particular channel conditions. Parameters are adjusted based on a specific value transmitted to the AN by an AT called a data rate control (DRC) value or index. The AT determines the DRC index based on a measurement of channel conditions, namely, CNIR. The DRC index is coded with the Walsh code assigned for the particular sector which the AT deems the strongest.

Consider that a number of ATs are active within a given sector. By "active" we mean that the AT has a valid MAC-ID for each member of its active set. (Note that an AT is said to have a DO connection when it has been assigned MAC-IDs for its active set and other identifiers not discussed here.) At the highest level the forward link transmission process is as follows:

1. Each AT in the given sector periodically measures its CNIR, maps the CNIR to a DRC value, and reports its channel condition by sending the DRC value covered by the Walsh code assigned for the sector. Reporting may be as often as every slot (1.67 ms), but the

reporting rate can be reduced in powers of 2 to once every 8×1.67 ms = 13.33 ms, depending on load within the sector.

2. The AN receives DRC values from all active ATs in the sector. A scheduler uses the data received from each AT and a wide range of other criteria to determine which AT is to receive the next downlink transmission. We discussed the concept of a scheduler as it related to HSDPA. Here, too, the standards do not mention a scheduling element, much less discuss how it determines which AT is to be served next. As stated earlier, infrastructure manufacturers may distinguish their system's performance-based features and the performance of their scheduler. As a consequence, the operation of a scheduler is carefully guarded proprietary information.
3. Once the scheduler determines the AT to be served, it inserts the MAC-ID in the form of a preamble at the beginning of the data packet and transmits it using the data format dictated by the DRC value received from the AT. The AN may not unilaterally change the transmission format determined by the DRC value.
4. Every active AT in the sector looks for a preamble containing its MAC-ID at the beginning of each slot. When it sees a packet containing its preamble, it performs a parity check to determine if errors are present. If not, it sends an ACK. Otherwise, it sends a NACK.

Associated with each DRC value there are seven radio configuration parameters that govern the format of the transmission. Table 8.5 summarizes the parameters for each DRC value.

Table 8.5 Summary of Data Rate Control Parameters

DRC Value	Data Rate (kbits/s)	Packet Length (bits)	Span (slots)	Turbo Coding Rate	Modulation	Preamble Length (bits)	Repetition Factor
1	38.4	1024	16	1/5	QPSK	1024	9.6
2	76.8	1024	8	1/5	QPSK	512	4.8
3	153.6	1024	4	1/5	QPSK	256	2.4
4	307.2	1024	2	1/5	QPSK	128	1.2
5	307.2	2048	4	1/3	QPSK	128	2.04
6	614.4	1024	1	1/3	QPSK	64	1
7	614.4	2048	2	1/3	QPSK	64	1.02
8	921.6	3072	2	1/3	8-PSK	64	1.02
9	1228.8	2048	1	1/3	QPSK	64	0.5
10	1228.8	4096	2	1/3	16-QAM	64	1.02
11	1843.2	3072	1	1/3	8-PSK	64	0.5
12	2457.6	4096	1	1/3	16-QAM	64	0.5

(Note that a DRC value is a 4-bit number. Values 13–15 are not defined in Rev. 0. DRC 0 is the null rate, implying that the AT is experiencing extremely poor channel conditions.) Note that the "span" refers to the number of slots that will be used to transmit the packet of given length. The preamble is nothing more than the 6-bit MAC-ID encoded to 32 bits and repeated. Note that for low DRC values the encoded MAC-ID is repeated many times (32 times for DRC 1) to ensure addressing integrity.

A brief discussion of Table 8.5 may bring further insight into the rate control mechanism. For example, DRC 1 dictates that the data packet size will be 1024 bits and is to be transmitted over 16 slots. Given that a slot is 1.67 ms, then the data rate, simply being the number of bits to be transmitted divided by the time it takes to transmit them, is $1024 \div (16 \times 1.67 \text{ ms}) = 38.4 \text{ kbits/s}$. For each 1.67 ms slot, however, 1600 chips will be transmitted, so there are 16 slots × 1600 chips/slot = 25, 600 chips that will be used to transmit the packet. The preamble will occupy the first 1024 chips of the first slot, so that the remaining 23,576 chips will be used to convey the payload. The payload is constructed as follows: The packet of length 1024 bits is encoded using a rate-1/5 turbo coder. For each input bit there are 5 output bits. The operation of the turbo coder is such that the input bits are output first followed by the added code bits. In this case the first 1024 bits out of the turbo coder are identically the input bits and the succeeding 4096 bits are the added code bits. Thus there is a total of $5 \times 1024 = 5120$ coded bits out of the turbo coder. Next the coded bits are interleaved, which merely rearranges the bit ordering to reduce the effects of burst errors. The coded bits are then QPSK modulated so that every 2 coded bits yield one modulation symbol, so that there are modulation symbols to be transmitted at the chip rate of 1.2288 Mchips/s. Note that of these 2560 modulation symbols, the first 1/5 (512) represent the data and the remaining 2048 represent the coding or protection bits. Given that there are 23,576 chips available to send, the 2560 modulation symbols representing the payload can be repeated $23{,}576 \div 2{,}560 = 9.6$ times. The complete transmission requires 16 slots or subpackets to send the preamble and 9.6 repeats. As a general rule, the packet length, span, number of repeats, and preamble length decrease, while the code rate and order of the modulation scheme increase, with increasing DRC value. The prescribed DRC configurations are expected to achieve no worse than a 1% PER for the given packet size and appropriate CNIR values. The mapping of CNIR to DRC value is not prescribed in the standard. Clearly, the mapping has strong performance implications. If CNIR measurements and mappings are not representative of the true channel conditions, average throughput will be less than optimum, and this may result in user dissatisfaction and lost revenue for the service provider if the discrepancy is large enough.

Note that for the higher DRC values, the repetition factor may be less than 1 when the span is one slot. For example, DRC 12 has a packet size of 4096 bits which when encoded and modulated requires 3072 modulation symbols to transmit the coded payload. Following the preamble, however, there are only 1536 chips remaining in the slot to transmit the payload modulation symbols. Thus half the modulation symbols must be eliminated. Recall that the information bits always emerge from the turbo coder first, so that the first 1/3 (1024) of the 3072 payload modulation symbols represent the original input data. The remaining 2048 modulation symbols represent the added turbo coder coding bits. To accommodate the dictates of the format, the code is punctured so that only 512 turbo code symbols are actually transmitted. This corresponds to an effective code rate of 2/3.

All of the packet sizes indicated in the table include 16 CRC parity bits and 6 encoder tail (or initialization) bits. Each DRC format is such that the contents of the first slot include the preamble and *all* of the modulation symbols representing the input packet, regardless of the span. Whatever space might remain is filled with modulation symbols representing the turbo coder bits. When an AT receives a packet, it performs a parity check on the information bits. If the packet is received error-free, it sends an ACK to the system and the remaining subpackets are not sent. If the packet has an error, the AT sends a NACK and the next subpacket is sent. The next subpacket may contain information symbols, coding symbols, or both. The contents of the newly received subpacket are processed with those already received in an attempt to obtain an error-free transmission. The transmission of subpackets ends when the information symbols can be decoded error-free or when the span of slots has been reached. If the span is reached and the errors cannot be resolved, the system takes no further action at the physical layer. This scheme is a form of incremental redundancy called hybrid ARQ, or simply H-ARQ. The efficacy of H-ARQ is based on the fact that channel conditions change and there is some finite probability that a packet will be received error-free regardless of the channel conditions. Considering DRC 1 that has a 16-slot span, the effective instantaneous data rate is 16×38.4 kbits/s $= 614.4$ kbits/s if the fist subpacket is received error-free.

Note that the DO forward link is shared among users and under direct control of the AN. Even though an AT is active, it occupies the spectrum only when there is an active transmission. Signaling and other reverse link transmissions (such as DRC values) are needed to support the forward link, but unproductive spectrum occupancy is minimized. Other resource allocations such as MAC-ID are quickly terminated when data transmission is terminated in either direction. Connections can be reestablished very rapidly once an AT has established a presence in the system.

The DO reverse link is based on power control and variable spreading rate. Reverse link rates are determined based on the amount of redundancy/protection provided for the payload. For a higher data rate protection bits are traded for payload and vice versa. Several channels are transmitted simultaneously within a single frame. Channels are separated by a specific Walsh code associated with the channel. Traffic channels are encoded with a four-bit Walsh code. The reverse link is limited to BPSK modulation and rate-1/2 or rate-1/4 convolutional encoding. Table 8.6 summarizes the reverse link formats.

All reverse link transmissions are conveyed within a 26.67 ms frame, regardless of the packet length. As can be seen from the table, longer packets can be transmitted over the same frame length as a small packet by removing some protection bits. This requires an increase in transmitted power, resulting in greater interference. The reverse link rate is autonomously determined by the AT using a prescribed randomization scheme. Rate limits, sector loading, and randomization parameters are provided to each AT by the AN. An AT may readjust its data frame to frame, but only in one-step increments. For each frame an AT determines the loading for each member of its active set by monitoring a reverse activity bit (RAB), which indicates whether or not a sector has too much interference (busy). If any one of the members of its active set

Table 8.6 Summary of DO Reverse Link Transmission Parameters

Data Rate (kbits/s)	Packet Length (bits)	Coding Rate	Repetitions	Processing Gain (PN chips/bit)*
9.6	256	1/4	8	128
19.2	512	1/4	4	64
38.4	1024	1/4	2	32
76.8	2046	1/4	1	16
153.6	4096	1/4	1	8

* Processing gain includes a four-bit Walsh code spreading factor to separate the traffic channel.

indicates that it is busy, the system is busy, and the AT may be required to lower its rate. If all of the members of an AT's active set report that they are not busy, the system is idle and an AT may be able to increase its rate. For each frame, an AT draws a random number and compares it to a prescribed threshold value called a **transition probability** associated with its current rate and system loading. If the drawn value is above the transition probability, the AT changes its rate by one step up or down, depending on loading. If the drawn value is less than the transition probability, the AT does not change its rate. In any event, an AT must always choose a rate associated with the minimum packet length required to carry the packet. Reverse rate control for DO Rev. 0 is considered a significant weak point. It uses randomization to prevent external events from triggering a widespread change in reverse link interference. It provides very loose power control and does not allow ATs to adjust to changes in user demand. Rev. A incorporated major changes to the reverse link, improving both its performance and the system's ability to manage interference.

DO Rev. 0 has several deficiencies that were addressed in Rev. A. Among them are the large imbalance between forward and reverse link peak data rates; poor packing efficiency in the forward link, especially for small packets; little support for real-time/low-latency applications; no support for multiple information flows requiring different QoS treatment; and restrictive limitations on the number of users that can share the forward link traffic channel.

In Rev. A the forward link peak rate was increased to 3.072 Mbits/s and the reverse rate was increased to 1.2 Mbits/s mandatory (or 1.8 Mbits/s optionally by incorporating an 8PSK modulator in the AT). The reverse link peak rate, by incorporating QPSK and 8PSK (optional) modulation schemes, a top-down revamping of the rate control mechanism, improved interference control management and allowed the incorporation of reverse link H-ARQ.

The DO Rev. 0 forward link suffers from poor packing efficiency because of the fixed packet lengths associated with the DRC formats. For example, voice over IP (VoIP) applications have relatively small packets, typically on the order of 150–200 bits. The packet lengths for Rev. 0

are defined in increments of 1024 bits. Since the forward link must respond with a prescribed format, packets are packed with zeros to obtain the prescribed length. In Rev. A smaller packet sizes were prescribed for most of the DRC formats, allowing the scheduler to more closely match the payload size. ATs use blind rate detection to determine the appropriate packet length based on the DRC value they reported. A multiuser transmission scheme was also devised to allow up to eight small packets to be combined and addressed to different ATs using a single transmission format. These enhancements dramatically improved efficiency and latency for small packets. Facilities were also added to allow ATs to predict handoffs, thereby reducing the time required to redirect forward link data flow between different base stations.

Other improvements include doubling the number of MAC-IDs to support up to 115 users, enhanced control channels and higher-rate access channels, and support for multiple simultaneous flows with varying QoS requirements. In the forward direction, the scheduler was afforded greater flexibility in selecting an appropriate transmission format to meet the QoS objectives for a given information flow. In the reverse link, ATs may request power allocations for each information flow independently and may manage power utilization for each flow separately. Additional modes and facilities were added to provide greater flexibility to meet low-latency or real-time applications.

OFDM: An Architecture for the Fourth Generation

It is not difficult to imagine that the quest for higher wireless data rates will continue until some yet unknown criteria are met, if not just for the foreseeable future. The advent of cellular communications introduced the world to the concept of being able to communicate almost anywhere and at any time. Because of the advancements made over the past two and a half decades, the public has come to expect wireless access to the same information and services they receive from the Internet, their cable or satellite company, or their provider of plain old telephone service. A look at history, even predating cellular to the time of the telegraph, suggests that the demand for wireless capabilities and services typically follows the trends set by the wired world. The recent deployment of high-rate data services is indeed demonstrating the existence of market demand for wireless access to virtually any information, application, or service that can be obtained by wired connection. Indications are that the trend will continue if it doesn't expand and accelerate. It is expected that the availability of relatively inexpensive high-rate data will, in all probability, spawn new real-time services outside the realm of wired communications, further increasing the need for higher rates and more capabilities. In all respects the emphasis of the fourth generation is on dramatically higher data rates, and much greater efficiency in utilizing the spectrum is justified. In fact, industry forums are looking at evolutionary frameworks that target gigabit-per-second data rates as the next frontier, with fourth-generation systems as the first stepping-stone. Although no one can really answer the question "How fast is fast enough?" with any great certainty, the general objectives for fourth generation aim at well over an order of magnitude increase in peak data rate with a similar increase in spectral efficiency.

In the earlier sections of this chapter we briefly reviewed the basic physical layer architectures and underlying principles that have allowed wireless systems to advance to their present state of development. Each generation required a change in fundamental approach followed by incremental improvements as the demands of the marketplace evolved. Here again, the challenges of the fourth generation cannot be met by incremental improvements in the approaches used in past system generations. We first look at some of the motivations influencing the choice of orthogonal frequency-division multiplexing (OFDM) as the worldwide choice for 4G architectures.

In view of all of our discussions it should be a familiar principle by now that peak data rate is limited by the system bandwidth and the required error rate. So on first consideration, the key to increasing the data rate is to increase the bandwidth while maintaining E_b/N_0. Of course, if we wish to maintain the error rate objective, we must also increase the peak transmitter power to maintain the same bit energy, E_b. Such an approach may be appropriate for incremental changes, but significant difficulties arise if we seek to obtain dramatic rate increases. Several of these difficulties are related to technology. As the pulse width decreases beyond certain limits, the difficulty, complexity, and cost of producing shaped pulses become prohibitive. Higher symbol rates require more stable references, and synchronization becomes more difficult. Finally, increasing peak power has significant implications for the entire RF chain, and in particular for the handset.

In Chapter 3 we discussed the coherence bandwidth as a channel parameter related to the environment. We defined the coherence bandwidth as an approximate measure of the range over which all frequencies of a transmitted symbol experience the same microscopic fading. We also related it to the concept of delay spread. As the symbol bandwidth becomes greater than the coherence bandwidth, nonuniform or frequency-selective fading will corrupt the signal spectrum, causing a "smearing" or spreading of the pulse, and therefore an increase in intersymbol interference (ISI). When ISI becomes significant, complex equalization techniques must be incorporated to estimate the channel and compensate for its nonuniformities.

In view of these considerations, it should be evident that a strategy to increase data rate by simply decreasing the symbol width becomes impracticable at some point. So, given sufficient bandwidth, we need a different strategy for enabling higher data rates. One strategy that immediately comes to mind is to divide the spectrum into some number of separate channels, each channel using a distinct carrier frequency. In this approach a serial bit stream is converted (after modulation) into some number of parallel symbol streams, one for each channel. The overall data rate is therefore increased directly by a factor equal to the number of channels. If we choose the single-channel bandwidth appropriately, we can eliminate both the effects of frequency-selective fading and the implementation difficulties of which we spoke earlier. This "multicarrier" approach is by no means novel. It is commonly used in cable systems, fiber-optic systems, and fixed long-haul telecommunications systems. In fact, the approach intended for cdma2000 $3 \times$ RTT is a multicarrier approach. There are inherent inefficiencies and complications associated with multicarrier systems, however. Let us take a closer look.

We described the spectrum of a rectangular pulse in Chapter 5. Consider such a pulse with amplitude A, pulse width T_p, and carrier frequency f_c. Figure 8.1 is a plot of the amplitude

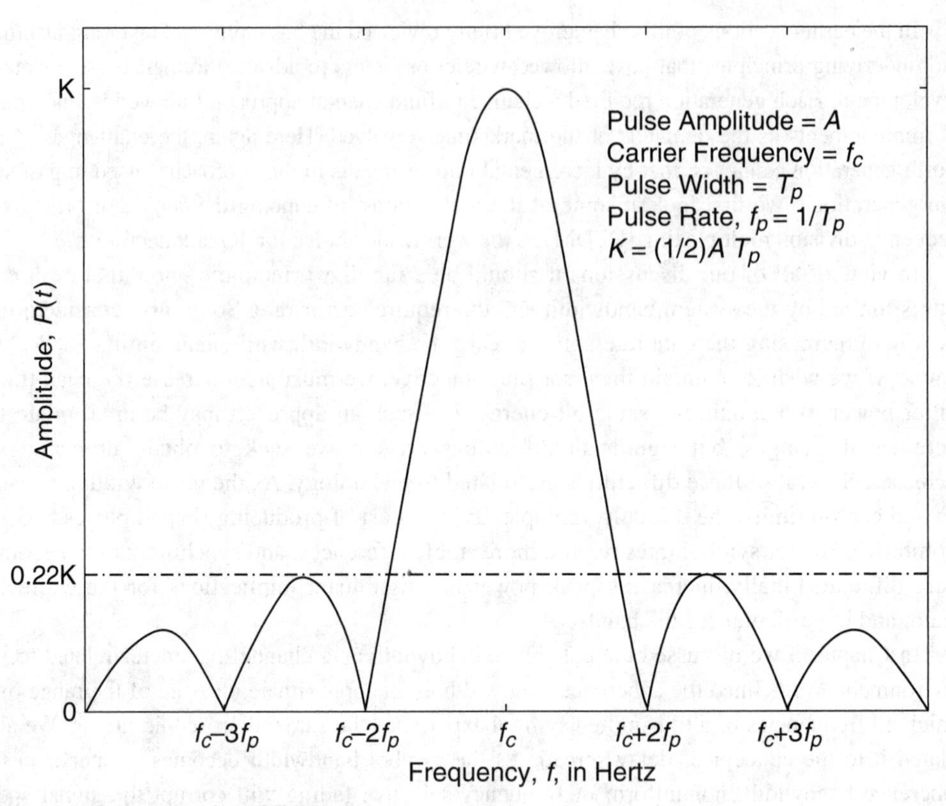

Figure 8.1 Spectrum of an RF Pulse

spectrum of this pulse on a linear scale. Note that nulls occur at multiples of the pulse repetition rate (or the inverse of the pulse width), specifically,

$$f_{nulls} = f_c \pm n\frac{1}{T_p}, n = 1, 2, \ldots. \tag{8.1}$$

Because a rectangular-pulse time function contains discontinuities, its power spectrum rolls off at a rate of $1/f^2$, or 6 dB per octave. Direct computation reveals that the first side lobe peak is only 13.15 dB below the peak value.

Consider now that we wish to transmit two symbol streams simultaneously using different carrier frequencies, that is, using a multicarrier approach. For a systems engineer the question that should immediately come to mind is: How close can we space the carrier frequencies so that the required total bandwidth is minimized? To help answer this question we look at the amplitude spectrum of two rectangular pulses with arbitrary frequency separation as shown in Figure 8.2 The figure suggests that interference between the spectra of the two pulses is the key factor to consider. Pulse shaping and filtering can be used to reduce interference, but the cost of such techniques increases very quickly as the required spacing is reduced. Protection in the form of FEC coding,

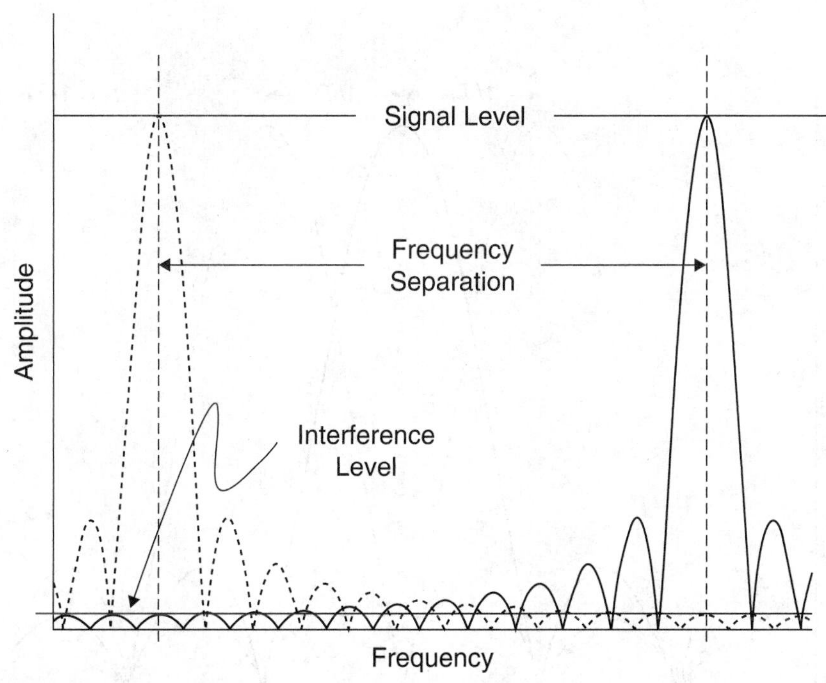

Figure 8.2 Two-Carrier Pulse Spectrum

orthogonal coding, and so on, can also be added to mitigate the effects of interference, but at the further expense of bandwidth. In any event, traditional multicarrier approaches require the use of guard bands between carriers to reduce interference. As a consequence, multicarrier systems are not typically deployed where spectrum efficiency is an important system constraint.

So on the basis of our first thoughts it does not appear that a multicarrier approach will afford us the dramatic improvements in efficiency we seek, although it will mitigate the effects of frequency-selective fading. If we look a little closer at Figure 8.2, however, we observe the periodic nulls, which are characteristic of time-limited waveforms. The nulls are at fixed frequencies and spaced at multiples of the pulse repetition rate. This suggests that we might be able to design our multicarrier system to exploit the absence of frequency content at the nulls. Such an approach is shown in Figure 8.3, which depicts a three-carrier system with a fixed frequency separation equal to the reciprocal of the pulse width T_p. Clearly, there is no intercarrier interference. Furthermore, we can add an arbitrarily large number of carriers, as many as the overall bandwidth will allow, without increasing the intercarrier interference. The implications are rather promising if not fascinating. Before proceeding with further discussion of the merits of this approach, however, let us briefly consider how one might construct a waveform having such a spectrum. In the following discussion we will use the term *subcarrier* instead of *carrier* to distinguish the information-bearing part of the transmitted signal from the channel center frequency. In other words, we will use the term *subcarriers* when referring to the baseband signal streams.

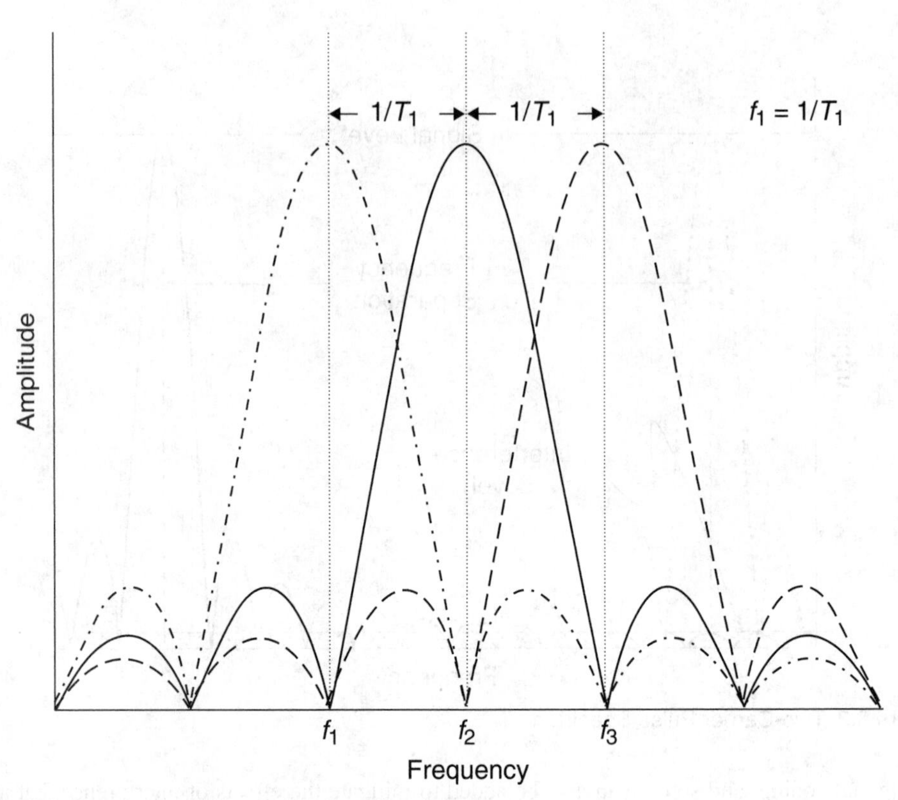

Figure 8.3 Spectra for Three Orthogonal Carriers

As Figure 8.3 shows, each subcarrier must use pulses of equal width, and the separation between carriers must be exactly equal to the reciprocal of the pulse width. Thus, the set of carrier frequencies must be chosen such that

$$f(n,t) = \cos(n\omega_0 t), \text{ where } n = \pm 1, 2, 3, \ldots, N, \text{ and } \omega_0 = \frac{2\pi}{T_p}. \tag{8.2}$$

You may recall from your studies of Fourier series that the frequencies represented by Equation (8.2) are orthogonal; that is, for any two integers m and n,

$$\int_0^{T_p} \cos(n\omega_0 t)\cos(m\omega_0 t)\,dt = \begin{cases} T_p/2, & \text{for } n = m \\ 0, & \text{for } n \neq m. \end{cases} \tag{8.3}$$

Therefore, a multicarrier system with minimum bandwidth is constructed from a set of orthogonal, equally spaced subcarriers, such that the subcarrier frequencies are integer multiples of the fundamental frequency $\omega_0 = 2\pi/T_p$, where T_p is the pulse or transmitted symbol width. It is easy to show that changing the amplitude or phase of any subcarrier does not introduce intercarrier interference, so the orthogonality does not depend on the type of pulse modulation used.

This strategy has dramatic implications for the ability to devise a highly flexible, spectrally efficient system. Multiple users can be supported on the basis of both frequency and/or time division. In such a system a resource allocation may consist of some number of carriers for some number of consecutive pulse periods. Several users can be supported during a single pulse interval, and the allocation can vary from pulse to pulse. A multicarrier system based on orthogonal subcarriers is called an **orthogonal frequency-division multiplex** (OFDM) system. When such a system supports a multitude of users it is referred to as an **orthogonal frequency-division multiple-access** (OFDMA) system. Even more opportune is the fact that such a system is easily grown as bandwidth becomes available. For example, consider an OFDM system that has an overall bandwidth of 5 MHz with 10 kHz subcarrier spacing. There are 500 subcarriers available for such a system. Assuming that the symbol rate for each carrier is equal to the bit rate, the system is capable of a peak symbol rate of 500×10 kbits/s $= 5$ Mbits/s. If another 5 MHz of spectrum becomes available, 500 more subcarriers can be added to double the peak data rate to 10 Mbits/s. Of course this implies modifying the RF structure to accommodate the wider bandwidth and power requirements. For this reason the system is often referred to as a **scalable-OFDM** (SOFDM) system.

As in any system design, there are trade-offs. Decreasing the subcarrier spacing provides greater granularity in allocating resources and better performance against frequency-selective fading; however, closer carrier spacing is more susceptible to Doppler frequency shift due to vehicle motion and requires greater stability in frequency-reference oscillators. On the other hand, increasing subcarrier spacing is limited by the coherence bandwidth of the channel. Current 4G system designs use subcarrier spacing on the order of 10 kHz, which appears to be a fairly good choice for most environments.

Although the attributes of OFDM/OFDMA are quite appealing, implementation appears problematic at first glance if one considers a traditional approach using modulators and local oscillators for each symbol stream. Such an approach is clearly untenable for a system with hundreds if not thousands of subcarriers. Fortunately, digital signal-processing techniques provide the basis for a rather simple and cost-effective approach.

For purposes of generality, let us assume that each information stream employs quadrature amplitude modulation (QAM) of some order. (Note that BPSK, QPSK, and 8PSK may be considered special cases of QAM so our assumption applies broadly.) This requires that the amplitude and the phase of each carrier be adjusted to form a symbol representing some combination of information bits. For any symbol interval, we can write an expression for the composite OFDM time signal in terms of the modulated carriers at baseband. Note that we can assume a symbol interval $0 \le t < T_p$ without loss of generality. The composite signal is given by

$$f(t) = \sum_{k=0}^{N-1} A_k \cos(k\omega_0 t + \varphi_k), \text{ for } 0 \le t < T_p, \tag{8.4}$$

where A_k and φ_k are the QAM amplitude and phase values for the time symbol of the kth stream, and $\omega_0 = \frac{2\pi}{T_p}$. The baseband symbol waveform is completely specified by Equation (8.4) and A_k and φ_k are completely determined by the bit stream. Our approach, then, is to create the time signal defined by Equation (8.4) and up-convert to the channel center frequency f_c.

The most direct approach is to digitally compute $f(t_i)$, for $i = 0,1,2,\ldots N-1$ at N equally spaced points in time and convert the sample values to a continuous-time waveform using a D/A converter. The digital evaluation is equivalent to finding the inverse discrete Fourier transform (IDFT). Given that there may be hundreds or even thousands of frequency components, the computation of Equation (8.4) quickly becomes intractable for such a straightforward approach. Therefore, we employ fast Fourier transform (FFT)[3] techniques to render the computation manageable. Recall that the FFT or inverse FFT (IFFT) is nothing more than a prescribed algorithm for dramatically reducing the amount of computation in finding a DFT or IDFT. The general approach to implementation, then, is to use an IFFT algorithm to create the baseband OFDM waveform and then to up-convert to the RF carrier frequency.[4] An OFDM receiver performs the inverse operations. The incoming waveform is sampled at an appropriate rate. The samples are processed with an FFT algorithm to recover the subcarriers. The modulation coefficient of each subcarrier is converted (demodulated) to recover the binary stream.

As mentioned earlier, the outstanding efficiency of OFDM/OFDMA is a consequence of the granularity that can be used in assigning resources. Users can be assigned one or more subcarriers for some number of pulses, depending on the amount of data that must be conveyed and the required QoS. Furthermore, as can be seen from Equation (8.4), the choice of modulation can be different for each carrier. Thus the channel parameters, including modulation order, can be adjusted to accommodate channel conditions. One other significant attribute of OFDM is that it simplifies the implementation of **multiple-input-multiple-output** (MIMO) and other multiple-antenna techniques. The reasons for the simplification may require a depth of discussion beyond our intended treatment, but suffice it to say that they are related to the ability of OFDM to adjust to the fading environment and to the rapid signaling allowed between the transmitters and receivers. In general, multiple-antenna techniques and systems are used to improve some dimension of system performance. In the next few paragraphs we briefly describe three broad classes of antenna system: those that adapt the antennas' radiation pattern; those that provide diversity in space, time, or both; and those that add more channels through spatial multiplexing.

The concept of using adaptable antenna systems is by no means new. In fact they have been used in military applications, such as phased-array radar, as far back as the mid-1900s. Their application to cellular communications is more recent, that is, within the past two decades. By **adaptable antenna system** we mean a class of multiple-antenna systems that are used to form and/or point an antenna beam, with the intent of enhancing the desired signal level relative to the surrounding

3. A. V. Oppenheim, R. W. Schafer, and J. R. Buck, *Discrete-Time Signal Processing,* 2nd ed. (Upper Saddle River, NJ: Prentice Hall, 1999), Chapter 9.
4. M. Schwartz, *Mobile Wireless Communications* (Cambridge: Cambridge University Press, 2005), 129–34.

interference. Specifically, they attempt to increase the signal energy in the direction of a specific user, or group of users, while reducing the energy in other directions. They range in complexity from simple switched-beam antennas to sophisticated (by today's standards) space-division multiple access (SDMA) systems. In switched-beam systems, several narrow-beam antennas are used to cover a region of a given sector. The signal is switched to the antenna having the strongest signal as the mobile unit moves within the sector. SDMA systems attempt to direct a narrow beam in the specific direction of each mobile unit as the unit moves throughout the sector. The more sophisticated techniques may require collaboration (signaling) between the transmitter and receiver.

Another class of antenna system is used to provide diversity gain. Spatial diversity, for example, may employ two or more antennas that are sufficiently separated that the probability of all antennas being in a fade is substantially reduced. Some form of combining, the simplest of which is to simply pick the strongest signal, is used obtain a diversity gain. We have seen this approach in CDMA systems that employ soft handoff, which is a form of macro diversity. The term *macro* refers to a very large separation between antennas. Another approach is to send a delayed replica of a given signal on a spatially separated antenna. Again, with appropriate combining techniques, this space-time diversity approach can be used to obtain even greater signal enhancements. Similarly, space-frequency diversity can be obtained by sending a signal using different frequencies and spatially separated antennas. Here, too, some of the techniques may require signaling and collaboration between the transmitter and receiver.

A third class of smart antenna system is referred to as a spatially multiplexing system. In particular, MIMO techniques create independent channels using spatially separated antennas and channel estimation techniques. Each independent channel can be used to communicate different OFDM symbols simultaneously, thereby increasing the overall peak data rate. The number of transmit and receive antennas may differ, but typically the number N of receive antennas is greater than or equal to the number M of transmit antennas. The peak data rate increases directly with the minimum of the number of transmit or receive antennas, that is, with (M, N). The simplified block diagram of Figure 8.4 illustrates the concept for a system employing two transmit and two receive antennas, that is, a 2×2 system. For this case, the input data stream is first demultiplexed into two parallel streams. Each stream is processed to form two OFDM symbols, $s_1(t)$ and $s_2(t)$, which are transmitted simultaneously on antennas 1 and 2 respectively. Note that the signal obtained from each receive antenna comprises signals from both transmit antennas, but the signals arrive over different paths. If the signals received at antennas 1 and 2 are $r_1(t)$ and $r_2(t)$ respectively, we can write

$$\begin{aligned} r_1(t) &= \mathbf{w}_{11}s_1(t) + \mathbf{w}_{12}s_2(t) \\ r_2(t) &= \mathbf{w}_{21}s_1(t) + \mathbf{w}_{22}s_2(t), \end{aligned} \tag{8.5}$$

where the $\mathbf{w}_{ij}$ refer to the complex path or channel coefficients that have been estimated for the path from transmit antenna i to receive antenna f. Note that the channel coefficients vary with time, but we have made the assumption that they are constant over a symbol period, which is on the order of 100 μs for typical implementations.

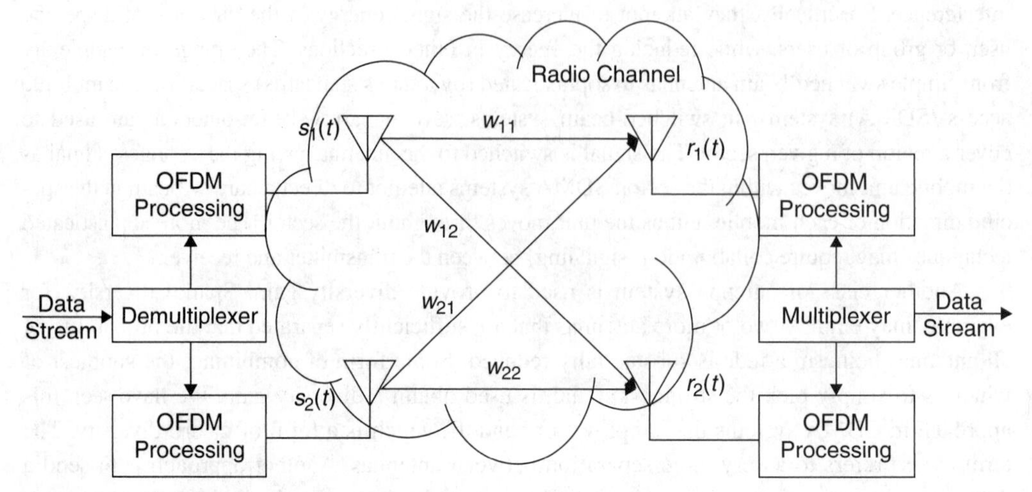

Figure 8.4 Example of a 2 × 2 MIMO System

The terms w_{ij} are of the form

$$w_{ij} = a_{ij}e^{j\theta_{ij}}. \tag{8.6}$$

They represent the amplitude and phase change over the path from antenna i to antenna j. If the paths are independent, the complex linear equations are independent and can be solved easily to find $r_1(t)$ and $r_2(t)$. For larger numbers of antennas, matrix techniques can be used to solve the system of equations. The efficacy of the approach depends on the accuracy of the channel estimation and the independence of the paths. Channel estimation may be inferred in an open-loop manner, or the transmitting and receiving ends may collaborate in a closed-loop fashion. All current proposals for 4G evolution support at least a 4 × 4 MIMO implementation as well as one or more of the aforementioned smart antenna techniques.

As of the writing of this book, there are three major proposals for 4G systems, all of which are based on OFDM/OFDMA. Two have been developed and evolved by what might be called the traditional or incumbent industry players, UMTS under 3GPP, and cdma2000 under 3GPP2. The UMTS evolution is called LTE for Long-Term Evolution, and the system evolving from cdma2000 is termed UMB for Ultra-Mobile Broadband. A third candidate system has arisen from the IEEE 802.16e standard, which is termed Mobile WiMAX for Mobile Worldwide Interoperability for Microwave Access. The fact that there are three system proposals suggests that the 4G marketplace may be fragmented further. Until service providers finalize their 4G strategies, however, it is not clear which or how many standards will survive. All three approaches have similar performance characteristics; some of the major differences are associated with how resources are allocated and the amount of overhead needed to derive the full benefits of OFDMA.

We stated earlier that OFDM/OFDMA systems are spectrally efficient. A common measure of a system's spectral efficiency is the data rate that can be achieved divided by the bandwidth required to support it. Therefore, the spectral efficiency has units of bits per second per hertz. Clearly, the spectrum efficiency is related to the assumption one makes about the data rate, be it peak theoretical rate, the peak rate actually implemented, or some other criterion. But for purposes of comparison, we derive the spectral efficiency of several of the systems we have described in this chapter, stating our assumptions. For HSCSD, a single user can achieve up to a 57.6 kbit/s data rate using a maximum of four slots. Since there are eight slots per channel, the maximum channel rate is 115.2 kbits/s in a channel bandwidth of 200 kHz. This yields a spectral efficiency of 0.576 bits/s/Hz. For GPRS the maximum theoretical rate is 171.2 kbits/s, assuming that all eight slots are allocated to a single user. This results in a spectral efficiency of 0.856 bits/s/Hz. For EDGE, a theoretical peak data rate of 473.6 kbits/s can be achieved, assuming that eight slots can be aggregated. Thus the maximum channel efficiency is 2.37 bits/s/Hz. Although the cdma2000 architecture was capable of higher peak rates, the systems that were actually deployed were limited to a maximum of 153.6 kbits/s, yielding an efficiency of 0.77 bits/s/Hz for the 1.25 MHz bandwidth. Given the bandwidth of W-CDMA, its maximum theoretical efficiency is 0.4 bits/s/Hz for a 2 Mbits/s peak data rate. Even HSDPA, assuming the maximum theoretical rate of 14 Mbits/s, is limited to 2.8 bits/s/Hz. Cdma2000 1 × EV-DO Rev. A has a maximum efficiency of 2.5 bits/s/Hz given its peak rate of 3072 Mbits/s and bandwidth of 1.25 MHz. By comparison, UMB has a peak theoretical data rate of about 280 Mbits/s using 20 MHz of spectrum. This corresponds to a spectral efficiency of 14 bits/s/Hz, a little over half an order of magnitude improvement over 1 × EV-DO. The peak data rate of UMB relative to 1 × EV-DO is based on 64-QAM modulation and 4 × 4 MIMO.

One common characteristic of all 4G standards is the absence of a circuit-switched core network. Adjunct servers that provide subscriber, gateway, and application services are replacing circuit-switched elements such as the mobile switching center (MSC), the home location register (HLR), and the visitor location register (VLR). This implies that all communications including voice will be conveyed using packet switching. The radio and core network architectures are nonhierarchical so that all base stations and system elements communicate over an Internet Protocol (IP) backbone. Handoffs and allocation of resources are handled at the base station. Base stations can communicate directly with each other so that data and signaling can be routed through the network seamlessly. Packet-switched interworking elements are being defined that will allow seamless roaming between different technologies, be they indoors or outdoors, on a global scale.

We conclude this section with a rather interesting and curious observation. The first cellular systems were analog and based on FDMA techniques. Analog systems strive to preserve the signal shape. Second-generation systems were digital and predominantly TDMA. Digital systems use analog techniques to represent discrete information, but they do not require the preservation of signal shape. The introduction of 2G cdmaOne drove the global evolution of 3G toward CDMA. In order to support true packet data operation efficiently and meet the ever increasing demand for higher peak data rates, 3G architectures began an evolution back to TDMA as we saw in 1 × EV-DO. In 4G we have fully reverted back to FDM in the form of OFDM/OFDMA.

In fact, OFDM requires that each symbol shape be preserved, a characteristic associated more with analog systems than with digital systems. Advancements in technology often make old ideas attractive and sometimes make them dramatically better than one could have imagined. Surely this is the case for OFDMA. Although the evolutionary path of cellular system architectures is rather interesting in and of itself, it is even more fascinating to wonder what 5G might bring, and whether or not the path already taken has portent for the future. There is no doubt that whatever the path, it will be an adventure for those who wish to contribute to the birth of the next generations.

Conclusions

In this chapter we began with a recap of our problem statement and a summary of the major principles and applications we discussed throughout the text. In the remainder of the chapter we sought to describe the major factors that have motivated the evolution of cellular systems from the first-generation analog systems, AMPS, through 3G and including a description of the next-generation OFDM/OFDMA systems. We have endeavored to couch our discussions in terms of evolving market needs and the systems-engineering concepts and architectures that were developed to meet those needs. Although our presentation has been more a historical than an analytical dissertation, we hope that it stimulated a realistic sense of the complexity of modern systems and provided insights into the vast array of disciplines that must come together to compete and succeed in this rapidly evolving field of communications. The fourth generation offers undreamed-of opportunities for developing new applications and services that will continue to have a major impact on the way people think, share, and communicate with each other. These systems will be deployed within the current decade. History suggests that 5G systems supporting gigabit data rates and unimagined opportunity for new applications and services will not be far behind. Indeed, we hope that we have stimulated the reader's interest in pursuing advanced studies and being a part of the professional community that will develop them.

APPENDIX A

Statistical Functions and Tables

The Normal Distribution

For a normally distributed random variable z with mean m and standard deviation σ, the probability that z is less than or equal to some arbitrary value x is given by

$$\text{Prob}\{z \le x\} = F_z(x) = \frac{1}{\sqrt{2\pi}\sigma}\int_{-\infty}^{x} \exp\left(-\frac{[\alpha - m]^2}{2\sigma^2}\right) d\alpha. \tag{A.1}$$

Likewise,

$$\text{Prob}\{z > x\} = \frac{1}{\sqrt{2\pi}\sigma}\int_{x}^{\infty} \exp\left(-\frac{[\alpha - m]^2}{2\sigma^2}\right) d\alpha. \tag{A.2}$$

For any valid probability distribution,

$$\frac{1}{\sqrt{2\pi}\sigma}\int_{-\infty}^{\infty} \exp\left(-\frac{[\alpha - m]^2}{2\sigma^2}\right) d\alpha = 1. \tag{A.3}$$

Therefore,

$$\text{Prob}\{z > x\} = 1 - F_z(x). \tag{A.4}$$

The function $F_z(x)$ is a normal or Gaussian probability distribution function. The integrand of c Equation (A.1) is the corresponding probability density function, $p_z(\alpha)$:

$$p_z(\alpha) = \frac{1}{\sqrt{2\pi}\sigma} \exp\left(-\frac{(\alpha - m)^2}{2\sigma^2}\right). \tag{A.5}$$

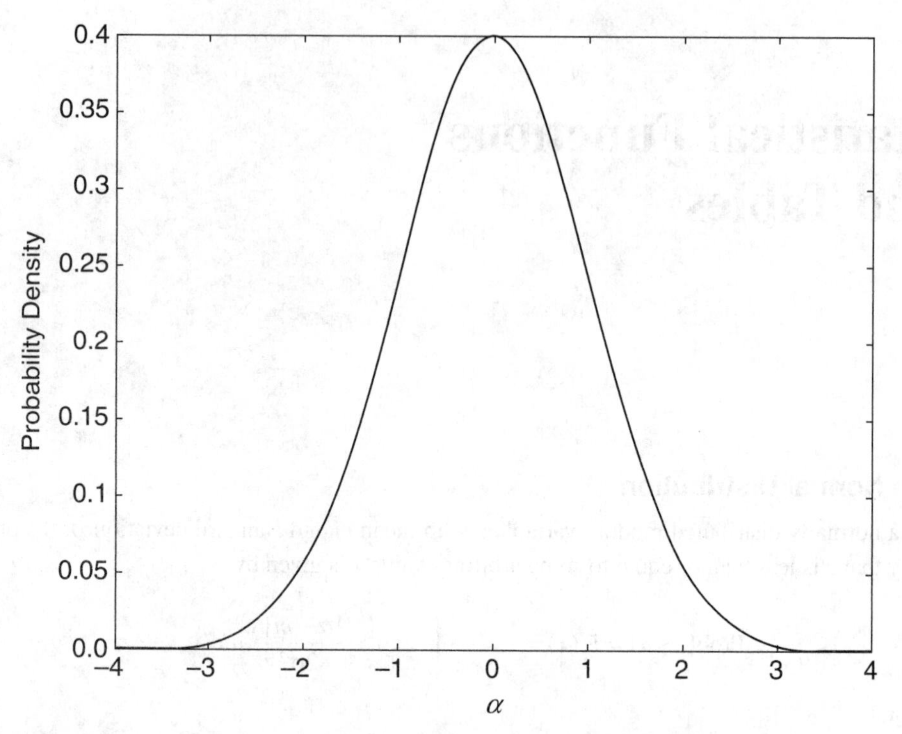

Figure A.1 Probability Density Function of a Normal Random Variable

Figure A.1 and Figure A.2 illustrate the probability density and probability distribution functions for a normal random variable with zero mean and unity standard deviation. For this case, the distribution and density functions are

$$\text{Prob}\{z \le x\} = F_z(x) = \frac{1}{\sqrt{2\pi}} \int_{-\infty}^{x} \exp\left(-\frac{\alpha^2}{2}\right) d\alpha, \tag{A.6}$$

$$p_z(\alpha) = \frac{1}{\sqrt{2\pi}} \exp\left(-\frac{\alpha^2}{2}\right). \tag{A.7}$$

We can obtain a result of the same form as Equation (A.6) by performing a change of variables for Equation (A.1).

Equation (A.6) is similar in form to the error function, erf(y), which has the form

$$\text{erf}(y) = \frac{2}{\sqrt{\pi}} \int_{0}^{y} \exp(-\beta^2) d\beta. \tag{A.8}$$

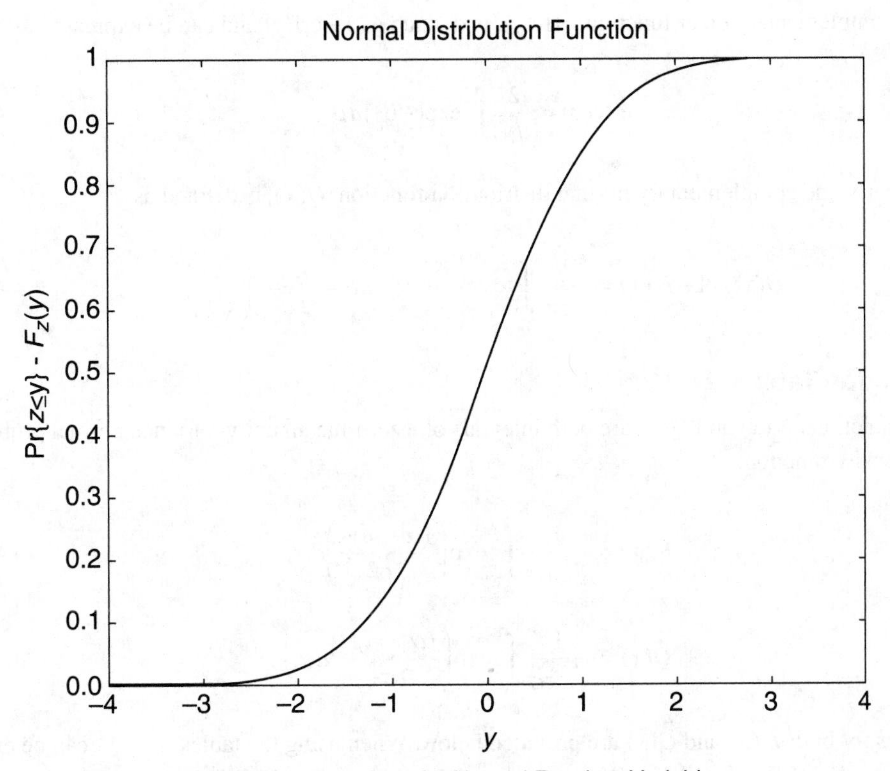

Figure A.2 Probability Distribution Function of a Normal Random Variable

Note that the erf is odd symmetric about the origin, that is, erf(−y) = −erf(y). Furthermore, erf(∞) = 1 so that erf(−∞) = −1. The simple change of variable $\beta = \alpha/\sqrt{2}$ yields

$$\frac{1}{2}\operatorname{erf}\left(\frac{x}{\sqrt{2}}\right) = \frac{1}{\sqrt{\pi}}\int_0^{\frac{x}{\sqrt{2}}} \exp(-\beta^2)d\beta = \frac{1}{\sqrt{2\pi}}\int_0^x \exp\left(-\frac{\alpha^2}{2}\right)d\alpha. \tag{A.9}$$

Given its odd symmetry, we can express Equation (A.6) in terms of the erf:

$$\text{Prob}\{z \le x\} = \frac{1}{\sqrt{2\pi}}\int_{-\infty}^x \exp\left(-\frac{\alpha^2}{2}\right)d\alpha = \frac{1}{2}\left(1 + \operatorname{erf}\left(\frac{x}{\sqrt{2}}\right)\right). \tag{A.10}$$

Simplifying the notation, the normal distribution function, $F_z(x)$, can be expressed in terms of the error function:

$$F_z(x) = \frac{1}{2}\left(1 + \operatorname{erf}\left(\frac{x}{\sqrt{2}}\right)\right). \tag{A.11}$$

The complementary error function, erfc(y), is defined as 1– erf(y) and can be expressed as

$$\mathrm{erfc}(y) = \frac{2}{\sqrt{\pi}} \int_y^{\infty} \exp(-\beta^2)\, d\beta. \tag{A.12}$$

Similarly, the complementary normal distribution function, $Q(x)$, is defined as

$$Q(x) = 1 - F_z(x) = \frac{1}{\sqrt{2\pi}} \int_x^{\infty} \exp\left(-\frac{1}{2}\alpha^2\right) d\alpha = \frac{1}{2}\mathrm{erfc}\left(\frac{x}{\sqrt{2}}\right). \tag{A.13}$$

Function Tables

The functions $F_z(x)$ and $Q(x)$ are both integrals of a zero-mean, unity-variance normal probability density function:

$$F_z(x) = \frac{1}{\sqrt{2\pi}\sigma} \int_{-\infty}^{x} \exp\left(\frac{(\alpha - m)^2}{2\sigma^2}\right) d\alpha, \tag{A.14}$$

$$Q(x) = \frac{1}{\sqrt{2\pi}\sigma} \int_{x}^{\infty} \exp\left(\frac{(\alpha - m)^2}{2\sigma^2}\right) d\alpha. \tag{A.15}$$

Tables for both $F_z(x)$ and $Q(x)$ are provided below. When using the tables, $F_z(-x)$ can be evaluated by using $Q(x)$. Similarly, $Q(-x)$ can be evaluated by using $F_z(x)$.

The following MATLAB functions, Ffun(x) and Qfun(x), compute $F_z(x)$ and $Q(x)$. They can be used for both positive and negative values of x.

```
%
% ********** Ffun(x) ***************
%
% This function computes the F-function F(x).
%
function Ffun(x)
sqrt2=sqrt(2);
F=1-0.5*erfc(x/sqrt2)

%
% ************ QFun ***************
%
% This function computes the Q-function Q(x).
%
function Qfun(x)
sqrt2=sqrt(2);
Q=0.5*erfc(x/sqrt2)
```

To evaluate either function for a distribution with mean m and standard deviation σ using the tables, use the forms

$$F_z\left(\frac{x-m}{\sigma}\right) \quad \text{and} \quad Q\left(\frac{x-m}{\sigma}\right). \tag{A.16}$$

The following MATLAB functions, Ffun2(*mean, std_dev, x*) and Qfun2(*mean, std_dev, x*), compute $F_z(x-m/\sigma)$ and $Q(x-m/\sigma)$. They can be used for both positive and negative values of x.

```
%
% ********** Ffun2(mean, std_dev,x) ***************
%
% This function computes the F-function F((x-mean)/std_dev)).
%
function Qfun2(mean, std_dev,x)
y=(x-mean)/std_dev;
sqrt2=sqrt(2);
F=1-0.5*erfc(y/sqrt2)

%
% ********** Qfun2(mean, std_dev,x) ***************
%
% This function computes the Q-function Q((x-mean)/std_dev)).
%
function Qfun2(mean, std_dev,x)
y=(x-mean)/std_dev;
sqrt2=sqrt(2);
Q=0.5*erfc(y/sqrt2)
```

x	F(x)									
	0.00	**0.01**	**0.02**	**0.03**	**0.04**	**0.05**	**0.06**	**0.07**	**0.08**	**0.09**
0.0	0.5000	0.5040	0.5080	0.5120	0.5160	0.5199	0.5239	0.5279	0.5319	0.5359
0.1	0.5398	0.5438	0.5478	0.5517	0.5557	0.5596	0.5636	0.5675	0.5714	0.5753
0.2	0.5793	0.5832	0.5871	0.5910	0.5948	0.5987	0.6026	0.6064	0.6103	0.6141
0.3	0.6179	0.6217	0.6255	0.6293	0.6331	0.6368	0.6406	0.6443	0.6480	0.6517
0.4	0.6554	0.6591	0.6628	0.6664	0.6700	0.6736	0.6772	0.6808	0.6844	0.6879
0.5	0.6915	0.6950	0.6985	0.7019	0.7054	0.7088	0.7123	0.7157	0.7190	0.7224
0.6	0.7257	0.7291	0.7324	0.7357	0.7389	0.7422	0.7454	0.7486	0.7517	0.7549
0.7	0.7580	0.7611	0.7642	0.7673	0.7704	0.7734	0.7764	0.7794	0.7823	0.7852
0.8	0.7881	0.7910	0.7939	0.7967	0.7995	0.8023	0.8051	0.8078	0.8106	0.8133
0.9	0.8159	0.8186	0.8212	0.8238	0.8264	0.8289	0.8315	0.8340	0.8365	0.8389
1.0	0.8413	0.8438	0.8461	0.8485	0.8508	0.8531	0.8554	0.8577	0.8599	0.8621
1.1	0.8643	0.8665	0.8686	0.8708	0.8729	0.8749	0.8770	0.8790	0.8810	0.8830
1.2	0.8849	0.8869	0.8888	0.8907	0.8925	0.8944	0.8962	0.8980	0.8997	0.9015
1.3	0.9032	0.9049	0.9066	0.9082	0.9099	0.9115	0.9131	0.9147	0.9162	0.9177
1.4	0.9192	0.9207	0.9222	0.9236	0.9251	0.9265	0.9279	0.9292	0.9306	0.9319
1.5	0.9332	0.9345	0.9357	0.9370	0.9382	0.9394	0.9406	0.9418	0.9429	0.9441
1.6	0.9452	0.9463	0.9474	0.9484	0.9495	0.9505	0.9515	0.9525	0.9535	0.9545
1.7	0.9554	0.9564	0.9573	0.9582	0.9591	0.9599	0.9608	0.9616	0.9625	0.9633
1.8	0.9641	0.9649	0.9656	0.9664	0.9671	0.9678	0.9686	0.9693	0.9699	0.9706
1.9	0.9713	0.9719	0.9726	0.9732	0.9738	0.9744	0.9750	0.9756	0.9761	0.9767
2.0	0.9772	0.9778	0.9783	0.9788	0.9793	0.9798	0.9803	0.9808	0.9812	0.9817

2.1	0.9821	0.9826	0.9830	0.9834	0.9838	0.9842	0.9846	0.9850	0.9854	0.9857
2.2	0.9861	0.9864	0.9868	0.9871	0.9875	0.9878	0.9881	0.9884	0.9887	0.9890
2.3	0.9893	0.9896	0.9898	0.9901	0.9904	0.9906	0.9909	0.9911	0.9913	0.9916
2.4	0.9918	0.9920	0.9922	0.9925	0.9927	0.9929	0.9931	0.9932	0.9934	0.9936
2.5	0.9938	0.9940	0.9941	0.9943	0.9945	0.9946	0.9948	0.9949	0.9951	0.9952
2.6	0.9953	0.9955	0.9956	0.9957	0.9959	0.9960	0.9961	0.9962	0.9963	0.9964
2.7	0.9965	0.9966	0.9967	0.9968	0.9969	0.9970	0.9971	0.9972	0.9973	0.9974
2.8	0.9974	0.9975	0.9976	0.9977	0.9977	0.9978	0.9979	0.9979	0.9980	0.9981
2.9	0.9981	0.9982	0.9982	0.9983	0.9984	0.9984	0.9985	0.9985	0.9986	0.9986
3.0	0.9987	0.9987	0.9987	0.9988	0.9988	0.9989	0.9989	0.9989	0.9990	0.9990
3.1	0.9990	0.9991	0.9991	0.9991	0.9992	0.9992	0.9992	0.9992	0.9993	0.9993
3.2	0.9993	0.9993	0.9994	0.9994	0.9994	0.9994	0.9994	0.9995	0.9995	0.9995
3.3	0.9995	0.9995	0.9995	0.9996	0.9996	0.9996	0.9996	0.9996	0.9996	0.9997
3.4	0.9997	0.9997	0.9997	0.9997	0.9997	0.9997	0.9997	0.9997	0.9997	0.9998
3.5	0.9998	0.9998	0.9998	0.9998	0.9998	0.9998	0.9998	0.9998	0.9998	0.9998

	Q(x)									
x	**0.00**	**0.01**	**0.02**	**0.03**	**0.04**	**0.05**	**0.06**	**0.07**	**0.08**	**0.09**
0.0	0.5000	0.4960	0.4920	0.4880	0.4840	0.4801	0.4761	0.4721	0.4681	0.4641
0.1	0.4602	0.4562	0.4522	0.4483	0.4443	0.4404	0.4364	0.4325	0.4286	0.4247
0.2	0.4207	0.4168	0.4129	0.4090	0.4052	0.4013	0.3974	0.3936	0.3897	0.3859
0.3	0.3821	0.3783	0.3745	0.3707	0.3669	0.3632	0.3594	0.3557	0.3520	0.3483
0.4	0.3446	0.3409	0.3372	0.3336	0.3300	0.3264	0.3228	0.3192	0.3156	0.3121
0.5	0.3085	0.3050	0.3015	0.2981	0.2946	0.2912	0.2877	0.2843	0.2810	0.2776
0.6	0.2743	0.2709	0.2676	0.2643	0.2611	0.2578	0.2546	0.2514	0.2483	0.2451
0.7	0.2420	0.2389	0.2358	0.2327	0.2296	0.2266	0.2236	0.2206	0.2177	0.2148
0.8	0.2119	0.2090	0.2061	0.2033	0.2005	0.1977	0.1949	0.1922	0.1894	0.1867
0.9	0.1841	0.1814	0.1788	0.1762	0.1736	0.1711	0.1685	0.1660	0.1635	0.1611
1.0	0.1587	0.1562	0.1539	0.1515	0.1492	0.1469	0.1446	0.1423	0.1401	0.1379
1.1	0.1357	0.1335	0.1314	0.1292	0.1271	0.1251	0.1230	0.1210	0.1190	0.1170
1.2	0.1151	0.1131	0.1112	0.1093	0.1075	0.1056	0.1038	0.1020	0.1003	0.0985
1.3	0.0968	0.0951	0.0934	0.0918	0.0901	0.0885	0.0869	0.0853	0.0838	0.0823
1.4	0.0808	0.0793	0.0778	0.0764	0.0749	0.0735	0.0721	0.0708	0.0694	0.0681
1.5	0.0668	0.0655	0.0643	0.0630	0.0618	0.0606	0.0594	0.0582	0.0571	0.0559
1.6	0.0548	0.0537	0.0526	0.0516	0.0505	0.0495	0.0485	0.0475	0.0465	0.0455
1.7	0.0446	0.0436	0.0427	0.0418	0.0409	0.0401	0.0392	0.0384	0.0375	0.0367
1.8	0.0359	0.0351	0.0344	0.0336	0.0329	0.0322	0.0314	0.0307	0.0301	0.0294
1.9	0.0287	0.0281	0.0274	0.0268	0.0262	0.0256	0.0250	0.0244	0.0239	0.0233
2.0	0.0228	0.0222	0.0217	0.0212	0.0207	0.0202	0.0197	0.0192	0.0188	0.0183

2.1	0.0179	0.0174	0.0170	0.0166	0.0162	0.0158	0.0154	0.0150	0.0146	0.0143
2.2	0.0139	0.0136	0.0132	0.0129	0.0125	0.0122	0.0119	0.0116	0.0113	0.0110
2.3	0.0107	0.0104	0.0102	0.0099	0.0096	0.0094	0.0091	0.0089	0.0087	0.0084
2.4	0.0082	0.0080	0.0078	0.0075	0.0073	0.0071	0.0069	0.0068	0.0066	0.0064
2.5	0.0062	0.0060	0.0059	0.0057	0.0055	0.0054	0.0052	0.0051	0.0049	0.0048
2.6	0.0047	0.0045	0.0044	0.0043	0.0041	0.0040	0.0039	0.0038	0.0037	0.0036
2.7	0.0035	0.0034	0.0033	0.0032	0.0031	0.0030	0.0029	0.0028	0.0027	0.0026
2.8	0.0026	0.0025	0.0024	0.0023	0.0023	0.0022	0.0021	0.0021	0.0020	0.0019
2.9	0.0019	0.0018	0.0018	0.0017	0.0016	0.0016	0.0015	0.0015	0.0014	0.0014
3.0	0.0013	0.0013	0.0013	0.0012	0.0012	0.0011	0.0011	0.0011	0.0010	0.0010
3.1	0.0010	0.0009	0.0009	0.0009	0.0008	0.0008	0.0008	0.0008	0.0007	0.0007
3.2	0.0007	0.0007	0.0006	0.0006	0.0006	0.0006	0.0006	0.0005	0.0005	0.0005
3.3	0.0005	0.0005	0.0005	0.0004	0.0004	0.0004	0.0004	0.0004	0.0004	0.0003
3.4	0.0003	0.0003	0.0003	0.0003	0.0003	0.0003	0.0003	0.0003	0.0003	0.0002
3.5	0.0002	0.0002	0.0002	0.0002	0.0002	0.0002	0.0002	0.0002	0.0002	0.0002

Traffic Engineering

Grade of Service and the State of the Switch

Consider a telephone switch that shares K trunks among a large community of users. From time to time, a call from one of the users arrives at the switch, and the switch removes a trunk from the trunk pool and assigns the trunk to the user. From time to time, a call in progress terminates. When this happens the trunk associated with that call is released and is returned to the pool. We can define the **state** of the switch as the number k of calls in progress at any given moment. Figure B.1 suggests the way the state might vary with time as calls arrive and terminate.

If the switch is in state $k = K$, then all the trunks are in use and a call arriving at the switch will be **blocked**. There are two policies for dealing with blocked calls. In the **blocked calls cleared** policy, the incoming call is shunted to a busy-signal generator. The user must terminate the attempt and may try again later. In the **blocked calls delayed** policy, the incoming call is held on a queue until an in-progress call terminates and a trunk becomes free. Queued calls are served on a first-come, first-served basis. The blocked calls cleared policy is familiar, as this policy is used by the public switched telephone network. The blocked calls delayed strategy is also familiar as the "please hold for the next available operator" routine used by the customer service departments of many organizations. In this appendix we will consider only the blocked calls cleared policy. The reader is referred to the literature for blocked calls delayed.

The **grade of service** (GOS) that the switch provides is the probability P_b that an incoming call will be blocked. Telephone systems are typically designed to provide a grade of service of a few percent during the weekly busy hour. We will assume that during the busy hour the average user places λ_{user} calls per minute and holds each call an average of $\bar{H}$ minutes. Our task in this appendix is to relate the call parameters λ_{user} and $\bar{H}$, the grade of service P_b, and the number of trunks K to the number N of users that the system can handle. As a result of our analysis we will

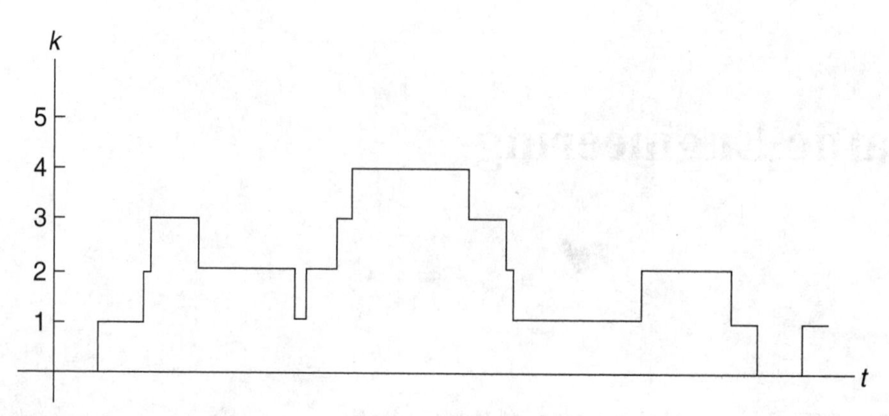

Figure B.1 The State of the Switch

be able to determine for a given number of users how many trunks will be needed to guarantee any desired grade of service. It should be noted that the call parameters λ_{user} and $\bar{H}$ are empirical parameters that are not produced by a theoretical calculation, but are obtained from measurement and experience.

A Model for Call Arrivals

Whenever a call arrives at the switch, the state of the switch increases by one. For the derivation to follow we will need a statistical model for call arrivals. We will begin with four assumptions.

1. If $t_1 \leq t < t_2$ and $t_3 \leq t < t_4$ are disjoint intervals of time, then the number of calls that arrive in the interval $t_1 \leq t < t_2$ is statistically independent of the number of calls that arrive in the interval $t_3 \leq t < t_4$.
2. If δt is a small interval of time, then the probability that one call arrives in the interval $t_0 \leq t < t_0 + \delta t$ is approximately $\lambda \delta t$, where t_0 is arbitrary and λ is a parameter called the "average call arrival rate," measured in calls per minute. The approximation becomes better as δt becomes smaller.
3. If δt is a small interval of time, then the probability that no calls arrive in the interval $t_0 \leq t < t_0 + \delta t$ is approximately $1 - \lambda \delta t$, where the approximation becomes better as δt becomes smaller.
4. Finally, the probability that more than one call arrives in the interval $t_0 \leq t < t_0 + \delta t$ is negligible and vanishes faster than δt as $\delta t \to 0$.

An arrival process that satisfies these four assumptions is called a "Poisson" process in the literature. It turns out that a Poisson process is an excellent model for call arrivals, especially when there is a large community of users, all of whom act independently. Figure B.2 shows the call arrivals from the example of Figure B.1. It should be noted that the call arrivals include new calls as well as calls that are retries of previously blocked calls. These all look the same to the switch!

Figure B.2 Call Arrivals from Figure B.1

A classical property of Poisson processes, derivable from the assumptions given above, is that if Δt is an arbitrary time interval, not necessarily small, then the probability of k arrivals in Δt seconds is given by

$$P_k = \frac{e^{-\lambda\Delta t}(\lambda\Delta - t)^k}{k!}, \; k = 0,1,2,\ldots. \tag{B.1}$$

Now suppose we begin observation at some arbitrary time t_0, and we wonder how long we will have to wait for the next call to arrive. Let T represent waiting time, and let τ be a particular value of T. Then

$$\begin{aligned} P[T < \tau] &= 1 - P[\text{no arrivals for } \tau \text{ seconds}] \\ &= 1 - P_0 \\ &= 1 - \frac{e^{-\lambda\tau}(\lambda\Delta - t)^0}{0!} \\ &= 1 - e^{-\lambda\tau}. \end{aligned} \tag{B.2}$$

This expression is the cumulative probability distribution function for the waiting time T. We can find the probability density function for T by differentiating with respect to τ:

$$\begin{aligned} p_T(\tau) &= \frac{dP[T < \tau]}{d\tau} \\ &= \frac{d(1 - e^{-\lambda\tau})}{d\tau} \\ &= \lambda e^{-\lambda\tau}. \end{aligned} \tag{B.3}$$

Now let us use this probability density function to find the average waiting time $\bar{T}$.

$$\begin{aligned} \bar{T} &= \int_0^\infty \tau p_T(\tau) d\tau \\ &= \int_0^\infty \tau\lambda e^{-\lambda\tau} d\tau \\ &= 1/\lambda. \end{aligned} \tag{B.4}$$

This is one way to justify our interpretation of λ as the average number of calls per second. It is interesting, and counterintuitive, to note that the average waiting time does not depend on when we start waiting. The start time t_0 is arbitrary and may be the time of the previous call arrival, or not.

A Model for Holding Time

Suppose we come upon a call in progress. How much longer will the call continue? Let H represent the remaining duration of the call, known as the **holding time**. To model the holding time we will take our inspiration from the preceding calculation for the time we must wait for the next call to arrive. The holding time H is modeled as having an exponential distribution

$$p_H(\eta) = \mu e^{-\mu\eta},\ 0 \le \eta, \tag{B.5}$$

where μ is a parameter to be identified shortly, and η is a dummy variable. Statistics gathered on the duration of actual calls have verified that the exponential distribution is indeed a very accurate model for holding time, with the exception that there are fewer very short calls than the model predicts. It turns out that relatively few calls are shorter than the time it takes to say "Hello."

Given the model in Equation (B.5), we can calculate the average holding time $\bar{H}$, that is,

$$\begin{aligned} \bar{H} &= \int_0^\infty \eta p_H(\eta) d\eta \\ &= \int_0^\infty \eta \mu e^{-\mu\eta} d\eta \\ &= 1/\mu. \end{aligned} \tag{B.6}$$

This identifies the holding time parameter μ as the reciprocal of the average holding time. As noted previously, average holding time is generally measured in minutes.

In the work to follow we will need to know the probability that a call in progress terminates in a very short interval of time. Suppose δt is such an interval, where $\delta t \ll \bar{H}$. Suppose we begin observing a call in progress at time t_0, and we want the probability that the call terminates on or before $t_0 + \delta_t$. Now from Equation (B.5)

$$\begin{aligned} P[H \le \delta t] &= \int_0^{\delta t} p_H(\eta) d\eta \\ &= \int_0^{\delta t} \mu e^{-\mu\eta} d\eta \\ &= -e^{-\mu\eta} \Big|_0^{\delta t} \\ &= 1 - e^{-\mu\delta t} \\ &\cong \mu\,\delta t = \frac{\delta t}{\bar{H}}, \end{aligned} \tag{B.7}$$

where in the last line we have used the Taylor series for the exponential, $e^{-\mu\delta t} \cong 1 - \mu\,\delta t$ for $\mu\,\delta t \ll 1$ or $\delta t \ll \bar{H}$.

The Switch State Probabilities

Let $p(k;t)$, $k = 0,\dots,K$ represent the probability that the switch is in state k at time t. We want to consider how these state probabilities evolve in time. As a way to get started, consider what can happen between time t_0 and time $t_0 + \delta t$. In particular, the switch can be in state $k = 0$ at time $t_0 + \delta t$ in two ways. Either the switch was in state $k = 1$ at time t_0 and the call in progress ended, or the switch was in state k = 0 at time t_0 and no new calls arrived. Therefore,

$$p(0;t_0 + \delta t) = p(1;t_0)\frac{\delta t}{\overline{H}} + p(0;t_0)(1 - \lambda\,\delta t), \tag{B.8}$$

where we have taken the probability that a call in progress ends from Equation (B.7) and the probability that no new calls arrive from assumption 3 of our Poisson model for call arrivals. Rearranging Equation (B.8) gives

$$\frac{p(0;t_0 + \delta t) - p(0;t_0)}{\delta t} = \frac{1}{\overline{H}}p(1;t_0) - \lambda p(0;t_0). \tag{B.9}$$

If we take the limit as $\delta t \to 0$, we obtain

$$\frac{\partial p(0;t)}{\partial t} = \frac{1}{\overline{H}}p(1;t) - \lambda p(0;t), \tag{B.10}$$

where the subscript has been dropped from the time variable in the interest of generality.

Now let us consider how the switch might be in state $k = 1$ at time $t_0 + \delta t$. First, the switch might have been in state $k = 0$ at time t_0 and a new call arrived. Remember that the probability that more than one new call arrives in a short interval of time is negligible. Second, the switch might have been in state $k = 2$ at time t_0 and one or the other of the two calls in progress ended. The probability that both calls end is on the order of $\left(\frac{\delta t}{\overline{H}}\right)^2$ and is therefore negligible. Finally, the switch might have been in state $k = 1$ at time t_0 and no new calls arrived and the existing call did not terminate. Putting these possibilities together gives us

$$p(1;t_0 + \delta t) = p(0;t_0)\lambda\,\delta t + p(2;t_0)\left(2\frac{\delta t}{\overline{H}}\right) + p(1;t_0)(1 - \lambda\,\delta t)\left(1 - \frac{\delta t}{\overline{H}}\right). \tag{B.11}$$

Rearranging Equation (B.11) gives

$$\frac{p(1;t_0 + \delta t) - p(1;t_0)}{\delta t} = \lambda p(0;t_0) + \frac{2}{\overline{H}}p(2;t_0) - \left(\lambda + \frac{1}{\overline{H}} - \frac{\lambda\,\delta t}{\overline{H}}\right)p(1;t_0). \tag{B.12}$$

Letting $\delta t \to 0$ then gives

$$\frac{\partial p(1;t)}{\delta t} = \lambda p(0;t_0) + \frac{2}{\overline{H}}p(2;t_0) - \left(\lambda + \frac{1}{\overline{H}}\right)p(1;t_0). \tag{B.13}$$

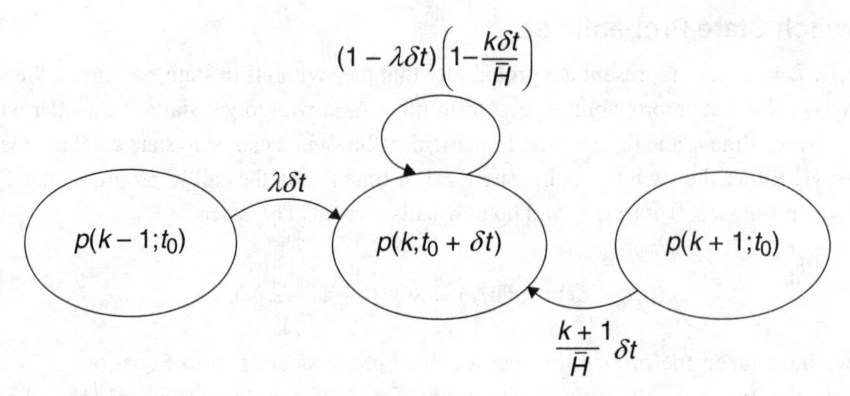

Figure B.3 How the Switch Can Get into State k at Time $t_0 + \delta t$

We can continue this argument state by state. For state k we find

$$\frac{\partial p(k;t)}{\delta t} = \lambda p(k-1;t) + \frac{k+1}{\overline{H}} p(k+1;t) - \left(\lambda + \frac{k}{\overline{H}}\right) p(k;t). \tag{B.14}$$

Figure B.3 shows the flow of probability leading to Equation (B.14). For the highest state $k = K$ the results are a little different, as there are only two ways of reaching state $k = K$ at time $t_0 + \delta t$. We obtain

$$\frac{\partial p(K;t)}{\delta t} = \lambda p(K-1;t) - \frac{K}{\overline{H}} p(K;t). \tag{B.15}$$

Now suppose we start the switch off in state zero at time zero and let it run (that is, $p(0;0) = 1$, and $p(k;0) = 0$ for $k = 1,\ldots,K$). After a while the solutions to the differential equations given in Equations (B.10), (B.14), and (B.15) will approach steady-state values. Let us call these steady-state probabilities $P(k),\ k = 0,\ldots,K$ (i.e., $P(k) = p(k;\infty)$). Before we solve for these steady-state probabilities, let us take a moment to discuss what the steady-state probabilities mean. Suppose we have a large number, say 1000, of identical switches that we can operate under identical conditions. Suppose we start them all off in state zero at the same time and let them run. After a long time we come back and examine the switches. At the moment we examine the switches we will find that a certain fraction of them happen to be handling k calls. This fraction is $P(k)$. If we go away and return later, we will find that calls have terminated in some of the switches and new calls have arrived at some of the switches. If we check to see which switches are handling exactly k calls, we may find a somewhat different group of switches from those we identified previously. Nevertheless, the fraction of the total number of switches that have k calls in progress will be about the same, $P(k)$.

Now in the steady state the probabilities $p(k;t)$ are constant with time. This means that the derivatives $\frac{\partial p(k;t)}{\partial t}$ will be zero for $k = 0,\ldots,K$. Our differential equations then become

$$0 = \frac{1}{\bar{H}}P(1) - \lambda P(0) \tag{B.16}$$

from Equation (B.10),

$$0 = \lambda P(k-1) + \frac{k+1}{\bar{H}}P(k+1) - \left(\lambda + \frac{k}{\bar{H}}\right)P(k),\ k = 1,\ldots,K-1 \tag{B.17}$$

from Equation (B.14), and

$$0 = \lambda P(K-1) - \left(\lambda + \frac{K}{\bar{H}}\right)P(K) \tag{B.18}$$

from Equation (B.15).

Beginning with Equation (B.16) and proceeding iteratively gives, first,

$$P(1) = \lambda\bar{H}P(0). \tag{B.19}$$

Then from Equation (B.17)

$$0 = \lambda P(0) + \frac{2}{\bar{H}}P(2) - \left(\lambda + \frac{1}{\bar{H}}\right)P(1)$$

$$0 = \lambda P(0) + \frac{2}{\bar{H}}P(2) - \left(\lambda + \frac{1}{\bar{H}}\right)\lambda\bar{H}P(0) \tag{B.20}$$

$$P(2) = \frac{(\lambda\bar{H})^2}{2}P(0),$$

and

$$0 = \lambda P(1) + \frac{3}{\bar{H}}P(3) - \left(\lambda + \frac{2}{\bar{H}}\right)P(2)$$

$$0 = \lambda(\lambda\bar{H})P(0) + \frac{3}{\bar{H}}P(3) - \left(\lambda + \frac{2}{\bar{H}}\right)\frac{\left(\lambda\bar{H}\right)^2}{2}P(0) \tag{B.21}$$

$$P(3) = \frac{(\lambda\bar{H})^3}{3\cdot 2}P(0),$$

and so on. The general term is

$$P(k) = \frac{(\lambda \bar{H})^k}{k!} P(0),\ k = 2, \ldots, K-1, \tag{B.22}$$

which the reader can easily verify by substituting into Equation (B.17). Finally, Equation (B.18) gives

$$\begin{aligned} 0 &= \lambda P(K-1) - \frac{K}{\bar{H}} P(K) \\ P(K) &= \frac{\lambda \bar{H}}{K} P(K-1) \\ P(K) &= \frac{\lambda \bar{H}}{K} \frac{(\lambda \bar{H})^{K-1}}{(K-1)!} P(0) \\ P(K) &= \frac{(\lambda \bar{H})^K}{(K)!} P(0). \end{aligned} \tag{B.23}$$

Blocking Probability, Offered Load, and Erlang B

Putting together Equations (B.19), (B.22), and (B.23) gives us

$$P(k) = \frac{(\lambda \bar{H})^k}{k!} P(0),\ k = 1, \ldots, K. \tag{B.24}$$

It seems that we would know all of the steady-state probabilities if we only knew $P(0)$. There is one more equation, however, and this will allow us to find $P(0)$ and then all of the $P(k)$. The additional equation represents the fact that the switch must be in *some* state, that is,

$$P(0) + P(1) + \cdots + P(K) = 1. \tag{B.25}$$

Substituting Equation (B.24) in Equation (B.25) gives

$$\sum_{n=0}^{K} \frac{(\lambda \bar{H})^n}{n!} P(0) = 1. \tag{B.26}$$

Solving for $P(0)$,

$$P(0) = \frac{1}{\displaystyle\sum_{n=0}^{K} \frac{(\lambda \bar{H})^n}{n!}}. \tag{B.27}$$

Finally, substituting back into Equation (B.24) gives us the steady-state probabilities that we seek:

$$P(k) = \frac{\dfrac{(\lambda\bar{H})^k}{k!}}{\displaystyle\sum_{n=0}^{K} \frac{(\lambda\bar{H})^n}{n!}}, \quad k = 0,\ldots,K. \tag{B.28}$$

We began this derivation with the intent of relating the blocking probability to the call statistics, the number of trunks, and the number of users. The goal is now within sight! If the switch is in state $k = K$ when a call arrives, then the call will be blocked, as all of the trunks will be in use. Thus $P[blocking] = P_b = P(K)$. We can now see from Equation (B.28) that the probability that an arriving call will find the switch in state $k = K$ is

$$P_b = \frac{\dfrac{(\lambda\bar{H})^K}{K!}}{\displaystyle\sum_{n=0}^{K} \frac{(\lambda\bar{H})^n}{n!}}. \tag{B.29}$$

There is now just one loose end to clean up. The parameter $\lambda\bar{H}$ that appears in Equation (B.29) is called the **offered load** and is represented by the symbol A. Offered load is described as a measure of the traffic intensity presented to the switch. It is the product of the average number of calls per unit time arriving at the switch and the average holding time. If λ is measured in calls per minute and $\bar{H}$ is measured in minutes, then A will be a dimensionless quantity. According to tradition, A is given the unit of **erlangs** after the Danish mathematician Agner Erlang (1878–1929), who initially developed the theory presented here. Now A is the aggregate offered load as seen by the switch. Each user presents an individual offered load given by $A_{user} = \lambda_{user}\bar{H}$. Recall that λ_{user} is the average number of calls per unit time generated by a single user. We can relate the aggregate offered load to the individual offered load by the simple relation

$$A = NA_{user}, \tag{B.30}$$

where N is the number of users. Substituting A into Equation (B.29) gives us the result we seek, the celebrated **Erlang B** formula,

$$P_b = \frac{\dfrac{(A)^K}{K!}}{\displaystyle\sum_{n=0}^{K} \frac{(A)^n}{n!}}, \tag{B.31}$$

where P_b is the blocking probability, A is the offered load, and K is the number of trunks.

Computational Techniques for the Erlang B Formula

The Erlang B formula contains terms of the form

$$\frac{A^K}{K!}. \tag{B.32}$$

When evaluating Equation (B.32), the factorial term may exceed the maximum floating-point number that a host computer can support for large values of K. For the computer that was used to generate the tables in this appendix, the upper limit for the factorial term was reached for $K = 170$. Likewise, the value of A^K may exceed the maximum number for large A and K. A typical switching office, however, may possess several thousand serving elements, thereby eliminating the possibility of computation of the GOS by direct evaluation of Equation (B.31). These computational difficulties can be eliminated by a simple rearrangement of the formula.

Starting with the Erlang B formula of Equation (B.31), we divide both numerator and denominator by the numerator. This yields

$$P_b = \frac{1}{\frac{K!}{A^K}\sum_{n=0}^{K}\frac{A^n}{n!}}. \tag{B.33}$$

Reversing the summation so that the terms range from K to 0 and are listed from left to right, we can expand the denominator to obtain

$$Den = 1 + \frac{K}{A} + \frac{K}{A}\cdot\frac{(K-1)}{A} + \frac{K}{A}\cdot\frac{(K-1)}{A}\cdot\frac{(K-2)}{A} + \cdots + \frac{K!}{A^K}. \tag{B.34}$$

A simple approach to evaluating *Den* is to form the vector of elements:

$$ratio = \left[1, \frac{K}{A}, \frac{(K-1)}{A}, \frac{(K-2)}{A}, \ldots, \frac{1}{A}\right]. \tag{B.35}$$

Next, compute a vector, *prods,* containing the cumulative products of the vector *ratio* such that the nth element is the product of the first n elements of *ratio.* The vector *prods* contains the elements

$$prods = \left[1, \frac{K}{A}, \frac{K}{A}\cdot\frac{(K-1)}{A}, \frac{K}{A}\cdot\frac{(K-1)}{A}\cdot\frac{(K-2)}{A}, \ldots, \frac{K!}{A^K}\right]. \tag{B.36}$$

The elements of *prods* are then summed to form *Den.*

For $A > K$ the values of the terms decrease from left to right. Therefore, *Den* converges to a limit for any practical value of K. For $K > A$ the values of the terms increase from left to right. If the ratio K/A is sufficiently large, *Den* may grow beyond the largest floating-point value that the host computer can support. For most practical situations, however, the ratio K/A and the absolute value of K are such that the limit is not reached. As an example, the largest floating-point number that can be represented by the PC used to generate the tables in this appendix is 1.7977e + 308. For situations that do exceed the maximum floating-point number limit a direct evaluation of Equation (B.31) using the techniques described above can be used effectively.

The following MATLAB function, "Erlangb.m," computes the blocking probability for a given number of servers, *K,* and offered load, *A,* in erlangs using the approach described above.

```
%
% **************** Function "ErlangB"   ***************
%
% This function evaluates the Erlang B formula to provide the GOS or
% blocking probability for a specified number of servers, K, and
% an offered load, A.
%
function Pb=ErlangB(A,K)
%
% Ensure that A is positive.
%
if A<0
    A=realmin;
end
%
% The numerator of the Erlang B formula contains terms of the form A^K/K!.
% The factorial term is limited to about K=170 depending on the maximum
% positive number that the host computer supports.
% To overcome this limitation we use the approach described in the notes.
%
% Form a vector of length equal to the rounded value K with
% each element value equal to A.
%
AtoK=ones(1,round(K))*A;
%
% Form a vector with element values from the rounded value of K to 1.
%
Kfac=[round(K):-1:1];
%
% Form the vector whose elements are [K/A, (K-1)/A, K-2)/A, ..., 1/A]
%
ratio=Kfac./AtoK;
%
% Augment the ratio vector with a "1" for the i=K term
%
%
ratio=[1 ratio];
%
%
% Form the vector "prods" whose elements are the cumulative products of the
% elements
% of the vector "ratio". The nth element of "prods" is the product of the first
% n elements of "ratio".
%
prods=cumprod(ratio);
%
%
% Add the elements of "prods" one at a time testing the fractional percentage
% change
%
% in the cumulative sum. When the change is smaller than the number of floating-point
% digits carried by the host computer, the summation is ended.
sum=0;
for i=1:K+1
    oldsum=sum;
    sum=oldsum+prods(i);
    delta=prods(i)/sum;
   if delta < eps
       break
   end
end
%
```

```
% The blocking probability is the reciprocal of the sum.
%
Pb=1/sum;
```

The MATLAB routine "GOS" below uses the function "ErlangB" to provide a convenient evaluation tool. The routine requests the number of servers available and the offered traffic to compute the GOS or blocking probability.

```
% *************** Routine GOS ***********************%
%
% This program computes the grade of service or blocking probability
% for a specified number of servers, K, and a specified offered load, A.
%
% The routine has two loops.  The outer loop requests the number of servers
% as an input.  The inner loop requests the offered load as an input.
% A blank or negative input breaks either loop.
%
clear
K=realmin;
while K>0
    K=input('How many servers are in the system?: ');
    A=realmin;
    if K<=0
        break;
    end
    while A>0 & K>0
       A=input('What is the offered load in Erlangs?: ');
       if A>0
           Grade_of_Service=ErlangB(A,K)
       end
   end
end
```

As its name implies, the routine "Offered_traffic" computes the maximum amount of offered traffic that can be supported by a specified number of servers at a specified GOS. The number of servers and desired GOS are inputs requested by the routine. Since the offered traffic cannot be expressed explicitly in terms of blocking probability and number of servers, this routine does an iterative search to find the value satisfying the input parameters. It uses functions "DeltaPb" and "ErlangB."

```
%
% *************** Routine "DeltaPb" *********************
%
% This function calculates the difference between a given
% blocking probability and the value returned by "ErlangB".
%
function f=DeltaPb(x)
global K pb
f=pb-ErlangB(x,K);
```

```
%
% **************   Routine "Offered_traffic" ***********
%
% This routine computes the offered traffic given the GOS
% and the number of servers.  Since the Erlang B formula
% cannot be expressed explicitly in terms of the offered
% traffic, A, it uses MATLAB function "fzero" to perform a
% search around an initial guess. It uses functions "DeltaPb"
% and "ErlangB".
%
clear
global K pb
Pb=eps;
opt=[];
while Pb>0
    Pb=input('What is the desired GOS in Percent? ');
    pb=Pb/100;
    K=eps;
    A=eps;
    while K>0 & Pb>0
        K=input('How many servers? ');
        if K>0
%
% As an initial guess assume that the offered traffic is
% 90% occupancy.
%
            x=.9*K;
%
% Perform a search using the initial guess and the input values
% of Pb and K which are passed to "DeltaPb" using global variables.
% "DeltaPb" is a function that computes the difference between the
% input "Pb" and the value of Pb computed using the initial guess
% for the offered traffic, "x", and the input number of servers, "K".
% "DeltaPb" uses function "ErlangB" to compute each value
% of Pb.
%
            A=fzero(@DeltaPb,x)
        end
    end
end
%
```

These routines and functions were used to compute the Erlang B tables that follow.

Erlang B Table

The following provides offered traffic for a range of values for the number of servers and the grade of service.

	Grade of Service or Blocking Probability in Percent									
	1%	2%	3%	4%	5%	6%	7%	8%	9%	10%
K	A, Offered Load in Erlangs									
1	0.010	0.020	0.031	0.042	0.053	0.064	0.075	0.087		0.111
2	0.153	0.223	0.282	0.333	0.381	0.427	0.470	0.513	0.555	0.595
3	0.455	0.602	0.715	0.812	0.899	0.980	1.057	1.131	1.202	1.271
4	0.869	1.092	1.259	1.399	1.525	1.640	1.748	1.851	1.949	2.045
5	1.361	1.657	1.875	2.057	2.218	2.366	2.504	2.634	2.760	2.881
6	1.909	2.276	2.543	2.765	2.960	3.139	3.305	3.462	3.613	3.758
7	2.501	2.935	3.250	3.509	3.738	3.946	4.139	4.322	4.497	4.666
8	3.128	3.627	3.987	4.283	4.543	4.779	4.999	5.207	5.405	5.597
9	3.783	4.345	4.748	5.080	5.370	5.634	5.879	6.111	6.333	6.546
10	4.461	5.084	5.529	5.895	6.216	6.506	6.776	7.031	7.275	7.511
11	5.160	5.842	6.328	6.727	7.076	7.393	7.687	7.965	8.231	8.487
12	5.876	6.615	7.141	7.573	7.950	8.292	8.610	8.910	9.197	9.474
13	6.607	7.402	7.967	8.430	8.835	9.202	9.543	9.865	10.173	10.470
14	7.352	8.200	8.803	9.298	9.730	10.121	10.485	10.828	11.156	11.473
15	8.108	9.010	9.650	10.174	10.633	11.048	11.434	11.799	12.147	12.484
16	8.875	9.828	10.505	11.059	11.544	11.983	12.390	12.776	13.144	13.500
17	9.652	10.656	11.368	11.952	12.461	12.924	13.353	13.759	14.147	14.522
18	10.437	11.491	12.238	12.850	13.385	13.870	14.321	14.747	15.154	15.548
19	11.230	12.333	13.115	13.755	14.315	14.822	15.294	15.740	16.166	16.579
20	12.031	13.182	13.997	14.665	15.249	15.779	16.271	16.737	17.182	17.613

21	12.838	14.036	14.885	15.581	16.189	16.740	17.253	17.738	18.202	18.651
22	13.651	14.896	15.778	16.500	17.132	17.705	18.238	18.743	19.225	19.692
23	14.470	15.761	16.675	17.425	18.080	18.674	19.227	19.751	20.252	20.737
24	15.295	16.631	17.577	18.353	19.031	19.646	20.219	20.762	21.281	21.784
25	16.125	17.505	18.483	19.284	19.985	20.622	21.215	21.775	22.313	22.833
26	16.959	18.383	19.392	20.219	20.943	21.600	22.212	22.792	23.347	23.885
27	17.797	19.265	20.305	21.158	21.904	22.582	23.213	23.811	24.384	24.939
28	18.640	20.150	21.221	22.099	22.867	23.566	24.216	24.832	25.423	25.995
29	19.487	21.039	22.140	23.043	23.833	24.552	25.221	25.855	26.464	27.053
30	20.337	21.932	23.062	23.990	24.802	25.540	26.228	26.881	27.506	28.113
31	21.191	22.827	23.987	24.939	25.773	26.531	27.238	27.908	28.551	29.174
32	22.048	23.725	24.914	25.890	26.746	27.524	28.249	28.937	29.597	30.237
33	22.909	24.626	25.844	26.844	27.721	28.518	29.262	29.967	30.645	31.301
34	23.772	25.529	26.776	27.800	28.698	29.515	30.277	31.000	31.694	32.367
35	24.638	26.435	27.711	28.758	29.677	30.513	31.293	32.033	32.745	33.434
36	25.507	27.343	28.647	29.718	30.657	31.513	32.311	33.069	33.797	34.503
37	26.378	28.254	29.585	30.680	31.640	32.514	33.330	34.105	34.850	35.572
38	27.252	29.166	30.526	31.643	32.624	33.517	34.351	35.143	35.904	36.643
39	28.129	30.081	31.468	32.608	33.609	34.521	35.373	36.182	36.960	37.715
40	29.007	30.997	32.412	33.575	34.596	35.527	36.396	37.222	38.016	38.787

(*Continued*)

	Grade of Service or Blocking Probability in Percent									
	1%	2%	3%	4%	5%	6%	7%	8%	9%	10%
K	A, Offered Load in Erlangs									
41	29.888	31.916	33.357	34.543	35.584	36.534	37.421	38.263	39.074	39.861
42	30.771	32.836	34.305	35.513	36.574	37.542	38.446	39.306	40.133	40.936
43	31.656	33.758	35.253	36.484	37.565	38.551	39.473	40.349	41.192	42.011
44	32.543	34.682	36.203	37.456	38.557	39.562	40.501	41.393	42.253	43.088
45	33.432	35.607	37.155	38.430	39.550	40.573	41.529	42.439	43.314	44.165
46	34.322	36.534	38.108	39.405	40.545	41.586	42.559	43.485	44.376	45.243
47	35.215	37.462	39.062	40.381	41.540	42.599	43.590	44.532	45.439	46.322
48	36.109	38.392	40.018	41.358	42.537	43.614	44.621	45.580	46.503	47.401
49	37.004	39.323	40.975	42.336	43.534	44.629	45.653	46.628	47.568	48.481
50	37.901	40.255	41.933	43.316	44.533	45.645	46.687	47.678	48.633	49.562
51	38.800	41.189	42.892	44.296	45.533	46.663	47.721	48.728	49.699	50.644
52	39.700	42.124	43.852	45.278	46.533	47.681	48.755	49.779	50.765	51.726
53	40.602	43.060	44.813	46.260	47.534	48.699	49.791	50.830	51.833	52.808
54	41.505	43.997	45.776	47.243	48.536	49.719	50.827	51.883	52.900	53.891
55	42.409	44.936	46.739	48.228	49.539	50.739	51.864	52.935	53.969	54.975
56	43.315	45.875	47.703	49.213	50.543	51.760	52.901	53.989	55.038	56.059
57	44.222	46.816	48.669	50.199	51.548	52.782	53.940	55.043	56.107	57.144
58	45.130	47.758	49.635	51.185	52.553	53.805	54.978	56.097	57.177	58.229
59	46.039	48.700	50.602	52.173	53.559	54.828	56.018	57.153	58.248	59.315
60	46.950	49.644	51.570	53.161	54.566	55.852	57.058	58.208	59.319	60.401

61	47.861	50.589	52.539	54.150	55.573	56.876	58.098	59.265	60.391	61.488
62	48.774	51.534	53.508	55.140	56.581	57.901	59.140	60.321	61.463	62.575
63	49.688	52.481	54.478	56.131	57.590	58.927	60.181	61.378	62.535	63.663
64	50.603	53.428	55.450	57.122	58.599	59.953	61.223	62.436	63.608	64.750
65	51.518	54.376	56.421	58.114	59.609	60.979	62.266	63.494	64.681	65.839
66	52.435	55.325	57.394	59.106	60.619	62.007	63.309	64.553	65.755	66.927
67	53.353	56.275	58.367	60.100	61.630	63.034	64.353	65.612	66.829	68.016
68	54.272	57.226	59.341	61.093	62.642	64.062	65.397	66.672	67.904	69.106
69	55.191	58.177	60.316	62.088	63.654	65.091	66.442	67.731	68.979	70.196
70	56.112	59.129	61.291	63.083	64.667	66.120	67.487	68.792	70.054	71.286
71	57.033	60.082	62.267	64.078	65.680	67.150	68.532	69.852	71.129	72.376
72	57.956	61.036	63.244	65.074	66.694	68.180	69.578	70.913	72.205	73.467
73	58.879	61.990	64.221	66.071	67.708	69.211	70.624	71.975	73.282	74.558
74	59.803	62.945	65.199	67.068	68.723	70.242	71.671	73.037	74.358	75.649
75	60.728	63.900	66.177	68.066	69.738	71.273	72.718	74.099	75.435	76.741
76	61.653	64.857	67.156	69.064	70.753	72.305	73.765	75.161	76.513	77.833
77	62.579	65.814	68.136	70.063	71.769	73.337	74.813	76.224	77.590	78.925
78	63.506	66.771	69.116	71.062	72.786	74.370	75.861	77.287	78.668	80.018
79	64.434	67.729	70.096	72.062	73.803	75.403	76.909	78.351	79.746	81.110
80	65.363	68.688	71.077	73.062	74.820	76.436	77.958	79.414	80.825	82.203

(*Continued*)

	Grade of Service or Blocking Probability in Percent									
	1%	2%	3%	4%	5%	6%	7%	8%	9%	10%
K	A, Offered Load in Erlangs									
81	66.292	69.647	72.059	74.062	75.838	77.470	79.007	80.479	81.904	83.297
82	67.222	70.607	73.041	75.063	76.856	78.504	80.057	81.543	82.983	84.390
83	68.152	71.568	74.024	76.065	77.874	79.539	81.107	82.608	84.062	85.484
84	69.084	72.529	75.007	77.067	78.893	80.574	82.157	83.672	85.141	86.578
85	70.016	73.490	75.990	78.069	79.912	81.609	83.207	84.738	86.221	87.672
86	70.948	74.452	76.974	79.071	80.932	82.644	84.258	85.803	87.301	88.767
87	71.881	75.415	77.959	80.075	81.952	83.680	85.309	86.869	88.381	89.861
88	72.815	76.378	78.944	81.078	82.972	84.716	86.360	87.935	89.462	90.956
89	73.749	77.342	79.929	82.082	83.993	85.752	87.411	89.001	90.543	92.051
90	74.684	78.306	80.915	83.086	85.014	86.789	88.463	90.068	91.624	93.146
91	75.620	79.271	81.901	84.091	86.035	87.826	89.515	91.134	92.705	94.242
92	76.556	80.236	82.888	85.096	87.057	88.863	90.568	92.201	93.786	95.338
93	77.493	81.201	83.875	86.101	88.079	89.901	91.620	93.269	94.868	96.434
94	78.430	82.167	84.862	87.106	89.101	90.939	92.673	94.336	95.950	97.530
95	79.368	83.133	85.850	88.112	90.123	91.977	93.726	95.404	97.032	98.626
96	80.306	84.100	86.838	89.119	91.146	93.015	94.779	96.472	98.114	99.722
97	81.245	85.068	87.826	90.125	92.169	94.054	95.833	97.540	99.196	100.819
98	82.184	86.035	88.815	91.132	93.193	95.093	96.887	98.608	100.279	101.916

99	83.124	87.003	89.804	92.140	94.216	96.132	97.941	99.676	01.362	03.013
100	84.064	87.972	90.794	93.147	95.240	97.171	98.995	100.745	102.445	104.110
101	85.005	88.941	91.784	94.155	96.265	98.211	100.049	101.814	103.528	05.207
102	85.946	89.910	92.774	95.163	97.289	99.251	101.104	102.883	104.611	106.304
103	86.888	90.880	93.765	96.172	98.314	100.291	102.159	103.952	105.695	107.402
104	87.830	91.850	94.756	97.180	99.339	101.331	103.214	105.022	106.778	108.500
105	88.773	92.821	95.747	98.190	100.364	102.372	104.269	106.092	107.862	109.598
106	89.716	93.791	96.738	99.199	101.390	103.413	105.325	107.161	108.946	110.696
107	90.660	94.763	97.730	100.208	102.415	104.454	106.380	108.231	110.030	111.794
108	91.604	95.734	98.722	101.218	103.441	105.495	107.436	109.302	111.114	112.892
109	92.548	96.706	99.715	102.228	104.468	106.536	108.492	110.372	112.199	113.991
110	93.493	97.678	100.708	103.239	105.494	107.578	109.549	111.442	113.284	115.089
111	94.438	98.651	101.701	104.249	106.521	108.620	110.605	112.513	114.368	116.188
112	95.384	99.624	102.694	105.260	107.548	109.662	111.662	113.584	115.453	117.287
113	96.330	100.597	103.688	106.271	108.575	110.704	112.718	114.655	116.538	118.386
114	97.277	101.571	104.682	107.283	109.602	111.746	113.775	115.726	117.623	119.485
115	98.223	102.545	105.676	108.294	110.630	112.789	114.832	116.797	118.709	120.584
116	99.171	103.519	106.670	109.306	111.658	113.832	115.890	117.869	119.794	121.684
117	100.118	104.493	107.665	110.318	112.685	114.875	116.947	118.940	120.880	122.783
118	101.066	105.468	108.660	111.331	113.714	115.918	118.005	120.012	121.965	123.883

(Continued)

	Grade of Service or Blocking Probability in Percent									
	1%	2%	3%	4%	5%	6%	7%	8%	9%	10%
K	A, Offered Load in Erlangs									
119	102.015	106.443	109.655	112.343	114.742	116.961	119.063	121.084	123.051	124.983
120	102.964	107.419	110.651	113.356	115.771	118.005	120.120	122.156	124.137	126.082
121	103.913	108.395	111.646	114.369	116.799	119.048	121.179	123.228	125.223	127.182
122	104.862	109.371	112.642	115.382	117.828	120.092	122.237	124.300	126.309	128.282
123	105.812	110.347	113.639	116.396	118.857	121.136	123.295	125.373	127.396	129.383
124	106.762	111.323	114.635	117.409	119.887	122.181	124.354	126.445	128.482	130.483
125	107.713	112.300	115.632	118.423	120.916	123.225	125.412	127.518	129.569	131.583
126	108.664	113.278	116.629	119.437	121.946	124.269	126.471	128.591	130.656	132.684
127	109.615	114.255	117.626	120.451	122.976	125.314	127.530	129.664	131.742	133.784
128	110.566	115.233	118.623	121.466	124.006	126.359	128.589	130.737	132.829	134.885
129	111.518	116.211	119.621	122.480	125.036	127.404	129.648	131.810	133.916	135.986
130	112.470	117.189	120.619	123.495	126.066	128.449	130.708	132.883	135.003	137.087
131	113.423	118.167	121.617	124.510	127.097	129.494	131.767	133.957	136.091	138.188
132	114.376	119.146	122.615	125.525	128.128	130.540	132.827	135.030	137.178	139.289
133	115.329	120.125	123.614	126.541	129.159	131.585	133.887	136.104	138.265	140.390
134	116.282	121.104	124.612	127.556	130.190	132.631	134.946	137.178	139.353	141.491
135	117.236	122.084	125.611	128.572	131.221	133.677	136.006	138.252	140.440	142.592
136	118.190	123.063	126.610	129.588	132.252	134.723	137.067	139.326	141.528	143.694
137	119.144	124.043	127.610	130.604	133.284	135.769	138.127	140.400	142.616	144.795

138	120.099	125.023	128.609	131.620	134.315	136.815	139.187	141.474	143.704	145.897
139	121.054	126.004	129.609	132.637	135.347	137.861	140.248	142.548	144.792	146.999
140	122.009	126.984	130.609	133.653	136.379	138.908	141.308	143.623	145.880	148.100
141	122.964	127.965	131.609	134.670	137.411	139.954	142.369	144.697	146.968	149.202
142	123.920	128.946	132.609	135.687	138.443	141.001	143.430	145.772	148.057	150.304
143	124.876	129.928	133.610	136.704	139.476	142.048	144.491	146.847	149.145	151.406
144	125.832	130.909	134.610	137.722	140.508	143.095	145.552	147.921	150.233	152.508
145	126.789	131.891	135.611	138.739	141.541	144.142	146.613	148.996	151.322	153.610
146	127.746	132.873	136.612	139.756	142.574	145.189	147.674	150.071	152.411	154.713
147	128.703	133.855	137.613	140.774	143.606	146.237	148.735	151.146	153.499	155.815
148	129.660	134.837	138.615	141.792	144.640	147.284	149.797	152.222	154.588	156.917
149	130.618	135.820	139.616	142.810	145.673	148.332	150.858	153.297	155.677	158.020
150	131.576	136.803	140.618	143.828	146.706	149.379	151.920	154.372	156.766	159.122
151	132.534	137.786	141.620	144.847	147.739	150.427	152.982	155.448	157.855	160.225
152	133.492	138.769	142.622	145.865	148.773	151.475	154.044	156.523	158.944	161.328
153	134.451	139.752	143.624	146.884	149.807	152.523	155.106	157.599	160.033	162.430
154	135.409	140.736	144.627	147.902	150.840	153.571	156.168	158.674	161.122	163.533
155	136.368	141.720	145.629	148.921	151.874	154.620	157.230	159.750	162.212	164.636
156	137.328	142.704	146.632	149.940	152.908	155.668	158.292	160.826	163.301	165.739
157	138.287	143.688	147.635	150.959	153.943	156.716	159.354	161.902	164.391	166.842
158	139.247	144.672	148.638	151.979	154.977	157.765	160.417	162.978	165.480	167.945

(Continued)

	Grade of Service or Blocking Probability in Percent									
	1%	2%	3%	4%	5%	6%	7%	8%	9%	10%
K	A, Offered Load in Erlangs									
159	140.207	145.657	149.641	152.998	156.011	158.814	161.479	164.054	166.570	169.048
160	141.167	146.641	150.644	154.018	157.046	159.862	162.542	165.130	167.660	170.151
161	142.128	147.626	151.648	155.037	158.080	160.911	163.604	166.207	168.749	171.255
162	143.088	148.611	152.651	156.057	159.115	161.960	164.667	167.283	169.839	172.358
163	144.049	149.596	153.655	157.077	160.150	163.009	165.730	168.360	170.929	173.461
164	145.010	150.582	154.659	158.097	161.185	164.058	166.793	169.436	172.019	174.565
165	145.972	151.567	155.663	159.117	162.220	165.108	167.856	170.513	173.109	175.668
166	146.933	152.553	156.667	160.138	163.255	166.157	168.919	171.589	174.199	176.772
167	147.895	153.539	157.672	161.158	164.291	167.206	169.982	172.666	175.289	177.876
168	148.857	154.525	158.676	162.179	165.326	168.256	171.046	173.743	176.380	178.979
169	149.819	155.511	159.681	163.199	166.361	169.306	172.109	174.820	177.470	180.083
170	150.781	156.498	160.686	164.220	167.397	170.355	173.172	175.897	178.560	181.187
171	151.744	157.484	161.690	165.241	168.433	171.405	174.236	176.974	179.651	182.291
172	152.707	158.471	162.695	166.262	169.469	172.455	175.299	178.051	180.741	183.394
173	153.669	159.458	163.701	167.283	170.504	173.505	176.363	179.128	181.832	184.498
174	154.633	160.445	164.706	168.304	171.540	174.555	177.427	180.205	182.922	185.602
175	155.596	161.432	165.711	169.326	172.576	175.605	178.491	181.282	184.013	186.706
176	156.560	162.419	166.717	170.347	173.613	176.655	179.555	182.360	185.104	187.810
177	157.523	163.407	167.723	171.369	174.649	177.706	180.618	183.437	186.194	188.915

178	158.487	164.395	168.728	172.391	175.685	178.756	181.682	184.514	187.285	190.019
179	159.451	165.382	169.734	173.412	176.722	179.806	182.747	185.592	188.376	191.123
180	160.416	166.370	170.740	174.434	177.758	180.857	183.811	186.670	189.467	192.227
181	161.380	167.358	171.747	175.456	178.795	181.908	184.875	187.747	190.558	193.332
182	162.345	168.347	172.753	176.478	179.832	182.958	185.939	188.825	191.649	194.436
183	163.310	169.335	173.759	177.501	180.869	184.009	187.004	189.903	192.740	195.541
184	164.275	170.323	174.766	178.523	181.905	185.060	188.068	190.981	193.831	196.645
185	165.240	171.312	175.773	179.545	182.942	186.111	189.133	192.058	194.923	197.750
186	166.205	172.301	176.779	180.568	183.980	187.162	190.197	193.136	196.014	198.854
187	167.171	173.290	177.786	181.590	185.017	188.213	191.262	194.214	197.105	199.959
188	168.136	174.279	178.793	182.613	186.054	189.264	192.327	195.292	198.196	201.063
189	169.102	175.268	179.800	183.636	187.091	190.315	193.391	196.371	199.288	202.168
190	170.068	176.258	180.808	184.659	188.129	191.367	194.456	197.449	200.379	203.273
191	171.035	177.247	181.815	185.682	189.166	192.418	195.521	198.527	201.471	204.378
192	172.001	178.237	182.822	186.705	190.204	193.470	196.586	199.605	202.562	205.482
193	172.968	179.226	183.830	187.728	191.241	194.521	197.651	200.684	203.654	206.587
194	173.934	180.216	184.838	188.751	192.279	195.573	198.716	201.762	204.746	207.692
195	174.901	181.206	185.845	189.775	193.317	196.624	199.781	202.840	205.837	208.797
196	175.868	182.196	186.853	190.798	194.355	197.676	200.846	203.919	206.929	209.902
197	176.835	183.187	187.861	191.822	195.393	198.728	201.912	204.997	208.021	211.007
198	177.803	184.177	188.869	192.845	196.431	199.780	202.977	206.076	209.113	212.112
199	178.770	185.168	189.878	193.869	197.469	200.832	204.042	207.155	210.205	213.217
200	179.738	186.158	190.886	194.893	198.507	201.884	205.108	208.233	211.296	214.323

Acronyms

1G	first generation
1 × EV-DO	1× evolution for data only
	1× evolution for data optimized (Rev A)
1 × EV-DV	1× evolution for data and voice
1 × RTT	1× radio transmission technology
2G	second generation
3G	third generation
3GPP	Third Generation Partnership Project
3GPP2	Third Generation Partnership Project 2
3xRTT	3x radio transmission technology
4G	fourth generation
8PSK	8-ary phase-shift keying
ACK	acknowledgment
AM	amplitude modulation
AMPS	Advanced Mobile Phone Service
AN	access network
ANSI	American National Standards Institute
ARQ	automatic repeat request
AT	access terminal
AWGN	additive white Gaussian noise
BER	bit error rate
BLER	block error rate
BPF	bandpass filter
BPSK	binary phase-shift keying
BS	base station
CDM	code-division multiplexing
CDMA	code-division multiple access
CDVCC	coded digital verification color code
CEPT	Conference of European Postal and Telecommunications Administrations
CFR	Code of Federal Regulations
CIR	carrier-to-interference ratio
CNIR	carrier-to-noise-and-interference ratio
CQI	channel quality indicator
CRC	cyclic redundancy check
CS	coding scheme

CSMA	carrier-sense multiple access
CSMA/CD	carrier-sense multiple access with collision detection
D-AMPS	Digital Advanced Mobile Phone Service
dB	decibels
dBm	dB relative to 1 mW
dBW	dB relative to 1 W
DCCH	dedicated control channel
DFT	discrete Fourier transform
DPCM	differential pulse-code modulation
DPSK	differential binary phase-shift keying
DQPSK	differential quadrature phase-shift keying
DR	data rate
DRC	data rate control
DSP	digital signal processing
DSSS	direct-sequence spread spectrum
DTC	digital traffic channel
DTMF	dual-tone multiple frequency
ECSD	Enhanced Circuit Switched Data
EDGE	Enhanced Data rates for Global Evolution
EGPRS	Enhanced General Packet Radio Service
EIRP	effective isotropic radiated power
ERP	effective radiated power
ETACS	European Total Access Communications System
	Extended Total Access Communications System
ETSI	European Telecommunications Standards Institute
FACCH	fast associated control channel
FCC	Federal Communications Commission
FCH	fundamental channel
FDD	frequency-division duplex
FDM	frequency-division multiplexing
FDMA	frequency-division multiple access
FEC	forward error correction
FFT	fast Fourier transform
FHSS	frequency-hopping spread spectrum
FM	frequency modulation
FPLMTS	Future Public Land Mobile Telecommunications System
FPR	functional product requirements
FSK	frequency-shift keying
GFSK	Gaussian frequency-shift keying
GMSK	Gaussian minimum-shift keying
GOS	grade of service

GPRS	General Packet Radio Service
GSM	Groupe Spéciale Mobile
	Global System for Mobile Communications
H-ARQ	hybrid automatic repeat request
HDR	high data rate
HLR	home location register
HSCSD	High-Speed Circuit-Switched Data
HSDPA	High-Speed Downlink Packet Access
HSPA	High-Speed Packet Access
HSUPA	High-Speed Uplink Packet Access
IDFT	inverse discrete Fourier transform
IF	intermediate frequency
IFA	intermediate frequency amplifier
IFFT	inverse fast Fourier transform
IMT	International Mobile Telecommunications
IMTS	Improved Mobile Telephone Service
IR	incremental redundancy
IS	interim standard
ISDN	Integrated Services Digital Network
ISI	intersymbol interference
ITU	International Telecommunication Union
LEO	low Earth orbit
LNA	low-noise amplifier
LPC	linear predictive coding
LPF	lowpass filter
LTE	long-term evolution
MAC	medium access control
MCS	modulation and coding scheme
MIMO	multiple input multiple output
MSC	mobile switching center
MSK	minimum-shift keying
MTSO	mobile telephone switching office
MU	mobile unit
NACK	negative acknowledgment
NA-TDMA	North American Time-Division Multiple Access
NRZ	non-return-to-zero
NTIA	National Telecommunications and Information Administration
NTSC	National Television System Committee
OFDM	orthogonal frequency-division multiplexing
OFDMA	orthogonal frequency-division multiple access
OQPSK	offset quadrature phase-shift keying

OVSF	orthogonal variable spreading factor
PA	power amplifier
PCM	pulse code modulation
PCS	personal communication services (system)
PDC	Pacific Digital Cellular System
PER	packet error rate
PN	pseudonoise
	pseudorandom number
POTS	plain old telephone service
PPSDN	public packet-switched data network
PRMA	packet-reservation multiple access
PSDN	packet-switched data network
PSK	phase-shift keying
PSTN	public switched telephone network
QAM	quadrature amplitude modulation
QoS	quality of service
QPSK	quadrature phase-shift keying
RAB	reverse activity bit
RACH	reverse access channel
RC	resistor-capacitor
	radio configuration
R-DCCH	reverse dedicated control channel
RF	radio frequency
R-FCH	reverse fundamental channel
RMS	root mean square
RoTT	rise-over-thermal threshold
R-PCSCH	reverse power control subchannel
RS	rate set
R-SCH	reverse supplementary channel
RTT	radio transmission technology
Rx	receiver
RZ	return-to-zero
SACCH	slow associated control channel
SCCH	supplemental code channel
SCH	supplementary channel
SDMA	space-division multiple access
SF	spreading factor
SIR	signal-to-interference ratio
SNIR	signal-to-noise-and-interference ratio
SNR	signal-to-noise ratio
SOFDM	scalable OFDM

SQNR signal-to-quantization-noise ratio
TC turbo coder
TDM time-division multiplexing
TDMA time-division multiple access
TIA Telecommunications Industry Association
TTI TDM time interval
TTL transistor-transistor logic
Tx transmitter
UE user element
UHF ultra-high frequency
UMB Ultra-Mobile Broadband
UMTS Universal Mobile Telecommunications Service (System)
USDC U.S. Digital Cellular system
UTRA Universal Terrestrial Radio Access
VBR variable bit rate
VLR visitor location register
VoIP voice over IP
VSELPC vector-sum excited linear predictive coder
W-CDMA Wideband Code-Division Multiple Access
WiFi Wireless Fidelity
WiMAX Worldwide Interoperability for Microwave Access

Index

1×EV-DO (1× evolution for data only) Rev. 0. *See* DO (data only) standard.
1×EV-DV (1× evolution for data and voice). *See* DV (data and voice) standard.
1G (first generation), 399–400
1-persistent CSMA, 330–331
1×RTT (1×radio transmission technology), 421–426
2G (second generation), 400–405
2.5G (second generation)
  CNIR (carrier-to-noise-and-interference ratio), 407
  EDGE (Enhanced Data for Global Evolution), 410–411
  GPRS (General Packet Radio Service), 408–410
  HSCSD (High-Speed Circuit-Switched Data), 407–408
  puncturing, 408
  RACH (reverse access channel), 409
2B1Q line code, 207
3G (third generation)
  overview, 405–407
  OVSF (orthogonal variable spreading factor), 413
  PPSDN (public packet-switched data network), 411–412
  W-CDMA (Wideband Code-Division Multiple Access), 411–414
3GPP (Third Generation Partnership Project), 406
3GPP2 (Third Generation Partnership Project 2), 406
8-ary phase-shift keying (8PSK), 279, 410
8PSK (8-ary phase-shift keying), 279, 410
16-QAM modulation, 279
64-QAM modulation, 279

A

A priori probabilities, 209
Access channels, 291
Access network (AN), 427
Access terminal (AT), 427–431
ACK (acknowledgment), 417
Active neighbor list, 404
Active set, 404
Adaptive DPCM, 371
Additive white Gaussian noise (AWGN), 204
Adequate received signal level, 133
Adjacent-channel interference, 171–173, 296
Advanced Mobile Phone Service (AMPS), 12, 297
Air interface. *See* RF links.
A-law compression, 367
Aliasing, 235, 358
Allocating frequency bands. *See* Spectrum allocation.
Allocation, 64
All-zero path, 388
Aloha protocol
  backoff procedure, 327
  efficiency, estimating, 327–328
  overview, 326
  slotted Aloha, 328–330
AM broadcasting
  adjacent-channel interference, 296
  AMPS (Advanced Mobile Phone Service), 297
  channel spacing, 298–299
  clear channels, 296
  cochannel interference, 296
  duplexers, 300
  efficiency, 298–299
  frequency-division duplexing, 299–300
  frequency-division multiplexing, 296–297
  full-duplex, 299–300
  grouping channels, 297
  groups, 297
  half-duplex, 299–300
  jumbo groups, 297
  mastergroups, 297
  overview, 296
  signal distance, 296
  single full-duplex, 300
  SIR (signal-to-interference ratio),298–299
  spectral shaping, 298–299
  supergroups, 297
  for telephone networks, 296–298

Amplifiers
designing, 65–66
LNA (low-noise amplifier), 56–57
power spectrum, 49
source of thermal noise, 44–45
AMPS (Advanced Mobile Phone Service), 12, 297
AN (access network), 427
Analog filters, 171
Analog signals, converting to digital. *See* PCM (pulse code modulation).
Angle-modulated signals, 254
ANSI (American National Standards Institute), 401
ANSI/TIA/EIA-136, 401
Antennas
adaptable, 438–439
beamwidth
azimuth plane, 25
definition, 25
effective aperture, relation to, 26–27
elevation plane, 25
first null-to-null, 26
and gain, 26–27
half-power, 26
physical dimensions, 25–26
planes, 25
definition, 4
dipole, 24–25
directional, sectoring, 175–179
dish, 26
diversity gain, 439
effective aperture. *See also* Range equation.
beamwidth, relation to, 26–27
calculating, example, 27–28
gain, relation to, 27
power factors, 21–22
efficiency, and physical dimensions, 19
EIRP (effective isotropic radiated power), 29, 65–66. *See also* ERP (effective radiated power).
ERP (effective radiated power), 66. *See also* EIRP (effective isotropic radiated power).
gain. *See also* Range equation.
and beamwidth, 26–27
calculating, example, 28
definition, 26
effective aperture, relation to, 27
increasing power, 28
maximum range, example, 33–34
reciprocity, 28
ground reflections, 79–86
half-wave dipole. *See* Dipole.
height, propagation modeling, 79–86
isotropic, 20–22
maximum range, example, 33–34. *See also* Range equation.
optimizing systems, 66
path loss, 10. *See also* Range equation.
path-loss exponent, 32
power factors
beamwidth, 25–27
distance between receiver and transmitter, 21–22
effective aperture, 21–22
gain, 26–27
radiation patterns
azimuth plane, 25
bandwidth planes, 25
beamwidth, 25–27
dipole antennas, 24–25
elevation plane, 25
far-field radiation region, 22
Fraunhofer region, 22
main beam, 22–23
main lobe, 22–23
plot of, 24
power pattern, 22
side lobes, 22–23
range, calculating, 28–34
range equation, 28–34
receiver sensitivity, 30
receiving, 19
size, relation to efficiency, 8
spatially multiplexing, 439
system loss, 31
transmitting, 19
Antialiasing filter, 358
ARQ (automatic repeat request), 376–377
AT (access terminal), 427–431
Autocorrelation, 217–220
Available gain, 47–49
Average call duration, 189
Average holding time, 189

Average power, 37–41
AWGN (additive white Gaussian noise), 204
Azimuth plane, 25

B

Backoff procedure, 327
Bandlimited sources, 6–7
Bandpass filter (BPF), 237
Bandwidth
  carrier-based signaling
    aliasing, 235
    definition, 229
    frequency-domain effect, 231
    minimum, calculating, 231–233
    Nyquist bandwidth, 233–235
    Nyquist equivalent spectrum, 232–235
    raised-cosine pulses, 234
    rolloff parameter, 234
    sampling theorems, 235
  carrier-based signaling modulation, 229–235
  definition, 229
  frequency-domain effect, 231
  minimum, calculating, 231
  Nyquist bandwidth, 233–235
  Nyquist equivalent spectrum, 232–235
  sharing. *See* Multiple user channels.
Bandwidth planes, 25
Base station (BS), 153–155
Baseband signaling. *See also* Carrier-based signaling; Spread-spectrum signaling.
  2B1Q line code, 207
  architecture, 204–207
  autocorrelation, 217–220
  BER (bit error rate), 208
  block diagram, 205
  convolution, 209, 219
  correlation, 216–220
  correlation receiver, 220–222
  correlators, 216–220, 222
  decision statistic, 208
  filters, 208
  lag, 216–220
  line code, 205
  Manchester code, 206
  matched filters, 212–215, 222
  NRZ (non-return-to-zero) line code, 205–206
  optimum threshold value, 210
  a priori probabilities, 209
  probability of error, 208–211
  probability of error, and received power, 223–226
  pulse detection, 207–212
  receiver performance, 222–226
  RZ (return-to-zero) line code, 206
  samplers, 208
  SNR (signal-to-noise ratio), 223–226
  symbol period, 205
  threshold comparators, 208
  waterfall curves, 223–226
  Wiener-Khinchine theorem, 217–218
Baseband signals, 7
Beamwidth
  and antenna physical dimensions, 25–26
  azimuth plane, 25
  definition, 25
  effective aperture, relation to, 26–27
  elevation plane, 25
  first null-to-null, 26
  and gain, 26–27
  half-power, 26
  planes, 25
BER (bit error rate), 61–62, 208, 351
BFSK (binary frequency-shift keying), 254–261
BLER (block error rate), 416
Blocked calls, 188, 453
Blocked calls cleared model, 190–191, 453
Blocked calls delayed model, 190–191, 453
Blocking probability, 460–461
Boundary coverage, 133
BPF (bandpass filter), 237
BPSK (binary phase-shift keying), 236–239
Brick wall filter, 171
BS (base station), 153–155
Burst error, 402
Burstiness, 351
Bursting *vs.* streaming, 354–355
Butterworth filter, 50

C

Call arrival model, 454–455
Call response, 292–294
Calls
  average duration, 189
  average holding time, 189
  blocked, 188, 453
  blocked calls cleared model, 190–191, 453
  blocked calls delayed model, 190–191, 453
  initiating, 292–293
Carried load, 190
Carried traffic intensity, 190
Carrier frequency
  path loss predictions, 101
  propagation modeling. *See* Hata model.
Carrier-based signaling. *See also* Baseband signaling; Spread-spectrum signaling.
  bandwidth
    aliasing, 235
    definition, 229
    frequency-domain effect, 231
    minimum, calculating, 231–233
    Nyquist bandwidth, 233–235
    Nyquist equivalent spectrum, 232–235
    raised-cosine pulses, 234
    rolloff parameter, 234
    sampling theorems, 235
  basic blocks, 227–228
  block diagram, 228
  channels, 227
  modulation
    8PSK (8-ary phase-shift keying), 279
    16-QAM, 279
    64-QAM, 279
    angle-modulated signals, 254
    bandwidth, 229–235
    BFSK (binary frequency-shift keying), 254–261
    BPSK (binary phase-shift keying), 236–239
    Carson's rule, 257
    coherent systems, 238
    continuous-phase FSK, 254–256
    definition, 226
    DPSK (differential binary phase-shift keying), 239–243
    DQPSK (differential quadrature phase-shift keying), 289
    frequency sensitivity, 255–256
    FSK (frequency-shift keying), 254–261
    Gaussian filters, 262–264
    GFSK (Gaussian frequency-shift keying), 262–264
    GMSK (Gaussian minimum-shift keying), 267
    in-phase components, 238–239
    instantaneous frequency, 254–255
    keying, 229
    linear, 229
    modulated carrier architecture, 227–228
    modulation index, 256–257
    MSK (minimum-shift keying), 264–267
    noncoherent systems, 238
    nonlinear, 229
    OQPSK (offset quadrature phase-shift keying), 251–254
    orthogonal signals, 260–261
    overview, 226–227
    peak frequency deviation, 256–257
    phase-locked receivers, 238
    principles, 229–235
    PSK (phase-shift keying), 229, 244
    QAM (quadrature amplitude modulation), 279
    QPSK (quadrature phase-shift keying), 243–251
    quadrature components, 238–239
    raised-cosine pulse, 237
    root-raised-cosine pulse, 237
    spectral efficiency, 237
  receiver architecture, 227–228
  recovering information, 227
  transmitter architecture, 227–228
Carriers, 8, 19
Carrier-sense multiple access (CSMA), 330–335
Carrier-sense multiple access with collision detection (CSMA/CD), 335
Carrier-to-interference ratio (CIR), 407, 415–416
Carrier-to-noise-and-interference ratio (CNIR), 407

Carson's rule, 257
CDM (code-division multiplexing), 415
CDMA (code-division multiple access)
collisions, 308–311
DSSS (direct-sequence spread spectrum)
Hadamard matrices, 317
multiple users, 315–316
nearly orthogonal spreading codes. *See* PN (pseudonoise) spreading codes.
orthogonal spreading codes, 316–318
overview, 311–315
PN (pseudonoise) spreading codes, 318–325
SNIR (signal-to-noise-and-interference ratio), 320–322
Walsh functions, 317
FHSS (frequency-hopping spread spectrum), 307–311
handoffs, 186–187
history of, 12
hits, 308–311
maximum subscribers, 311, 323–234
overview, 306–307
reuse factor, 401
spread-spectrum modulation, 306–307
cdma2000 system
history of cellular systems, 289
standards. *See* RTT (radio transmission technology).
cdmaOne system, 289, 402–405
CDVCC (coded digital verification color code), 303
Cell layout
available channels, calculating, 163
cell splitting, 179–183
circular regions, 157
clusters of cells
definition, 163
minimum size, calculating, 165–166,169
minimum size, limitations, 167
number of, determining, 165–166
reuse distance, 163–164
size, trade-offs, 173–175
contiguous cells. *See* Clusters.
hexagonal grid, 158–160
most efficient pattern, 158
reuse distance, 163–164
separation distance, calculating, 161–163
square grid, 158–160
triangular grid, 158–160
Cells
clusters
definition, 163
minimum size, calculating, 165–166, 169
minimum size, limitations, 167
number of, determining, 165–166
reuse distance, 163–164
size, trade-offs, 173–175
contiguous. *See* Clusters.
definition, 153
radius
reducing, 181–183
trade-offs, 173–175
splitting, 179–183
Cellular concept, history of, 11–12
Cellular digital packet data, 335
Central office, 187
CEPT (Conference of European Postal and Telecommunications Administrations), 400
CFR (Code of Federal Regulations), 150
Channel quality indicator (CQI), 416
Channel sets, 155–156
Channels
access methods
access channels, 291
call initiation, 292–293
call response, 292–294
code division. *See* CDMA (code-division multiple access).
control channels, 291
frequency division. *See* FDMA (frequency-division multiple access).
overview, 290–294
paging channels, 291
reverse control channels, 291
sync channels, 291
time division. *See* TDMA (time-division multiple access).
available
calculating, 152, 163
and number of subscribers, 152–153
carrier-based signaling, 227
cellular system, diagram, 291

Channels (*Continued*)
  definition, 4, 19, 291
  dividing by
    code. *See* CDMA (code-division multiple access).
    frequency. *See* FDMA (frequency-division multiple access).
    time. *See* TDMA (time-division multiple access).
  dynamic assignment, 185
  grouping, 297
  jumbo groups, 297
  mastergroups, 297
  multiple user access. *See* Multiple user channels.
  organizing into groups. *See* Channel sets.
  partitioning, 173
  selection, history of, 11
  spacing, AM broadcasting, 298–299
  supergroups, 297
Checksums, 376
Chips, spread-spectrum signaling, 271
CIR (carrier-to-interference ratio), 407, 415–416
Circular cell regions, 157
Clear channels, 296
Clusters of cells
  definition, 163
  minimum size, calculating, 165–166, 169
  minimum size, limitations, 167
  number of, determining, 165–166
  reuse distance, 163–164
  size, trade-offs, 173–175
CNIR (carrier-to-noise-and-interference ratio), 407
Cochannel interference
  AM broadcasting, 296
  cluster size, effect on, 167
  definition, 153
  predicting, 167–171
Code of Federal Regulations (CFR), 150
Codecs, 366
Coded digital verification color code (CDVCC), 303
Code-division multiple access (CDMA). *See* CDMA (code-division multiple access).
Code-division multiplexing (CDM), 415
Coding for error correction
  ARQ (automatic repeat request), 376–377
  checksums, 376
  convolutional codes. *See also* Forward error correction.
    all-zero path, 388
    constraint length, 378
    decoding, 380–383
    definition, 377
    encoding, 378–379
    Hamming distance, 381, 388
    maximum likelihood rule, 382–383
    minimum free distance, 388
    path metrics, 382
    performance, 387–389
    trellis diagrams, 379–380
    Viterbi algorithm, 383–387
  forward error correction, 376–377. *See also* Convolutional codes.
  parity check, 376
  triple redundancy, 377
Coding scheme (CS), 409
Coherence bandwidth, 101, 115–120
Coherence time, 101, 121
Coherent systems, 238
Collision probability
  calculating, 328–330
  reducing. *See* CSMA (carrier-sense multiple access).
Collisions, CDMA, 308–311
Compandors, 366
Components, 13
Compression
  A-law, 367
  codecs, 368
  compandors, 368
  compression/expansion, 364–366
  compressors, 364
  expanders, 364
  images, 350
  μ-law, 367
  quantization, 364–366
Compressors, 364
Conference of European Postal and Telecommunications Administrations (CEPT), 400
Connection layers, 398

Constraint length, 378
Contention-based methods
1-persistent CSMA, 330–331
Aloha protocol, 326–328
backoff procedure, 327
collision probability, calculating, 328–330
CSMA (carrier-sense multiple access), 330–335
nonpersistent CSMA, 330–331
overview, 325–326
packets, 326
packet-switching, 326
p-persistent CSMA, 331
PRMA (packet-reservation multiple access), 331
random numbers, 327
slotted Aloha, 328–330
streaming *vs.* bursting, 325–326
Contiguous cells. *See* Clusters.
Continuous-phase FSK, 254–256
Control channels, 291
Convolution, 209, 219
Convolutional codes
all-zero path, 388
constraint length, 378
decoding, 380–383
definition, 377
encoding, 378–379
Hamming distance, 381, 388
maximum likelihood rule, 382–383
minimum free distance, 388
path metrics, 382
performance, 387–389
trellis diagrams, 379–380
Viterbi algorithm, 383–387
Correction for
antenna height, 92–93
errors. *See* Coding for error correction.
terrain, 93–95
Correlation, 216–220
Correlation receiver, 220–222
Correlators, 216–220, 222
Coverage area
adequate received signal level, 133
boundary coverage, 133
calculating, 134–137
definition, 133
designing, 173–175
separation distance (geographic), 156
CQI (channel quality indicator), 416
CRC (cyclic redundancy check), 402
CS (coding scheme), 409
CSMA (carrier-sense multiple access), 330–335
CSMA/CD (carrier-sense multiple access with collision detection), 335
Cumulative probability distribution function, 107–108

D

D-AMPS (Digital Advanced Mobile Phone Service), 401
Data
BER (bit error rate), 351
burstiness, 351
delays, 352
QoS (quality of service), 353–354
source characteristics, 351
subscriber requirements, 351–352
throughput, 351
Data and voice (DV) standard, 425–426
Data only (DO) standard
Rev. 0, 425–432
Rev. A, 441
Data rate (DR), 414
Data rate control (DRC), 427–432
Data transmission
BER (bit error rate), 351
burstiness, 351
delays, 352
source characteristics, 351
subscriber requirements, 351–352
throughput, 351
Datagrams, 354
Data-sense multiple access, 335
dBm (dB relative to 1 mW), 31–33
dBW (dB relative to 1 W), 31
DCCH (dedicated control channel), 422–423
Decision levels, 359–361
Decision statistic, 208
Delay budget, speech sources, 347
Delay jitter, 347, 350–351

Delays
  data transmission, 352
  speech sources, 347–348
  video conferencing, 350–351
Delta modulation, 369–371
Demodulation, 7–9. *See also* Modulation.
Designing systems
  cell radius
    reducing, 181–183
    trade-offs, 173–175
  cell splitting, 179–183
  cluster size, 173–175
  directional antennas, 175–179
  dynamic channel assignment, 185
  FDM (frequency-division multiplexing), 151
  FDMA (frequency-division multiple access), 151
  frequency band allocation. *See* Spectrum allocation.
  geographic coverage, 173–175
  GOS (grade of service), 187–194
  handoffs
    CDMA (code-division multiple access), 186–187
    mobile assisted, 186
    overview, 185–187
    process description, 186–187
    purpose of, 185
  MSC (mobile switching center), 184
  MTSO (mobile telephone switching office), 184
  PSTN (public switched telephone network), 184
  for QoS (quality of service). *See* Optimizing systems.
  requirements assessment, 150–153
  sectoring, 175–179, 193–194
  subscriber density, 173–175
  system architecture, 150–153
  trade-offs, 173–175
  traffic engineering, 187–194
  trunking, 187–194
Development processes, 14–15
DFT (discrete Fourier transform), 428
Differential binary phase-shift keying (DPSK), 239–243
Differential PCM (DPCM). *See* DPCM (differential PCM).
Differential quadrature phase-shift keying (DQPSK), 289
Digital Advanced Mobile Phone Service (D-AMPS), 401
Digital signal processing (DSP), 8
Digital signaling. *See* Baseband signaling; Carrier-based signaling; Spread-spectrum signaling.
Digital signals, converting from analog. *See* PCM (pulse code modulation).
Digital traffic channel (DTC), 303
Digitizing speech. *See also* PCM (pulse code modulation).
  formants, 371
  LPC (linear predictive coder), 373–376
  model of human voice, 371–373
  overview, 355–356
  source coding. *See* LPC (linear predictive coder); Vocoders.
  vocoders, 371–376
  VSELPC (vector-sum excited linear predictive coder), 375
  waveform coding. *See* PCM (pulse code modulation).
Dipole antennas, 24–25
Directional antennas, 175–179
Direct-sequence spread spectrum (DSSS). *See* DSSS (direct-sequence spread spectrum).
Dish antennas, 26
Distortion, aliasing, 235, 358
Diversity gain, 404
DO (data only) standard
  Rev. 0, 425–432
  Rev. A, 441
Documentation, system-level, 15
Doppler spread, 121–123
Downlink, 66–70
DPCM (differential PCM)
  adaptive, 371
  delta modulation, 369–371
  performance, 371
  sampling, 367–369
  slope overload noise, 370
  vs. LPC (linear predictive coder), 374
DPSK (differential binary phase-shift keying), 239–243
DQPSK (differential quadrature phase-shift keying), 289
DR (data rate), 414

DRC (data rate control), 427–432
DSP (digital signal processing), 8
DSSS (direct-sequence spread spectrum)
  Hadamard matrices, 317
  multiple users, 315–316
  narrowband interference, 275–278
  orthogonal spreading codes, 316–318
  overview, 311–315
  PN (pseudonoise) spreading codes, 318–325
  spread-spectrum signaling, 271–278
  Walsh functions, 317
DTC (digital traffic channel), 303
DTMF (dual-tone multiple frequency), 303
Duplexers, 300
DV (data and voice) standard, 425–426
Dynamic channel assignment, 185
Dynamic range, 347, 364
  speech sources, 347

E

ECSD (Enhanced Circuit Switched Data), 410
EDGE (Enhanced Data for Global Evolution), 410–411
Effective aperture, antennas
  beamwidth, relation to, 26–27
  calculating, example, 27–28
  gain, relation to, 27
  power factors, 21–22
Effective input-noise temperature
  calculating, example, 58–61
  calculating output-noise power spectrum, 57
  definition, 51
  vs. noise figure, 54–55
EGPRS (Enhanced General Packet Radio Service), 410
8-ary phase-shift keying (8PSK), 279, 410
8PSK (8-ary phase-shift keying), 279, 410
EIRP (effective isotropic radiated power), 29, 65–66
Elements, 13
Elevation plane, 25
Empirical models, 86–95
End users. *See* Subscribers.
Endpoints, 6
Engineering systems. *See* Designing systems.
Environmental factors, common characteristics, 5
Equalization, 118
Erlang, A. K., 189, 461
Erlang B formula, 460–465
Erlang B table, 465–475
Erlang formula, 189
erlangs, 189, 461
ERP (effective radiated power), 66
Error correction. *See* Coding for error correction.
Error function, 98
ETACS (European Total Access Communications System), 399
ETSI (European Telecommunications Standards Institute), 400
Expanders, 364
Exponential probability distribution, 112

F

FACCH (fast associated control channel), 303
Fade margin, 100
Fading
  link budget, 137–139
  propagation modeling
    coherence bandwidth, 101, 115–120
    coherence time, 101
    equalization, 118
    fast, 127–128, 131–132
    flat, 104, 118
    frequency changes, 104
    frequency-selective, 104, 118
    large-scale, 100
    log-normal, 100
    macro-scale, 100
    macroscopic, 100
    micro-scale, 100–106
    microscopic, 100
    multipath propagation, 100
    Rayleigh fading, 101
    RMS (root mean square) delay spread, 101, 118–120
    slow, 127–128, 131–132
    small-scale, 100
    two-ray model, stationary receiver, 102–106
Far-field radiation region, 22

Fast fading, 127–128, 131–132
Fast frequency hoppers, 269
FCC (Federal Communications Commission)
communications regulations, 150
creation date, 150
definition, 8
history of cellular systems, 11
spectrum allocation, 179
Web site, 296
FCH (fundamental channel), 421
FDD (frequency-division duplex), 299–300, 400
FDM (frequency-division multiplexing), 151, 296–297
FDMA (frequency-division multiple access), 151, 295
FEC (forward error correction), 402, 407–409
FFT (fast Fourier transform), 438
FHSS (frequency-hopping spread spectrum), 268–270, 307–311
Filters
analog, 171
antialiasing, 358
baseband pulse detection, 208
baseband signaling, 208
BPF (bandpass filter), 237
brick wall, 171
Butterworth, 50
Gaussian, 262–264
interference, 171–173
LPF (lowpass filter), 237
matched, 212–215, 222
selective bandpass filter, 171
Fingers, 404
First generation (1G), 399–400
First null-to-null, 26
Flat fading, 104, 118
Flicker, video, 350
FM (frequency modulation), 348
Formants, 371
Forward error correction, 376–377
Forward error correction (FEC), 402, 407–409
Forward link, 417, 421
FPLMTS (Future Public Land Mobile Telecommunications System), 405
FPR (functional product requirements), 15
Frame preamble, 301
Frames, TDMA, 301
Fraunhofer region, 22
Free-space loss. *See* Path loss.
Frequency
fading, 104
modulation, 255–256
relation to wavelength, 8
reuse ratio
cluster size, 174–175
definition, 155
SIR (signal-to-interference ratio), 169
sensitivity, 255–256
Frequency bands
allocating. *See* Spectrum allocation.
laws and regulations, 150
Frequency modulation (FM), 348
Frequency-division duplex (FDD), 299–300, 400
Frequency-division multiple access (FDMA), 151, 295
Frequency-division multiplexing (FDM), 151, 296–297
Frequency-hopping spread spectrum (FHSS), 268–270, 307–311
Frequency-selective fading, 104, 118
FSK (frequency-shift keying), 254–261
Full-duplex, 299–300
Full-scale range quantizer, 361
Function tables, 446–451
Functional product requirements (FPR), 15
Fundamental channel (FCH), 421
Future Public Land Mobile Telecommunications System (FPLMTS), 405

G

Gain, antennas
and beamwidth, 26–27
calculating, example, 28
definition, 26
effective aperture, relation to, 27
increasing power, 28
maximum range, example, 33–34
reciprocity, 28
Gaussian filters, 262–264
Gaussian frequency-shift keying (GFSK), 262–264
Gaussian minimum-shift keying (GMSK), 267

General Packet Radio Service (GPRS), 408–410
Generational descriptions. *See also* History of cellular systems.
  1G (first generation), 399–400
  2G (second generation), 400–405
  2.5G (second generation)
    CNIR (carrier-to-noise-and-interference ratio), 407
    EDGE (Enhanced Data for Global Evolution), 410–411
    GPRS (General Packet Radio Service), 408–410
    HSCSD (High-Speed Circuit-Switched Data), 407–408
    puncturing, 408
    RACH (reverse access channel), 409
  3G (third generation). *See also* HSPA (High-Speed Packet Access).
    overview, 405–407
    OVSF (orthogonal variable spreading factor), 413
    PPSDN (public packet-switched data network), 411–412
    W-CDMA (Wideband Code-Division Multiple Access), 411–414
  3GPP (Third Generation Partnership Project), 406
  3GPP2 (Third Generation Partnership Project 2), 406
  4G (fourth generation). *See* OFDM (orthogonal frequency-division multiplexing).
Geographic coverage. *See* Coverage area.
Geographic location of subscribers, 156
GFSK (Gaussian frequency-shift keying), 262–264
GMSK (Gaussian minimum-shift keying), 267
GMSK modulation, 289–290
Gold codes, 413
GOS (grade of service). *See also* QoS (quality of service).
  average call duration, 189
  average holding time, 189
  blocked calls, 188
  blocked calls cleared model, 190–191
  blocked calls delayed model, 190–191
  carried load, 190
  carried traffic intensity, 190
  central office, 187
  definition, 188–189
  offered load, 189
  offered traffic intensity, 189
  Poisson model, 190
  sectoring, 193–194
  subscriber calling habits, 189
  throughput, 190
  traffic engineering, 453–454
  trunking, 187–194
  trunking efficiency, 192–194
  trunking theory, 188
  trunks, 188
GPRS (General Packet Radio Service), 408–410
Ground reflections, 79–86
Groups
  AM broadcasting channels, 297
  cells. *See* Clusters.
  channel sets, 155–156
  jumbo groups, 297
  mastergroups, 297
  supergroups, 297
GSM (Global System for Mobile Communications)
  history of, 12, 289–290
  overview, 304–305
GSM (Groupe Spéciale Mobile), 400
Guard interval, 301–302

H

Hadamard matrices, 317
Half-duplex, 299–300, 306
Half-power beamwidth, 26
Half-wave dipole antenna. *See* Dipole antennas.
Hamming distance, 381, 388
Handoffs
  CDMA (code-division multiple access), 186–187
  definition, 153
  hard, 404
  mobile assisted, 186
  overview, 185–187
  process description, 186–187
  purpose of, 185
  soft, 404

Handovers. *See* Handoffs.
Hard handoffs, 404
H-ARQ (hybrid automatic repeat request), 417, 430
Hata model
link budget, 137–139
path loss predictions, 87–90, 95, 98–100
HDR (high data rate), 425
Hertz, Heinrich, 226
Hexagonal grid, 158–160
Hidden stations, 335
High-Speed Circuit-Switched Data (HSCSD), 407–408
High-Speed Downlink Packet Access (HSDPA), 414
High-Speed Packet Access (HSPA). *See* HSPA (High-Speed Packet Access).
High-Speed Uplink Packet Access (HSUPA), 414, 418–419
History of cellular systems. *See also* Generational descriptions.
access methods, 289–290
AMPS (Advanced Mobile Phone Service), 12
automatic channel selection, 11
CDMA (code-division multiple access), 12
cdma2000 system, 289
cdmaOne system, 289, 402–405
cellular concept, 11–12
CEPT (Conference of European Postal and Telecommunications Administrations), 400
D-AMPS (Digital Advanced Mobile Phone Service), 401
ETACS (European Total Access Communications System), 399
ETSI (European Telecommunications Standards Institute), 400
evolution of radio, 10
FDD (frequency-division duplex), 400
GMSK modulation, 289–290
GSM (Global System for Mobile Communications), 12, 289–290
GSM (Groupe Spéciale Mobile), 400
IMTS (Improved Mobile Telephone Service), 152
IS-95, 289
mobile telephones, 10–11
Morse code, 226
multiple user channels, 289–290
NA-TDMA (North American Time-Division Multiple Access), 401
QPSK modulation, 289–290
Qualcomm, Inc., 12, 289–290, 401
radio, 226
roaming automation, 400
spectrum shortages, 12
spread-spectrum techniques, 12, 401
telegraphy, 226
UMTS (Universal Mobile Telecommunications System), 289
USDC (U.S. Digital Cellular) system, 289–290
W-CDMA (Wideband Code-Division Multiple Access), 289
Hits. *See* Collisions.
HLR (home location register), 441
Hold time model, 456
Hopping away from interference, 270
HSCSD (High-Speed Circuit-Switched Data), 407–408
HSDPA (High-Speed Downlink Packet Access), 414
HSPA (High-Speed Packet Access). *See also* RTT (radio transmission technology).
BLER (block error rate), 416
CQI (channel quality indicator), 416
forward link, 417
H-ARQ (hybrid automatic repeat request), 417
HSDPA (High-Speed Downlink Packet Access), 414
HSUPA (High-Speed Uplink Packet Access), 414, 418–419
PER (packet error rate), 416
protocol enhancements, 415
RoTT (rise-over-thermal threshold), 418
scheduler, 417
UEs (user elements), 416
HSUPA (High-Speed Uplink Packet Access), 414, 418–419
Human voice, model of, 371–373
Hybrid automatic repeat request (H-ARQ), 417, 430

I

IDFT (inverse discrete Fourier transform), 428
IF (intermediate frequency), 56
IFFT (inverse fast Fourier transform), 438
Images
  QoS (quality of service), 353
  source characteristics, 349
  subscriber requirements, 349
IMT-2000 (International Mobile Telecommunications 2000), 405
IMTS (Improved Mobile Telephone Service), 152
Incremental redundancy (IR), 410
Information sinks. *See* Sinks.
Information sources. *See* Sources of information.
In-phase components, 238–239
Input-noise temperature. *See* Effective input-noise temperature.
Instantaneous frequency, 254–255
Integrated Services Digital Network (ISDN), 207, 355
Interference. *See also* Noise.
  adjacent-channel, 295–296
  AM broadcasting, 296, 298–299
  FDMA (frequency-division multiple access), 295
  hopping away from, 270
  spreading. *See* Spread-spectrum signaling.
Interference prediction
  adjacent-channel interference, 171–173
  analog filters, 171
  brick wall filter, 171
  channel partitioning, 173
  channel sets, 155–156
  cochannel interference, 167–171
  customer density management, 156
  filters, 171–173
  frequency reuse ratio
    cluster size, 174–175
    definition, 155
    SIR (signal-to-interference ratio), 169
  geographic location of subscribers, 156
  interference-limited systems, 154
  noise-limited systems, 154
  power control, 173
  selective bandpass filter, 171
  separation distance (geographic), 156
  sharp cutoff filters, 171
  SIR (signal-to-interference ratio), 155, 169
  SNIR (signal-to-noise-and-interference ratio), 154
Interference reduction factor. *See* Frequency, reuse ratio.
Interference-limited systems, 153–154
Interlaced scanning, 350
Interleaved scanning, 350
Intermediate frequency (IF), 56
Inverse discrete Fourier transform (IDFT), 428
Inverse fast Fourier transform (IFFT), 438
IR (incremental redundancy), 410
IS (interim standard), 400
IS-95 standard, 12, 289
IS-95A standard, 401
IS-95B standard, 401
ISDN (Integrated Services Digital Network), 207, 355
ISI (intersymbol interference), 433
Isotropic antennas, 20–22
Isotropic radiation, 20–22
ITU (International Telecommunication Union), 405

J

Johnson, J. B., 38
Johnson noise, 38–39
Joint probability density function, 107
Jumbo groups, 297

K

Keying
  8PSK (8-ary phase-shift keying), 279
  BFSK (binary frequency-shift keying), 254–261
  BPSK (binary phase-shift keying), 236–239
  definition, 229
  DPSK (differential binary phase-shift keying), 239–243
  DQPSK (differential quadrature phase-shift keying), 289
  FSK (frequency-shift keying), 254–261
  GFSK (Gaussian frequency-shift keying), 262–264
  GMSK (Gaussian minimum-shift keying), 267
  MSK (minimum-shift keying), 264–267
  OQPSK (offset quadrature phase-shift keying), 251–254

Keying (*Continued*)
PSK (phase-shift keying), 229, 244
QPSK (quadrature phase-shift keying), 243–251
Kirchhoff's laws, 19

L

Lag, 216–220
Large-scale fading, 100
Laws and regulations, 8, 150. *See also* Standards.
Layers, 397–398
Lee model, 90–95
LEO (low Earth orbit), 405
Line codes, 205
Line drivers, 9
Linear modulation, 229
Linear predictive coder (LPC). *See* LPC (linear predictive coder).
Link budget
fading, 137–139
Hata model, 137–139
path loss, 137–139
radio links, 66–70
Links. *See* RF links.
LNA (low-noise amplifier), 56–57
Log-normal fading, 100
Log-normal shadowing, 95–100
Lossy transmission lines, 57–58
LPC (linear predictive coder)
output of, 374
overview, 373–374
vs. DPCM (differential PCM), 374
VSELPC (vector-sum excited linear predictive coder), 375
LPF (lowpass filter), 237
LTE (long-term evolution), 440
Lumped-element analysis, 19

M

MAC (medium access control), 397–398
MacDonald, V. H., 159
MAC-ID, 427–432
Macro diversity, 404
Macro-scale fading, 100
Macroscopic fading, 100
Main beam, 22–23
Main lobe, 22–23
Manchester code, 206
Marconi, Guglielmo, 226
Mastergroups, 297
Matched filters, 212–215, 222
Maximum likelihood rule, 382–383
Maxwell's equations, 18–19
MCS (modulation and coding scheme), 410
Mean-square quantization noise, 363
Median signal attenuation, calculating, 87
Micro-scale fading, 100–106
Microscopic fading, 100
Midriser quantizer, 360–361
Midtread quantizer, 360–361
MIMO (multiple input multiple output), 438
Minimum free distance, 388
Minimum-shift keying (MSK), 264–267
μ-law compression, 367
Mobile switching center (MSC), 184
Mobile telephone switching office (MTSO), 184
Mobile telephones, history of, 10–11
Mobile unit (MU), 196
Mobile-assisted handoffs, 186
Modulated carrier architecture, 227–228
Modulation. *See also* Demodulation.
in block diagrams, 7–9
carrier-based signaling
8PSK, 279
16-QAM, 279
64-QAM, 279
angle-modulated signals, 254
bandwidth, 229–235
BFSK (binary frequency-shift keying), 254–261
BPSK (binary phase-shift keying), 236–239
Carson's rule, 257
coherent systems, 238
continuous-phase FSK, 254–256
definition, 226
DPSK (differential binary phase-shift keying), 239–243
frequency sensitivity, 255–256
FSK (frequency-shift keying), 254–261
Gaussian filters, 262–264
GFSK (Gaussian frequency-shift keying), 262–264
GMSK (Gaussian minimum-shift keying), 267

in-phase components, 238–239
instantaneous frequency, 254–255
keying, 229
linear, 229
modulated carrier architecture, 227–228
modulation index, 256–257
MSK (minimum-shift keying), 264–267
noncoherent systems, 238
nonlinear, 229
OQPSK (offset quadrature phase-shift keying), 251–254
orthogonal signals, 260–261
overview, 226–227
peak frequency deviation, 256–257
phase-locked receivers, 238
principles, 229–235
PSK (phase-shift keying), 229, 244
QAM (quadrature amplitude modulation), 279
QPSK (quadrature phase-shift keying), 243–251
quadrature components, 238–239
raised-cosine pulse, 237
root-raised-cosine pulse, 237
signal constellation, 238–239
spectral efficiency, 237
converting analog to digital. *See* PCM (pulse code modulation).
DQPSK (differential quadrature phase-shift keying), 289
Modulation and coding scheme (MCS), 410
Modulation index, 256–257
Modulator, 9
Morse, Samuel B., 226
Morse code, 226
MSC (mobile switching center), 184
MSK (minimum-shift keying), 264–267
MTSO (mobile telephone switching office), 184
MU (mobile unit), 196
Multipath propagation
fading, 100–106
propagation modeling, 100–102
statistical models
coherence bandwidth, 115–120
cumulative probability distribution function, 107–108
exponential probability distribution, 112
joint probability density function, 107
probability density function, 108
probability distribution function, 108
Rayleigh density function, 109–113
Rayleigh fading, 106–113
region of integration, 109
Ricean probability distribution, 113–114
Multiple input multiple output (MIMO), 438
Multiple user channels. *See also* Subscriber density.
AM band. *See* AM broadcasting.
cellular digital packet data, 335
code division. *See* CSMA (carrier-sense multiple access).
contention-based methods
1-persistent CSMA, 330–331
Aloha protocol, 326–328
backoff procedure, 327
collision probability, calculating, 328–330
collision probability, reducing. *See* CSMA (carrier-sense multiple access).
CSMA (carrier-sense multiple access), 330–335
nonpersistent CSMA, 330–331
overview, 325–326
packets, 326
packet-switching, 326
p-persistent CSMA, 331
PRMA (packet-reservation multiple access), 331
random numbers, 327
slotted Aloha, 328–330
streaming *vs.* bursting, 325–326
CSMA/CD (carrier-sense multiple access with collision detection), 335
data-sense multiple access, 335
frequency division. *See* FDMA (frequency-division multiple access).
hidden stations, 335
history of, 289–290
multiplexing *vs.* multiple-access, 288
spread-spectrum-based code-division. *See* CDMA (code-division multiple access).
throughput, calculating, 334–335
time division. *See* TDMA (time-division multiple access).

Multiplexing *vs.* multiple-access, 288
Music
  QoS (quality of service), 353
  source characteristics, 348
  subscriber requirements, 348–349

N

NACK (negative acknowledgment), 410
Narrowband interference
  DSSS (direct-sequence spread spectrum), 275–278
  FHSS (frequency-hopping spread spectrum), 269–270
NA-TDMA (North American Time-Division Multiple Access), 401
Noise. *See also* Interference.
  bandwidth, 49–50
  definition, 34
  floor, 35
  relation to signal. *See* SNR (signal-to-noise ratio).
  system, 9
  temperature, 45–46
  from thermal motion of electrons. *See* Thermal noise.
Noise factor. *See* Noise figure.
Noise figure
  calculating, example, 58–61
  overview, 50–54
  vs. effective input-noise temperature, 54–55
Noise-limited systems, 152, 154
Noncoherent systems, 238
Nonlinear modulation, 229
Nonlinear quantization, 364–367
Nonpersistent CSMA, 330–331
Nonuniform quantization, 364
Normal distribution, 443–446
Normalized power. *See* Average power.
NRZ (non-return-to-zero) line code, 205–206
NTIA (National Telecommunications and Information Administration), 8
NTSC (National Television System Committee), 349
Null-to-null frequency interval, 105–106
Null-to-null time interval, 127
Nyquist bandwidth, 233–235
Nyquist equivalent spectrum, 232–235

O

OFDM (orthogonal frequency-division multiplexing)
  4G standards, 441–442
  4G system proposals, 440–441
  adaptable antennas, 438–439
  antenna systems, 438–439
  definition, 437
  diversity gain antennas, 439
  increasing data rates, 433–437
  MIMO (multiple input multiple output), 438
  OFDMA (orthogonal frequency-division multiple access), 437
  overview, 432–433
  spatially multiplexing antennas, 439
  trade-offs, 437–438
OFDMA (orthogonal frequency-division multiple access), 437
Offered load, 189, 460–461
Offered traffic intensity, 189
Offset quadrature phase-shift keying (OQPSK), 251–254
Okumura model, 87
1 × EV-DO (1× evolution for data only) Rev. 0. *See* DO (data only) standard.
1 × EV-DV (1× evolution for data and voice). *See* DV (data and voice) standard.
1G (first generation), 399–400
1-persistent CSMA, 330–331
1 × RTT (1× radio transmission technology), 421–426
Operating frequency, specifying, 64
Optimizing systems
  allocation, 64
  amplifier design, 65–66
  antennas, 66
  BER (bit error rate), 61–62
  design considerations, 64–66
  link budget, example, 66–70
  operating frequency, specifying, 64
  QoS (quality of service), subscriber perceptions, 61–62
  range equation, 63–66
  receiver sensitivity, 62–63
  requirement flowdown, 64
  SNR (signal-to-noise ratio), 62

system-level design, 61–62
top-level design, 63–66
OQPSK (offset quadrature phase-shift keying), 251–254
Orthogonal
definition, 317
frequency division. *See* OFDM (orthogonal frequency-division multiplexing).
signals, 260–261
spreading codes
overview, 316–318
OVSF (orthogonal variable spreading factor), 413
PN (pseudonoise), 318–325
Orthogonal frequency-division multiple access (OFDMA), 437
Orthogonal frequency-division multiplexing (OFDM). *See* OFDM (orthogonal frequency-division multiplexing).
Orthogonal variable spreading factor (OVSF), 413
OVSF (orthogonal variable spreading factor), 413

P

PA (power amplifier), 65
Pacific Digital Cellular System (PDC), 406
Packet error rate (PER), 416
Packet-reservation multiple access (PRMA), 331
Packets, 326
Packet-switched data network (PSDN), 398
Packet-switching, 326
Paging channels, 291
Parity check, 376
Partitioning channels, 173
Passive systems, source of thermal noise, 39–44
Path loss
definition, 10
link budget, 137–139
propagation modeling
carrier frequency, 101
correction for antenna height, 92–93
correction for terrain, 93–95
error function, 98
fade margin, 100
Hata model, 87–90, 95, 98–100
Lee model, 90–95
probability density function, 96–100
range of loss, 96–100
rural environment, 88
shadowing losses, 96–100
suburban environment, 88
transmitter distance, 96–100
urban environments, 88, 101
Path metrics, 382
Path-loss exponent, 32, 79
PCM (pulse code modulation)
aliasing, 358
antialiasing filter, 358
coding, 361
decision levels, 359–361
distortion, 358
DPCM (differential PCM)
adaptive, 371
delta modulation, 369–371
performance, 371
sampling, 367–369
slope overload noise, 370
vs. LPC (linear predictive coder), 374
overview, 356
performance, 362–364
quantization
A-law compression, 367
codecs, 366
compandors, 366
compression/expansion, 364–366
compressors, 364
dynamic range, 364
error, 362–364
expanders, 364
full-scale range quantizer, 361
intervals, 359–361
levels, 359–361
mean-square quantization noise, 363
midriser quantizer, 360–361
midtread quantizer, 360–361
μ-law compression, 367
nonlinear, 364–367
nonuniform, 364
overview, 359–361
SQNR (signal-to-quantization-noise ratio), 362–364
sampling, 356–359
PCS (personal communication services), 4, 16

PDC (Pacific Digital Cellular System), 406
Peak frequency deviation, 256–257
PER (packet error rate), 416
Perceptual measures of quality. *See* Subscriber requirements.
Phase-locked receivers, 238
Phase-shift keying (PSK), 229, 244
Physical layers, 397–398
Plane wave incident, 79–80
Planning systems. *See* Designing systems.
PN (pseudorandom number), 268
PN (pseudonoise) sequences, 402
PN (pseudonoise) spreading codes, 318–325. *See also* Orthogonal, spreading codes.
Poisson model, 190
POTS (plain old telephone service), 9
Power, measuring radiated
  EIRP (effective isotropic radiated power), 29, 65–66
  ERP (effective radiated power), 66
Power factors, antennas
  beamwidth, 25–27
  distance between receiver and transmitter, 21–22
  effective aperture, 21–22
  gain, 26–27
Power spectrum
  amplifiers, 49
  noise, calculating, 36–38, 41–43, 46
p-persistent CSMA, 331
PPSDN (public packet-switched data network), 411–412
Preamble interval, 301–302
PRMA (packet-reservation multiple access), 331
Probability density function, 96–100, 108
Probability distribution function, 108
Probability of error
  digital signal performance, 208–211
  a priori probabilities, 209
  and received power, 223–226
Propagation modeling
  carrier frequency. *See* Hata model; Lee model.
  large-scale fading, 100
  macroscopic fading, 100
  microscopic fading, 100
  path loss prediction
    rural environments, 88
    suburban environments, 88
    urban environments, 88, 101. *See also* Hata model.
  small-scale fading, 100
Propagation of radio waves
  Maxwell's equations, 18–19
  physical circuits, effects of, 19
  range equation, 20
  in the real world. *See* Fading; Path loss; Propagation modeling.
  in space, 18–20
Protocol stack, 398
PSDN (packet-switched data network), 398
Pseudonoise (PN) sequences, 402
Pseudonoise (PN) spreading codes, 318–325. *See also* Orthogonal, spreading codes.
Pseudorandom number (PN), 268
PSK (phase-shift keying), 229, 244
PSTN (public switched telephone network), 184
Pulse code modulation (PCM). *See* PCM (pulse code modulation).
Pulse detection, 207–212
Puncturing, 408

Q

QAM (quadrature amplitude modulation), 279
QoS (quality of service). *See also* GOS (grade of service).
  common parameters, 352
  data, 353–354
  designing for. *See* Optimizing systems.
  images, 353
  for information sources, 353–354
  music, 353
  perceptual measures. *See* Subscriber requirements.
  speech, 353
  subscriber requirements, 61–62
  toll quality speech communication, 167
  user expectations, 344–345
  video, 353
QPSK (quadrature phase-shift keying), 243–251
QPSK modulation, 289–290
Quadrature components, 238–239
Qualcomm, Inc., 12, 289–290, 401

Quantization
A-law compression, 367
codecs, 366
compandors, 366
compression/expansion, 364–366
compressors, 364
dynamic range, 364
error, 362–364
expanders, 364
full-scale range quantizer, 361
intervals, 359–361
levels, 359–361
mean-square quantization noise, 363
midriser quantizer, 360–361
midtread quantizer, 360–361
μ-law compression, 367
nonlinear, 364–367
nonuniform, 364
overview, 359–361
SQNR (signal-to-quantization-noise ratio), 362–364
Quantization error, 362–364
Quantization intervals, 359–361
Quantization levels, 359–361

R

RAB (reverse activity bit), 430
RACH (reverse access channel), 409
Radiation patterns, antennas
azimuth plane, 25
bandwidth planes, 25
beamwidth, 25–27
beamwidth, physical dimensions, 25–26
dipole antennas, 24–25
elevation plane, 25
far-field radiation region, 22
Fraunhofer region, 22
main beam, 22–23
main lobe, 22–23
plot of, 24
power pattern, 22
side lobes, 22–23
Radio, history of, 10, 226
Radio configuration (RC), 412
Radio transmission technology (RTT). *See* RTT (radio transmission technology).
Raised-cosine pulses, 234, 237
RAKE receivers, 404
Random numbers, collision prevention, 327
Range
antennas, calculating, 28–34
mobile unit to base station, 66–70
Range equation
alternate forms, 30
EIRP (effective isotropic radiated power), 29
example, 29
Friis transmission form, 30
optimizing systems, 63–66
overview, 28–34
propagation in free space, 20
variables, 63–66
Range of loss, 96–100
Raster images, 349
Rate control, 429
Rate set (RS), 422
Rayleigh density function, 109–113
Rayleigh fading, 101, 106–113
RC (radio configuration), 412
R-DCCH (reverse dedicated control channel), 424
Received power, and probability of error, 223–226
Receiver (Rx). *See* Rx (receiver).
Receiving antennas, 19
Reciprocity, antenna gain, 28
Recovering information, 227
Region of integration, 109
Regulations. *See* Laws and regulations.
Requirement flowdown, 64
Requirements assessment, 150–153
Resistors, source of thermal noise, 35–40
Return-to-zero (RZ) line code, 206
Reudink curves, 137
Reuse distance, 163–164
Reverse access channel (RACH), 409
Reverse activity bit (RAB), 430
Reverse control channels, 291
Reverse dedicated control channel (R-DCCH), 424
Reverse link configurations, 424
Reverse power control subchannel (R-PCSCH), 423
Reverse supplementary channel (R-SCH), 424

RF links
 definition, 17
 designing. *See* Optimizing systems.
 interference. *See* Noise.
 power transmission. *See* Antennas.
 range, calculating. *See* Range equation.
R-FCH (reverse fundamental channel), 424
Rice, S. O., 113
Ricean probability distribution, 113–114
Rise-over-thermal threshold (RoTT), 418
RMS (root mean square)
 calculating, 38
 delay spread, 101, 118–120
Roaming automation, 400
Rolloff parameter, 234
Root mean square (RMS)
 calculating, 38
 delay spread, 101, 118–120
Root-raised-cosine pulse, 237
RoTT (rise-over-thermal threshold), 418
R-PCSCH (reverse power control subchannel), 423
RS (rate set), 422
R-SCH (reverse supplementary channel), 424
RTT (radio transmission technology). *See also* HSPA (High-Speed Packet Access).
 1 × EV-DO (1× evolution for data only) Rev. 0. *See* DO (data only) standard.
 1 × EV-DO (1× evolution for data optimized) Rev. A, 441
 1 × EV-DV (1× evolution for data and voice). *See* DV (data and voice) standard.
 1 × RTT (1× radio transmission technology), 421–426
 DO (data only) standard, 425–432
 DV (data and voice standard), 425–426
 FCH (fundamental channel), 421
 forward link configurations, 421
 H-ARQ (hybrid automatic repeat request), 430
 MAC index. *See* MAC-ID.
 MAC-ID, 427–432
 overview, 420–421
 rate control, 429
 R-DCCH (reverse dedicated control channel), 424
 reverse link configurations, 424
 R-FCH (reverse fundamental channel), 424
 R-PCSCH (reverse power control subchannel), 423
 R-SCH (reverse supplementary channel), 424
 SCCH (supplemental code channel), 421
 SCH (supplementary channel), 421–422
 transition probability, 431
Rural environments, path loss, 88
Rx (receiver)
 definition, 10
 performance, 222–226
 sensitivity, 30, 62–63
 thermal noise analysis. *See* Thermal noise.
RZ (return-to-zero) line code, 206

S

SACCH (slow associated control channel), 303–305
Samplers, 208
Sampling, 356–359, 367–369
Sampling theorems, 235
Scalable OFDM (SOFDM), 437
SCCH (supplemental code channel), 421
SCH (supplementary channel), 421–422
Scheduler, 417
SDMA (space-division multiple access), 439
Second generation (2G), 400–405
Second generation (2.5G)
 CNIR (carrier-to-noise-and-interference ratio), 407
 EDGE (Enhanced Data for Global Evolution), 410–411
 GPRS (General Packet Radio Service), 408–410
 HSCSD (High-Speed Circuit-Switched Data), 407–408
 puncturing, 408
 RACH (reverse access channel), 409
Sectoring, 175–179, 193–194
Selective bandpass filter, 171

Sensitivity, 34
Separation distance (AM channels), 298–299
Separation distance (geographic), 156, 161–163
SF (spreading factor), 267
Shadowing losses, 96–100
Shannon, Claude, 227
Sharing bandwidth. *See* Multiple user channels.
Sharp cutoff filters, 171
Side lobes, 22–23
Signaling. *See* Baseband signaling; Carrier-based signaling; Spread-spectrum signaling.
Signal-processing functions, 7
Signal-to-quantization-noise ratio (SQNR), 362–364
Single full-duplex, 300
Sinks, 6–7
SIR (signal-to-interference ratio), 155, 169, 298–299
16-QAM modulation, 279
64-QAM modulation, 279
Slope overload noise, 370
Slotted Aloha, 328–330
Slow associated control channel (SACCH), 303–305
Slow fading, 127–128, 131–132
Slow frequency hoppers, 269
Small-scale fading, 100
Snell's law, 79–80
SNIR (signal-to-noise-and-interference ratio), 154
SNR (signal-to-noise ratio)
  optimizing systems, 62–63
  speech sources, 347
  waterfall curves, 223–226
SOFDM (scalable OFDM), 437
Soft handoffs, 404
Source coding. *See* LPC (linear predictive coder); Vocoders.
Sources of information
  in block diagrams, 6–7
  data
    BER (bit error rate), 351
    burstiness, 351
    delays, 352
    QoS (quality of service), 353–354
    source characteristics, 351
    subscriber requirements, 351–352
    throughput, 351
  datagrams, 354
  error correction. *See* Coding for error correction.
  images
    QoS (quality of service), 353
    source characteristics, 349
    subscriber requirements, 349
  music
    QoS (quality of service), 353
    source characteristics, 348
    subscriber requirements, 348–349
  QoS (quality of service)
    common parameters, 352
    for information sources, 353–354
    perceptual measures. *See* Subscriber requirements.
  streaming *vs.* bursting, 354–355
  video
    QoS (quality of service), 353
    source characteristics, 350
    subscriber requirements, 350–351
Sources of information, speech
  delay budget, 347
  delay jitter, 347
  delays, 347–348
  digitizing. *See also* PCM (pulse code modulation).
    formants, 371
    LPC (linear predictive coder), 373–376
    model of human voice, 371–373
    overview, 355–356
    source coding. *See* LPC (linear predictive coder); Vocoders.
    vocoders, 371–376
    VSELPC (vector-sum excited linear predictive coder), 375
    waveform coding. *See* PCM (pulse code modulation).
  dynamic range, 347
  QoS (quality of service), 353
  SNR (signal-to-noise ratio), 347
  source characteristics, 347
  streaming signals, 347
  subscriber requirements, 347–348

Space-division multiple access (SDMA), 439
Spectral efficiency
BFSK (binary frequency-shift keying), 258
cost of, 279
definition, 237
FSK (frequency-shift keying), 257
increasing, 279
QPSK (quadrature phase-shift keying), 2 43–245
Spectral shaping, 298–299
Spectrum allocation. *See also* Traffic engineering.
available channels, and number of subscribers, 152–153
cochannel interference, 153. *See also* Interference prediction.
distinct channels available, 152
example, 151
FDM (frequency-division multiplexing), 151
FDMA (frequency-division multiple access), 151
interference-limited systems, 153
laws and regulations, 150–151
noise-limited systems, 152
NTIA chart of, 8
Spectrum shortages, history of, 12
Speech
delay budget, 347
delay jitter, 347
delays, 347–348
digitizing
formants, 371
LPC (linear predictive coder), 373–376
model of human voice, 371–373
overview, 355–356
vocoders, 371–376
VSELPC (vector-sum excited linear predictive coder), 375
dynamic range, 347
QoS (quality of service), 353
SNR (signal-to-noise ratio), 347
source characteristics, 347
streaming signals, 347
subscriber requirements, 347–348
Spreading factor (SF), 267
Spreading interference. *See* Spread-spectrum signaling.
Spread-spectrum modulation, 306–307
Spread-spectrum signaling. *See also* Baseband signaling; Carrier-based signaling; CDMA (code-division multiple access).
chips, 271
DSSS (direct-sequence spread spectrum), 271–278
fast hoppers, 269
FHSS (frequency-hopping spread spectrum), 268–270
hopping away from interference, 270
narrowband interference, 269–270, 275–278
necessary conditions, 267
overview, 267–268
SF (spreading factor), 267
slow hoppers, 269
spreading interference, 276
Spread-spectrum techniques, 12, 401
Spread-spectrum-based code-division. *See* CDMA (code-division multiple access).
SQNR (signal-to-quantization-noise ratio), 362–364
Square grid, 158–160
Standards. *See also* Laws and regulations; RTT (radio transmission technology).
ANSI (American National Standards Institute), 401
ANSI/TIA/EIA-136, 401
cdma2000. *See* RTT (radio transmission technology).
D-AMPS (Digital Advanced Mobile Phone Service), 401
GPRS (General Packet Radio Service), 408–410
GSM (Global System for Mobile Communications), 12
HSPA, 414
HSUPA, 414
IS-95, 12
IS-95A, 401
IS-95B, 401
ITU (International Telecommunication Union), 405
NA-TDMA (North American Time-Division Multiple Access), 401

TIA/IS-41 Interim, 400
TIA/IS-54, 401
USDC (U.S. Digital Cellular), 401
Statistical model, moving receiver, 129–132
Streaming signals
speech sources, 347
vs. bursting, 325–326, 354–355
Subscriber density. *See also* Multiple user channels.
cell radius, effects of, 173–175
growth. *See* Cells, splitting; Sectoring.
managing, 156
Subscriber requirements. *See also* GOS (grade of service); QoS (quality of service).
data, 351–352
designing for, 61–62
images, 349
music, 348–349
speech sources, 347–348
video, 350–351
Subscribers
calling habits, 189
end users, definition, 6
maximum, CDMA, 311, 323–234
maximum, TDMA, 300
Subsystems, 13
Suburban environments, path loss, 88
Supergroups, 297
Supplemental code channel (SCCH), 421
Supplementary channel (SCH), 421–422
Switch state, 453–454
Switch state probabilities, 457–460
Symbol period, 205
Sync channels, 291
Sync interval, 301–302
System architecture, 150–153. *See also* Designing systems.
System loss, 31
System noise, 9. *See also* Noise.
System-level design, 61–62
Systems, definition, 13
Systems engineers
applicable disciplines, 14
development processes, 14–15
responsibilities, 15
role of, 12–15
system documentation, 15

T

TC (turbo coder), 413
TDM (time-division multiplexing), 415–416
TDM time interval (TTI), 416–417, 420
TDMA (time-division multiple access)
efficiency, 301
frame preamble, 301
frames, 301
GSM (Global System for Mobile Communications), 304–305
guard interval, 301–302
maximum subscribers, 300
overview, 300–302
preamble interval, 301–302
sync interval, 301–302
time-division duplexing, 305–306
USDC (U.S. Digital Cellular), 302–304
Telecommunications Industry Association (TIA), 400
Telegraphy, history of cellular systems, 226
Television, NTSC scans, 349
Thermal noise. *See also* Noise.
average power, 37–41
definition, 34
Johnson noise, 38–39
mobile telephone receiver, example, 58–60
noise floor, 35
output-noise power spectrum
average power, 44–45
calculating, 42, 56–57
LNA (low-noise amplifier), 56–57
power spectrum, 36–38
sources of
amplifiers, 44–45
passive systems, 39–44
resistors, 35–40
RMS (root mean square), calculating, 38
typical values, 60
white noise
available power, 46
definition, 36
noise temperature, 45–46
power spectrum, 36–37
Thermal noise, two-ports
available gain, 47–49
Butterworth filter, 50

Thermal noise, two-ports (*Continued*)
cascade of, 54–57
definition, 47
effective input-noise temperature
calculating, example, 58–61
calculating output-noise power spectrum, 57
definition, 51
vs. noise figure, 54–55
LNA (low-noise amplifier), 56–57
lossy transmission lines, 57–58
noise bandwidth, 49–50
noise figure
calculating, example, 58–61
overview, 50–54
vs. effective input-noise temperature, 54–55
Third generation (3G)
overview, 405–407
OVSF (orthogonal variable spreading factor), 413
PPSDN (public packet-switched data network), 411–412
W-CDMA (Wideband Code-Division Multiple Access), 411–414
Third Generation Partnership Project (3GPP), 406
Third Generation Partnership Project 2 (3GPP2), 406
Threshold comparators, 208
Throughput
calculating, 334–335
data transmission, 351
definition, 190
TIA (Telecommunications Industry Association), 400
TIA/IS-54 standard, 400
Time dispersive channels, 118
Time-division duplexing, 305–306
Time-division multiple access (TDMA). *See* TDMA (time-division multiple access).
Time-division multiplexing (TDM), 415–416
Time-varying channels, 121
Titanic, wireless system, 1
Toll quality speech communication, 167
Top-level design, 63–66
Trade-offs, 173–175
Traffic engineering
average call duration, 189
average holding time, 189
blocked calls, 188, 453
blocked calls cleared model, 190–191, 453
blocked calls delayed model, 190–191, 453
blocking probability, 460–461
call arrival model, 454–455
carried load, 190
carried traffic intensity, 190
definition, 188–189
Erlang B formula, 460–465
Erlang B table, 465–475
GOS (grade of service), 189, 453–454
hold time model, 456
offered load, 189, 460–461
offered traffic intensity, 189
Poisson model, 190
state of the switch, 453–454
subscriber calling habits, 189
switch state probabilities, 457–460
throughput, 190
trunking efficiency, 192–194
trunking theory, 188
trunks, 188
Transducers, 19. *See also* Antennas.
Transferring calls. *See* Handoffs.
Transition probability, 431
Transmitter (Tx). *See* Tx (transmitter).
Transmitter distance, path loss, 96–100
Transmitting antennas, 19
Trellis diagrams, 379–380
Triangular grid, 158–160
Triple redundancy, 377
Trunking, 187–194
Trunking efficiency, 192–194
Trunking theory, 188
Trunks, 188
TTI (TDM time interval), 416–417, 420
TTL (transistor-transistor logic), 205
Turbo coder (TC), 413
Turbo coders, 413
2G (second generation), 400–405
2.5G (second generation)
CNIR (carrier-to-noise-and-interference ratio), 407
EDGE (Enhanced Data for Global Evolution), 410–411
GPRS (General Packet Radio Service), 408–410
HSCSD (High-Speed Circuit-Switched Data), 407–408

puncturing, 408
RACH (reverse access channel), 409
2B1Q line code, 207
Two-ports. *See* Thermal noise, two-ports.
Two-ray model, 102–106, 121–129
Tx (transmitter)
definition, 9
line drivers, 9
signal-processing functions, 7
wired systems, 9
wireless systems, 9

U

UEs (user elements), 416
UHF (ultra-high frequency), 8
UMB (Ultra-Mobile Broadband), 440–441
UMTS (Universal Mobile Telecommunications System), 289, 406
UMTS Release 5. *See* HSDPA (High-Speed Downlink Packet Access).
UMTS Release 99, 414
Unlicensed radio services, 150
Uplink, 66–70
Urban environments, path loss predictions, 88, 101. *See also* Hata model.
U.S. Digital Cellular (USDC) standard. *See* USDC (U.S. Digital Cellular) standard.
USDC (U.S. Digital Cellular) standard
2G (second generation) systems, 401
history of, 289–290
overview, 302–304
Users. *See* Subscribers.
UTRA (Universal Terrestrial Radio Access), 406

V

VBR (variable bit rate), 403
Video
QoS (quality of service), 353
source characteristics, 350
subscriber requirements, 350–351
Video conferencing, 350–351
Viterbi, A. J., 383
Viterbi algorithm, 383–387
VLR (visitor location register), 441
Vocoders
DPCM *vs.* LPC, 374
formants, 371
LPC (linear predictive coder), 373–376
model of human voice, 371–373
VSELPC (vector-sum excited linear predictive coder), 375
VoIP (voice over IP), 415, 431
VSELPC (vector-sum excited linear predictive coder), 375

W

Walsh functions, 317, 402–403
Waterfall curves, 223–226
Waveform coding. *See* PCM (pulse code modulation).
Wavelength, relation to frequency, 8
W-CDMA (Wideband Code-Division Multiple Access), 289, 406, 411–414
White noise
available power, 46
definition, 36
noise temperature, 45–46
power spectrum, 36–37
Wiener-Khinchine theorem, 217–218
WiFi (Wireless Fidelity), 25
WiMAX (Worldwide Interoperability for Microwave Access), 440
Wired systems
transmitters, 9
vs. wireless, 5
Wireless systems
definition, 4
overview, 4–6
vs. wired systems, 5

LearnIT at InformIT

Go Beyond the Book

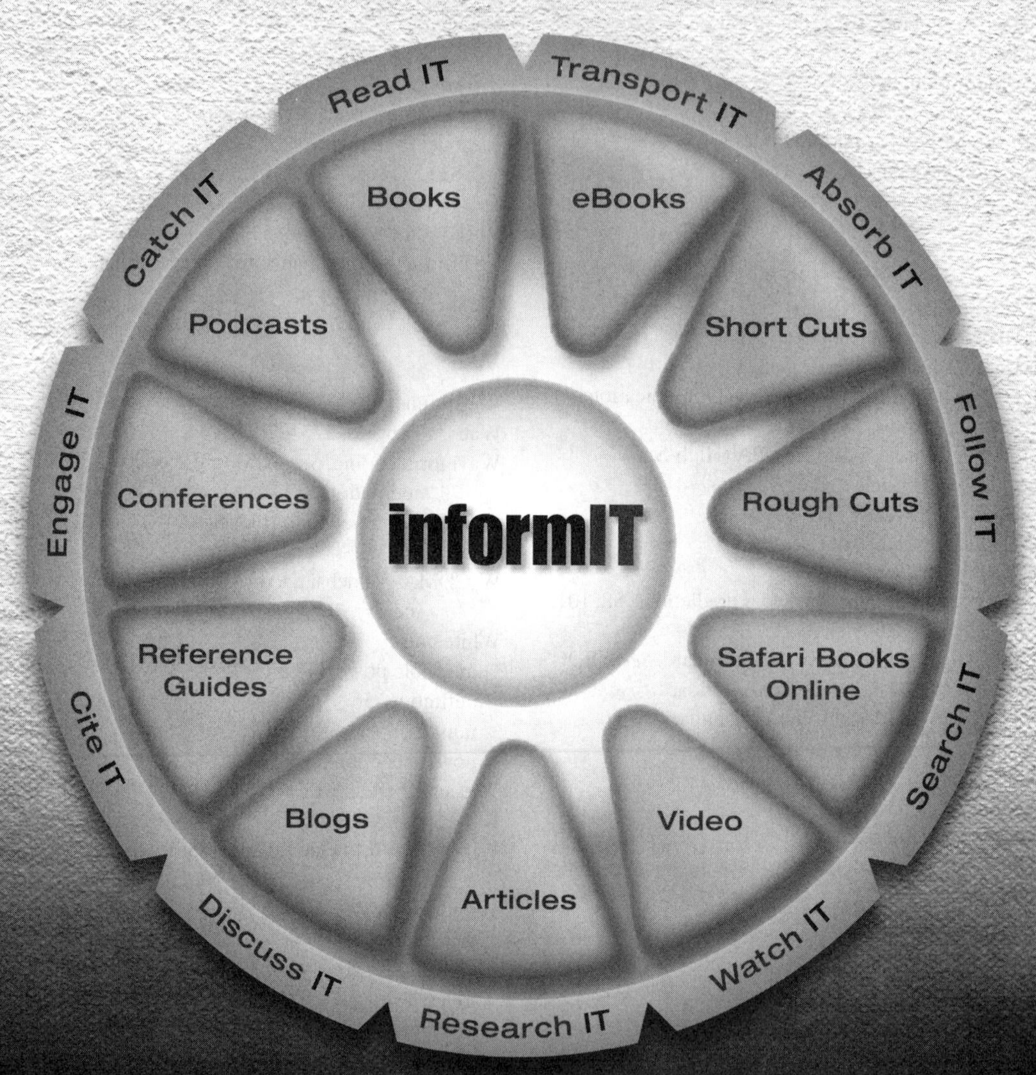

11 WAYS TO LEARN IT at **www.informIT.com/learn**

The online portal of the information technology publishing imprints of Pearson Education

Cisco Press EXAM/CRAM IBM QUE

SAMS